T0399598

Emerging Topics in Ecotoxicology

Principles, Approaches and Perspectives

Volume 3

Series Editor

L.R. Shugart and Associates, Oak Ridge, TN, USA

For further volumes:
http://www.springer.com/series/7360

John E. Elliott · Christine A. Bishop
Christy A. Morrissey
Editors

Wildlife Ecotoxicology

Forensic Approaches

Editors
John E. Elliott
Environment Canada
Science and Technology Branch
Pacific Wildlife Research Centre
5421 Robertson Road Delta
British Columbia, V4K 3N2, Canada
john.elliott@ec.gc.ca

Christy A. Morrissey
Department of Biology
University of Saskatchewan
112 Science Place, Saskatoon
Saskatchewan, S7N 5E2, Canada
christy.morrissey@usask.ca

Christine A. Bishop
Environment Canada
Science and Technology Branch
Pacific Wildlife Research Centre
5421 Robertson Road Delta
British Columbia, V4K 3N2, Canada
cab.bishop@ec.gc.ca

The chapters in this book reflect the views of the authors and not necessarily those of Environment Canada or other government agencies.

ISSN 1868-1344 e-ISSN 1868-1352
ISBN 978-0-387-89431-7 e-ISBN 978-0-387-89432-4
DOI 10.1007/978-0-387-89432-4
Springer New York Dordrecht Heidelberg London

Library of Congress Control Number: 2011930754

Printed on acid-free paper

Springer is part of Springer Science+Business Media (www.springer.com)

Lindsay Oaks and Munir Virani examining a dead vulture in Pakistan, 2000

This book is dedicated to the memory of J. Lindsay Oaks, a friend and colleague to many, and a great wildlife forensic toxicologist.

Preface

Toxicology and forensics are very historical applications of science, going back even as far as the Roman Empire; and, forensic toxicology has been the subject of many a mystery thriller. Yet, forensic *ecotoxicology* as an applied science is quite new. In fact, the word *ecotoxicology* was not coined, as we know it today, as an applied branch of toxicological science until 1969 by R. Truhaut (see B. A. Rattner, 2009, History of wildlife toxicology, *Ecotoxicology* 18:773–783). And, as an applied field of ecology and conservation biology, forensic ecotoxicology and related policy applications have come into their own only in the last few decades, and its growth and development has been interesting to say the least. Most of the work described in this book, as well as other similar toxics work, is not forensics per se, but in essence it is forensic in nature. In our specialty, forensic ecotoxicology and policy applications today ideally follow a hypothesis-centered process that must lead to problem-solving and also builds strongly on known information from past work (see Fig. 1).

Using as an example, the DDE-induced (DDE is one of the recalcitrant, persistent DDT metabolites) eggshell-thinning phenomenon, a long series of investigations typically went along the following steps:

1. Describe the pattern, extent, and timing of the phenomenon of eggshell thinning.
2. Hypothesize and then determine causation and relationships to suspected environmental factors (DDE, PCBs, other POPs, stress, nutrition, etc.).
3. Identify relationships to individual and population health.
4. Test field hypotheses with controlled experiments and hypothesis-testing field designs.
5. Determine physiological and biochemical mechanisms of action.
6. Develop predictive ability through models.
7. Translate causation to policy and regulation (this seems to be the most difficult part).

None of that was easy, and just for DDE alone, the process took 25–30 years and many hundreds, more likely thousands, of scientific studies involving also thousands of scientists and technicians. Additional effects, causes, and complications still continue to be found as research has progressed (for example: A. N. Iwaniuk

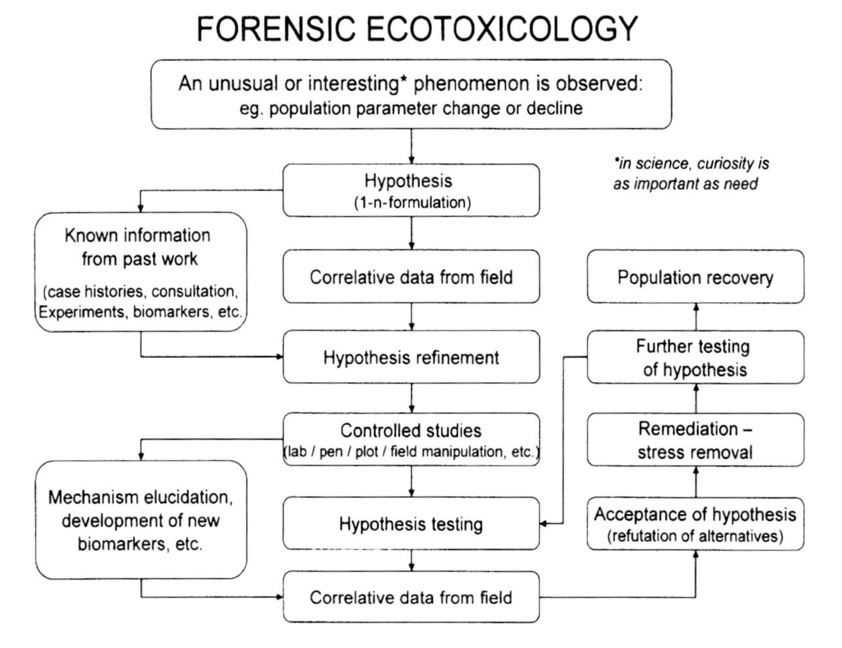

Fig. 1 Idealized, step-wise process in forensic ecotoxicology, in the case of conservation, leading to population, species, or system restoration, and based largely on scientific hypothesis-testing throughout (modified from D. W. Anderson, 1998, Evaluation and impact of multiple stressors on ecosystems: four classic case histories, *in* Cech, Wilson, and Crosby, eds., Multiple stresses in ecosystems, Lewis Publishers)

et al., 2006, The effects of environmental exposure to DDT on the brain of a songbird: Changes in structure associated with mating and song, *Behavioural Brain Research* 173:1–10) Nonetheless, there are some people trying to challenge the results and especially the subsequent policy/regulation results. And of course, the listed steps did not occur perfectly in the order listed, and often due to conservation urgency, regulation and policy are sometimes (but not very often) enacted early in the process, once causation has been reasonably well determined (a "better safe than sorry" or "precautionary principle" philosophy). R. W. Risebrough (1986, Pesticides and bird populations, *Current Ornithology* 3:397–427) briefly summarized the DDE/eggshell thinning phenomena, and I borrowed the seven steps described above from my class notes in *Wildlife Ecotoxicology* (UC Davis). I think this entire process represents one of the important early "forensic successes," although some may disagree.

Of course, a process such as that described above applied to many ecological circumstances, and many species, and involved multiple-stressors in almost every instance. The process is more clear and straightforward when applied to individual studies, such as described by C. J. Henny, L. J. Blus, E. J. Kolbe, and R. E. Fitzner (1985, Organophosphate insecticide [famphur] topically applied to cattle kills magpies and hawks, *Journal of Wildlife Management* 49:648–658). In contrast, just how complex such a diagnostic approach can become, has been nicely illustrated for Bald Eagles in the Great Lakes (D. A. Best et al., 2010, Productivity, embryo and eggshell characteristics, and contaminants in bald eagles from the Great Lakes, USA, 1986–2000, *Environmental Toxicology and Chemistry* 29:1581–1592), leading to scientifically sound conclusions and supporting policy and regulation.

Going back again to the persistent organic pollutants (POPs), I think also that DDT and its introduction during WWII, as a *secret* weapon nobody knew about at first, nonetheless, came to represent the beginnings of widespread concern for the individual and environmental effects of unwanted toxic compounds on ecosystems as well as their components (species and populations). This led to greater understanding of the causes for degradation of biodiversity (it started with exploitable fish and wildlife species), and then for finding solutions to hopefully alleviate these unwanted ecological factors. Today, it just seems like "common sense." In the early days, it almost seemed like heresy to even question the use of pest-control chemicals, no matter how indiscriminate, because they were part of the "green revolution" and the need to feed a rapidly growing human population.

This is not to say that concern and fledgling approaches to (what we know now as) ecotoxicological approaches did not go back even as far as the early 1900s, especially involving compounds like rodenticides, predator-control agents, and chemically simple, inorganic poisonous compounds (see review by Rattner 2009; and R. L. Rudd and R. E. Genelly, 1956, *Pesticides: their use and toxicity in relation to wildlife*, CA Department of Fish and Game, *Game Bull.* 7). Rudd and Genelly (1956) provided a critical "early-step" in modern forensic ecotoxicology, but times were simpler then in that *spray-count* era. (I also recommend J. O. Keith, 1991, Historical perspectives, *in* T. J. Peterle, Wildlife toxicology, Van Nostrand Reinhold). In 1946, when DDT and then soon other POPs were introduced into general use by the overall public, society was *intoxicated*, not by chemicals, but by the euphoria that technology was going to solve all of mankind's problems (albeit, many of them self-inflicted). But even as early as 1946 and 1951, *The Journal of Wildlife Management* published a series of papers (*JWM* Vols. 10 and 15) on the potential effects of chemicals applied directly to the open environment. Then, of course, came Rachel Carson (1962, *Silent Spring*, Houghton-Mifflin, Boston) and Robert L. Rudd (1964, *Pesticides and the living landscape*, University of Wisconsin Press, Madison), which really got us all started. As surely as most biologists are the "intellectual children" of Darwin, ecotoxicologists today are, in essence, the intellectual children of Rachel Carson and Robert Rudd, too.

In forensic ecotoxicology, I think, the field has also evolved through at least three scientific "generations" of scientists in about 65 years of existence. Most people in our field first started out as non-toxicological ecologists or conservationists, people out there studying ecological relationships, demography, behavior, applied ecology, wildlife management, or whatever. Yes, some were physiological ecologists whose field of study, especially involving secondary plant compounds, would eventually lead directly into ecotoxicology. And, some were traditional toxicologists at first. Most ecologists at the time were faced with the conundrum of describing and trying to alleviate newly discovered, man-induced toxicological problems – not that natural systems are not evolutionarily involved in their own ecotoxicological challenges, resulting in coevolved ecological systems, and biochemical pathways that could involve evolutionarily unique toxicants to some degree (see the early classic, *Herbivores: their interaction with secondary plant compounds*, edited by G. A. Rosenthal and D. H. Janzen, 1979, Academic Press). But, most of the "first-generation" scientists were ecologists and conservationists who were led into ecotoxicology by the circumstances. Something was "messing up" their ecological studies, and they had to find out what it was. So ecologists had to do some sleuthing, and that is where forensics first came in: they had to confirm their hypotheses as best they could through the scientific, forensic method. The very resources that these ecologists were studying and trying to understand were "going down." By the way, habitat loss is still the most significant factor resulting in the loss of biodiversity today – but after all, pollution by toxicants is simply one of the human-related factors that is degrading the quality of habitats.

I scanned through J. J. Hickey's book (1965, *Peregrine Falcon populations: their biology and decline*, University of Wisconsin Press) and that of T. J. Cade and W. Burnham (2003, *Return of the Peregrine: a North American saga of tenacity and teamwork*, The Peregrine Fund, Boise) and pulled the following notable names related to the forensics and regulation centered about just *one* bird species, the Peregrine Falcon. Similar stories were being told around all kinds of species of wildlife, such as the Osprey, Bald Eagle, Brown Pelican, White Pelican, Black-crowned Night Heron and other ardeids, various waterfowl and game birds, and many others. This "peregrine-list" reads like a "who's who" of ecologically oriented pioneers in wildlife forensic ecotoxicology (just a few examples are given): Tom J. Cade, Joseph J. Hickey, Ian Newton, David Peakall, Ian Prestt, Derek A. Ratcliffe, R. W. Risebrough, L. F. Stickel, W. H. Stickel, and many others. This list could go on for pages. As a graduate student invited to attend the meeting, Hickey had setup that resulted in Hickey (1965), cited above, one had to have been sufficiently impressed with such a field of experts and it became apparent to me that there was no other more exciting field to devote one's professional life! Of course, similar "stories" like this were developing all over the world with diverse species and ecosystems, and my bias is obviously North American and for wildlife in particular.

The second generation still comprised some taxonomic or subfield specialists, but now included people who obtained additional specialized training in and emphasizing toxicology and physiological ecology. Those were mostly graduate students and/or mentees of "first-generation" forensic ecotoxicologists, such as mentioned above. Then, the current generation seems to be quite a mix of organism plus discipline-oriented field and laboratory specialists, but most have primarily a strong background in the ecological, chemical, and physical sciences and might now be called "true" ecotoxicological specialists, much as we now have "conservation biologists" and "restoration ecologists." Many have developed ancillary expertise in some taxonomic group or subfield, such as reptiles and amphibians, systems ecology, demography, and other specialties. These ecotoxicologists importantly apply ecology, chemistry, physics, biochemistry and molecular biology, physiology, systems ecology, population biology, modeling, statistics, and other disciplines to their specialties; and, they use the forensic approach to apply their derived knowledge often to mathematical and statistical modeling and then to policy and regulation.

This book is a stimulating and quite comprehensive compendium of many of the currently relevant ecotoxicological problems and their histories and their solutions, written by many of today's leaders in the field. Yet, the work itself continues-on, so much in fact, that this book could easily become the start of an important series. I have told my students in that university class I mentioned above (Wildlife Ecotoxicology) that it is the ultimate objective of ecotoxicologists to work ourselves out of our jobs. I am afraid that is a long time off.

Davis, California, USA Daniel W. Anderson

Contents

Reviewers

Thomas Augspurger United States Fish and Wildlife Service, Raleigh, NC, USA

Alain Baril Environment Canada, Science and Technology Branch, National Wildlife Research Centre, Ottawa, ON, Canada

Gian Basili St. Johns River Water Management District, Palatka, FL, USA

Anne Fairbrother Exponent Engineering and Scientific Consultants, Bellevue, WA, USA

Kim Fernie Environment Canada, Science and Technology Branch, Canada Centre for Inland Waters, Burlington, ON, Canada

Michael Gilbertson Guelph, ON, Canada

Anders Goksøyr Department of Biology, University of Bergen, Bergen, Norway

Jay Harrington United States Fish and Wildlife Service, Jacksonville, FL, USA

Megan Harris Lorax Consulting, Charlottetown, PEI, Canada

William Hopkins Department of Fisheries and Wildlife Sciences, Virginia Tech, Blacksburg, VA, USA

Peter Kille School of Biosciences, Cardiff University, Cardiff, UK

Pamela Martin Environment Canada, Science and Technology Branch, Canada Centre for Inland Waters, Burlington, ON, Canada

Rafael Mateo Instituto de Investigación en Recursos Cinegéticos, Ciudad Real, Spain

Miguel Mora Department of Wildlife and Fisheries Sciences, Texas A&M University, College Station, TX, USA

Barnett Rattner United States Geological Survey, Patuxent Wildlife Research Center, Beltsville, MD, USA

Peter Ross Fisheries and Oceans Canada, Institute of Ocean Sciences, Sidney, BC, Canada

Judit Smits Faculty of Veterinary Medicine, University of Calgary, Calgary, AB, Canada

Donald Tillitt United States Geological Survey, Columbia Environmental Research Center, Columbia, MO, USA

NicoVan den Brink Alterra Research, Wageningen, Netherlands

Shari Weech Minnow Consulting, Victoria, BC, Canada

Daniel Welsh United States Fish and Wildlife Service, Sacramento, CA, USA

Lisa Williams United States Fish and Wildlife Service, East Lansing, MI, USA

David Hoffman United States Geological Survey, Patuxent Wildlife Research Center, Beltsville, MD, USA

Don Sparling Cooperative Wildlife Research Laboratory, Southern Illinois University, Carbondale, IL, USA

Michelle Boone Department of Zoology, Miami University, Oxford, OH, USA

Alexandra van Geel Industrial Economics, Incorporated, Cambridge, MA, USA

Sherry Krest United States Fish and Wildlife Service, Chesapeake Bay Field Office, Annapolis, MD, USA

Christopher Rowe University of Maryland Center for Environmental Science, Solomons, MD, USA

Contributors

Christine A. Bishop Environment Canada, Science and Technology Branch, Pacific Wildlife Research Centre, Delta, BC, V4K 3N2, Canada
cab.bishop@ec.gc.ca

Roxanne Conrow St. Johns River Water Management District, Palatka, FL, 32178, USA
rconrow@sjrwmd.com

Michael F. Coveney St. Johns River Water Management District, Palatka, FL, 32178, USA
mcoveney@sjrwmd.com

Christine M. Custer US Geological Survey, Upper Midwest Environmental Sciences Center, 2630 Fanta Reed Rd., La Crosse, WI, 54603, USA
ccuster@usgs.gov

Jeffrey T. Edson Edson Ecosystems LLC, Boulder, CO, 80302, USA
jeffedson@comcast.net

John E. Elliott Environment Canada, Science and Technology Branch, Pacific Wildlife Research Centre, Delta, BC, V4K 3N2, Canada
John.elliott@ec.gc.ca

Amy L. Filby School of Biosciences, Hatherly Laboratories, University of Exeter, Prince of Wales Road, Exeter, Devon, EX4 4PS, UK
a.l.filby@exeter.ac.uk

Timothy S. Gross University of Florida, Environmental Resource Consultants, Gainesville, FL, 32611, USA
tsgross@cox.ne

Tyrone B. Hayes Department of Integrative Biology, University of California, Berkeley, CA, 94720, USA
tyrone@berkeley.edu

James V. Holmes Stratus Consulting Inc., Boulder, CO, 80302, USA
jholmes@stratusconsulting.com

Tim Lougheed Science Writer, Ottawa, ON, K1K 2A5, Canada
Stormchild@sympatico.ca

Edgar F. Lowe St. Johns River Water Management District, P.O. Box 1429, Palatka, FL, 32178, USA
elowe@sjrwmd.com

Greg Masson Division of Environmental Quality, United States Fish and Wildlife Service, 4401N. Fairfore, Room 810k, Arlington, VA, 22203, USA
Greg_Masson@fws.gov

Christy A. Morrissey Department of Biology, University of Saskatchewan, 112 Science Place, Saskatoon, SK, S7N 5E2, Canada
christy.morrissey@usask.ca

J. Lindsay Oaks Department of Veterinary Microbiology and Pathology, College of Veterinary Medicine, Washington State University, Box 647040, Pullman, WA, 99164–7040, USA
loaks@vetmed.wsu.edu

Harry M. Ohlendorf CH2M Hill, Sacramento, CA, 95833-2937, USA
Harry.Ohlendorf@CH2M.com

H. Franklin Percival US Geological Survey, Florida Cooperative Fish and Wildlife Research Unit, University of Florida, P.O. Box 110485, Gainesville, FL, 32611, USA
percivaf@ufl.edu

R. Heath Rauschenberger US Fish and Wildlife Service, North Florida Ecological Services Field Office, 7915 Bay Meadows Way, Suite 200, Jacksonville, FL, 32256, USA
Heath_Rauschenberger@fws.gov

Rick A. Relyea Department of Biological Sciences, University of Pittsburgh, Pittsburgh, PA, 15260, USA
relyea@pitt.edu

Kenneth G. Rice US Geological Survey, Southeast Ecological Science Center, 7920 NW 71 Street, Gainesville, FL, 32653-3701, USA
krice@usgs.gov

A.M. Scheuhammer Environment Canada, National Wildlife Research Centre, Carleton University, Ottawa, ON, K1A 0H3, Canada
Tony.Scheuhammer@ec.gc.ca

Vernon G. Thomas Department of Integrative Biology, University of Guelph, Guelph, ON, N1G 2W1, Canada
VThomas@uoguelph.ca

Charles R. Tyler School of Biosciences, Hatherly Laboratories, University of Exeter, Prince of Wales Road, Exeter, Devon, EX4 4PS, UK
c.r.tyler@exeter.ac.uk

Robert Vernon Agriculture and Agri-foods Canada, Agassiz, BC, V0M 1A0, Canada
vernonbs@agr.gc.ca

Richard T. Watson The Peregrine Fund, Boise, ID, 83709, USA
rwatson@peregrinefund.org

Laurie K. Wilson Canadian Wildlife Service, Pacific Wildlife Research Centre, Environment Canada, Delta, BC, V4K 3N2, Canada
laurie.wilson@ec.gc.ca

Allan R. Woodward Fish and Wildlife Research Institute, Florida Fish and Wildlife Conservation Commission, Gainesville, FL, 32601, USA
allan.woodward@myfwc.com

Chapter 1
Wildlife Ecotoxicology: Forensic Approaches

John E. Elliott, Christine A. Bishop, and Christy A. Morrissey

Abstract This introductory chapter provides an overview of the book and some discussion of the emergent themes. The nature of forensic ecotoxicology is considered, and a definition proposed. We reflect on the experiences of some authors in trying to translate scientific evidence of toxicant effects into regulatory or non-regulatory action. We further examine the problem of bias in data interpretation, and consider some of the dispute resolution processes discussed by the various authors.

Introduction

The field of wildlife toxicology emerged in response to the use, misuse, and ecological mishaps associated with the explosion of commercial chemical use in the twentieth century (Rattner 2009). Many of the earliest studies, and those that really define the field, and in a sense its mythology, were exercises in detective or forensic science. Numerous publications, including chapters in this book, cite "Silent Spring" (Carson 1962) and its influence on widening the awareness of the hazards of environmental contaminants. Yet even as Carson was finishing and publishing her book, one of the most interesting and compelling scientific narratives in wildlife ecotoxicology was just unfolding. In the spring of 1961, Derek Ratcliffe, a biologist with the British Trust for Ornithology, organized a survey of peregrine falcon (*Falco peregrinus*) populations in Great Britain. Ironically, the survey was commissioned in response to calls by homing pigeon enthusiasts to remove legal protection for the falcon. It was alleged that peregrine populations were increasing and killing many of their homers. However, by the end of the survey in the summer of 1961, it was apparent

J.E. Elliott (✉)
Environment Canada, Science and Technology Branch,
Pacific Wildlife Research Centre, Delta, BC, V4K 3N2, Canada
e-mail: John.elliott@ec.gc.ca

J.E. Elliott et al. (eds.), *Wildlife Ecotoxicology: Forensic Approaches*,
Emerging Topics in Ecotoxicology 3, DOI 10.1007/978-0-387-89432-4_1,

that British peregrine populations were in a state of serious decline or even a crash. Ratcliffe describes the ensuing investigations as follows:

> The search for causes had begun in 1961 when the first serious symptoms of decline became clear. It was case of detective work with few clues.
>
> Ratcliffe 1980, page 306

When Ratcliffe began his detective work and search for clues, he had no operating paradigm of contaminant effects on wildlife and limited supporting forensics, such as analytical chemical methods. There were no published accounts linking agricultural or industrial chemicals and declines in wildlife population size or occurrences. Peregrines in particular were associated with wilder spaces, certainly not with agricultural or urban activities. Fifty years later, we live in a world where the awareness and knowledge of the spread and potential hazard of toxic chemicals is one of the acknowledged "risks of modernity" (Beck 1992). The field of environmental toxicology and chemistry now constitutes a virtual industry employing thousands of toxicologists, ecologists, chemists, veterinarians, hydrologists, soil scientists, statisticians, risk assessors, regulators, other specialists, and many students. We believe that group is the primary audience for this book along with a broader readership of people interested in environmental pollution problems, and the human element behind the work.

The book is about investigating the cause and effect relationships between environmental toxicants and vertebrate wildlife populations. We have used a case study format. Many of the cases began as a conservation problem in which a detective-like or forensic investigation was initiated to determine the cause of the health effects in the animals. Some of the studies are more chemical based. They examine the evidence for effects of a given compound or element on a specific taxonomic group, which is how most environmental toxicology now proceeds. Researchers conduct dose response studies in the laboratory or collect environmental samples and quantify the presence of a chemical. That may lead to an assessment of exposure, hazard, or risk. Alternately there may be an attempt to look for correlative evidence of effects in the field using biomarkers or ecological techniques. In essence, it is the reverse of traditional scientific approach; the putative cause is known and the search is for an effect.

Most of the significant advances in the wildlife ecotoxicology, as in other fields, occur when biologists or other scientists observe unusual phenomena in nature, and wonder about the cause. Such phenomena have included declining populations or unusual patterns of deformed offspring, whether as overt morphologies or more subtle physiological features. Whatever approach was taken, the authors of each chapter have first defined the problem and proceeded to describe the ensuing scientific investigations. In each case they have gone on to depict the broader regulatory or remedial aspects of the problem, and in some cases to expand on their personal experiences.

The concept for the book arose from a session "Wildlife Toxicology: Forensic Approaches" at the Society of Environmental Toxicology and Chemistry (SETAC) World Congress held in November of 2004 in Portland, Oregon. Definitions of the word, forensic, are varying. Forensic science involves the application of scientific disciplines to resolving medical and legal problems. Forensic toxicology is an important part of that field and focuses on identifying injury or death caused by poisoning. More recently the field of environmental forensics has emerged, which has its own society and journal, and focuses mainly on applications of environmental chemistry and associated disciplines to identifying sources of contamination. Forensic toxicologists, who largely work with human cases, define their work as the application of toxicology to the purposes of the law (e.g. Cravey and Baselt 1981). Thus for forensic ecotoxicology, we could extend that definition to legal applications of ecotoxicology. Investigations of an "eco" nature normally would not involve criminal law, but rather regulatory or non-regulatory efforts to attenuate the impact of pollutants and contaminants on organisms, or ideally on ecosystems, both small and large. We think a definition of forensic which invokes the apparent original connotation of the word might be more appropriate here. It applies to investigations which invoke public debate or discussion: "of, relating to, or used in public debate or argument" (Princeton University 2010). Thus, forensic ecotoxicology could be defined as: "the investigation of causal linkages between source(s) and presence of a chemical or mixture, and biological effects, with the goal of reducing impact via regulatory or non-regulatory interventions".

The examples presented here are selective, and were based on developing a list of possible subjects and issues, then inviting people to write those stories including their experiences in doing the research and the regulatory process. There are many more topics and examples that could have been described. We have tried to provide an overview covering a variety of classes of compounds, sources, types of forensic investigations, and regulatory aspects. The spatial scope of the issues varies in many of the examples, but those localized cases are useful in describing what worked and did not work, and can provide guidance and understanding for related problems elsewhere.

Two themes emerge repeatedly from these case studies. We see that, despite all the advances in knowledge and technology since Ratcliffe's day, establishing cause and effect in the field, remains a daunting challenge. It is worth noting that in the wildlife toxicology field, the most successful examples from a regulatory point of view involved "simple lethality". Among the chapters, those include organochlorine pesticides and waterbirds, lead shot and pesticides in birds of prey, and diclofenac and vultures. To that list we could add the classic studies by Ian Newton and colleagues, not included here but often cited, on the impact of dieldrin and other cyclodienes on raptors in Britain and on bats and other species elsewhere (Newton et al. 1992; Blus 2003). In most of those cases, the species of concern were long lived and K-selected for which survival of breeding adults was the crucial demographic parameter. Such studies commonly took place over extended temporal and spatial scales often involving a number of collaborators as well as a broader network of volunteers and naturalists, or they were highly focused local studies of marked populations also carried out over longer time periods. When the evidence is in the form of more subtle physiological or reproductive endpoints, establishing cause and effect linkages, particularly at the population level, is even more difficult. In a number of the chapters here, the authors describe attempts to make those connections, including the three case studies on reptiles, and amphibians and fish. Other studies have relied primarily on a risk assessment approach, establishing only exposure in the field. For many wildlife top predators, practical and ethical constraints preclude anything but a risk assessment strategy.

The second theme emerging is the challenge of trying to mitigate the identified threat to wildlife, whether through regulation, remediation, or other means. It is encouraging to see how many were successful in bringing about positive change, despite the many obstacles encountered. Under the broader definition of forensics, and thus invoking the social and regulatory implications of the ecotoxicological science, it is interesting to compare that context in Ratcliffe's time with that of today. Ratcliffe describes the response of various agencies and institutions at the time to the evidence accumulating against the organochlorine insecticides:

> It was not an easy time. Some of us had our first experience of scientists playing politics, and we learned how vicious a vested interest under pressure can be. It was clearly in many people's interests, one way or another, to believe that the wildlife conservationists were talking nonsense, and they left no stone unturned in trying to establish this. Every new paper with more evidence was dissected and gone over with a fine tooth comb, to see what flaws could be found...Tactics at times resembled those of a courtroom rather than the scientific debating chamber. There were tedious arguments about the nature of proof, and the validity of circumstantial evidence. The attempts to deny effects of pesticides on wild raptors descended now and then into obscurantism.
>
> Ratcliffe 1980, page 331

Given the financial implications, some opposition from affected parties to new chemical regulations or cleanup actions would be expected. Corporations are formed to make money, thus whether it is loss of market share from banning of a pesticide or other product, upgrades in pollution control technology, clean up and restoration of a contaminated site, financial resources have to be diverted to the problem, and so away from profit or other investment needs. Differences in resource access and use are the source of most human conflicts. Such conflicts sparked the original conservation activism in North America, and eventually the environmental movement and laws of the 1970s such as the U.S. Endangered Species Act (McCormick 1989). Resource conflicts have lead to widespread civil disobedience and even violence (Marr-Liang and Severson-Baker 1999; Amster 2006), and in the extreme can result in societal breakdown and civil war (Le Billon 2001), sometimes leading to further environmental degradation (Dudley et al. 2002).

As dispute resolution mechanisms, many jurisdictions employ science advisory boards or ecological risk panels, composed of a range of scientists and other specialists from opposing perspectives, usually industry, government, and academia (EPA 2002). Such bodies are charged with examining the data and determining whether there is evidence, for example, for a commercial chemical such as a pesticide to cause significant damage to biota.

A number of the chapters in this book recount the authors' experiences with such decision making tribunals. Opposing sides will differ on not only the interpretation of the pertinent data, but also on study design and methods. As scientists, at least initially we tend to harbor myths about objectivity and simplicity and purity of truth seeking and scientific endeavor (Lackey 2002). In reality, science has always been about testing competing hypotheses and variations in interpretation of results. The debates often become even more contentious when they widen into the social and political realms. Both scientists and the broader society continually debate questions of evolution, nature versus nurture, theories of cosmology and most recently climate change. Fundamental to quality control and "resolving" scientific arguments is the need for repeatability of results and the peer-review process. Even repeatability of experiments can get bogged down, however, in the complexities of experimental design. It has even been asserted that some governments have legislated the requirement for quality assurance procedures so stringent as to prevent publication of data that is critical of important commercial interests. That topic is discussed in the chapter by Lougheed and by others in this book. Even the peer review process, which like democracy is one of those "worst imaginable but better than the rest" systems, has recently been questioned (Brahic 2010; New Scientist 2010; Sieber 2006; Smith 2006).

Data interpretation is unavoidably a function of perspective, which is influenced by background, education and experience, but can also be affected by biases of the financial and career nature. Accepting some degree of bias as unavoidable, the goal is to avoid conflicts of interest that are primarily of a financial nature (e.g., Barrow and Conrad 2006). It is a delicate topic, but such distinctions can become murky, when we consider that most industry advocates are paid consultants, whose continued employment depends in part on their success in defending the interests of their clients. Similarly, careers can be advanced and, therefore, salaries increased for government scientists whose findings provide the evidence needed for regulatory decisions. The academic community is not isolated from the problem of conflicts of interest. Regardless of the consideration that many academics also act as paid consultants to industry or government, university based researchers can also see their funding and careers advanced if they work on high profile conservation issues. Thus, there is the potential of direct gain from over- or under-stating the implications of their findings.

There is some recent literature examining the problems associated with interpretation of contentious data and how to deal with bias and conflict of interest (Rosenstock 2002; Hayes 2004; Barrow and Conrad 2006; Huss et al. 2007; Rohr and McCoy 2010). Recommendations include the need for: complete and transparent disclosure of funding sources in all reports, presentations and publications; balanced makeup of review boards and panels; improved education of both professionals and the broader public about environmental ethics and conflicts of interest.

The implications of the choices made in environmental policy and regulatory decisions cannot be overstated. Such decisions on use of pesticides or industrial chemicals, on large commercial and industrial land developments, and forestry, agricultural and fishery habitat stewardship clearly have profound consequences. The outcomes can lead, on the one hand, to excessive environmental damage and loss of biodiversity and ecological services. On the other hand, the decision may result in unwarranted restrictions on resource development or protection leading to economic hardship or even poverty in local communities. That can in turn contribute to even greater losses of biodiversity (Bradshaw et al. 2009).

As some authors recount here, opposition to change or action can also come within government agencies. It can stem from intervention at the political level or even from corruption. However, new actions can be stifled by simple bureaucratic inertia, and the tendency for large hierarchical organizations to suppress scientific dialog and in particular, dissenting views (Bella 1992, 2004).

Resolution of environmental disputes sometimes has to move beyond the scientific panel or board to be settled by a judicial process even by civil litigation. For example, the USA has the Natural Resource Damage Assessment (NRDA) process to assess injury and economic costs from oil and chemical contamination (Ofiara 2002). Among the cases discussed here, a number had to resort to litigation such as those described by Christine Custer, the chapter by Edson et al. on the Rocky Mountain Arsenal, and the original ban on lead shot for waterfowl hunting in the US, discussed in the chapter by Thomas and Scheuhammer.

However, it also emerges how often the dialogs and eventual solutions were devised in more open and conciliatory ways, with efforts by both parties to devise a consensus solution. Ratcliffe and other colleagues in Britain and across the Atlantic in North America began to make a case against broad scale organochlorine pesticide usage in the early 1960s. It took almost a decade for the first significant regulatory restrictions. That contrasts with the experience 40 years later of the multinational team investigating the crashing Asian vulture populations. Lindsay Oaks and colleagues faced the same skepticism that a chemical, in that case a veterinarian drug, could in any way be poisoning vultures on a continental scale and pushing them to the brink of extinction. Until they completed their work in Pakistan, most of the experts involved in studying the problem were convinced that it must have been a disease or some other undefined factor. But, once faced with the carefully compiled evidence against diclofenac, the governments and industry in Pakistan and India responded relatively quickly to try and address the problem by banning and restricting use of the drug. The question now is whether identification of the problem and attempts to address came quickly enough and are adequate and enforceable. Whether all or any of the vulture populations recover like the peregrines will only be apparent with time.

The first chapters in this book deal mainly with contamination by persistent organic pollutants, the so-called legacy 'POP's. There has been a great deal of success in regulating these compounds, which now proceeds under the Stockholm Convention on Persistent Organic Pollutants. We considered possible chapters on the newer POPs issues, particularly the brominated flame retardants and perfluorinated compounds. However, evidence for a potential problem from those compounds has come primarily from monitoring of temporal and spatial patterns of contamination, rather than evidence of effects in the field (Hites 2004). The process has also been broad and diffused without a defining focal narrative. How the perfluorinated contaminant issue was identified and addressed by John Giesy and colleagues is covered briefly in the chapter by Tim Lougheed, which discusses the social and political context of regulating environmental contaminants.

On the persistent organic pollutant theme the opening chapter examines dioxin pollution from the forest industry. That was problem for many jurisdictions, and is detailed here in the context of first reports of significant wildlife exposure and effects in waterbirds and other wildlife on the Pacific coast of Canada. The main ongoing problem with POPs is the legacy of contaminated sites from manufacturing, waste disposal or intensive use. As a consequence of the degree of economic and industrial development, the US likely has among the largest inventory of such sites, and probably the best characterized and addressed (http://www.epa.gov/superfund/index.htm). Many past and ongoing investigations in the US have been conducted under the Natural Resource Damage Assessment (NRDA) process. Four chapters address issues associated with that legacy contamination. The chapter by Chris Custer takes a species approach rather than a problem based approach demonstrating how a range of contaminant sources and types can be addressed using a versatile indicator species, the tree swallow (*Tachycineta bicolor*), at NRDA sites across the United States. From there, the focus narrows to an NRDA case study of the Rocky Mountain Arsenal, which we selected because of the important role that wildlife poisoning had on identifying and defining the extent of the contamination at that site. We follow that with two forensic studies about South Florida's Lake Apopka. One involves the efforts by Alan Woodward and others to understand the causes of reproductive and developmental problems in alligators (*Alligator mississippiensis*). Apart from the complexity of establishing cause and effect in the wild, there emerges the need for baseline ecological monitoring to provide some understanding of what is "normal" in a wildlife population. The companion chapter by Roxanne Conrow and colleagues describes the efforts to rehabilitate Lake Apopka. It focuses on acute mortality of waterbirds from OC pesticides, an unexpected consequence of efforts to restore the ecological integrity of the lake following decades of anthropogenic insults.

The next series of chapters continues the pesticide theme but shifts to studies of current use compounds. They start with two terrestrial studies of birds. The first is in an agricultural setting and describes efforts to understand how, over a period of a decade, top predators were poisoned from a number of supposedly non-accumulative and non-persistent organophosphorus and carbamate insecticides. The second is in a forestry setting and addresses an unusual situation of a compound, monosodium methane arsenate (MSMA), originally deployed as herbicide, but later developed as a systemic insecticide to suppress bark beetles (*Dendroctonus* spp.) in an epidemic outbreak. It is, to our knowledge, the only ecotoxicological study of woodpeckers (*Picoides* spp.). They are followed by investigations of two of the major herbicides on a global scale, and the experiences of Rick Relyea and Tyrone Hayes, respectively, dealing with the awakening reality for many that impacts on amphibians were overlooked when assessing the risks of these chemicals in the environment. The experiences of those two scientists, as recounted in their chapters, appear a lot more like those of Derek Ratcliffe than some of the more conciliatory approaches taken by industry as described in other chapters.

Two classic problems in wildlife toxicology are the focus of the chapters by Harry Ohlendorf on selenium and by Vernon Thomas and Tony Scheuhammer on continuing issues caused by poisoning from lead projectiles. Ohlendorf's is a classic case study using typical forensic approaches, which started with the finding of deformed ducks in areas subject to runoff from agricultural drainage water. Most readers would already have some awareness and, therefore, would not be surprised at the challenges associated with resolving water related issues in California. The chapter by Thomas and Scheuhammer addresses the ongoing problem of lead, and whether we should be continuing to use lead projectiles when safe and affordable options are available.

The final case studies examined two very different aspects of environmental contamination by pharmaceutical compounds. Great progress has been made in reducing eutrophication and industrial pollution resulting in recovery on lakes and rivers in many countries. However, with growing human populations and increasing complexity and variety of pharmaceutical and other commercial chemicals, we have seen the emergence of new sorts of challenges such as those chronicled by Charles Tyler and Amy Filby on developmental abnormalities in fish from British rivers. The second is the story of secondary poisoning of vultures through the use of a veterinary drug widely used to treat joint and muscle problems in cattle and other livestock. It is a compelling example of Beck's risks of modernity. The plight of the vultures poignantly demonstrates that despite our new awareness of the risks of using drugs and other commercial chemicals, and all the testing and safeguards that we can put in place, there will likely be other such far-reaching and profound consequences from the deployment of chemical technologies.

The final chapter by Tim Lougheed examines the broader context of environmental contamination, and compares the political and regulatory approaches adopted in various countries.

The authors of these 14 case studies have described the science behind each topic, the relations to regulatory actions and the human stories behind the science. They have each tried to make the topics accessible to a broader audience. There are also some insights from the various personal experiences on how the systems worked and how attempts were made, successfully or not, to make changes to improve conditions for the affected wildlife. We hope that the readers find the chapters useful to their own research or related activities, and can appreciate the magnitude of work involved from identifying a contaminant-related problem to achieving a resolution.

References

Amster R (2006) Perspectives on ecoterrorism: catalysts, conflations, and casualties. Contemp Justice Rev 9:287–301

Barrow CS, Conrad JW (2006) Assessing the reliability and credibility of industry science and scientists. Environ Health Perspect 114:153–155

Beck U (1992) Risk society: towards a new modernity. Sage, London, p 252

Bella DA (1992) Ethics and the credibility of applied science. In: Reeves GH, Bottom DI, Brookes MA (eds) Ethical questions for resource managers. U.S. Dept Agriculture, Forest Service, General Technical Report PNW-GTR-288. Portland, OR, USA. pp 19–32

Bella DA (2004) Salmon and complexity: challenges to assessment. Hum Ecol Risk Assess 8:55–73

Blus LJ (2003) Organochlorine pesticides. In: Hoffman DJ, Rattner BA, Burton GA, Cairns J (eds) Handbook of ecotoxicology. CRC, Boca Raton, FL, pp 313–339

Bradshaw CJA, Sodhi NS, Brook BW (2009) Tropical turmoil: a biodiversity tragedy in progress. Front Ecol Environ 7:79–87

Brahic C (2010) Climategate scientist breaks his silence. New Sci 207(2771):10–11

Carson RL (1962) Silent spring. Houghton Mifflin, Boston, MA, p 368
Cravey RH, Baselt RC (1981) The science of forensic toxicology. In: Cravey RH, Baselt RC (eds) Introduction to forensic toxicology. Biomedical, Davis, CA, USA
Dudley JP, Ginsberg JR, Plumptre AJ, Hart JA, Campos LC (2002) Effects of war and civil strife on wildlife and wildlife habitats. Conserv Biol 16:319–329
Hites RA (2004) Polybrominated diphenyl ethers in the environment and in people: a meta-analysis of concentrations. Environ Sci Technol 38:945–956
Huss A, Egger M, Hug K, Huwiler-Muntener K, Roosle M (2007) Source of funding and results of studies of health effects of mobile phone use: systemic review of experimental studies. Environ Health Perspect 115:1–4
Lackey RT (2002) Values, policy and ecosystem health. Bioscience 51:437–453
Le Billon P (2001) The political ecology of war: natural resources and armed conflicts. Polit Ecol 20:561–584
Marr-Liang T, Severson-Baker C (1999) Beyond eco-terrorism: the deeper issues affecting Alberta's oilpatch. Pembina Institute, Drayton Valley, AB, p 26
McCormick J (1989) Reclaiming paradise. Indiana University Press, Bloomington, ID, USA, p 259
New Scientist (2010) End dirty tactics in the climate war: editorial. New Sci 207(2771):3
Newton I, Wyllie I, Asher A (1992) Mortality from the pesticides aldrin and dieldrin in British Sparrowhawks and Kestrels. Ecotoxicology 1:31–44
Ofiara D (2002) Natural resource damage assessments in the United States: rules and procedures for compensation from spills of hazardous substances and oil in waterways under US jurisdiction. Mar Pollut Bull 44:96–110
Princeton University (2010) Wordnet. http://wordnet.princeton.edu/.
Ratcliffe D (1980) The peregrine falcon. T & AD Poyser, Calton, UK, p 416
Rattner BA (2009) History of wildlife toxicology. Ecotoxicology 18:773–783
Rohr JR, McCoy KA (2010) Preserving environmental health and scientific credibility: a practical guide to reducing conflicts of interest. Conserv Lett 3:143–150
Rosenstock L (2002) Attacks on science: the risks to evidence-based policy. Am J Public Health 92:14–18
Sieber JE (2006) Quality and value: how can we research peer review? Nature. doi:10.1038/nature05006
Smith R (2006) Peer review: a flawed process at the heart of science and journals. J R Soc Med 99:178–182
U.S. EPA (United States Environmental Protection Agency) (2002) Overview of the panel formation process at the Environmental Protection Agency Science Advisory Board. EOA-SAB-EC-02-010, U.S. Environmental Protection Agency, Washington, DC

Chapter 2
Dioxins, Wildlife, and the Forest Industry in British Columbia, Canada

John E. Elliott

J.E. Elliott (✉)
Environment Canada, Science and Technology Branch,
Pacific Wildlife Research Centre, Delta, BC, V4K 3N2, Canada
e-mail: John.elliott@ec.gc.ca

J.E. Elliott et al. (eds.), *Wildlife Ecotoxicology: Forensic Approaches*,
Emerging Topics in Ecotoxicology 3, DOI 10.1007/978-0-387-89432-4_2,

On the Pacific coast of Canada and neighboring American states, the resident great blue heron subspecies, Ardea herodias fannini, commonly nests in colonial aggregations high in spindly alder trees. During spring, the colony is a cacophony of noise from adult birds and chicks, while the air is ripe with the smell of feces, rotting fish and carcasses of chicks displaced by siblicide. So it was unsettling in late June of 1987 to enter a once thriving colony at Crofton on Vancouver Island to find a silent place devoid of herons. The residual smells of the heronry still mingled with the equally pungent odors of a kraft pulp mill. The occasional calls of corvids and gulls broke the silence, while trucks rumbled along the gravel causeway to and from the mill. The truck noise and the smell of the cooking and bleaching of wood were a reminder that viewed from above, the heronry was on a small island of green beside a massive industrial complex. Data from previous years showed that heron eggs from this site contained some of the highest concentrations of chlorinated dioxin contaminants reported for eggs of birds. Given that the mill was discharging large amounts of chlorinated waste into the local marine environment, it did not take a lot of imagination to wonder if there was a connection with the silent colony. The failure of that heron colony provoked a series of investigations that continued for the next ten years. Working together, biologists and chemist used ecological, toxicological and chemical analytical methods to investigate the health of wildlife and to address the possible role of persistent pollutants from pulp mills and other forest industry activities. Those results were then used by engineers, regulators and, industry officials to devise solutions to the contamination problems and to monitor their effectiveness.

Abstract The exposure and effects of wildlife to persistent pollutants from forest industry sources were studied over the period 1986–2008 in British Columbia, Canada. Elevated concentrations of specific polychlorinated dibenzo-*p*-dioxin and furan (PCDD/PCDF) congeners were measured in a variety of aquatic and predatory birds and mustelid mammals, and related to sources including chlorine bleaching of wood pulp and use of chlorophenolic herbicides. Exposure was correlated to biochemical, physiological, and morphological variables in various species, and to reproductive success in great blue herons (*Ardea herodias*) and bald eagles (*Haliaeetus leucocephalus*) at specific study sites impacted by pulp mill effluents. From the late 1980s into the early 1990s, changes to the bleaching process and regulatory restrictions on chlorophenol use produced significant reductions in ambient contamination and wildlife exposure to PCDD/Fs, as well as improvements in physiological responses. My experiences with both the science and the regulatory processes are described, along with a reassessment of the wildlife effects data and some consideration of the lessons that might be learned from this work.

Introduction and Background

During the 1950s and 1960s millions of poultry in the USA died from a condition referred to as chick-edema disease (Friedman et al. 1959). Following many years of scientific investigation, 'dioxins', primarily the compound, 1,2,3,7,8,9-hexachlorodibenzo-*p*-dioxin (HxCDD), were identified as the cause of that disease outbreak

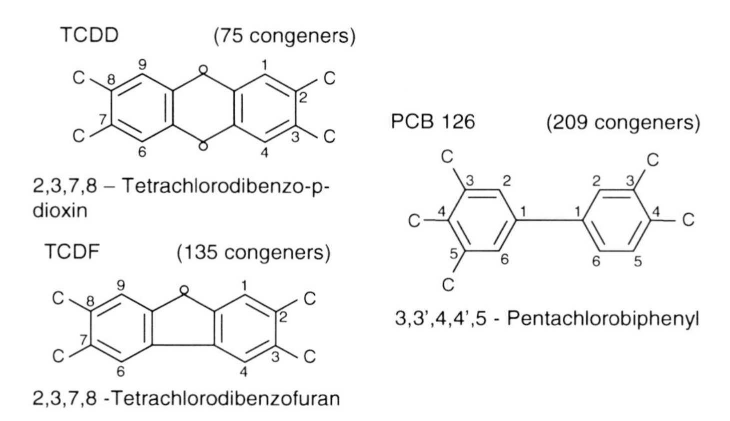

Fig. 2.1 Structure of 2,3,7,8-TCDD and related polychlorinated dibenzofurans (PCDFs) and polychlorinated biphenyls (PCBs). Nomenclature is based on presence of chlorine atoms on the molecule at sequentially numbered carbon atoms. For PCDDs and PCDFs the carbons are numbered in a clockwise order around the molecule. For PCBs, each ring is numbered separately, using a *prime* (′) designation for the second ring

(Higginbotham et al. 1968; Firestone 1973). The structure of 1,2,3,7,8,9-HxCDD can be deduced from Fig. 2.1, which shows the generic polychlorinated dibenzo-*p*-dioxin (PCDD), dibenzofuran (PCDF), and the related polychlorinated biphenyl (PCB) molecules. PCDDs and PCDFs were contaminants in the widely used chlorophenolic biocides, particularly penta- and tetrachlorophenol. Chlorophenolics along with their dioxin contaminants had entered the poultry food chain in fatty acid feed supplements obtained from waste oils and fats produced by the hide-tanning industry, where they had been used as preservatives.

By the early 1970s, the word 'dioxin' began to diffuse into the broader consciousness as something to be vaguely fearful of, something akin to a plague or other disease. *Silent Spring* (Carson 1962) and the subsequent dissemination of the book's information and had created public awareness that, like microbes, agricultural and industrial chemicals could contaminate air, water, and food. Such wider implications of technological change and the social perspectives have been referred to by some writers as the "Risks of Modernity" (Beck 1992).

Further comprehension of the potential hazards posed by dioxins began with the evidence they were formed during the manufacture of the widely used phenoxyacid herbicides, 2,4-dichlorophenoxyacetic acid (2,4-D) and 2,4,5-trichlorophenoxyacetic acid (2,4,5-T). Agent Orange, a mixture of 2,4,-D and 2,4,5-T, became synonymous with dioxin. An estimated 400 kg of 2,3,7,8-TCDD was sprayed by the US military onto the forests of Indochina as an impurity in the 20 million kg of 2,4,5-T applied as the Agent Orange chemical warfare agent (Huff and Wassom 1974). Concerns over the impact on health of Vietnamese people, particularly children, and the American Vietnam War veterans involved in the application, as well as the Vietnamese ecosystem continue to this day (Minh et al. 2008). During the 1970s, dioxin contamination associated with chlorophenolic manufacturing and disposal sites and with industrial accidents in Times Beach Missouri (Powell 1984), Seveso Italy (Fanelli et al. 1980a, b) and Love Canal in the USA drew increased public attention to dioxins (Pohl et al. 2002).

The Great Lakes Story

This narrative really begins during the early 1970s and with the colonial waterbirds of the North American Great Lakes. There were reports at that time of unusual reproductive problems and anomalies at colonies of cormorants (*Phalacrocorax* spp.), gulls (*Larus* spp.) and terns (*Sterna* spp.) nesting on the Great Lakes of Ontario and Michigan. At some islands on eastern Lake Ontario, nests were abandoned during mid-breeding season; eggs contained dead embryos, often with deformities. At other colonies, nests contained chicks also with various deformities, particularly of the bill, eyes, and feet (Gilbertson 1974, 1975). Failed eggs were salvaged during nest visits, and Canadian Wildlife Service biologists, such as Michael Gilbertson, aware of similar scenarios at seabird colonies in Britain, looked for a laboratory to analyze the eggs for residues for industrial and agricultural chemicals. Gilbertson worked with Lincoln Reynolds at the Ontario Research Foundation to adapt and apply methods for chlorinated hydrocarbons, such as DDE and PCBs for the analysis of wildlife tissues (Gilbertson and Reynolds 1972). Mean PCB concentrations, using methodology of the time, and expressed as a ratio of two commercial Aroclor[1] mixtures 1254:1260, were much higher than were being reported elsewhere. Eggs of herring gull (*Larus argentatus*) had 142 μg/g total PCBs at a colony in Lake Ontario and 92 μg/g, both on a wet weight basis, at a Lake Michigan colony (Gilman et al. 1977).

In 1975, the Canadian Wildlife Service opened the National Wildlife Research Centre on the grounds of a former agricultural research facility in Hull, Quebec. The problems with fish-eating birds and other wildlife on the Great Lakes had attracted wide public and political attention, which helped to shift the paradigm defining the Great Lakes Water Quality Agreement. The governments of Canada and the USA began to place priority on persistent toxic substances, rather than the original emphasis on eutrophication (IJC 1978). There was new funding to

[1] Aroclors are technical mixtures of PCBs manufactured by Monsanto Corporation from the 1930s to the 1970s. Those mixtures of PCB congeners were named according to their chlorine content, e.g. Aroclor 1254 contains 54% chlorine by weight and Aroclor 1260 contains 60%. Quantification of PCBs in environmental samples posed a challenge to analytical chemists for many years. Until analytical standards became widely available in the mid 1980s, chemists used one or two major chromatographic peaks considered representative of the major Aroclor mixtures to estimate PCBs on an Aroclor basis.

address the problem of environmental toxicants in the Great Lakes and elsewhere in Canada. New research and technical staff were hired, many of who focused on the Great Lakes. However, the establishment of cause-effect linkages between the observed health problems and specific compounds or groups of compounds proved to be challenging. During the period when signs of toxicity were overt, egg contents of fish-eating birds contained elevated concentrations of a complex mixture of halogenated aromatic contaminants in addition to PCBs, including DDTs, mirex, hexachlorobenzene and other compounds, later discussed in a paper by Peakall and Fox (1987) and examined retrospectively by Hebert et al. (1994).

Given the findings of reproductive failure among the fish-eating birds of Lake Ontario, particularly the mortality of embryos and the presence of deformed chicks, the presence of potent toxicants such as 2,3,7,8-TCDD in the Great Lakes food chains had been hypothesized by Michael Gilbertson, but could not be established with analytical methods employed at that time (Bowes et al. 1973). By the early 1980s, advances in analytical methodology and the availability of high resolution mass spectrometry combined with gas chromatography (GC/MS) enabled the quantification of PCDDs and PCDFs in tissue samples at toxicologically significant concentrations, such as the low parts per trillion range or less than 10 pg/g. A method was developed and eggs were analyzed from colonies of herring gulls in each of the Great Lakes, which demonstrated that PCDDs and PCDFs were present in food-chains from all the lakes, but with hotspots evident in Lakes Ontario, Michigan, and Huron. The temporal nature of the dioxin contamination was made possible by retrospective analysis of herring gull egg samples stored in the Canadian Wildlife Service National Specimen Bank, also one of the earliest demonstrations of the value of environmental specimen banks (Elliott 1984; Elliott et al. 1988a). The results showed that in 1971 eggs from a colony in Lake Ontario contained mean concentrations of 2,3,7,8-TCDD > 1,000 pg/g, which had decreased to about 100 pg/g by 1980 (Stalling et al. 1985). At the time of that work, the data base was very limited for comparative toxicity of dioxins in birds. Work on chickens found that LD_{50}s for 2,3,7,8-TCDD injected into chicken eggs were as low as 200 parts per trillion (pg/g, Verrett 1970).

Casting a Wider Net

Dioxins were present in wildlife from the Great Lakes, but what about other Canadian ecosystems? In 1983, the new high resolution GC/MS method was applied to a survey of wildlife samples from across Canada available from the National Specimen Bank (Norstrom and Simon 1983). Included in that survey were eggs of the Great Blue Heron collected from a colony breeding on the grounds of the University of British Columbia, and which foraged in the nearby Fraser River estuary. Along with the usual mix of halogenated contaminants, such as PCBs and organochlorine insecticides, those heron eggs contained 50 pg/g (wet weight) of

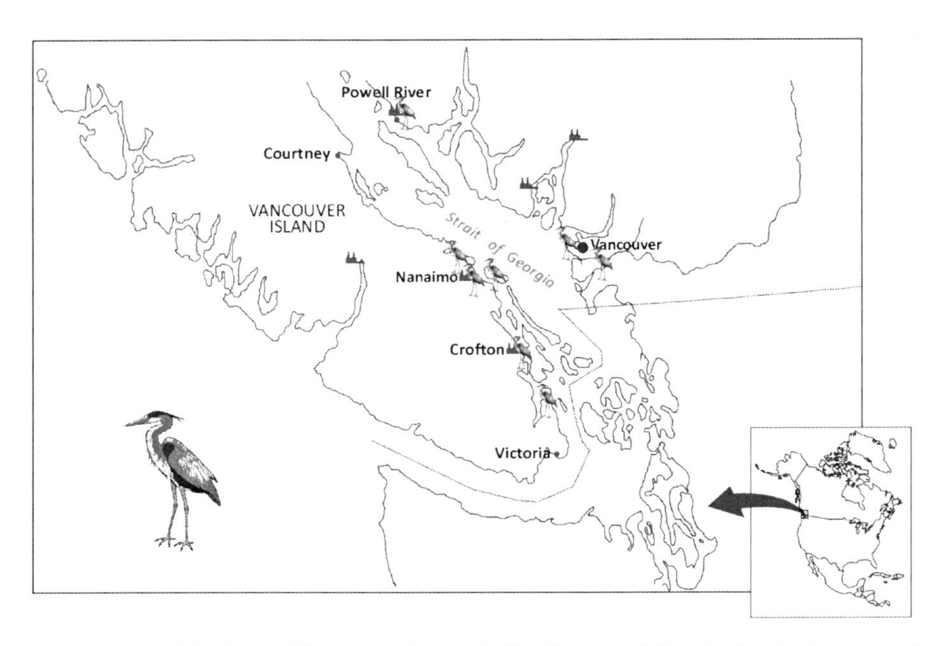

Fig. 2.2 Map of the Strait of Georgia region on the Pacific coast of Canada showing locations of pulp mills and heron colony study sites

2,3,7,8-TCDD. That amount was considered surprising, given the prevailing conception at the time that ecosystems on the west coast of Canada were relatively pristine, at least in comparison to the heavily industrialized Great Lakes. To follow up on those results, in 1983, great blue heron eggs were collected from a number of breeding colonies around the southern Strait of Georgia region on the coast of British Columbia (Fig. 2.2). The survey included two colonies on or near Vancouver Island, one on Gabriola Island located only a few kilometers across from a large bleached-kraft pulp and paper mill and other associated forest industry activities at Harmac near Nanaimo and a colony near Crofton, located on a small island immediately adjacent to a large bleached-kraft mill. The results continued to surprise and are summarized in Fig. 2.3.

The data from 1983 were interesting, particularly the high concentrations of 1,2,3,7,8-pentachlorodibenzo-*p*-dioxin (PnCDD) and 1,2,3,6,7,8-hexachlorodibenzo-*p*-dioxin (HxCDD) in the eggs from the Crofton colony. Environment Canada biologists with support from the Department of Fisheries began to collect fish, crab, and other marine organisms around the contaminated heron colonies (Harding and Pomeroy 1990). At that time, although there had been some speculative writing and proposals, there were no published links between production of wood pulp and formation of dioxins. The putative links between the dioxin contamination of the heron eggs and the forest industry had focused on the use of chlorophenolic biocides. Chlorophenols were widely used at that time in the range of millions of kilogram per year in British Columbia as wood preservatives (PCP, pentachlorophenol) and

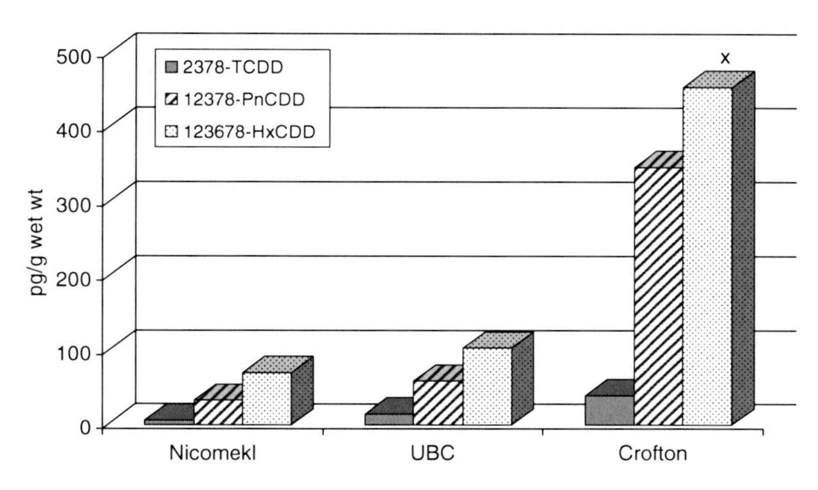

Fig. 2.3 PCDDs in eggs of great blue heron collected in 1983 from colonies around the southern Strait of Georgia, British Columbia, Canada

anti-sapstain agents (TCP, tetrachlorophenol) considered necessary to stop fungal staining of wood piled outside during the cool wet winters.[2] In 1984, the contaminants biologist, Phil Whitehead, took an acting management position and as a result there was no one in biologist position to undertake or supervise any follow-up field work that year.

In November of 1984, there was a federal election in Canada, the government of the day was defeated and a new one under Brian Mulroney came to power. Mdm Suzanne Blais-Grenier was appointed Minister of the Environment, where she made cuts to the departmental budget, particularly the Canadian Wildlife Service. The Research and Interpretation Division of the Canadian Wildlife Service was eliminated along with approximately 25% of the fulltime positions. Those budget reductions affected many projects and virtually all of the regional toxic chemical biologists across Canada, including the Pacific and Yukon Region. In Ottawa, the toxic chemicals program staff at the National Wildlife Research Centre were not made redundant, although most of their colleagues involved with research on animal health issues, including veterinarians, pathologist, and parasitologists were given 90 days notice of job termination. For months, the building was threatened with closure and transfer of the staff and facilities to the Canada Centre for Inland Waters in Burlington, Ontario. Needless to say, those changes

[2]Chlorophenolic biocides refer to the penta- and tetrachlorophenols widely used in many commercial applications to preserve wood and the U.S. Pacific Northwest and British Columbia to prevent mold and slime formation on lumber stored outside during the wet winters. Chlorophenols are unavoidably contaminated with a range of polychlorinated dioxins and furans; however, certain congeners are considered indicative of chlorophenol sources. See for example: Elliott et al. 1998b, b paper on ospreys for background references.

meant that very little new research was undertaken in 1985. However, by the following year, the budgetary and staffing situation had stabilized, and we developed a plan to study the dioxin contamination of the west coast herons.

The proposed focus of the work was the breeding biology of the herons at Crofton, and other sites in relation to the contaminant exposure. A former interpretive biologist, Rob Butler, who had been affected by the 1984 cuts to that program, was interested in undertaking a PhD as a work–study assignment. Thus, in the late winter of 1986, Butler began his study of heron biology.

At that time little was known about the ecology of the west coast subspecies of the great blue heron. There was a need for new techniques to assess breeding success and study other important topics such as diet and feeding behavior. We needed the most basic information about the local population, such as how many heron colonies were there on the south coast of B.C., where were they located in proximity to pollution sources, what was normal reproductive success, what factors could contribute to colony establishment, failure and relocation? Much of the new findings were later recounted in a book on the great blue heron (Butler 1997).

Eggs collected in 1986 confirmed the results obtained 3 years earlier, that the Crofton herons were exposed to relatively high concentrations of PCDDs and PCDFs, and in a particular pattern of the tetra:penta:hexa 2,3,7,8-subtstituted congeners Fig. 2.3. The estimated breeding success of the 64 nesting heron pairs was about 1.24 chicks per active nest, comparable to the rates measured at another dioxin contaminated colony in the Fraser River estuary (UBC) and a putative reference colony on the Nicomekl River (Harris et al. 2003a). Thus, we tentatively concluded that during the 1986 breeding season the dioxin contamination was not preventing the herons from producing what seemed to be a normal complement of fledglings.

We initially planned a 3-year study of heron reproduction and contamination at Crofton and three other colonies. The scene at the heron colony in the spring of 1987 is partially described in the introduction to the chapter. Of the estimated 200 or so eggs laid earlier by the female herons, many still lay in the nests and others were on the ground beneath the colony. Most had been pecked open and the contents eaten, probably by crows. Rob Butler was the first to enter the colony in April of 1987 and find it abandoned (Fig. 2.4). He returned immediately by car ferry across the Strait of Georgia to the Canadian Wildlife Service office in Delta. Phil Whitehead had a private pilot's license and owned a small airplane. Concerned about losing all the eggs and, therefore, evidence, he called a professional tree climber and asked him to meet them at the local Boundary Bay airport in Delta. The three of them then flew immediately to the nearest airport at Nanaimo, rented a car, and drove to the site of the Crofton heronry.

Whitehead and Butler salvaged a number of intact eggs and the contents were removed and couriered to the National Wildlife Research Centre. In the laboratory, the eggs were analyzed for chemical clues related to the colony failure. Later that summer the results came back and the concentrations of 2,3,7,8-TCDD had increased threefold from 70 to 210 parts per trillion. Interestingly, TCDD also increased about twofold over that same period in the eggs of herons at the UBC colony (Fig. 2.5).

Fig. 2.4 The bleached kraft pulp mill at Crofton, B.C., spring of 1990. The heronry was located on the island in the right foreground. Just to the right of where the causeway road bends to the left can be seen the viewing tower used to observe the herons. The new secondary treatment system was under construction at the time and is evident from the half finished tanks on the foreshore below the chip piles. The effluent pipe can also be discerned snaking under the water from near those tanks

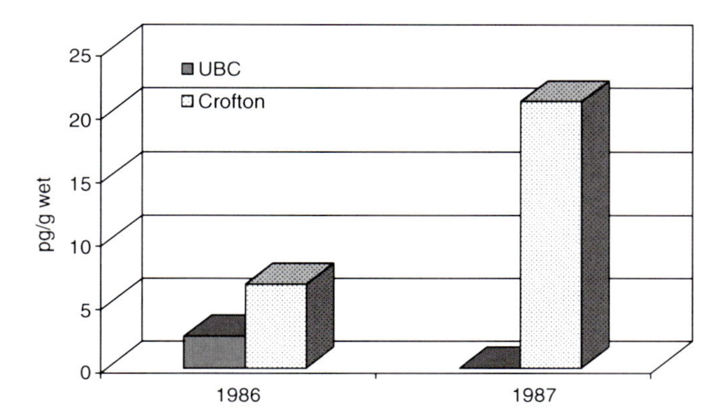

Fig. 2.5 Change in mean concentration of 2,3,7,8-TCDD in heron eggs at Crofton and UBC colonies between 1986 and 1987. Sample size was ten eggs in all cases (Elliott et al. 1989)

The Forensic Phase: I. Tracking the Sources

The steep increase in concentrations of TCDD in the eggs salvaged from Crofton forced us to confront two inter-related problems. Firstly, to reduce or ideally eliminate the contamination, we had to track and identify the source(s). Progress on source identification required better information on the environmental chemistry of the PCDD and PCDF compounds, and on the feeding ecology of the herons. We also had to engage in further dialog with regulatory officials and engineers knowledgeable about pulp mill operations and use of chlorophenolic compounds. Secondly, we needed to investigate whether the TCDD increase was coincident or causally linked with the failure of the colony. Thus, we had to understand more about the general toxicity of the dioxin and furan contaminants in birds, particularly herons, and dose-response relations between dioxins and various aspects of reproduction, ultimately whether there was the potential to cause reproductive failure at the colony.

It now seems obvious in hindsight that cooking wood, a ready source of complex organic molecules, at high temperatures with elemental chlorine would lead to the production of complex chlorinated byproducts. However, at the time, there was no clear published data linking contamination of the ambient environment of a pulp mill with significant elevated concentrations of TCDD and TCDF. Canadian and international scientists and regulators concerned with dioxin sources were focused on waste incineration, chlorophenols, and landfills (Czuczwa and Hites 1984; Interdepartmental Committee on Dioxins 1985), and on atmospheric transport (Czuczwa et al. 1984). Whereas, we knew that the tetra- and penta-chlorophenols were heavily used as wood preservatives on the B.C. coast, they had mainly been linked to contamination with higher chlorinated PCDDs and PCDFs (Bright et al. 1999; Macdonald et al. 1992), not 2,3,7,8-TCDD or -TCDF. During the early 1980s, greater concentrations of PCDDs in human adipose tissue samples from British Columbia relative to elsewhere in Canada had been attributed to chlorophenol use (Ryan et al. 1985). Those findings and other reports of exposure and contamination of people, even industrial workers, also had been down played by some authors as of little consequence for human health (Tschirley 1986).

By March 1986, Canadian regulators were aware of substantial 2,3,7,8-TCDD residues in fish sampled in the international Rainy River drainage in Northern Ontario.[3] A sludge sample from the upstream bleached kraft pulp mill at International Falls, Minnesota contained 414 pg/g of 2,3,7,8-TCDD. By May of the same year there was high level concern at the US EPA that PCDD contamination of effluents and sludges from the pulp and paper industry could be a widespread and generic problem. Those concerns centered on the practice of applying sludges as fertilizer on forestry and agricultural land (Swanson 1988). PCDD contamination of sediments from a Wisconsin river had already been reported (Kuehl et al. 1987). In response, industry had sponsored the avian egg injection studies to assess the risk of PCCD/F contamination of forest fertilizer sludge to wild birds (Thiel et al. 1988). In February 1987, shortly before the colony failure at Crofton, a U.S. National Dioxin

[3]Memo from K. Shikaze, Regional Director, Ontario Region, Conservation and Protection Branch, Environment Canada to P. Higgins, Director General, Environmental Pollution and Protection Division, Environment Canada, Ottawa, March 12, 1986 on the subject "Dioxins – Pulp and Paper Mills". Attached is a memo from T. Tseng, Investigations of sources of dioxin in pulp and paper mills, dated March 10, 1986.

Study reported that PCDDs and PCDFs exceeded background concentrations in fish collected downstream of a number of pulp and paper mills, where no other sources of dioxins had been identified (Amendola et al. 1989). Greenpeace also produced a report in 1987 criticizing government and industry for not releasing information on dioxin contamination from the pulp and paper industry (Van Strum and Merrell 1987). Suddenly, everything was coming together fast, and it was increasingly clear that pulp mills were associated with the specific formation of 2,3,7,8,-TCDD and -TCDF.

As government scientists located in a small regional wildlife centre, we were in no isolated ivory tower. As awareness of the issue widened, political and economic factors came increasingly into play. We tried to focus on the intricacies of dioxin chemistry and toxicology interfaced with heron population and foraging ecology, while trying to absorb explanations on how chlorophenols were used and pulp mills operated. Meanwhile, we went to numerous meeting with managers and officials from our own and other departments, prepared summary and briefing notes for parliamentary question period, and answered questions and requests for interviews and presentations from the media and the public. As with most similar problems, the questions distilled down to the main issue of the significance of the problem in relation to costs of reducing the release of pollutants to the environment. That quickly became a public and political debate waged in many forums and with many competing voices.

Throughout that period there was considerable public pressure on the government and industry, evidenced by the widespread media coverage. Much of the public had grasped the idea that dioxins were not something that you wanted in your bacon or sushi, or even washing around in the water at your local beach. In British Columbia the publicity escalated rapidly after January 5, 1988 when Greenpeace released data on PCDD and PCDF contamination of sediments collected near the Harmac bleached kraft pulp mill at Nanaimo. Consistent with the importance of the forest industry to the provincial economy, the Crofton herons had already received full color press on the front page of the Vancouver Sun. In 1989, data on the increasing trend of 2,3,7,8-TCDD in herons at the UBC colony was featured in national radio and television coverage. Lobbyists for industry and their political allies countered in the media, hiring respected scientists such as Bruce Ames to speak at industry sponsored forums. In media reports, Ames contended there was no elevated threat of cancer from the dioxins in pulp effluent and products, but missed the point that wildlife concerns were not driven by cancer, but by affecting reproductive and basic physiological processes.

The Canadian pulp and paper industry did respond quickly to evidence that their operations were producing and releasing dioxin contamination into local marine and aquatic environments. PAPRICAN (Pulp and Paper Research Institute of Canada) researchers identified the role of elemental chlorine bleaching in TCDD and TCDF formation; subsequently, chemical defoaming agents were also found to be a significant source of TCDF in particular (Berry et al. 1989). The specific pattern in wildlife from the Strait of Georgia, where 1,2,3,7,8-PnCDD and 1,2,3,6,7,8-HxCDD were so prevalent (Elliott et al. 2001a), was traced to the pulping of chlorophenol contaminated wood chips. Luthe et al. (1993) later reported how HxCDDs in particular were produced from a condensation reaction when polychlorinated phenoxyphenol impurities in anti-sapstain treated wood chips were cooked in a pulp mill digester.

It was clear that the Crofton mill in particular had a dioxin problem. A bleached kraft and chemi–thermo–mechanical pulp mill had operated since 1935 at Crofton, processing a variety of local wood species, and discharging liquid effluent into adjacent Stuart Channel. Effluents were discharged under 1971 federal regulations, which exempted mills, such as Crofton, built before the implementation of those regulations. In the spring of 1988, the mill began to test their feedstock chips for chlorophenols, and to divert non-compliant materials. Later that year, they converted first their 'A' bleach plant, and then their 'B' plant to 50% chlorine dioxide, which reduced production of the most toxic, 2,3,7,8-substituted dioxins and furans (Berry et al. 1989). In October 1989, the mill began using defoamers reformulated with purified oils free of precursor dioxins and furans, as had been recommended by researchers at Paprican (Berry et al. 1989). By early 1990, the bleaching sequence was modified by introducing molecular chlorine first and chlorine dioxide last, also in response to the Paprican research.

Those process and product changes had rapid effects on contamination levels in the great blue herons, which are detailed in articles by Elliott et al. (2001a) and Harris et al. (2003a). Within a year of the changes at the mill and over the period 1989–1991, concentrations of 2,3,7,8-substituted TCDD, PnCDD, and HxCDD all dropped significantly. Concentrations of 2,3,7,8-TCDF in heron eggs also declined, although more slowly. We turn from chemistry to ecology to understand why the contaminants dropped so fast in the heron eggs. Diet studies had shown that herons ate mainly small fish between 1 and 2 years of age (Harfenist et al. 1993). Once the tap was essentially turned off, the mill effluents were cleaner, the active tidal and current movements flushed the marine system and the newly hatched fish were growing in a much cleaner environment than had the parent fish. Thus, it was not surprising that a reduced uptake of contaminants in herons quickly followed reductions in contaminant discharges.

Different dioxin congeners did show different patterns of decline, likely due to varying principal sources within and around the pulp mill at Crofton. Throughout the 1980s, the gradual decline of 1,2,3,6,7,8-HxCDD was consistent with the earlier theory that the pervasive contamination of the Georgia Basin with that compound originated from chlorophenol use, which was being regulated and declining by the mid-1980s. At Crofton, the major source of HxCDD in heron eggs was probably contaminated wood chips in the pulp-mill, but local sawmills may have also contributed to contamination. The increased contamination of eggs with TCDD and, to a lesser extent TCDF, in the early 1980s was consistent with increased production and sales of bleached pulp over the decade at that mill. The dramatic declines after 1990 were clearly linked to process changes, as pulp production continued to increase during that time. Because several changes to bleaching technologies and defoamer use were implemented within a short time span, it is virtually impossible to sort out which of those effected the most change in TCDD and TCDF production.

The Forensic Phase: II. Toxicology of Dioxins and Wildlife

By the time that chlorine bleaching of wood pulp was shown to cause formation of 2,3,7,8-substituted dioxins and furans, there had been some progress in understanding the basic toxicology of dioxins and related chemicals. The role of the cellular Ah or arylhydrocarbon receptor had been identified and widely investigated (Poland et al. 1976; Poland and Knutson 1982). The dioxin-like toxicity of some polychlorinated biphenyl (PCB) molecules had also been examined (Safe 1984). A scheme had been developed to rank the toxicity of the 75 PCDD, 135 PCDF, and 209 PCB congeners and produce a single number, the dioxin equivalent or TEQ, to estimate the toxicity of the complex mixtures commonly found in the environment (Bradlaw and Casterline 1979). The finding of a receptor led to a general theory of how dioxin-like chemicals caused toxicity, depicted in Fig. 2.6.

Following the 1987 failure of the Crofton heron colony and the increase in 2378-TCDD burden of the eggs, we initiated a toxicological study for the 1988 breeding season. The Crofton colony eventually failed again that year; however, partially incubated eggs were collected prior to the abandonment and at two other colonies, and transported to the Animal Science laboratories at the University of British Columbia. Using information on incubation conditions obtained from the San Diego zoo, the eggs were placed into incubators designed for chickens, and were tended carefully until they began to hatch. Under those controlled conditions in the laboratory, and

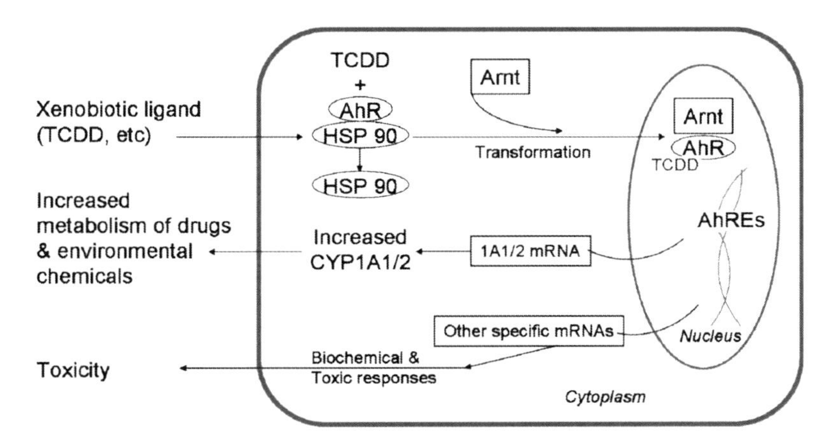

Fig. 2.6 Mode of action for dioxin-like chemicals at the cellular and molecular level. The xenobiotic ligand such as TCDD is able to move passively across the lipid rich cell membrane due to its fat solubility. Once inside the cell, it binds with the Ah-receptor in the process displacing a heat shock protein (HSP 90); it then further binds with the Arnt, a nuclear translocation protein, which facilitates transport into the nucleus where it interacts with responsive elements on the genome. A variety of messenger RNAs (mRNAs) are produced, which leave the nucleus and proceed to code for the production of various proteins, including cytochrome P450s (CYPs) such as CYP1A, involved in metabolism of xenobiotics, including contaminants and drugs. Similarly, other proteins are produced leading to various biochemical and toxic responses

Table 2.1 Result of great blue heron eggs incubated in the laboratory, 1988

Variable	Nicomekl	UBC	Crofton
Mean 2378-TCDD	8 pg/g	84 pg/g	170 pg/g
Fertility	100%	100%	100%
Hatchability	100%	92%	92%
Growth	Normal	Normal	Chicks significantly smaller than Nicomekl
Incidence of edema	0	15%	30%
EROD induction	Normal	Elevated	Significantly elevated over Nicomekl

thus without the influence of other factors that can affect the outcome of the eggs in the wild, it was clear that despite the dioxin burden, there were no significant differences in hatching success or fertility of the eggs among the different colonies (Table 2.1). Thus, a first and important element was determined; there was no overt toxicity to developing embryos. The colony was not failing because eggs were not hatching. There were, however, other physiological and biochemical effects in the heron embryos, which correlated significantly with TCDD exposure. In that study and in further assessments during 1990–1992, there were statistically significant correlations between concentrations of 2,3,7,8-TCDD and of dioxin toxic equivalents (TEQs) in eggs with activity of a liver enzyme referred to as ethoxyresorufin-0-deethylase or EROD, a known marker of exposure to dioxin-like chemicals (Bellward et al. 1990; Sanderson et al. 1994a, b; Harris and Elliott 2011). Exposure to 2,3,7,8-TCDD also correlated with several measures of chick growth, such as body, organ, and bone weights (Hart et al. 1991). Some chicks at Crofton and UBC had jelly-like deposits of fluid or edema under the skin of the neck, breast and leg regions, and one chick from UBC had a crossed bill. No gross abnormalities were observed in chicks from the reference colony at Nicomekl River.

The findings from those studies, including similarities and differences compared to what was reported in the Great Lakes water birds, often referred to as GLEMEDS (Great Lakes Embryo Mortality Edema and Deformities Syndrome, Gilbertson et al. 1991) were examined in more detail in the publications cited above and later in Elliott et al. (2001a, b), and Harris et al. (2003a, b). Among other items of importance, we learned that great blue herons, along with the other species studied, cormorants, bald eagles and ospreys, are not particularly sensitive to dioxin-like toxicity. In 1988, the dioxin-TEQs in heron eggs chicks averaged 370 pg/g at Crofton and 360 pg/g at UBC. Those TEQ tissue loads were exceeded in eggs collected from Crofton in all of the years up to 1991. At UBC, they were exceeded in 1985 and 1989. Comparable toxic burdens were also measured in 1989 at a second colony near near Crofton. We suspect, given the statistical relationships between TEQs and developmental effects in chicks, that there was at least EROD induction and possibly growth effects in chicks at those other sites. With LD_{50}s in the range of 200–250 pg/g, TEQs (Harris and Elliott 2011), chicken embryos would have died at the exposures commonly found in avian predators near pulp mills during the 1980s and to the mid-1990s.

In 1988 there was also an attempt to do an egg swap study. That approach had been applied successfully in contaminant studies of osprey (*Pandion haliaetus*), and involved switching eggs between locations (Wiemeyer et al. 1975; Spitzer et al. 1978; Woodford et al. 1998). The objective is to separate effects 'intrinsic to the egg' such as infertility or direct embryo toxicity from those caused by parental behaviors such as failing to adequately incubate eggs. Egg swap studies are logistically complex at the best of times; however, the approach did not work for great blue herons. There were two main problems; the first was the tendency of herons at some sites to abandon nests if disturbed early in incubation. The second problem was the lack of synchrony in breeding at various colonies, likely due to variation in local availability of food resources (Butler 1997), which made it very difficult to successfully switch eggs among colonies.

Overall, the productivity of great blue herons in the Strait of Georgia, although variable, was similar to that recorded for colonies in Washington and Oregon (Thomas and Anthony 1999). Poor reproductive success, where less than one chick was produced per active nest, occurred at Crofton (5 years), Holden Lake (1 year), Powell River (3 years), UBC (1 year), and Victoria (2 years). The first three colonies (comprising 75% of the cases) were located within 10 km of pulp mills, while the latter two colonies were located within urban centers. Breeding at Crofton and Powell River failed completely in several years, and none of the colonies surveyed around those mills were in apparent use from 1996 to 1999 (Moul et al. 2001). The poorest effort recorded for a more remote colony in our subset of data was one young per active nest (2.11 young per successful nest); however, colony failures have been recorded in rural sites, including Sidney Island in 1989 and 1990 (Butler 1997; Moul et al. 2001). Reproductive success is affected by a variety of ecological as well as anthropogenic factors; nonetheless, there is an apparent trend of lower nest success of great blue herons near areas of significant industrial and urban pollutant sources.

Assessments of Other Wildlife

Herons were not the only species on the BC coast using habitat affected by pollution from forest industry activities. Beginning in the early 1970s, eggs of two species of *cormorant*, the widely distributed, double-crested cormorant, *Phalacrocorax auritus*, and the smaller Pacific coast species, *P. pelagicus* or the pelagic cormorant, were monitored for the exposure to persistent contaminants. Double-crested cormorant eggs contained on average 50% greater concentrations of 2,3,7,8-TCDD than pelagics, while being in turn about 30% less on average than heron eggs from the same locations (Harris et al. 2003a, b). Trends in cormorant eggs followed the same temporal patterns as were measured in herons, falling dramatically in the early 1990s at the forest industry dominated sites. Although cormorants and herons share many ecological niche characteristics, there are important differences, which contributed to overall lower dioxin burdens in cormorants. Most importantly, there are marked differences in fine scale habitat use. Herons tend to feed in slow moving or even stagnant waters where particulate matter tends to settle out, often carrying adsorbed contaminants as compared to the open, faster moving waters where cormorants prefer to forage. Larger scale seasonal differences in feeding area are also likely important; a radio-telemetry study showed that double crested cormorants can move away from breeding areas, traveling in winter down into areas of Puget Sound (unpublished data).

We also conducted an artificial egg incubation experiment with double-crested cormorants. Hepatic EROD activity in hatchlings from three British Columbia colonies, including Crofton and a second pulp mill influenced colony, was significantly elevated as compared to hatchlings from a reference colony in Saskatchewan (Sanderson et al. 1994b). There were no gross abnormalities, edema, or changes in morphological measurements in hatchlings from the pulp mill colonies. Henshel (1998) did report that some hatchlings exhibited at least one type of brain asymmetry, possibly indicating neurotoxic effects of contaminants.

Waterfowl and waterbird populations from breeding sites across western North America overwinter in the Strait of Georgia and many other protected bays and estuaries of the west coast (Campbell et al. 1990). Because waterfowls are hunted for human consumption, their contamination by industrial and pulp mill effluents was a human health issue (Braune et al. 1999). There have also been recent conservation concerns for some North Pacific populations of eider and scoter species (Henny et al. 1995). A study of wood ducks (*Aix sponsa*) suggested that species, and possibly other waterfowl, were very sensitive to dioxin-like effects on deformities and reproduction (White and Segnak 1994), although that was not confirmed by later egg injection studies (Augspurger et al. 2008).

To address both conservation concerns and risk to human consumers, samples of seaducks and grebe species were collected in the late 1980s and early 1990s near pulp mills, and from reference sites, including the Yukon Territory. Sampling occurred in late winter or early spring and assumed exposure would be maximized after overwintering in the collection area (Elliott and Martin 1998). Generally, of all samples collected in 1989, those from near the bleached kraft pulp mill site at Port Alberni were the most contaminated with PCDDs and PCDFs (Vermeer et al. 1993). Piscivorous or fish-eating species, including western grebes (*Aechmophorus occidentalis*) and common mergansers (*Mergus merganser*), contained the greatest concentrations of all contaminants (Elliott and Martin 1998). Those samples also contained other chlorinated hydrocarbons such as DDE and PCBs. The only compound detected in all samples was 2,3,7,8-TCDF, although 2,3,7,8-TCDD was regularly present. International TCDD toxic equivalents (I-TEQs) in some western grebe samples were comparable to herons and cormorants (200–400 pg/g). Top predators, particularly bald eagles, resident near pulp mill sites on the British Columbia coast probably accumulated high levels of chlorinated hydrocarbons in part from feeding on fish-eating birds, including grebes and seaducks (Elliott et al. 1996b, c).

The greater liver PCDD/F concentrations in 1989 at Port Alberni prompted the Canadian federal health agency (Health Canada) to issue an advisory limiting the human consumption of common merganser liver to <50 g/week and of surf scoters liver to <75 g/week (Whitehead et al. 1990). Waterfowl samples were collected at the end of winter, potentially following maximum exposure to pulp mill effluents; therefore, it was supposed that by the time of the important fall hunting season, and a summer on less contaminated breeding grounds, tissue contaminant levels would be much lower, and less of a concern for human consumers.

Among the resident birds on the Pacific coast, the *bald eagle* is the largest predatory species. Eagles are generally considered to be top predators; however, we found that based on their stable isotope or $\delta^{15}N$ signature (Elliott et al. 2009), an independent measure of trophic level, that at least during the breeding season, eagles were similar to other piscivorous birds on the British Columbia coast. They appear to have been eating a variety of medium-sized forage fishes such as herring (*Clupea*), mackerel (*Scomber*), and midshipman (*Porichthys*) species (Gill and Elliott 2003; Elliott et al. 2005).

In 1990, I began a study of eagle biology in the context of contaminant exposure and effects on the British Columbia coast. In 1990 and 1991, eagle eggs from around Crofton and other pulp mill and reference sites had elevated PCDDs and PCDFs with patterns similar to those in other fish-eating birds (Elliott et al. 1996a). In 1992, we collected eggs for an artificial incubation study of the bald eagle. We measured significant hepatic EROD induction, and using molecular techniques found that specific toxicant metabolizing enzymes, including a cytochrome P450 or CYP1A-like enzyme was expressed (Elliott et al. 1996b). That year we also began to monitor productivity of eagles across a broad area of the British Columbia coast, and to use non-destructive blood sampling of nestling eagles, rather than egg collection. The results of the 4-year study revealed that nest success was significantly lower in the bald eagle nests within the dioxin fishery closure zone around the Crofton mill compared to nests outside that zone (Elliott and Norstrom 1998). However, productivity of bald eagles was also low at other areas of the coast that had low exposure to PCDDs and PCDFs. At those sites, such as areas of the west coast and north-eastern coasts of Vancouver Island, and the Queen Charlotte Islands, low production of young eagles was linked to low food availability during the breeding season (Elliott et al. 1998a).

Later we studied both contaminants and food supply more intensively in eagles from the Crofton area. At territories south of the mill, average productivity was 50% lower than territories north of the mill. At individual nests south of the mill, reproductive success was correlated to lower rates of prey delivery in comparison to nests north of the mill. That in turn was linked to topographic differences, which limited feeding areas south of the mill, compared to the habitat further north where there were greater shallows and mudflats. However, contaminants, particularly PCDDs and PCDFs also were higher at the territories south of the mill, and we recognized the potential for those contaminants to have acted in concert with food stress to affect eagle breeding success (Gill and Elliott 2003; Elliott et al. 2005). Most recently, we have shown that more than 30 years after regulatory bans on PCB use, that nestling bald eagles, similar to birds from many other locations, exhibit apparent thyroid hormone disruption caused by PCB exposure (Cesh et al. 2010).

Forest sector industrialization of the British Columbia landscape was not confined to the coast. Well before 1990, the major waterways of the interior of the province were dotted with numerous log sorts and lumber mills. At strategic locations, integrated pulp and paper installations were constructed. Moving east in British Columbia, the moderating influence of the Pacific wanes while successive mountain ranges create a diversity of climate and habitats and with it changes in the wildlife community, and some key life history strategies. Marine species such as pelagic cormorants disappear, and species common to the coast, such as the great blue heron or the bald eagle, may now be forced to migrate to avoid subzero winters with ice cover of rivers and lakes. To expand our study of forest industry pollution to the interior, we needed a different approach and sometimes different species.

The osprey (*Pandion haliaetus*) is an obligate fish-eating bird of prey with a nearly worldwide breeding distribution. Like the bald eagle, the osprey was particularly sensitive to DDE-induced effects on eggshell quality, and populations in many parts of the world declined during the organochlorine era (Poole 1989). Exposure and effects of chlorinated hydrocarbon and mercury pollutants has been widely studied in this species (Ames 1966; Spitzer et al. 1978; Wiemeyer et al. 1988; Steidl et al. 1991). Many features of osprey life history make it a useful species for monitoring and research of contaminants (Grove et al. 2009).

Over a number of years, we studied ospreys breeding up and downstream of bleached-kraft pulp mills on the Fraser and Columbia River drainage systems of British Columbia, Washington, and Oregon. We learned that mean concentrations of 2,3,7,8-TCDD were significantly higher in eggs collected at downstream compared to upstream nests near pulp mills at Kamloops and Castlegar, British Columbia (Elliott et al. 1998b). We monitored a number of sites over time, and found that, like the coastal birds by the late 1990s, concentrations of 2,3,7,8-TCDD and -TCDF were significantly lower than in the early 1990s, consistent with changes in bleaching technology and evidence from sediments (Elliott et al. 1998b; Macdonald et al. 1998).

We found an unusual pattern of higher chlorinated PCDDs and PCDFs in many of the osprey eggs collected in that study, with highest concentrations from nests on the Thompson River, a tributary of the Fraser River, and contrary to expectations, concentrations were generally higher at nests upstream of pulp mills. That contamination was related to use of chlorophenolic wood preservatives by lumber processors based on patterns of trace PCDFs in eggs, essentially a fingerprint particular to that source, and significant positive correlations between egg concentrations of pentachlorophenol. It had been widely thought that octachlorodioxin (OCDD) had a very low capacity to bioaccumulate. Bioavailability of chemicals in food chains is largely determined by their tendency to move or partition between water (aqueous) and octanol (lipid) phases, called the octanol:water partition coefficient, and expressed as a logarithm, log K_{OW}. It was thought that OCDD and other higher chlorinated compounds had such a high log K_{OW} that it inhibited their capacity efficiently bioaccumulate (Segstro et al. 1995). However, the theory does not apply to some benthic or bottom feeding food chains where, for example, fresh water clams or aquatic insects appear to efficiently bioaccumulate higher chlorinated PCDDs and PCDFs (Fig. 2.7). A connection with the ospreys and such food chains was made from the work of fishery biologist, Don McPhail, from the University of British Columbia who showed that some subspecies of mountain whitefish, that he referred to as "Pinocchios", because of their longer snouts, had adapted to specialize on benthic invertebrates (McPhail and Troffe 1998). Thus it appeared that those

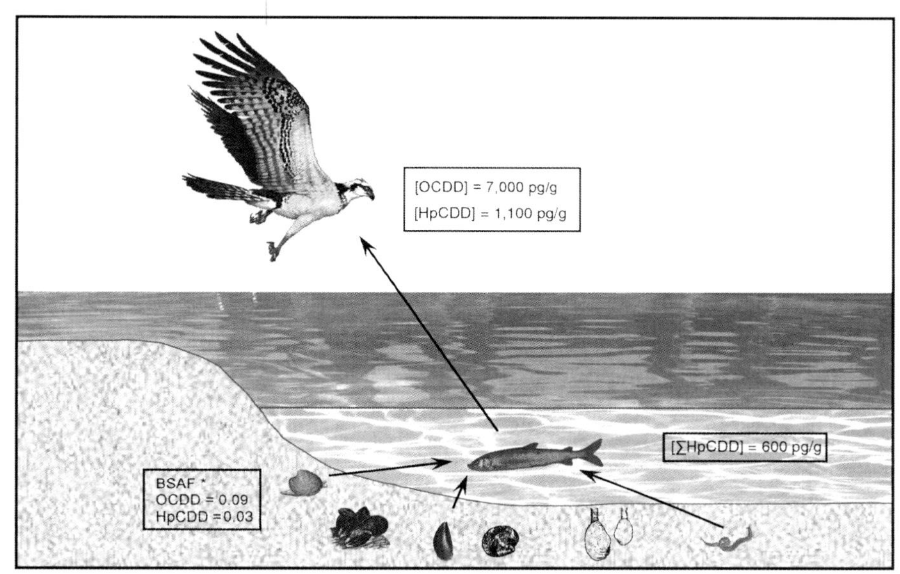

Fig. 2.7 Ospreys accumulated unusually high concentrations of supposedly non-bioaccumulative compounds, such as OCDD. Some ospreys probably fed on 'pinnochio' forms of the mountain whitefish, adapted to feed on the benthic filter-feeding invertebrates that are able to accumulate such chemicals

bottom feeding whitefish were accumulating OCDD and other higher chlorinated compounds (Nener et al. 1995) and transferring them to ospreys.

Similar to the studies with other birds, in 1995 and 1996, we collected 54 osprey eggs at seven sites along the Fraser and Columbia Rivers in BC, Washington and Oregon, USA and brought them to the laboratory for incubation (Elliott et al. 2001b). That work was done five years after the pulp mills had substantially reduced dioxin and furan contamination of effluent, and may have been in factor in our finding no differences in contaminant concentrations between eggs which hatched and those that did not. As with the other fish-eating birds, EROD and related proteins were induced with exposure to toxic equivalents and PCBs. Tissue concentrations of retinols or vitamin A compounds also varied among sites and correlated positively with concentrations of TEQs and PCBs, indicating there were subtle effects of those chemicals on the physiology of ospreys.

Although fish-eating birds are among the most visible top predators in temperate aquatic food webs, they may not be indicative of contaminant exposure and effects on other wildlife, such as mammalian predators. Pacific killer whales in the Strait of Georgia (*Orcinus orca*) have substantial body burdens of persistent contaminants, particularly PCBs (Ross et al. 2000). Harbor seals (*Phoca vitulina*) collected in the mid 1990s near the Strait of Georgia pulp mill sites showed the same distinctive patterns of PCDDs and PCDFs, but body burdens were lower than exhibited by their avian counterparts (Ross et al. 2004; Addison et al. 2005). *Mink* (*Mustela vison*) and *river otter* (*Lontra canadensis*) have a similar trophic status, but their home ranges are relatively small and seasonally more constant compared to their larger marine counterparts. In the 1960s, minks were identified as particularly sensitive to PCBs, when a connection was made between mink reproductive failure and diets of contaminated Great Lakes fish (Aulerich and Ringer 1977; Smit et al. 1994). Declines of European otter (*Lutra lutra*) populations (Macdonald and Mason 1988; Kruuk and Conroy 1966) and North American mink populations (Henny et al. 1981) have been linked tentatively to elevated environmental PCB concentrations. In a later study, mink exposed via the diet to bleached kraft mill effluent had sublethal effects on immune function (Smit et al. 1994).

During the early 1990s, I worked with Chuck Henny of the US Geological Survey in Corvallis, Oregon to study semiaquatic mammals on the Fraser and Columbia Rivers. (Elliott et al. 1999) Using carcasses obtained from commercial trappers, we measured chlorinated hydrocarbon residue concentrations in livers of American mink and river otter collected from the Fraser River and Columbia River watersheds. For the most part, significant amounts of the two congeners 2,3,7,8-TCDD and TCDF were not found in mink and otter. International TCDD toxic equivalent levels in mink (31 pg/g) and otter (93 pg/g) from the lower Columbia River approached toxicity thresholds for effects on reproduction in ranch mink.

Elevated concentrations of higher chlorinated PCDDs and PCDFs, probably resulting from use of chlorophenolic wood preservatives, were found in both species. Most samples also contained other contaminants including DDE and PCBs, although there was substantial variability in patterns and trends among neighboring samples.

More recently we have used a nonintrusive technique to study river otter populations, by collecting their feces at communal latrine sites (Elliott et al. 2008). We found that otter feces collected in 1998 from pulp mill sites at Crofton, Nanaimo, and Powell River did not contain elevated TCDD, although one latrine near the pulp mills at Powell River still had higher TCDF present. Elevated concentrations of higher chlorinated PCDDs and PCDFs were found at some sites, particularly in Victoria Harbour and Nanaimo. Based on a fingerprint pattern of specific PCDF compounds, the source was attributed to use of chlorophenolic biocides, as we had found elsewhere and in other species, most likely by previous forest industry operations. Recently, we have combined molecular DNA genotyping techniques with contaminant analyses in feces to show movement and population structuring of river otters in the Victoria area (Guertin et al. 2009).

Forensic Phase: III. Reappraising the Toxicological Data

With the benefit of hindsight and new information what can we now conclude about causal relationships between the forest industry contaminants and health effects in wildlife? Significant exposure was clearly evident; in habitats impacted by forest industry effluents, predatory and fish-eating wildlife accumulated much greater

PCDD and PCDF concentrations than at reference sites. The only exceptions were the mink and otter, which did not appear to forage extensively on the more contaminated main stems of those rivers (Elliott et al. 1999).

Local populations of three coastal bird species (herons, cormorants, and eagles) exhibited measurable and significant biochemical responses, including increased activity of the enzyme(s) measured by the EROD assay and in some cases specific induction of CYP1A-like enzymes. Interpreting the ecological or even basic toxicological significance of induction of CYP1 enzymes remains complex and somewhat unresolved (Nebert et al. 2004). In avian species, we now know that there are at least two forms of the enzyme associated with the measured EROD response, as well as more than one form of the Ah-receptor (reviewed by Harris and Elliott 2011). It may be that the CYP1 induction measured in those various birds was simply an adaptive physiological response to xenobiotic exposure – with no toxicologically significant consequences. However, if viewed as a biomarker of an Ah-receptor mediated response, CYP1A induction also indicates that other physiological processes were affected. Effects on immune function, for example, appear to be Ah-receptor mediated, but are more difficult to measure reliably in the field, but have potentially greater significance to the overall fitness of the bird. In the ospreys, we initiated the laboratory incubation study because breeding success downstream of some mills, such as Kamloops, was lower than upstream and not explained by food availability or other discernible factors (Elliott et al. 2001a). However, by the time of that study, PCDDs and PCDFs had declined significantly in those systems. Nevertheless, we detected a CYP1A type response which correlated with exposure to PCBs.

The population level parameters, such as the failure of the Crofton colony during the years of highest PCDD and PCDF contamination, and the lower nest success of bald eagles within the fishery closure zone are confounded by other ecological factors, including disturbance, weather and food supply. Thus it is always difficult to relate cause and effect, which is no different, however, from the complexity facing human epidemiological work, and is not a reason for avoiding attempts to understand the system. In 1988 and 1989 we used a basic ecological technique of intensive observation of individual birds. Observation of the Crofton heron colony during two breeding seasons revealed periodic disturbance and subsequent predation and scavenging by crows (*Corvus caurinus*) or ravens (*Corvus corax*) of the vacated nests. The proximal cause of the abandonment appeared to have been that disturbance, either from humans or bald eagles (Norman et al. 1989; Moul 1990). There is then the question of relative sensitivity of herons to disturbance, also not a simple topic. At some sites, herons are more easily provoked to abandon a breeding attempt, particularly early in the incubation cycle and before a lot of investment of time and energy. However, elsewhere, such as a colony located for years above the old Stanley Park zoo in Vancouver, herons bred successfully in trees overhanging the howler monkey cages (Butler 1997). The role of disturbance by bald eagles has also proven more complex than originally thought. Some authors considered the recovering bald eagle populations in the Pacific Northwest to pose a threat to long-term viability of heron populations by disturbance and predation at breeding colonies (Vennesland and Butler 2004). In my experience, however, during the breeding season most eagle pairs consume and feed their young on fish (Elliott et al. 1998a, 2005). Recent studies also showed the eagle–heron relationship to be more complicated. Herons which nested in close proximity to eagle nests appear to obtain protection from other foraging eagles in return for some loss of chicks (Jones 2009). That finding is consistent with a so-called predator protection or 'sleeping with the enemy' hypothesis (Quinn and Ueta 2008).

There remains the potential for a negative relationship between PCDD and PCDF contamination and the sensitivity of the herons' to disturbance by humans or predators. Although the physiological mechanisms are not understood, some chlorinated hydrocarbon compounds have been associated with aberrant parenting behavior in avian species in the laboratory (Peakall and Peakall 1973; McArthur et al. 1983) and in some field situations (reviewed in: Harris and Elliott 2011). Thus, contaminants in incubating adult herons may have affected their nest attentiveness and subsequent susceptibility to disturbances.

In addition to the 'realtime' exposure and effects of dioxins on adult herons, another mechanism may have been in play. In the early 1990s, research with laboratory rodents showed that exposure to very low concentrations of 2,3,7,8-TCDD during early development, in the womb or from maternal milk, could have significant consequences for later sexual development particularly of male animals (Mably et al. 1992; Peterson et al. 1993). Subsequent studies with a variety of PCBs and polybrominated diphenyl–ether flame retardant compounds have corroborated those findings (e.g. Dickerson and Gore 2007). Overall there is a lack of such research on birds. Avian work has focused primarily on the possible estrogenic or anti-androgenic effects of early exposure to DDE (Fry and Toone 1981; Helander et al. 2002; Holm et al. 2006; Iwaniuk et al. 2006) rather than delayed responses to PCBs or PCDDs. However, some more recent laboratory work with avian models has shown later effects on reproductive and related endpoints from in ovo exposure to PCBs (Fernie et al. 2001; Hoogesteijn et al. 2008).

The embryos of herons, cormorants, eagles, and ospreys collected during the 1980s and 1990s near any of the bleached kraft pulp mills had PDDD and PCDF TEQ concentrations in the range of hundreds of pg/g. Although speculative, it possible that male birds exposed during early development to pulp mill derived dioxins, later suffered such neuroendocrine impairments with subtle effects on reproductive behaviors during mating or later during incubation There are repeated hints from the data, such as the consistent poor reproductive success of herons near Crofton and other mill sites, and of the more highly exposed bald eagle south of the Crofton mill.

The Regulatory Process

I have often wondered if the evidence for wildlife alone would have been sufficient cause for the pulp and paper industry to have spent any money, never mind hundreds of millions of dollars, to reduce output of chlorinated waste, especially dioxins?[4] Perhaps it is a moot point, as the concerns for wildlife were only a part of the issue. Large areas of the coast were closed to commercial harvest of crab, shellfish, and fishes. The industry also faced consumer boycotts and 'Reach for Unbleached' campaigns. Regulators in many major markets in the USA and Europe were defining low limits on the amount of dioxins permitted in food contact products, such a coffee filters, diapers, and paper towels.

Involvement with the regulatory process meant many meetings with government and industry representatives to present and discuss the implications of our findings. Phil Whitehead and I made numerous presentations to technical and public forums. At the request of management, we published the first account of the heron contamination in a Canadian government technical forum (Elliott et al. 1988a, b). The contamination of herons at Crofton and by implication their food chain as evidenced by the fishery closures, and the intense publicity played some role in the pre-emptive changes made at that mill site, and soon after at all Canadian mills (see for example: Servos et al. 1996).

In 1992 the Pulp and Paper Regulatory Framework was made law, consisting of the *Pulp and Paper Effluent Regulations* (PPER) under the *Fisheries Act* and two regulations under the *Canadian Environmental Protection Act* (Government of Canada 1992). The PPER revoked and replaced an earlier set of regulations passed in 1971. The 1971 regulations were considered deficient in that they were legally binding only on new mills built after the regulation's November 2, 1971 promulgation date. That exempted many of the mills in British Columbia, such as Crofton, which were built before that date. The new regulations were designed to ensure that all mills were subject to regulatory requirements. The PPER set discharge limits for Biological Oxygen Demand (BOD) and Total Suspended Solids (TSS), and prohibited the discharge of effluent that was acutely lethal to rainbow trout, based on a standard assay.[5] The new regulations developed under CEPA specifically targeted the formation of chlorinated dioxins and furans during pulp bleaching. The *Pulp and Paper Mill Effluent Chlorinated Dioxins and Furans Regulations* prohibited the release of measurable concentrations of the 2,3,7,8-chlorinated dioxin and furan in effluent from mills that used chlorine or chlorine dioxide to bleach pulp. The *Pulp*

[4]Cost estimates for 96 Canadian pulp and paper mills to become compliant with new effluent regulations were 2.2 billion dollars at an average cost of $ 23.8 million per mill. Costs of meeting PCDD and PCDF regulations were estimated at $500 million for 47 affected mills (Environment Canada, 1990, Cleaning up pollution in the pulp and paper industry: an overview of the federal regulatory strategy. Environmental Protection Branch, Environment Canada, Ottawa, ON).

[5]For more details on treatment of pulp mill effluent, consult Biermann (1996). Processes are labeled as primary, secondary, and sometimes, tertiary treatment. Primary treatment, removes solids by allowing them to settle out was widely implemented in the 1950s. Currently, most mills also employee at least a secondary treatment process involving use of oxygen and microorganisms to remove oxygen-consuming materials, which significantly decreases toxicity of the effluent. In Canada the most common secondary steps are aerated stabilization basins and activated sludge. Some mills also have a tertiary process following the secondary treatment normally to reduce odor and color.

and Paper Mill Defoamer and Wood Chip Regulations, targeted the defoamers used in chlorine bleaching processes and prohibited the manufacture of pulp from wood chips treated with polychlorinated phenols.

In 1992, the British Columbia government moved to further regulate the industry by enacting the Pulp and Paper Mill Liquid Effluent Control Regulations, which provided a 10-year staged process to eliminate chlorinated organic compounds from effluent by requiring increasingly strict controls on AOX. Alberta had already set a high standard in its 1988 legislation, which required all mills to implement best available technologies to minimize the impact of effluent on receiving waters, a decision at that time based largely on application of the precautionary principle.

Implementation of those new regulations led to industry-wide improvements in the treatment processes. Significant changes in the chemical composition and toxicity of the final effluent and significant reductions in hazard to aquatic organisms were demonstrated (e.g. Bothwell 1992; Servos et al. 1996; Dubé and McLatchy 2000). Previous sections have discussed how PCDD and PCDF contamination of wildlife decreased both at coastal and interior aquatic sites. Those declines were also documented in sediments and other biota (Hagen et al. 1997; Macdonald et al. 1998; Bright et al. 1999; Yunker et al. 2002). In the late 1990s, studies were beginning to document early signs of recovery in fish reproductive parameters at a number of mills that had modernized their processes (Munkittrick et al. 1997). Those improvements to pulp mill effluent treatment have reportedly also reduced toxicity events and the effects of eutrophication on benthic invertebrate communities (Felder et al. 1998; Chambers et al. 2000; Culp et al. 2000a, b; Lowell and Culp 2002). In addition, the PPER required implementation of an Environmental Effects Monitoring (EEM) program at each mill in order to monitor the quality of mill effluent and to provide long-term data on the impacts on Canadian aquatic ecosystems. As a result of those changes, all marine and freshwater systems in Canada receiving pulp and paper effluents or runoff from sawmills and log sorts are now much cleaner than they were prior to the early 1990s.

Conservation Gains

Presently, on both the south coast and the interior rivers of British Columbia, the eggs and offspring of wildlife species, including herons, cormorants, grebes, seaducks, eagles, ospreys, mink, and otter are exposed to concentrations of PCDDs and PCDFs, which are in many cases several orders of magnitude lower than pre 1990s. In the case of the herons, we documented the improvement of the physiological and morphological responses in concert with the reduced exposure in the Crofton area (Sanderson et al. 1994a, b). It seems reasonable to suppose that other species would have exhibited similar improvements in those health parameters. Given variation in species sensitivity and lack of firm information about developmental and behavioral effects of those chemicals, it is difficult to confidently assess the degree of later injury caused by exposure during early development. Some impact at the population level is possibly evident from the consistent poor reproductive success of both herons and eagles in the number of the zones most heavily impacted by pulp mill pollution. Unfortunately, there is limited subsequent information other than reports that herons have returned to breed successfully in the Crofton area.

Overall, removal of that contamination reduced a significant anthropogenic stress on the populations. In the process, we learned a great deal about the basic ecology and reproductive biology, of some key species, which has aided in assessment of overall habitat loss and degradation.

Summary and Lessons Learned

These are complex societal issues eventually fought out, at least in democratic societies, in courts of public opinion and, therefore, at political levels (e.g. Harrison 2002; Lougheed 2009, this volume). Truth and facts may often be obscured in the arguments made by the various opponents and proponents of the given commercial activity causing the pollution (Elliott et al. Introduction to this volume). So, it is incumbent on the scientists working for both parties to make truthful and factual arguments, and to resist tendencies to over extend interpretation, whether in defense of the wildlife resource or the commercial client. Involvement in this and related issues has personally emphasized the need to focus on producing the best scientific data with the available resources. It may become difficult if the issue expands along with related and extraneous demands on one's time. It is also important, not to ignore, but to remain skeptical about threats such as job losses and damage to commerce. That is for the regulators and ultimately politicians to work out using cost-benefits models. In this case, most of the mills continue to operate, and profitably it seems. Some mills have closed for wider reasons, such as depleted timber supplies and market factors including competition from lower cost producers elsewhere, related to labor costs and environmental and other regulations. That is a complex topic in environmental economics and beyond the scope of this article (see, for example: Norberg-Bohm and Rossi 1998; Clarkson et al. 2004).

This work provides another lesson in how wildlife and particularly fish-eating and predatory birds make elegant and efficient sentinels of ecosystem contamination by persistent bioaccumulative contaminants. In the examples discussed here from British Columbia, but also from the Great Lakes, there was evidence of a contamination problem in wildlife many years before there was corroborating evidence from other levels of the foodchain, or from sampling of nonbiota, such as sediment. The work also further underscores the power of charismatic wildlife to attract public attention and galvanize people to respond to such evidence.

Looking ahead, there are a number of observations if we wish to maintain the capability for this type of investigation, and ways to improve. These include:

- The need for in-house expertise in government agencies, both in the laboratory to adapt and develop methods for new chemicals and other threats to wildlife health, and in the field to identify and address problems before they reach a critical stage.

- The need to apply that expertise to assess new and emerging chemical threats as was done here in the case of dioxins, and more recently in the case of the brominated flame retardant chemicals (e.g. Elliott et al. 2005; McKinney et al. 2006; Cesh et al. 2010).
- The fostering of close cooperation between researchers, risk assessors, and regulators in the evaluation and monitoring of new chemicals.
- The maintenance of tissue archives such as the National Specimen Bank.
- The continuous development and refinement of wildlife monitoring techniques, particularly the use of nondestructive and ideally non-intrusive techniques such as fecal sampling.
- The development and application of appropriate study designs, many of which, such as the sample egg technique, have been around for some time but are not always applied (e.g. Blus 1984; Custer et al. 1999; Henny et al. 2009), including use of blood, feather, feces, etc.
- Continuous awareness and adaptation of improvement in data analysis and statistical methods such as information theoretic, and probabilistic approaches (Burnham and Anderson 2001; and see examples: Elliott et al. 2009; Custer et al. 2010; Cesh et al. 2010; Best et al. 2010).
- The continued development and application of a landscape approach using spatial analysis methods.
- The development and refinement of biomarkers, particularly using new genomics technologies.
- The application of other molecular technologies, such a DNA genotyping, into population level assessments (e.g., Guertin et al. 2010).

Acknowledgments I am indebted to the many people who contributed to this work, in particular, Phil Whitehead. I would also like to thank: D. Bennet, G. Bellward, A. Breault, R. Butler, K. Cheng, C. Coker, M. Harris, L. Hart, D. Janz, S, Lee, I. Moul, R. Norstrom, M. Simon, T. Sanderson, T. Sullivan, L Wilson, and H. Won. Useful comments on an earlier version of the manuscript were made by T. Augspurger, M. Gilbertson, and P. Ross. S. Lee assisted with drafting figures.

References

Addison RF, Ikonomou MG, Smith TG (2005) PCDD/F and PCB in harbour seals (*Phoca vitulina*) from British Columbia: response to exposure from pulp mill effluents. Mar Environ Res 59:165–176

Amendola G, Barna D, Blosser R, LaFleur L, McBride A, Thomas F, Tiernan T, Whittemore R (1989) The occurrence and fate of PCDDs and PCDFs in five bleached kraft pulp and paper mills. Chemosphere 18:1181–1188

Ames PL (1966) DDT residues in the eggs of the osprey in the northeastern USA and their relation to nest success. J Appl Ecol 3:87–97

Augspurger TP, Tillitt DE, Bursian SJ, Fitzgerald SD, Hinton DE, Di Giulio RT (2008) Embryo toxicity of 2,3,7,8-tetrachlorodibenzo-p-dioxin to the wood duck (Aix sponsa). Arch Environ Contam Toxicol 55:659–669

Aulerich RJ, Ringer RK (1977) Current status of PCB toxicity to mink, and effect on their reproduction. Arch Environ Contam Toxicol 6:279–292

Beck U (1992) Risk society: towards a new modernity. Sage, London

Bellward GD, Norstrom RJ, Whitehead PE, Elliott JE, Bandiera SM, Dworschak C, Chang T, Forbes S, Cadario B, Hart LE, Cheng KM (1990) Comparison of polychlorinated dibenzodioxin levels with hepatic mixed-function oxidase induction in great blue herons. J Toxicol Environ Health 30:33–52
Berry RM, Fleming BI, Voss RH, Luthe CE, Wrist PE (1989) Towards preventing the formation of dioxins during chemical pulp bleaching. Pulp Paper Can 90:48–58
Best DA, Elliott KH, Bowerman WW, Shieldcastle M, Postupalsky S, Nye PE, Kubiak TJ, Tillitt DE, Elliott JE (2010) Productivity, embryo and eggshell characteristics and contaminants in bald eagles from the Great Lakes, U.S.A. Environ Toxicol Chem 29:1581–1592
Biermann CJ (1996) Pulping and paper making. Academic, San Diego, California
Blus LJ (1984) DDE in birds' eggs: comparison of two methods for estimating critical levels. Wilson Bull 96:268–276
Bothwell ML (1992) Eutrophication of rivers by nutrients in treated kraft pulp mill effluent. Water Pollut Res J Can 27:447–472
Bowes GW, Simoneit BR, Burlingame AL, de Lappe BW, Risebrough RW (1973) The search for chlorinated dibenzofurans and chlorinated dioxins in wildlife populations showing elevated levels of embryonic death. Environ Health Perspect 5:191–198
Bradlaw JA, Casterline JL (1979) Induction of enzyme activity in cell culture: a rapid screen for detection of planar polychlorinated organic compounds. J Assoc Off Anal Chem 62:904–916
Braune BM, Malone BJ, Burgess NM, Elliott JE, Garrity N, Hawkings J, Hines J, Marshall H, Marshall WK, Rodrigue J, Wakeford B, Wayland M, Weseloh DV, Whitehead PE (1999) Chemical residues in waterfowl and gamebirds harvested in Canada, 1987–1995. Canadian Wildlife Service, Technical Report Series, No 326, Ottawa ON, Canada
Bright DA, Cretney WJ, Macdonald RW, Ikonomou MG, Grundy SL (1999) Differentiation of polychlorinated dibenzo-p-dioxin and dibenzofuran sources in coastal British Columbia, Canada. Environ Toxicol Chem 18:1097–1108
Burnham KP, Anderson DR (2001) Kullback-Leibler information as a basis for strong inference in ecological studies. Wildl Res 28:111–119
Butler RW (1997) The great blue heron. UBC, Vancouver BC, Canada, p 167
Campbell RW, Dawe NK, McTaggert-Cowan I, Cooper JM, Kaiser GW, McNall MCE (1990) The birds of British Columbia, vol 1. Victoria, BC, Canada, 514 pp
Carson R (1962) Silent spring. Houghton Mifflin, New York
Cesh L, Elliott KH, McKinney M, Quade S, Maisonneuve F, Garcelon DK, Sandau CD, Letcher RJ, Williams TD, Elliott JE (2010) Polyhalogenated aromatic hydrocarbons and metabolites: relation to circulating thyroid hormone and retinol in nestling bald eagles (*Haliaeetus leucocephalus*). Environ Toxicol Chem 29:1301–1310
Chambers PA, Dale AR, Scrimgeour GC, Bothwell ML (2000) Nutrient enrichment of northern rivers in response to pulp mill and municipal discharges. J Aquat Ecosyst Stress Recov 8:53–66
Clarkson PM, Li Y, Richardson GD (2004) The market valuation of environmental capital expenditures by pulp and paper companies. Accounting Rev 79:329–353
Culp JM, Lowell RB, Cash KJ (2000a) Integrating mesocosm experiments with field and laboratory studies to generate weight-of-evidence risk assessments for large rivers. Environ Toxicol Chem 19:1167–1173
Culp JM, Podemski CL, Cash KJ (2000b) Interactive effects of nutrients and contaminants from pulp mill effluents on riverine benthos. J Aquat Ecosyst Stress Recov 8:67–75
Custer TW, Custer CM, Hines RK, Gutreuter S, Stromborg KL, Allen PD, Melancon MJ (1999) Organochlorine contaminants and reproductive success of double-crested cormorants from Green Bay, Wisconsin, USA. Environ Toxicol Chem 18:1209–1217
Custer TW, Custer CM, Gray BR (2010) Polychlorinated biphenyls, dioxins, furans, and organochlorine pesticides in spotted sandpiper eggs from the upper Hudson River basin, New York. Ecotoxicology 19:391–404
Czuczwa JM, Hites RM (1984) Environmental fate of combustion-generated polychlorinated dioxins and furans. Environ Sci Technol 18:444–450

Czuczwa JM, McVeety BD, Hites RA (1984) Polychlorinated dibenzo-p-dioxins and dibenzofurans in sediments from Siskiwit Lake, Isle Royale. Science 226:568–569

Dickerson SM, Gore AC (2007) Estrogenic environmental endocrine-disrupting chemical effects on reproductive neuroendocrine function and dysfunction across the life cycle. Rev Endocr Metab Disorders 8:143–159

Dubé MG, McLatchy DL (2000) Endocrine responses of *Fundulus heteroclitus* to effluent from a bleached-kraft mill before and after installation of reverse osmosis treatment of a waste stream. Environ Toxicol Chem 19:2788–2796

Elliott JE (1984) Collecting and archiving wildlife specimens in Canada. In: Lewis RA, Stein N, Lewis CW (eds) Environmental specimen banking and monitoring as related to banking. Martinus Nijhoff, The Hague, pp 45–64

Elliott JE, Martin PA (1998) Chlorinated hydrocarbons in grebes and seaducks wintering on the coast of British Columbia, Canada, 1988–93. Environ Monitor Assess 53:337–362

Elliott JE, Norstrom RJ (1998) Chlorinated hydrocarbon contaminants and productivity of bald eagle populations on the Pacific coast of Canada. Environ Toxicol Chem 17:1142–1153

Elliott JE, Butler RW, Norstrom RJ, Whitehead PE (1988b) Levels of polychlorinated dibenzo-dioxins and polychlorinated dibenzofurans in eggs in Great Blue Herons (*Ardea herodias*) in British Columbia, 1983–1987: possible effects on reproductive success. Canadian Wildlife Service Progress Notes, Ottawa, ON, No. 176, p 7

Elliott JE, Norstrom RJ, Kennedy SK, Fox GA (1988a) Trends and effects of environmental contaminants determined from analysis of archived wildlife samples. In: Wise SA, Zeisler GM (eds) Progress in environmental specimen banking. National Bureau of Standards (U.S.) Spec. Publ. No. 740:131–142

Elliott JE, Butler RW, Norstrom RJ, Whitehead PE (1989) Environmental contaminants and reproductive success of Great Blue Herons *Ardea herodias* in British Columbia, 1986–87. Environ Pollut 59:91–114

Elliott JE, Norstrom RJ, Lorenzen A, Hart LE, Philibert H, Kennedy SW, Stegeman JJ, Bellward GD, Cheng KM (1996a) Biological effects of polychlorinated dibenzo-*p*-dioxins, dibenzofurans, and biphenyls in bald eagle (*Haliaeetus leucocephalus*) chicks. Environ Toxicol Chem 15:782–793

Elliott JE, Norstrom RJ, Smith GEJ (1996b) Patterns, trends and toxicological significance of chlorinated hydrocarbons and mercury in bald eagle eggs. Arch Environ Contam Toxicol 31:354–367

Elliott JE, Wilson LK, Langelier KE, Norstrom RJ (1996c) Bald eagle mortality and chlorinated hydrocarbons in liver samples from the Pacific coast of Canada, 1988–1995. Environ Pollut 94:9–18

Elliott JE, Machmer MM, Henny CJ, Whitehead PE, Norstrom RJ (1998a) Contaminants in ospreys from the Pacific Northwest: I. Trends and patterns in polychlorinated dibenzo-p-dioxins and dibenzofurans in eggs and plasma. Arch Environ Contam Toxicol 35:620–631

Elliott JE, Moul IE, Cheng KM (1998b) Variable reproductive success of bald eagles on the British Columbia coast. J Wildl Manage 62:518–529

Elliott JE, Henny CJ, Harris ML, Wilson LK, Norstrom RJ (1999) Chlorinated hydrocarbons in livers of American mink (*Mustela vison*) and river otter (*Lutra canadensis*) from the Columbia and Fraser River basins, 1990–1992. Environ Monit Assess 57:229–252

Elliott JE, Harris ML, Wilson LK, Whitehead PE, Norstrom RJ (2001a) Monitoring temporal and spatial trends in polychlorinated dibenzo-*p*-dioxins (PCDDs) and dibenzofurans (PCDFs) in eggs of great blue heron (*Ardea herodias*) on the coast of British Columbia, Canada, 1983–19. Ambio 30:416–428

Elliott JE, Wilson LK, Henny CJ, Trudeau SF, Leighton FA, Kennedy SW, Cheng KM (2001b) Assessment of biological effects of chlorinated hydrocarbons in osprey chicks. Environ Toxicol Chem 20:866–879

Elliott KH, Gill CE, Elliott JE (2005) Influence of tides and weather on bald eagle provisioning rates. J Raptor Res 39:99–108

Elliott JE, Guertin DA, Balke JME (2008) Chlorinated hydrocarbon contaminants in faeces of coastal marine river otters. Sci Total Environ 397:58–71

Elliott KH, Cesh LS, Dooley JA, Letcher RJ, Elliott JE (2009) PCBs and DDE, but not PBDEs, increase with trophic level and marine input in nestling bald eagles. Sci Total Environ 407:3867–3875

Fanelli R, Bertoni MP, Castelli MG, Chiabrando C, Martelli GP, Noseda A, Garattini S, Binaghi C, Marazza V, Pezza F (1980a) 2,3,7,8-tetrachlorodibenzo-p-dioxin toxic effects and tissue levels in animals from the contaminated area of Seveso, Italy. Arch Environ Contam Toxicol 9:569–577

Fanelli R, Castelli MG, Martelli GP, Noseda A, Garattini S (1980b) Presence of 2,3,7,8-tetrachlorodibenzo-p-dioxin in wildlife living near Seveso, Italy: a preliminary study. Bull Environ Contam Toxicol 24:460–462

Felder DP, D'Surney SJ, Rodgers JH Jr, Deardorff TL (1998) A comprehensive environmental assessment of a receiving aquatic system near an unbleached kraft mill. Ecotoxicology 7:313–324

Fernie KJ, Smits JE, Bortolotti GR, Bird DM (2001) In ovo exposure to polychlorinated biphenyls: reproductive effects on second-generation American kestrels. Arch Environ Contam Toxicol 40:544–550

Firestone D (1973) Etiology of chick-edema disease. Environ Health Perspect 5:59–66

Friedman L, Firestone D, Horwitz W, Banes D, Anstead M, Shue G (1959) Studies of the chick edema disease factor. J Assoc Offic Agr Chem 42:129–140

Fry DM, Toone CK (1981) DDT-induced feminization of gull embryos. Science 213:922–924

Gilbertson M (1974) Pollutants in breeding herring gulls from the lower Great Lakes. Can Field Nat 88:273–280

Gilbertson M (1975) A Great Lakes tragedy. Nat Can 4:22–25

Gilbertson M, Reynolds L (1972) DDE and PCB in Canadian birds 1969 to 1972. Canadian Wildlife Service. Occasional Paper No. 19, p 17

Gilbertson M, Kubiak T, Ludwig J, Fox G (1991) Great Lakes embryo mortality, edema, and deformities syndrome (GLEMEDS) in colonial fish-eating birds: similarity to chick-edema disease. J Toxicol Environ Heath 33:455–520

Gill CE, Elliott JE (2003) Influence of food supply and chlorinated hydrocarbon contaminants on breeding success of bald eagles. Ecotoxicology 12:95–112

Gilman AP, Fox GA, Peakall DB, Teeple SM, Carroll TR, Haymes GT (1977) Reproductive parameters and egg contaminant levels of Great Lakes herring gulls. J Wildl Manage 41:458–468

Government of Canada (1992) Pulp and paper effluent regulations. Canada Gazette Part II, Registration SOR/92-269 126, 1967–2006

Grove RA, Henny CJ, Kaiser JL (2009) Osprey: worldwide sentinel species for assessing and monitoring environmental contamination in rivers, lakes, reservoirs, and estuaries. J Toxicol Environ Health B Crit Rev 12:25–44

Guertin DA, Ben-David M, Harestad A, Drouillard KG, Elliott JE (2009) Non-invasive fecal sampling links individual river otters to chlorinated hydrocarbon contaminant exposure. Environ Toxicol Chem 29:275–284

Guertin DA, Ben-David M, Harestad A, Drouillard KG, Elliott JE (2010) Non-invasive fecal sampling links individual river otters to chlorinated hydrocarbon contaminant exposure. Environ Toxicol Chem 29:275–284

Hagen ME, Colodey AG, Knapp WD, Samis SC (1997) Environmental response to decreased dioxin and furan loadings from British Columbia coastal pulp mills. Chemosphere 34:1221–1229

Harding LE, Pomeroy WM (1990) Dioxin and furan levels in sediments, fish and invertebrates from fishery closure areas of Coastal British Columbia. Environment Canada Regional Data Report 90-09. Environment Canada, Pacific and Yukon Region, North Vancouver, BC, Canada, p 77

Harfenist A, Whitehead PE, Cretney WJ, Elliott JE (1993) Food chain sources of polychlorinated dioxins and furans to great blue herons (*Ardea herodias*) foraging in the Fraser River Estuary, British Columbia. Canadian Wildlife Service (CWS) Technical Report Series No. 169. CWS, Pacific and Yukon Region, Environment Canada, Delta, BC, Canada

Harris ML, Elliott JE (2011) Polychlorinated biphenyls, dibenzo-*p*-dioxins and dibenzofurans in birds. In: Beyer WN, Meador J (eds) Environmental contaminants in wildlife – interpreting tissue concentrations. CRC, New York, NY, pp 471–522

Harris ML, Elliott JE, Wilson LK, Butler RW (2003a) Reproductive success and chlorinated hydrocarbon contamination of resident great blue herons (*Ardea herodias*) from coastal British Columbia, Canada, 1977 to 1998. Environ Pollut 121:207–227

Harris ML, Wilson LK, Norstrom RJ, Elliott JE (2003b) Egg concentrations of polychlorinated dibenzo-p-dioxins and dibenzofurans in double-crested (*Phalacrocorax auritus*) and pelagic (*P. pelagicus*) cormorants from the Strait of Georgia, Canada, 1973–1998. Environ Sci Technol 37:822–831

Harris ML, Wilson LK, Elliott JE (2005) An assessment of chlorinated hydrocarbon contaminants in eggs of double-crested (*Phalacrocorax auritus*) and pelagic (*P. pelagicus*) cormorants from the west coast of Canada, 1970 to 2002. Ecotoxicology 14:607–625

Harrison K (2002) Ideas and environmental standard-setting: a comparative study of regulation of the pulp and paper industry. Governance 15:65–96

Hart LE, Cheng KM, Whitehead PE, Shah RM, Lewis RJ, Ruschkowski SR, Blair RW, Bennett DC, Bandiera SM, Norstrom RJ, Bellward GD (1991) Dioxin contamination and growth and development in great blue heron embryos. J Toxicol Environ Health 32:331–344

Hebert CE, Norstrom RJ, Simon M, Braune BM, Weseloh DV, Macdonald CR (1994) Temporal trends and sources of PCDDs and PCDFs in the Great Lakes: herring gull egg monitoring, 1981–1991. Environ Sci Technol 28:1268–1277

Helander B, Olsson A, Bignert A, Asplund L, Litzén K (2002) The role of DDE, PCB, coplanar PCB and eggshell parameters for reproduction in the white-tailed sea eagle (*Haliaeetus albicilla*) in Sweden. Ambio 31:386–403

Henny CJ, Blus LJ, Gregory SV, Stafford CJ (1981) PCBs and organochlorine pesticides in wild mink and river otters from Oregon. In: Chapman JA, Pursley D (eds) Worldwide furbearer conference proceedings, August 1980, vol 3. Frostburg, MD, USA, pp 1763–1780

Henny CJ, Rudis DD, Roffe TJ, Robinson-Wilson E (1995) Contaminants and sea ducks in Alaska and the circumpolar region. Environ Health Perspect 103:41–49

Henny CJ, Kaiser JL, Grove RA, Johnson BL, Letcher RJ (2009) Polybrominated diphenyl ether flame retardants in eggs may reduce reproductive success of ospreys in Oregon and Washington, USA. Ecotoxicology 18:802–813

Henshel DS (1998) Developmental neurotoxic effects of dioxin and dioxin-like compounds on domestic and wild avian species. Environ Toxicol Chem 17:88–98

Higginbotham GR, Huang A, Firestone D, Verrett J, Ress J, Campbell AD (1968) Chemical and toxicological evaluations of isolated and synthetic chloro derivatives of dibenzo-p-dioxin. Nature 220:702–703

Holm L, Blomqvist A, Brandt I, Brunstrom B, Ridderstale Y, Berg C (2006) Embryonic exposure to o, p'-DDT causes eggshell thinning and altered shell gland carbonic anhydrase expression in the domestic hen. Environ Toxicol Chem 25:2787–2793

Hoogesteijn AL, Kollias GV, Quimby FW, De Caprio AP, Winkler DW, DeVoogd TJ (2008) Development of a brain nucleus involved in song production in zebra finches (*Taeniopygia guttata*) is disrupted by Aroclor 1248. Environ Toxicol Chem 27:2071–2075

Huff JE, Wassom JS (1974) Health hazards from chemical impurities: chlorinated dibenzodioxins and chlorinated dibenzofurans. Int J Environ Stud 6:1–17

Interdepartmental Committee on Dioxins (1985) Dioxins in Canada: the federal approach. Environmental Protection Branch, Ottawa ON, Canada

International Joint Commission (IJC) (1978) Great Lakes Water Quality Agreement of 1978. Agreement with annexes and terms of reference, between the United States of America and Canada. 22 November 1978, Ottawa, ON, Canada, p 52

Iwaniuk AN, Koperski DT, Cheng KM, Elliott JE, Smith LK, Wilson LK, Wylie DRW (2006) The effects of environmental exposure to DDT on the brain of a songbird: changes in structures associated with mating and song. Behav Brain Res 173:1–10

Jones I (2009) Bald eagle and great blue heron. Thesis, Masters of Science, Simon Fraser University, Department of Biological Sciences

Kruuk H, Conroy JWH (1966) Concentrations of some organochlorines in otters (*Lutra lutra* L.) in Scotland: implications for populations. Environ Pollut 92:165–171

Kuehl DW, Cook PM, Batterman AR, Lothenbach D, Butterworth BC (1987) Bioavailability of polychlorinated dibenzo-p-dioxins and dibenzofurans from contaminated Wisconsin River sediment to carp. Chemosphere 16:667–679

Lougheed T (2009) Outside looking in: understanding the role of science in regulation. Environ Health Perspect 117:105–110

Lowell RB, Culp JM (2002) Implications of sampling frequency for detecting temporal patterns during environmental effects monitoring. Water Qual Res J Can 37:119–132

Luthe CE, Berry RM, Voss RH (1993) Formation of chlorinated dioxins during production of bleached kraft pulp from sawmill chips contaminated with polychlorinated phenols. Tappi J 76:63–69

Mably TA, Bjerke DL, Moore RW, Gendron-Fitzpatrick A, Peterson RE (1992) In utero and lactational exposure of male rats to 2,3,7,8-tetrachlorodibenzo-p-dioxin. III. Effects on spermatogenesis and reproductive capability. Toxicol Appl Pharmacol 114:118–126

Macdonald SM, Mason CF (1988) Observations on an otter population in decline. Acta Theriol 33:415–434

Macdonald RW, Cretney WJ, Crewe N, Paton D (1992) A history of octachlorodibenzo-p-dioxin, 2,3,7,8-tetrachlorodibenzofuran, and 3,3′,4,4′-tetrachlorobiphenyl contamination in Howe Sound, British Columbia. Environ Sci Technol 26:1544–1550

Macdonald RW, Ikonomou MG, Paton DW (1998) Historical inputs of PCDDs, PCDFs, and PCBs to a British Columbia interior lake: the effect of environmental controls on pulp mill emissions. Environ Sci Technol 33:331–337

McArthur MLB, Fox GA, Peakall DB, Philogene BJR (1983) Ecological significance of behavioural and hormonal abnormalities in breeding ring doves fed an organochlorine chemical mixture. Arch Environ Contam Toxicol 12:343–353

McKinney MA, Cesh LS, Elliott JE, Williams TD, Garcelon DK, Letcher RJ (2006) Novel brominated and chlorinated contaminants and hydroxylated analogues among North American west coast populations of bald eagles (Haliaeetus leucocephalus). Environ Sci Tech 40: 6275–6281

McPhail JD, Troffe PM (1998) The mountain whitefish (Prosopium williamsoni): a potential indicator species for the Fraser System. DOE FRAP 1998-16 Report, Environment Canada, Vancouver, BC, Canada

Minh TB, Iwata H, Takahashi S, Viet PH, Tuyen BC, Tanabe S (2008) Persistent organic pollutants in Vietnam: environmental contamination and human exposure. Rev Environ Contam Toxicol 193:213–290

Moul IE (1990) Environmental contaminants, disturbance and breeding failure at a great blue heron colony on Vancouver Island. Unpubl. MSc. Thesis, University of British Columbia, Vancouver, BC, Canada

Moul IE, Vennesland RG, Harris ML, Butler RW (2001) Standardizing and interpreting nesting records for Great Blue Herons in British Columbia. Canadian Wildlife Service (CWS) Progress Note No. 217. CWS, Pacific and Yukon Region, Vancouver, BC, Canada. Available from http://www.cws-cf.ec.gc.ca/pub/pnotes/index_e.html

Munkittrick KR, Servos MR, Carey JH, Van Der Kraak GJ (1997) Environmental impacts of pulp and paper wastewater: evidence for a reduction in environmental effects at North American pulp mills since 1992. Water Sci Technol 35:329–338

Nebert DW, Dalton TP, Okey AB, Gonzalez FJ (2004) Role of aryl hydrocarbon receptor-mediated induction of the CYP1 enzymes in environmental toxicity and cancer. J Biol Chem 279:23847–23850

Nener J, Kieser D, Thompson JA, Lockhart WL, Metner DA, Roome R (1995) Monitoring of mountain whitefish (*Prosopium williamsoni*), from the Columbia River system near Castlegar, British Columbia: health parameters and contaminants in 1992. Can Tech Rep Fish Aquat Sci, No. 2036, p 89

Norberg-Bohm V, Rossi M (1998) The power of incrementalism: environmental regulation and technological change in pulp and paper bleaching in the US. Technol Anal Stateg Manage 10:225–245

Norman D, Breault AM, Moul IE (1989) Bald eagle incursions and predation at great blue heron colonies. Colonial Waterbirds 12:215–217

Norstrom RJ, Simon M (1983) Preliminary appraisal of tetra- to octachlorodibenzodioxin contamination in eggs of various species of wildlife in Canada. In: Miyamoto J (ed) IUPAC pesticide chemistry: human welfare and the environment. Pergamon, Oxford, UK, pp 165–170

Peakall DB, Fox GA (1987) Toxicological investigations on pollutant-related effects in Great Lakes gulls. Environ Health Perspect 71:187–193

Peakall DB, Peakall ML (1973) Effects of a polychlorinated biphenyl on the reproduction of artificially and naturally incubated dove eggs. J Appl Ecol 10:863–868

Peterson RE, Theobald HM, Kimmel GL (1993) Developmental and reproductive toxicity of dioxins and related compounds: cross-species comparisons. CRC Rev 23:283–335

Pohl HR, Hicks HE, Jones DE, Hansen H, De Rosa CT (2002) Public health perspectives on dioxin risks: two decades of evaluations. Hum Ecol Risk Assess 8:233–250

Poland A, Knutson JC (1982) 2,3,7,8-tetrachlorodibenzo-*p*-dioxin and related halogenated aromatic hydrocarbons: examination of the mechanism of action. Ann Rev Pharmacol Toxicol 22:517–554

Poland A, Glover E, Kende AS (1976) Stereospecific, high affinity binding of 2,3,7,8-tetrachlorodibenzo-p-dioxin by hepatic cytosol. J Biol Chem 251:4936–4946

Poole AF (1989) Ospreys: a natural and unnatural history. Cambridge University Press, New York

Powell RL (1984) Dioxin in Missouri: 1971–1983. Bull Environ Contam Toxicol 33:648–654

Quinn JL, Ueta M (2008) Protective nesting associations in birds. Ibis 150:146–167

Ross PS, Ellis GM, Ikonomou MG, Barrett-Lennard LG, Addison RF (2000) High PCB concentrations in free-ranging Pacific killer whales, *Orcinus orca*: effects of age, sex and dietary preference. Mar Pollut Bull 40:504–515

Ross PS, Jeffries SJ, Yunker MB, Addison RF, Ikonomou MG, Calambokidis JC (2004) Harbor seals (*Phoca vitulina*) in British Columbia, Canada, and Washington State, USA, reveal a combination of local and global polychlorinated biphenyl, dioxin, and furan signals. Environ Toxicol Chem 23:157–165

Ryan JJ, Lizotte R, Lau BPY (1985) Chlorinated dibenzo-p-dioxins and chlorinated dibenzofurans in Canadian human adipose tissue. Chemosphere 14:697–706

Safe S (1984) Polychlorinated biphenyls (PCBs) and polybrominated biphenyls (PBBs): biochemistry, toxicology, and mechanism of action. Crit Rev Toxicol 13:319–393

Sanderson JT, Elliott JE, Norstrom RJ, Whitehead PE, Hart LE, Cheng KM, Bellward GD (1994a) Monitoring biological effects of polychlorinated dibenzo-p-dioxins, dibenzofurans and biphenyls in great blue heron chicks. J Toxicol Environ Health 41:435–450

Sanderson JT, Norstrom RJ, Elliott JE, Hart LE, Cheng KM, Bellward GD (1994b) Biological effects of polychlorinated dibenzo-*p*-dioxins, dibenzofurans, and biphenyls in double-crested cormorant chicks (*Phalacrocorax auritus*). J Toxicol Environ Health 41:247–265

Segstro MD, Muir DCG, Servos MR, Webster GRB (1995) Long-term fate and bioavailability of sediment-associated polychlorinated dibenzo-*p*-dioxins in aquatic mesocosms. Environ Toxicol Chem 14:1799–1807

Servos ME, Munkittrick KR, Carey JH, van der Kraak GJ (1996) Environmental fate and effects of pulp and paper mill effluents. St. Lucie, Delray Beach, FL

Smit MD, Leonards PEG, van Hattum AGM, de Jongh AWJJ (1994) PCBs in European otter (*Lutra lutra*) populations, report nr 4-97/7. Institute for Environmental Studies, Vrije Universiteit, Amsterdam, p 45

Smit MD, Leonards PEG, van Hattum AGM, de Jongh AWJJ (1994) PCBs in European otter (Lutra lutra) populations. Report R-94/7. Institute for Environmental Studies, Vrije Universiteit, Amsterdam, The Netherlands

Spitzer PR, Risebrough RW, Walker W, Hernandez R, Poole A, Puleston D, Nisbet ICT (1978) Productivity of ospreys in Connecticut–Long Island increases as DDE residues decline. Science 202:333–335

Stalling DL, Norstrom RJ, Smith LM, Simon M (1985) Patterns of PCDD, PCDF and PCB contamination in Great Lakes fish and birds and their characterization by principal components analysis. Chemosphere 14:627–643

Steidl RJ, Griffin CR, Niles LJ (1991) Contaminant levels of osprey eggs and prey reflect regional differences in reproductive success. J Wildl Manage 55:601–608

Swanson SE (1988) Dioxins in the Bleach Plant. Ph.D. Thesis, Institute of Environmental Chemistry, Umea, Sweden

Thiel DA, Martin SG, Duncan JW, Lemke MJ, Lance WR, Peterson RE (1988) Evaluation of the effects of dioxin-contaminated sludges on wild birds. In: Proceedings of the 1988 Technical Association of Pulp and Paper Environmental Conference. Charleston, SC, p 145–148

Thomas CM, Anthony RG (1999) Environmental contaminants in great blue herons (*Ardea herodias*) from the lower Columbia and Willamette Rivers, Oregon and Washington, USA. Environ Toxicol Chem 18:2804–2816

Tschirley FH (1986) Dioxin. Sci Am 254:19–22

Van Strum C, Merrell PE (1987) No margin of safety: a preliminary report on dioxin pollution and the need for emergency action in the pulp and paper industry. Greenpeace, Toronto, ON

Vennesland RG, Butler RA (2004) Factors influencing Great Blue Heron nesting productivity on the Pacific coast of Canada from 1998 to 1999. Waterbirds 27:289–296

Vermeer K, Cretney WJ, Elliott JE, Norstrom RJ, Whitehead PE (1993) Elevated polychlorinated dibenzodioxin and dibenzofuran concentrations in grebes, ducks and their prey near Port Alberni, British Columbia, Canada. Mar Pollut Bull 26:431–435

Verrett MJ (1970) Effects of 2,4,-T on man and the environment. In: Hearings before the Subcommittee on Energy, Natural Resources and the Environment of the Committee on Commerce, US Senate, Serial 91-60. US Government Printing Office, Washington, DC

White DH, Seginak JT (1994) Dioxins and furans linked to reproductive impairment in wood ducks. J Wildl Manage 58:100–106

Whitehead PE, Elliott JE, Norstrom RJ, Vermeer K (1990) PCDD and PCDF contamination of waterfowl in the Strait of Georgia, British Columbia, Canada, 1989-90. Organohalogen Compounds 1:459–462

Wiemeyer SN, Spitzer PR, Krantz WC, Lamont TG, Cromartie E (1975) Effects of environmental pollutants on Connecticut and Maryland ospreys. J Wildl Manag 39:124–139

Wiemeyer SN, Bunck CM, Krynitsky AJ (1988) Organochlorine pesticides, polychlorinated biphenyls and mercury in osprey eggs – 1970–79 – and their relationships to shell thinning and productivity. Arch Environ Contam Toxicol 17:767–787

Woodford JE, Krasov WH, Meyer ME, Chambers L (1998) Impact of 2,3,7,8-TCDD exposure on survival, growth, and behaviour of ospreys breeding in Wisconsin, USA. Environ Toxicol Chem 17:1323–1331

Yunker MB, Cretney WJ, Ikonomou MG (2002) Assessment of chlorinated dibenzo-*p*-dioxin and dibenzofuran trends in sediment and crab hepatopancreas from pulp mill and harbour sites using multivariate- and index-based approaches. Environ Sci Technol 36:1869–1878

Chapter 3
Swallows as a Sentinel Species for Contaminant Exposure and Effect Studies

Christine M. Custer

It is always satisfying to see a tree swallow apparently materialize out of thin air to investigate a nest box you just erected. Because tree swallows are secondary cavity nesters and cavity limited in most situations, they seem to have a keen search image for a nesting cavity. I have seen this phenomenon many times over the past 15 years that I have worked with this species.

C.M. Custer (✉)
US Geological Survey, Upper Midwest Environmental Sciences Center,
2630 Fanta Reed Rd., La Crosse, WI, 54603, USA
e-mail: ccuster@usgs.gov

J.E. Elliott et al. (eds.), *Wildlife Ecotoxicology: Forensic Approaches*,
Emerging Topics in Ecotoxicology 3, DOI 10.1007/978-0-387-89432-4_3,

During that time, I have investigated exposure to, and effects of, various environmental contaminants under a variety of ecological situations across a broad swath of the U.S using swallows as the model species. By having multiple study sites with differing mixtures of contaminants, I have been able to parse out effects in field studies that would be impossible otherwise. I have also been fortunate to be able to do multi-year studies that have provided insights into both normal variability of the endpoints being measured, but also to see the effects of unusual physical or ecological situations that only occasionally occur. An example of the insights made across studies was the phenomenon of buried eggs. It was initially described as female swallows behaving abnormally and actively burying her eggs. In a study along the Woonasquatucket River we found that nest abandonment was occurring that was associated with dioxin contamination. Because our box occupancy at that site was relatively low, the eggs remained visible in the nest. During a subsequent study at another location, eggs began 'disappearing'. The eggs were not really disappearing, but were being covered up when a new female began using the nest box and started her normal nest building activity in preparation for egg laying. The behavioral effect was nest abandonment not burying eggs. Our working hypothesis to explain this nest abandonment is that dioxins and/or furans interfere with prolactin, the hormone that promotes incubation once egg laying is complete. This remains to be tested along with discovering whether both dioxins and furans cause this effect or if it is specific to one or the other of these two chemical classes. As it happens, nest abandonment as a result of dioxin and furan contamination was commented on many decades ago in Great Lakes' gulls and found in a study of wood ducks in the southeastern U.S. Why this effect did not get traction or notice is unknown. This points out the value of a series of integrated studies on the same species where hypotheses raised in one study can be tested in subsequent studies. After 15 years of working with swallows, I look forward to and know that some new wrinkle, some new discovery, or some new direction will result from the next study.

Abstract Tree swallows are an important model species to study the effects of contaminants in wild bird populations and have been used extensively in studies across North America. The advantages of swallows compared to other avian species are detailed. Three case histories are provided where swallows have been successfully used in Natural Resource Damage and Ecological Risk Assessments. The final two sections of this chapter are for individuals who want more in-depth information and include a summary of the chemical classes for which there are swallow data, including effect levels when known. Information provided in this section can be used to put exposure to most classes of contaminants into context with other sites across North America. Finally, commonly used endpoints, ranging from population-level down to cellular and genetic endpoints, are discussed including considerations and pitfalls, and when further work is needed to more fully understand the role of environmental and biological variation in interpreting these endpoints.

Introduction

Toxic substances originating from industrial, agricultural, and urban sources have contaminated portions of the landscape throughout North America. Even though we wish that contamination of the environment did not exist, the ultimate question for state, provincial, and federal government regulatory agencies is whether these contaminants are causing harm to the environment and specifically harm to the biota living in those habitats. The US Environmental Protection Agency (EPA), US Fish and Wildlife Service (USFWS), and National Oceanographic and Atmospheric Administration (NOAA) conduct either Ecological Risk Assessments (ERAs by EPA) or Natural Resource Damage Assessments (NRDAs by USFWS and NOAA). In Canada, Environment Canada (EC) establishes Federal Contaminated Sites Action Plans (FCSAPs). The goal of all of these are to assess whether there is risk for impairment or injury as a result of the contamination, and if so, what appropriate remediation or mitigation actions are needed. In both Canada and the USA, the philosophy is that the "the polluter pays." When that entity no longer exists, it falls to the government to pay the cost to repair and restore the environment. Birds are often used as part of these assessments perhaps because they were one of the first taxa where serious population-level effects due to environmental contaminants were brought to the public's attention (Carson 1962) and because birds have been protected for many decades under the Migratory Bird Treaty Act in the USA, Canada, and Mexico. Birds are protected by statute in most States and Provinces as well, so there are legal mandates for the protection of these natural resources. Once areas are remediated, swallows can also be used to great advantage to assess the efficacy of those actions undertaken as a result of NRDAs, ERAs, and FCSAPs. An example of this is work being undertaken as part of the Great Lakes Restoration Initiative which began in 2010. As part of this initiative, tree swallow boxes are being installed in areas undergoing remediation. Measurements on swallows will be taken before, during, and after remediation to quantify the success of these actions.

Over the past 25 years, tree swallows (*Tachycineta bicolor*) have become an important and more widely used study species to assess contaminant exposure and effects. To understand the widespread usage of swallows in these contaminant assessments and to provide information for those who might be considering their use, a brief overview will be presented of why swallows are a useful avian model compared to other bird species. This will be followed by several case studies where swallows were used in NRDAs and ERAs. The final two sections are for individuals who want more in-depth information and include a summary of the chemical classes for which there are swallow data, including effect levels when known. For well-studied contaminants, such as polychlorinated biphenyls (PCBs) and mercury, graphs of mean concentrations from across North America are provided as a quick reference guide. There are also descriptions of the numerous endpoints and metrics, from population level down to the subcellular level, that have been used for swallows including references to other cavity-nesting species where appropriate. These latter two sections provide a road map both for what we know, and also for where further work is needed to fully develop and understand the use of many of these endpoints. For a generalized discussion of avian biosentinels in contaminant studies, see Golden and Rattner (2003) and for tree swallow in particular, see McCarty (2001/2002) and Jones (2003). For a compilation of other contaminant-related information on avian biosentinels and a database of these publications, consult the USGS Contaminant Exposure and Effects – Terrestrial Vertebrates (CEE-TV) database web site at http://www.pwrc.usgs.gov/contaminants. See Robertson et al. (1992) for general information on tree swallows.

Why Swallows?

Many bird species have been used in contaminant studies during the past half century and information on many of these species has been presented in other chapters of this book. One group of species, which has been more recently and widely used for contaminants exposure and effects studies, are cavity nesting birds such as the tree swallow (*T. bicolor*) and to a lesser extent their noncavity or cavity nesting conspecifics, the barn swallow (*Hirundo rustica*), cliff swallow (*Petrochelidon pyrrhonota*), and cave swallow (*Pseudomonas fulva*). Other cavity nesting birds used in contaminant studies include the European starling (*Sturnus vulgaris*) and house wren (*Troglodytes aedon*) and in Europe, the blue tit (*Parus caeruleus*). Tree swallows, and to a lesser extent these other species are used as a surrogate for songbirds, which as a whole are poorly studied from a contaminant-effects perspective.

The reasons that species of the swallow family (Hirundinidae) have proven to be very useful for contaminant exposure and effects studies include the following:

1. Swallows have a wide breeding distribution encompassing most of North America from Alaska, throughout Canada, to Florida. This contrasts with some other commonly used avian species which are more regionally distributed. For example, herring gulls (*Larus argentatus*) breed primarily in the Great Lakes, and across Canada and Alaska, whereas another widely used species, the osprey (*Pandion haliaetus*) tends to have a coastal distribution or more northern breeding distribution, the central part of Canada and south into the USA.
2. Swallows are associated with a larger variety of aquatic habitats than other waterbird species, including rivers, lakes, other palustrine wetlands, and even highly urbanized aquatic areas where other waterbird species are rarely found. For example, American dippers (*Cinclus mexicanus*) are associated primarily with fast-flowing mountain streams in the West, while common loons (*Gavia immer*) are mainly associated with northern lakes. The habitat specificity of these other species limits their utility from both a habitat and range standpoint.
3. Some swallow species (tree swallows and purple martins [*Progne subis*]) are secondary cavity nesting species and are often cavity limited as a result. Because of this, they will readily use artificial nest boxes and can be easily attracted to areas where information is needed. The ability to attract tree swallows to an area where data are needed is critical. Unlike the tree swallow, cliff and barn swallows require a preexisting structure (e.g., a bridge, building, or culvert) to attach their nest. These species have been used as sentinels of contamination, but only if such structures were available for nesting in the appropriate locations. In Europe, the blue tit which is also a cavity-nesting species has been used similarly to the tree swallow to assess contaminant exposure and effects.
4. Because sample size must be sufficient to provide unequivocal, statistically valid results, tree swallows present a clear advantage over most other avian species. Many nest boxes can be erected in a specific location which nearly always ensures the presence of an adequate sample size. Other aquatic species, such as the belted kingfisher (*Ceryle alcyon*) or osprey, may be good indicators of environmental contamination and are widespread across the landscape, but they are

never locally abundant and sample sizes are often too small to demonstrate effects, especially if effects are slight or the contaminated area is small.

5. Swallows tend to forage near their nest (<1 km ±), although this depends on insect abundance and other factors, so contaminant concentrations in their eggs and nestlings reflect levels of contaminants from the local area. Because of this, swallows can be used to assess contaminant exposure over very small (<1 km) as well as large spatial scales. In contrast, many other waterbird species such as herons and egrets may fly 10–20 km to a feeding location making them less suitable for assessing contaminant exposure and effects at fine geographic scales.
6. The specific location where tree swallow boxes are erected will nearly always be indicative of contamination at that site. This contrasts with some other bird species where the congruence between nesting and feeding habitats may not be as high. The requirements for a good nesting site (an island which offers predator protection, for example) may be different than good feeding habitat (shallow water with adequate food). If those two life history requirements are not colocated, then the location of a nesting colony may not provide an assessment of the contamination present at that site.
7. During the breeding season swallows are obligate aerial insectivores and they will not switch to a different diet even if other food sources become more abundant. If the nest boxes are placed in proximity to aquatic systems, then there is a high probability that the majority of the bird's diet will be emergent, aquatic insects. Terrestrial insects can also be an important component of the diet, however, in certain situations and times. The benthic aquatic insects, which are subsequently consumed by the swallows, will have accumulated contaminants that are present and bioavailable in the sediments. This diet specificity contrasts with other birds that are generalists and will opportunistically switch their diet and feeding habitat to take advantage of abundant or easily captured prey. The black-crowned night-heron (*Nycticorax nycticorax*) is an example of a generalist. The lack of consistency in diet for many other waterbird species can make interpretation of exposure data more difficult.
8. The food chain for swallows is short and less complicated than many other species, which further facilitates the interpretation of contaminant data and demonstrating the link between sediment contaminants and a biological effect. In general, only three levels link sediment contamination to larval insects that inhabit those sediments, through their aerial stages which are then consumed by the swallows. This is important because the USFWS and EPA among others use sediment benchmarks for assessing contamination, setting cleanup levels, and other aspects of toxic substance evaluations. It must be recognized, however, that benthic aquatic insects have diverse life histories which may affect the movement of contaminants from sediments to swallows.
9. Swallows are a link between the aquatic and terrestrial food chains and can transfer contaminants from aquatic systems to the terrestrial ecosystem.
10. Swallow studies often require less equipment and fewer personnel than do studies of other aquatic species. This increases sampling efficiency and reduces costs. Once the nest boxes are erected, they are relatively simple to find, check, and sample. For example, data on hundreds of swallow nests can be collected

in a day by one person, whereas it may require several people and the same amount of time to excavate and check fewer than 5–10 kingfisher burrows or to climb 5–10 trees to sample raptor nests.

11. Compared to many species, collecting data from swallow boxes is safer. Standing on the ground to check a nest box that is 4-ft above the ground is less dangerous than climbing a tree or power pole to collect samples from bald eagle (*Haliaeetus leucocephalus*) or other raptor nests. Some raptor species will physically attack an intruder, which can pose a danger to the individual collecting data at that nest.
12. Tree swallows are docile and amenable to repeated handling and observation, allowing observers to collect data on egg and nestling contamination, food habits, behavior, site fidelity, and various other aspects of reproduction and survival with minimal observer effect or bias. Conversely, precocial species, such as shorebirds, terns, and ducks where young birds leave the nest shortly after hatching, make sampling nestlings much more difficult. It is often quite hard or time consuming to find and capture these mobile young and it may not be possible to associate them with specific nests. This often limits the types of data that can be collected.
13. Swallow nests can be protected from most predators. In contrast, when collecting data from some types of colonial waterbirds, depredation by gulls (*Larus* sp.) can interfere with or dictate the timing and duration of visits. Depredation may also reduce the available sample size. Excessive depredation can reduce the type, quantity, and quality of data that are possible to collect.
14. State and Federal collecting permits are usually easily obtained for most swallow species because swallows are widespread and common and have no special protection status. Other species, such as threatened and endangered species or those that are charismatic or nest in low densities, are more difficult to study because human activity, such as collection of eggs or other data, are heavily regulated or prohibited by management agencies. For example, tissue collections are often limited to addled eggs or only a few samples for eagles, loons, and kingfishers.
15. There already exists a large body of information on the natural history of swallows. Techniques have been developed for swallows that can be readily adapted for use in contaminant studies. This information allows swallow research to progress more rapidly and efficiently in many areas than it might progress for other less well-known or well-studied species.
16. Finally, large amounts of data on contaminant exposure and effects are already available for swallows (Fig. 3.1), which allows new information to be quickly put into context. This is of particular importance to EC, EPA, and USFWS because it can allow for an initial and quick assessment of a contaminated site. The distribution of contaminant concentrations in selected tissues for several well-studied chemicals will also be provided for this purpose in this chapter.

Although there are many advantages to using swallows as the study species for environmental contamination studies, there are some caveats which need to be recognized.

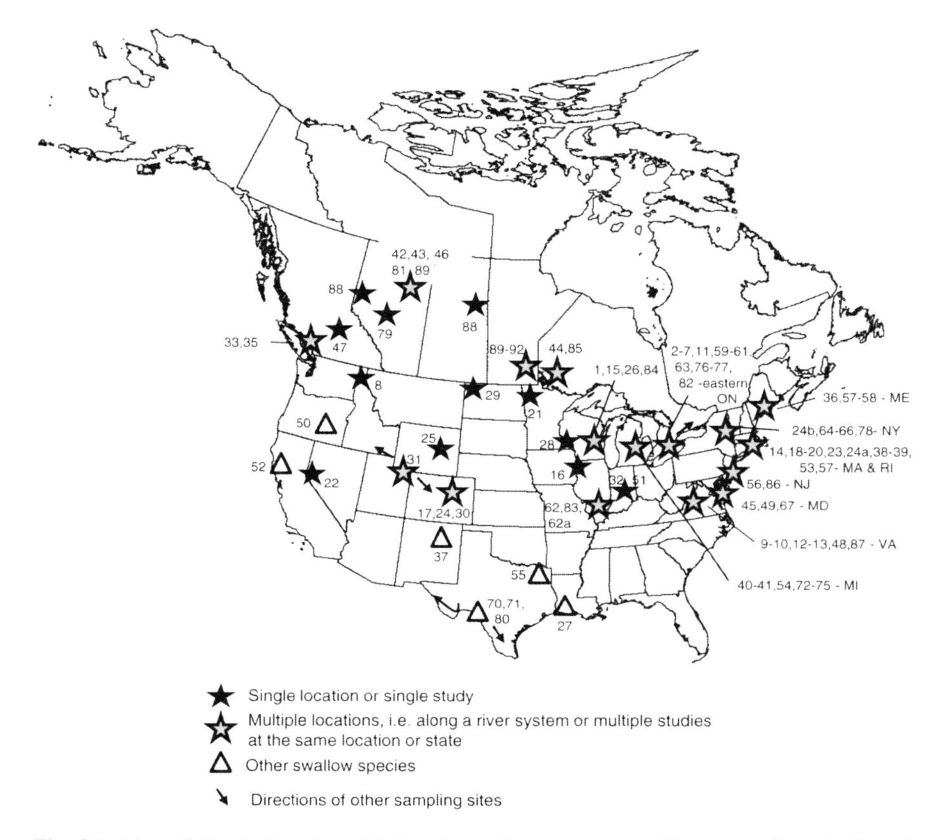

Fig. 3.1 Map of North America with locations of contaminant studies on swallows indicated. *Numbers* refer to numbers in brackets at the end of citations in the references section

1. The small sample mass of swallows (adult weight=22 g) especially their eggs (1.8 g) may require sampling multiple individuals from a nest. Although this is generally not an issue for tree swallows, because they lay 5–7 eggs per clutch, it can be more problematic for barn or cliff swallows that have smaller clutches (3–5 eggs/clutch). Recent advances in analytical chemistry have made low sample mass less of a concern.
2. Another consideration when using swallows is their trophic status. They are a mid-level consumer, rather than a top-level consumer. Because swallows are insectivores, they tend to not bioaccumulate organic contaminants to the extent that higher trophic level species, such as fish-eating birds can. There are some notable exceptions to this paradigm (C. Custer et al. 2010b), however, that are important to be cognizant of. For example, some dioxins and furans are either not accumulated or metabolized by fish (Opperhuizen and Sijm 1990), so an insectivorous species may actually accumulate higher dioxin and furan concentrations and be more likely to demonstrate effects than a piscivorous species. Also contrary to the typical

food-chain paradigm predictions, blood mercury levels were the same in belted kingfishers as in tree swallows nesting along the mercury-contaminated South River in Virginia (Cristol et al. 2008). There are some comparative data available for contaminant levels in tissues of swallows and fish-eating species at the same location, so generalized predictions and extrapolations to higher trophic levels can be made for some contaminants (T. Custer et al. 1999 for PCBs; Henny et al. 2002 and Cristol et al. 2008 for mercury).

3. Finally, swallows are migratory birds and adults could accumulate contamination on wintering grounds or during migration. This is a situation common to all contaminant studies on migratory birds, however. The use of nestling swallows and the calculation of accumulation rates avoids this potential problem. Calculation of accumulation rates (mass of a contaminant accumulated per day) and other techniques allow for the assessment of local versus previous exposure. See further discussions about accumulation rate calculations in the Section "Accumulation Rate." Additionally, as has been found for other migratory species especially those that are income breeders, the carryover of contamination into eggs from migration and wintering area exposure is often minimal; it is even less in nestlings. For example, the amount of 2,3,7,8-tetrachlorodibenzo-*p*-dioxin (TCDD) in eggs at two sites in Rhode Island varied by a factor of almost 50 even though they nested only 1.6 km apart (C. Custer et al. 2005). There is no reason to believe that those two groups of swallows wintered or migrated through different areas and hence the exposure was of local origin.

Other box-nesting passerine species have some significant shortcomings for contaminant assessments. Bluebirds (*Sialia* sp.) do not nest densely enough to provide an adequate sample size. There are generally only one to two pairs of bluebirds in an array of 30 nest boxes. House wrens, as well as bluebirds, tend to forage on more terrestrial insects so that they are less useful when assessing aquatic contamination. Wrens, however, can be useful when waste disposal and contamination of terrestrial ecosystems has occurred, such as locations affected by abandoned mine tailings. Additionally, house wren eggs are smaller than tree swallows and their nestlings are quite energetic making them less easy to handle and sample. Starlings have the benefit of being a non-native, invasive species, but have not been used as much recently in contaminant field studies (Arenal et al. 2004; Reynolds et al. 2004). In the 1970s and 1980s, however, the USFWS had a nationwide monitoring program that used starlings to assess exposure to organic contaminants on a national scale (Bunck et al. 1987 and for references to earlier results). These programs have been discontinued, and since then few studies have used starlings. Why starlings have not been used more is unclear. Prothonotary warblers (*Protonotaria citrea*) are an aquatic species that have been used only occasionally for contaminant studies (Reynolds et al. 2004), but would seem to share some of the positive characteristics of tree swallows. They are limited by range (southeastern USA mainly), although their range complements the range of tree swallows, limited by habitat (heavily wooded water bodies), and perhaps by having low nest densities.

Tree swallows, and where necessary barn and cliff swallows, are well-suited to assess contaminant exposure and effects. They provide information on transfer and bioaccumulation of contaminants from sediment to biota in a very direct manner, they are amenable to intensive research on reproductive and survival effect endpoints that require extensive nest visitations and handling, adequate sample sizes are nearly always achievable, and there is extensive natural history information to build upon. There has also been extensive work done to develop biochemical and physiological endpoints for use with swallows.

NRDA Case Histories

NRDAs recover money to restore habitats damaged by the release of hazardous materials, but also to compensate for lost services that may have resulted from those releases. NRDAs differ from risk assessments used by EPA, EC, and others, whose primary goal is to reduce the risk posed by the releases. An NRDA is ultimately a negotiation process where the results from various lines of evidence are used, including scientific studies, to reach an agreement with the principal responsible parties about the damages caused by the contamination. These cases can range from amicable and fairly quickly resolved to highly contentious and lasting for decades, and often only finally resolved through litigation, e.g., the Coeur D'Alene, Idaho case. The scientific research conducted as part of an NRDA can run the gamut from finding no evidence of injury to documenting serious injury.

For the NRDAs and ERAs where swallows have been used, swallows were but one part of the much larger assessment. There is generally a separate group of scientists who prepare and present the actual case documentation materials. This separation of responsibilities between the research component and the regulatory and management component is one mechanism to protect the scientific credibility and independence of the research and to reduce any perception of conflict of interest. In many, if not most of the NRDAs and ERAs, multiple lines of evidence are used to build the case including not only field research data, but also laboratory studies and extensive use of modeling. We have used tree swallows as parts of NRDAs and ERA in Bayou D'Inde, Louisiana; Green Bay, Wisconsin; Housatonic River, Massachusetts; upper Arkansas River, Colorado; and the Woonasquatucket River, Rhode Island. Case studies for three locations are provided below.

Upper Arkansas River and California Gulch, Colorado

The upper Arkansas River basin in central Colorado near Leadville has hundreds of abandoned mines, many miles of underground tunnels and shafts, and large tailing deposits dotting the landscape. Historically some 75 mills and 44 smelters were active in the area beginning in the late 1800s. From these various sources, an estimated 115,000 cubic yards of fluvial tailings were deposited in the flood plain of the Arkansas River. These fluvial tailings are, and have been, actively eroded by the river and result in the ongoing releases of metals. The metals of concern include zinc, lead, cadmium, and copper which seasonally exceed state aquatic life standards.

Fig. 3.2 Map of the upper Arkansas River in central Colorado with tree swallow study sites indicated

In order to assess possible injury to birds, tree swallow boxes were distributed in 1997 and 1998 along a 100-km stretch of the upper Arkansas River beginning above the historic mining district and extending downstream (Fig. 3.2). Egg and nestling samples were collected at these sites over a 2-year period and analyzed for a suite

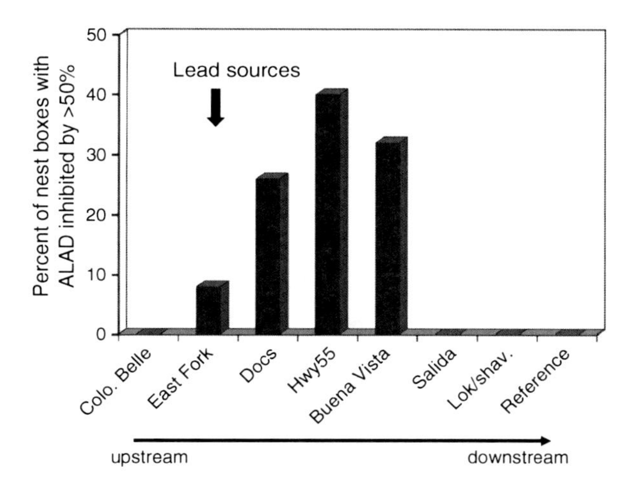

Fig. 3.3 Percent of tree swallow blood samples with ALAD inhibited by >50% at sites along the upper Arkansas River near and downstream from Leadville, Colorado in 1997 and 1998. Adapted from Fig. 28.5 in C. Custer et al. (2003a)

of 19 inorganic elements including lead. Swallow diet samples were also chemically analyzed to evaluate the association of contamination in avian tissue samples to the local aquatic environment via the food they consumed. Several metrics for documenting possible injury were used, including hatching success and a variety of biomarkers. One biomarker, δ-aminolevulinic acid dehydratase (ALAD), is according to Federal statute de facto evidence of injury in the NRDA context. There is injury if there is a decrease of ALAD activity >50% relative to a reference population. In this 2-year study, the presence of detectable quantities of lead in bird tissues followed an upstream to downstream gradient along the 100-km stretch of river and there was a significant proportion of the swallow population that had ALAD activity decreased by >50% of normal (Fig. 3.3, adapted from C. Custer et al. 2003a). There was also an association between cadmium concentrations in both carcass and liver tissues and genetic damage in red blood cells ($r=0.440$, $p=0.019$). It was the ALAD metric, however, that was most important for the NRDA case although the other data on exposure patterns and pathway were important to build the case as well.

These swallow data which documented exposure and effects from lead exposure were an integral part of the NRDA documentation (http://www.fws.gov/mountain-prairie/nrda/LeadvilleColo/CaliforniaGulch.htm). The tree swallow data were provided to the NRDA Trustees in the form of peer-reviewed publications and the Trustees were then responsible for "translating" the information for the NRDA case. The swallow data were used in conjunction with American dipper, another aquatic bird species in this assessment, and brown trout (*Salmo trutta*) data to present a compelling assessment of injury to wildlife in that system. These injuries, coupled with the surface and ground water injuries, as well as terrestrial injuries, have resulted in damages of >$50 million. Restoration plans are currently being written and implemented. Tree swallows have been proposed as one mechanism to evaluate the efficacy of the restoration actions on the upper Arkansas River. The proposal is to collect tree swallow data beginning in 2012 to assess lead and other trace element concentrations, and to quantify current ALAD activity levels. Those results will be compared to data from 1997 and 1998 to determine whether ALAD levels have decreased to within 50% of normal and whether contaminant exposure has declined.

Housatonic River, Massachusetts

The General Electric (GE) Company owned and operated a 254-acre facility in Pittsfield, Massachusetts, where PCBs were used in the manufacture of electrical transformers from the late 1930s to the 1970s. During this time period, hazardous substances were released from the GE facility to the Housatonic River and Silver Lake in Pittsfield (Federal Register 2005). These hazardous substances include PCBs, dioxins and furans (a contaminant of the PCBs), volatile and semivolatile organic compounds, and inorganic constituents (i.e., metals).

To assess the potential adverse effects of the PCB mixture as part of EPA's ERA, beginning in 1998 and for the next 2 years over 200 swallow next boxes were put up above and below the GE facility, and at a nearby reference lake unconnected to the Housatonic River. Egg and nestling samples, along with diet samples from the nestling's gastrointestinal (GI) tract, were collected for 3 years. These data provided information on exposure patterns as well as spatial and temporal variability of exposure. The primary "effect" endpoint was to establish whether there was an association between hatching success and contaminant concentrations (C. Custer et al. 2003b). Assessing hatching success is a time-consuming process. Nest boxes must be visited every 5–7 days beginning immediately prior to the beginning of egg laying. Nest visits then continue throughout incubation and until the nestlings reach 12 days of age. Because the fate of each egg in the nest must be known, it may be necessary to visit nest boxes daily for 2–3 days during the anticipated hatching period. It is essential to know exactly how many eggs hatched and why they did not hatch, if that was the case, in order to assess the possible effects of chemical contamination.

Concentrations of total PCBs along this stretch of the Housatonic River were some of the highest ever reported in birds. There was a statistically significant negative association with hatching success (Fig. 3.4). This was one of the first field studies to report such an association in swallows. This type of association is usually logistic, not linear, in shape. There is a point along the curve where the probability of reduced hatching success starts to increase significantly. This does not mean, however, that every egg with concentrations above that level fails to hatch. Along the Housatonic River, ~20 μg/g wet wt. (ww) total PCBs was the beginning of that inflection point. It is also important to note that the same negative association was present in both 1998 and 1999 providing a second level confidence that the negative association was real. See further discussion in Section "Hatching Success" about assessing reproductive success in relation to contaminant concentrations in eggs.

Like many other NRDAs and ERAs, the swallow data for the ERA on the Housatonic River were just one portion of the total assessment. Seven receptor taxa, in addition to the tree swallows, were used (EPA 2003) in that assessment. Within each of those seven receptor taxa there were often multiple endpoints and studies.

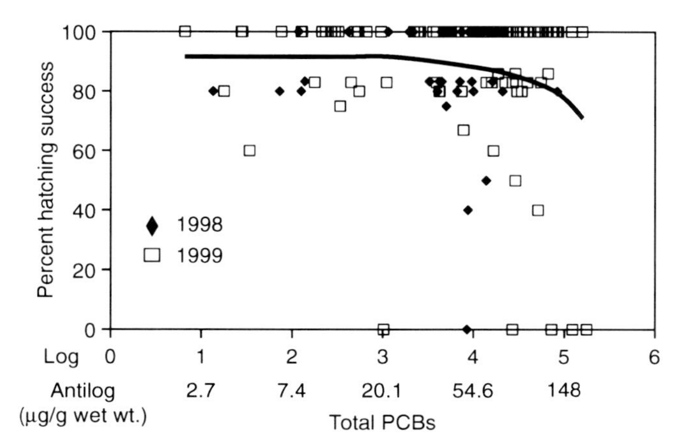

Fig. 3.4 Logistic regression of percent hatching and concentrations of total polychlorinated biphenyls (PCBs) in tree swallow eggs from the Housatonic River valley, Berkshire County, Massachusetts, in 1998 and 1999. Adapted from Fig. 5 in C. Custer et al. (2003b)

For example, within the amphibian taxa there were eight separate toxicity studies (different life stages and two different species), a community structure study, and a reproductive study on one frog species.

At this same location, but in a separate study which was not part of the ERA, adult female swallows were captured and banded each year for 5 years and the association with contaminant exposure and overwinter survival of adult birds was assessed (C. Custer et al. 2007b). There was on average a 5% reduction in adult female survival at the contaminated sites compared to the reference area. Overwinter survival, therefore, could also be an important NRDA and ERA population-level endpoint. Although neither the reproductive nor the survival effects seemed catastrophic to the local population, even a small but statistically significant diminution in these two important population-level endpoints is a cause for concern. Both the reproductive and adult survival studies on the Housatonic River were the first to definitively link population level effects with exposure to PCB mixtures in tree swallows. A population modeling effort now seems warranted to integrate these two data sets.

Woonasquatucket River, Rhode Island

The Centredale Manor Restoration Superfund site near Providence, Rhode Island, is a highly dioxin contaminated site. The dioxin contamination was most likely from a manufacturing plant that produced the antibacterial pesticide hexachlorophene. A common byproduct of the manufacture of hexachlorophene in the 1960s and 1970s was TCDD. As part of the ERA at this site, tree swallow boxes were in place in 2000 and 2001 at Allendale and Lyman Mill Ponds, two ponds within the Superfund site, and at an upstream reference location, Greystone Pond. A standard

Fig. 3.5 Pictures of the Woonasquatucket River and Centredale Manor Restoration Project activities near Providence, Rhode Island

tree swallow exposure and effects study was undertaken (Fig. 3.5). Egg and 12-day-old nestling samples, as well as, diet samples were collected and chemically analyzed for organic contaminants. The progress of nesting, the number of eggs laid and the number that hatched, was followed to assess whether there were adverse effects on reproductive success. The concentrations of TCDD in tree swallow eggs were some of the highest ever reported in bird eggs from anywhere in the world (C. Custer et al. 2005). Total PCBs and organic pesticides, however, were at background concentrations, which allowed for an assessment of dioxin effects without the confounding factor of other chemicals. Mean concentrations of TCDD were between 572 and 990 pg/g ww compared to average values of between 6 and 9 pg/g ww at the upstream reference locations. Concentrations of TCDD are often not detected in swallow egg samples from elsewhere in the USA. Additionally, TCDD was the dominant dioxin and furan congener, which is also unusual and occurred because of the very specific manufacturing process and source material. Reproductive success was reduced at the two sites with high dioxin contamination. Only 47 and 51% of the eggs hatched at Allendale Pond which is about half the normal hatching rate. The plot of hatching success and TCDD concentration in the eggs followed the expected logistic shape (Fig. 3.6) and was similar in both years of the study. Nest failures began at about 200 pg/g ww TCDD. Figure 3.6 also illustrates the large number of total nest failures. This rate of total nest failures is unusual for tree swallows, it occurred soon after egg laying was completed, and was associated with nest abandonment. Nest abandonment implies a strong behavior component

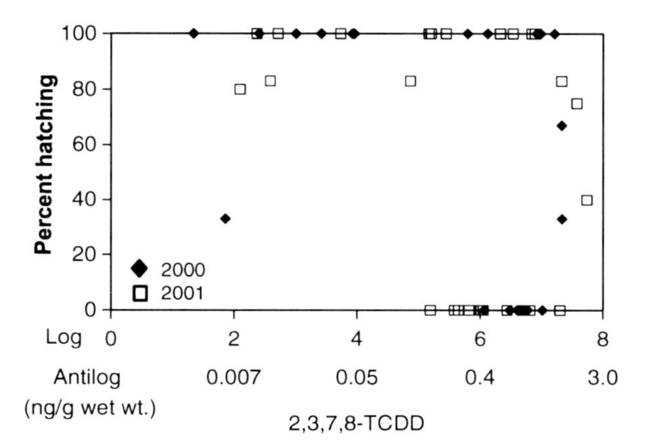

Fig. 3.6 Logistic regression of percent hatching and concentrations of 2,3,7,8-tetrachlorodibenzo-*p*-dioxin (TCDD) in tree swallow eggs from the Woonasquatucket River valley, Rhode Island in 2000 and 2001. Adapted from Fig. 5 in C. Custer et al. (2005)

rather than a direct toxicological effect within the eggs. Abandonment attributed to dioxin and furan contamination also occurred in Lake Ontario herring gulls (*L. argentatus*) in the early 1970s (Gilbertson 1983). Similarly, there were a higher percent of total nest failures in wood ducks (*Aix sponsa*) nesting in the most highly dioxin-contaminated reaches of Bayou Meto in Arkansas (White and Seginak 1994). Prolactin, a hormone associated with maintaining incubation behavior (el Halawani et al. 1980; Youngren et al. 1991), may have been insufficient to sustain incubation following exposure to dioxins and furans. It is unknown if all, or specific dioxin or furan congeners, cause this behavioral effect.

From a purely scientific standpoint, this study also allowed for an assessment of the World Health Organization's toxic equivalency factors (TEFs) for PCBs relative to TCDD. Concentrations of total PCBs averaged 0.64 μg/g ww in eggs on the Woonasquatucket River (C. Custer et al. 2005). This concentration is far below any effect level for wild birds and is less than, or equivalent to, total PCB concentrations at many locations across the USA. In spite of these background concentrations, PCB TEQs accounted for 31% of the total TEQs (220.9 pg/g TEQs for PCBs) compared to 69% for dioxin and furans (502 pg/g TEQs for dioxins and furans) at Allendale Pond. It is unlikely that there was any measurable toxicity associated with such low concentrations of PCBs and argues for a reassessment of various PCB congener TEFs as they relate to hatching success. Additionally, the lowest effect level for TCDD was approximately 250 pg/g (Fig. 3.6), which would imply that PCB TEQs were also at a level to affect the hatching success. This was clearly not the case because total PCBs averaged only 0.6 μg/g ww in eggs.

As for the two case studies above, the tree swallow information was only one part of a larger assessment. Along the Woonasquatucket River, aquatic community assessments and a 2-year census of amphibians were also done. In addition, laboratory toxicity tests were done to assess fish embryo and invertebrate survival. Currently, EPA is evaluating potential cleanup alternatives for addressing contamination in Allendale and Lyman Mill Ponds and results of this evaluation process are expected to be available to the public in 2010.

The remaining two sections provide a synopsis of swallow information by contaminant class (Section "Contaminants Measured in Birds and Effect Levels Where Known") and by endpoints (Section "Measurement Endpoints from Cellular to Population Level"). These data can be used to quickly put new data into context, help interpret the meaning of endpoints, and to identify where data gaps exist.

Contaminants Measured in Birds and Effect Levels Where Known

Metals and Other Inorganic Elements

There are approximately 14 inorganic elements, most of which are metals, but some like selenium and arsenic are metalloids or nonmetals, that are commonly analyzed in swallow tissues. Some of these elements, such as zinc, copper, and selenium, are essential in living organisms while others are either nonessential or in some cases hazardous (cadmium, mercury, and lead). The inorganic elements differ from many of the organic contaminants, especially the organic pesticides and contaminants such as PCBs, dioxins, furans, because the inorganic elements are naturally occurring in the environment some level of exposure is normal and birds have adapted to that exposure. Because of this, concentrations of the essential elements especially, are usually well regulated by the bird unless exposure is quite high and the normal homeostatic processes are overwhelmed. Other elements, such as cadmium, lead, and mercury, even though naturally occurring in the environment have no known biological purpose and adverse effects have been demonstrated in birds. An issue, therefore, with the inorganic contaminants is to separate normal, background, and essential concentrations from those that are excessive. Additionally, some elements are not accumulated in some tissues; therefore it is important to analyze the correct tissue for each of the inorganic elements. Most concentration data are present in the literature on a dry weight (dw) basis except for mercury where the basis is evenly split between ww and dw basis in the published literature.

Cadmium (Cd)

Liver and kidney tissues are preferred for diagnostic purposes. Although Cd is often found at higher concentrations in kidney and hence may be preferable to liver tissue for that reason (Scheuhammer 1987; Elliott et al. 1992), when kidney failure is imminent from Cd exposure, concentrations drop substantially in the kidney (Furness 1996). Low concentrations in the kidney, therefore, can be indicative of either very little exposure or extremely high exposure. Liver tissue, therefore, may provide a better overall measure of Cd exposure.

Kidney is the organ generally affected by cadmium based on studies of other bird species. Little data are available for Cd exposure and effects in swallows in part because it is rarely detected. Concentrations in swallow tissues are provided in C. Custer et al. (2003a, 2009), which showed elevated concentrations in swallows nesting along mining-impacted streams; little cadmium accumulation has been documented elsewhere (Blus et al. 1995; T. Custer et al. 2001, 2006, 2007; C. Custer 2005, 2006, 2007a). Cadmium is more likely to occur in marine food chains (see Elliott et al. 1992 for a review), but even in the marine environment adverse effects have rarely been documented. Along the upper Arkansas River in Colorado, however, Cd concentrations in liver tissue were correlated with genetic damage (C. Custer et al. 2003a). This effect has not been reported for any other avian species; that association needs further investigation.

Lead (Pb)

Liver tissue is most often used to assess exposure to Pb. Lead is generally not transferred to eggs unless exposure is quite high, therefore eggs are not considered a suitable diagnostic tissue. Sources of lead can be from mining activities, but also as a result of lead additives (since banned) in gasoline and the resultant deposition along roadsides from automotive exhaust. Data on Pb concentrations in eggs, livers, and whole carcasses are available in C. Custer et al. (2003a) and Grue et al. (1984). As with Cd, Pb is generally not detected in swallow tissues unless exposure is high. Concentrations of Pb in bone can be used to assess long term or historic exposure to Pb (Elliott et al. 1992; Franson 1996; Pain 1996); no data on bone concentrations are available for swallows, however.

Mercury (Hg)

Mercury concentrations are routinely assessed in egg and liver tissue, and more recently in blood and feathers of swallows. Effect levels have been established for hatchability in eggs predominantly from field studies (~5 μg/g dw, discussed in Longcore et al. 2007a) and from laboratory egg injection work (1.6 μg/g dw, Heinz et al. 2009). See Fig. 3.7 for ranges of mean Hg concentrations in swallow eggs and liver tissues from across North America. Graphed values represent one or two mean concentrations from various studies and are intended to provide a broad overview of egg and liver concentrations from across North America; consult individual publications referenced on the graph for complete details. Mean concentrations ranged as high as 9.2 μg/g dw in eggs and 4.2 μg/g in livers (C. Custer et al. 2007a). The highest mean concentration of Hg in adult blood was 3.56 μg/g ww on the Shenandoah River, Virginia (Brasso and Cristol 2008). In nestlings, the maximum mean concentration in blood was 0.23 μg/g ww at that same study area in Virginia.

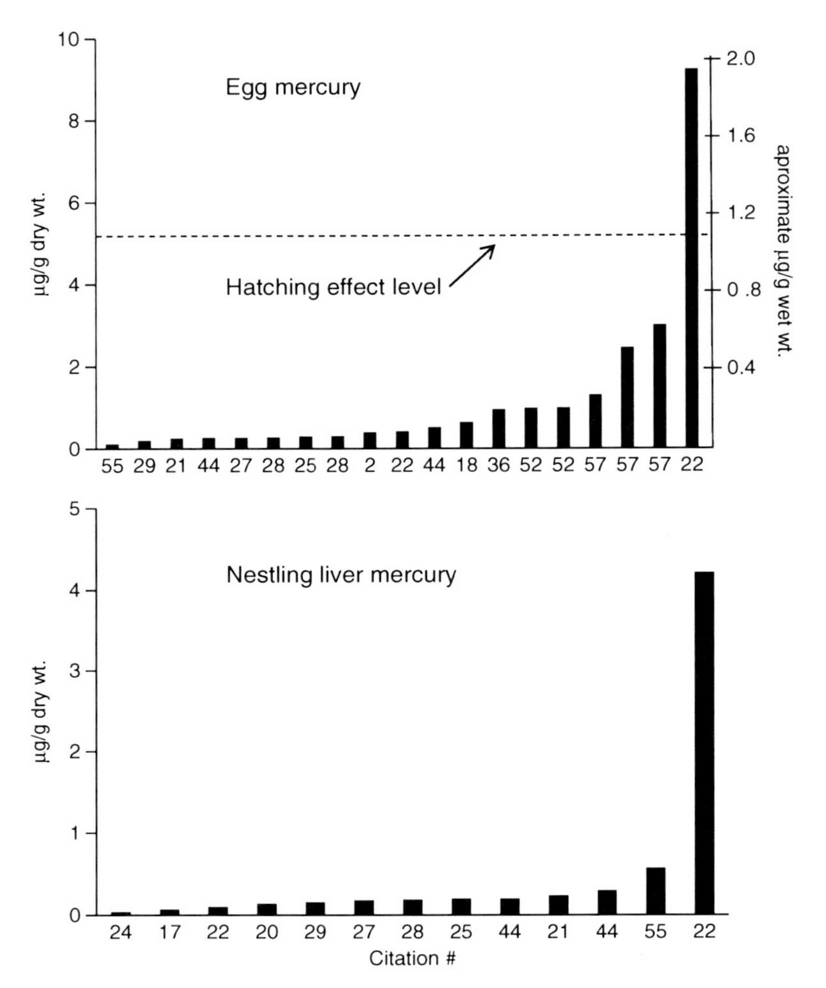

Fig. 3.7 Mean concentrations of mercury in swallow eggs (*upper panel*) and nestling liver tissue (*lower panel*) from studies across North America. Only 1 or 2 means, generally geometric means, per study are presented. Citation numbers referred to are in brackets at the end of citations in the references section

As in other birds, actively growing feathers can sequester Hg (Condon and Cristol 2009) and effectively remove it from circulation within the body. Reproductive effects from Hg exposure have been documented in field studies of swallows (C. Custer et al. 2007a; Brasso and Cristol 2008).

There are numerous forms of both organic (the methyl form predominates) and inorganic Hg in the environment, with the methyl form being the most toxic to birds (Thompson 1996). Nearly 100% of the Hg in avian blood, brain, muscle, feather, and egg tissues is methylmercury so there is little need to quantify Hg residues beyond total Hg concentrations. Henny et al. (2002) and Gerrard and St. Louis (2001), however, both postulated that methylmercury could be demethylated in the liver when exposure was very high. This may reduce the deleterious effects of mercury exposure; thus it might be advisable to analyze liver samples for both total and methyl Hg in certain situations.

Selenium (Se)

Selenium is an essential element for which there is a fine line between adverse effects from too little and adverse effects from too much Se. No study to date has found that Se has been sufficiently high or low to adversely affect tree swallows.

Organic Contaminants

Some organic contaminants accumulate in tissues and others do not. Because of this, methods used to document exposure and quantify effects differ for these two general groupings of organic chemicals. Each chemical in the following discussion is classified as accumulating or nonaccumulating. Concentrations of organics in all tissues are generally presented on a ww basis. See Peakall and Gilman (1979) for a discussion of the limitations of using a lipid-weight basis. Except in unique situations, and for specific and well-justified purposes, presenting data on a lipid wt. basis should be avoided.

For accumulating chemicals, the concentrations in tissues are correlated with the amount of exposure, i.e., high concentrations in an egg, carcass, or specific tissues indicate a high level of exposure. Nonaccumulating chemicals, however, preclude this standard tissue analysis and interpretation. These nonaccumulating chemicals are either quickly metabolized or not absorbed into the bird, but can cause effects in spite of that. For those chemicals, such as many of the petroleum or combustion products, the mere presence of a chemical in the target tissue is enough to indicate significant exposure. The concentrations are often quite low and there is no dose–response relationship. For the organophosphate (OP) insecticides, a physiological response indicates exposure. If chemical analyses are done, they are done on the food in the GI tract and not on a target tissue.

Aliphatic Hydrocarbons

Nonaccumulating. This is a group of 28+ related compounds that can be naturally occurring in the environment or can be of anthropogenic origin. Each aliphatic hydrocarbons (ALHs) congener differs in the number of carbon atoms and whether those carbon atoms are joined in straight or branched chains or in nonaromatic rings. The distribution of the various ALHs in a sample can be used to distinguish between a natural plant-based source and a petroleum source, as well as indicate chronic or current exposure to petroleum. Ratios >1.0 of pristine:n-C_{17} and phytane:n-C_{18} are indicative of chronic exposure to petroleum. A preponderance of odd-numbered ALHs is indicative of recent biological origin rather than petroleum or combustion sources.

The pristine:n-C_{17} and phytane:n-C_{18} ratios ranged between 0.10 and 0.25 in tree swallow carcasses on the North Platte River in central Wyoming indicating that those birds were not chronically exposed to petroleum (T. Custer et al. 2001). That conclusion was further confirmed by the similarity in concentration of the odd- and even-numbered ALHs.

Alkylphenol Ethoxylates

Nonaccumulating. These are nonionic surfactants sometimes present in municipal and industrial effluents. They are used in pulp and paper production, in crop protection chemicals, and in industrial and institutional cleaning products. One alkylphenol ethoxylate (AE), 4-nonylphenol (4-NP), is suspected of being an estrogen mimic. Although 4-NP was present in insects consumed by swallows, it was either not detected or detected at very low levels in nestling liver tissue (Dods et al. 2005). These ethoxylates are apparently quickly metabolized and not stored in tissues for any length of time.

There was no effect of AEs on hatching success at wastewater treatment plants contaminated with AEs near Vancouver, British Columbia, Canada; however, clutch size was reduced in 2 of 3 years at those site compared to a nearby reference area (Dods et al. 2005).

Organophosphate and Carbamate Insecticides

Nonaccumulating. These two classes of chemicals replaced many of the previously used organochlorine pesticides. The organophosphate and carbamate chemicals are less likely to accumulate in the environment and biota because they are rapidly degraded by hydrolysis upon exposure to sunlight, air, and soil. They also do not accumulate in biotic tissues. The OP pesticides, as well as the carbamate pesticides, are neurotoxins and can be acutely toxic to birds. Because they do not bioaccumulate or persist as long in the environment, the time period when they can cause adverse effects in birds is shorter than for the organochlorine insecticides that they replaced. Examples of OP chemicals are diazinon, parathion, phosmet, and many, many others.

Because the OPs do not bioaccumulate in biotic tissues, exposure to and effects from them must be assessed differently. Inhibition of cholinesterase activity (ChE) is one method of assessment. Cholinesterase activity can be measured in either brain tissue or blood plasma. Because of methodological variation, ChE levels generally need to be compared to reference material, analyzed at the same time in the laboratory.

Cholinesterase levels in the brain increase with age in nestling birds, so nestling age must be accounted for when doing these types of analyses (Grue and Hunter 1984). There is no effect of nestling age for plasma ChE (Burgess et al. 1999), however. Brain ChE activity offers a more precise measure than blood plasma sampling, but the former is lethal whereas the latter can be sampled nondestructively. If a bird does not die from exposure to the OP, the ChE activity returns to normal over the course of days or weeks. It is useful, therefore, when interpreting data to know something about the spray history before an assessment is made. Alternatively, or in conjunction with ChE analyses, GI tract contents can be analyzed for the presence of an OP or carbamate. If a bird is found dead, the brain ChE level is <50% of normal, and there is an OP present in the GI tract, then there is presumptive evidence of death from OP poisoning. A live bird can survive and have brain ChE levels <50% of normal, however, the percent inhibition is not definitive proof of anything other than exposure.

Cholinesterase levels in plasma of adult tree swallows were inhibited after apple orchards were sprayed with some, but not all OP applications (Burgess et al. 1999). On a mean basis, nestling plasma ChE activity did not vary between control and OP-treated locations; however, some individuals had as much as a 55% decrease in plasma ChE at sprayed locations (Burgess et al. 1999); this was indicative of exposure. It is unknown, however, what effects ChE depression may have on behavior or other physiological conditions. In 4 of 7 years, Bishop et al. (2000a) found a negative association between cumulative toxicity scores and daily egg and nestling survival; however, there were potentially effects from both OP and residual legacy organic pesticides. In starlings, Grue and Shipley (1984) found that nestlings were more sensitive than adults to OP exposure and that one mechanism for effects was a reduction in feeding rates by OP-exposed compared to reference starlings (Grue et al. 1982).

Perfluorinated Compounds

Accumulating. This is a group of approximately 14 chemicals that are used as stain repellents, in nonstick cookware and food packaging, as surfactants and wetting agents, and for other industrial uses. Their uses are determined by the number of fluorine atoms and the functional group that is attached. Perfluorooctanesulfonate (PFOS) is the form present in the highest concentration in birds (T. Custer et al. 2009), probably because many of the other perfluorinated compounds (PFCs) break down to this compound; PFOS appears to be quite stable and bioaccumulative. PFCs have been detected in tree swallow blood plasma (unpublished data). Work is ongoing to determine the best swallow tissues for sampling and to collect other baseline data needed to assess exposure and effects of this class of chemicals.

Polybrominated Diphenyl Ethers

Accumulating. This is a group of halogenated chemicals used as flame retardants in furniture, clothing, and hard plastics such as computer monitors and television housings.

Because polybrominated diphenyl ethers (PBDEs) are mixed into, rather than covalently bonded with the product, they can migrate out and enter the environment. They are now ubiquitous in the environment and, in contrast to many of the other organic contaminants, are increasing in the environment rather than declining. No work has been published to date on exposure or effects of PBDEs in swallows, however, adverse effects have been reported in other aquatic bird species in field situations (Henny et al. 2009). In laboratory studies, effects of PBDEs on behavioral, endocrine, developmental, and reproductive systems have been found in captive American kestrels (*Falco sparverius*, Fernie et al. 2008, 2009; McKernan et al. 2009) and starlings (Van den Steen et al. 2009). Field studies on exposure and effects in tree swallows seem warranted.

Polychlorinated Biphenyls

Accumulating. There are a total of 209 related chemicals (congeners) that comprise what are collectively called PCBs. Different combinations of these 209 congeners were used to formulate different mixtures, called Aroclors or by other trade names, which in turn had different industrial uses. Fewer than half of these PCB congeners are detected in bird tissues with any regularity (McFarland and Clarke 1989). PCBs became widely distributed in the environment through their widespread use as a nonflammable insulators and other industrial applications. Concentrations have been declining in the environment for the past 30 years in most areas. In tree swallows, PCBs are probably the mostly widely studied chemical and there are ample data on background, as well as, elevated levels (Fig. 3.8) in various tissues. The modes of action include effects on the aryl hydrocarbon receptor (cytochrome P450 enzymes), possible neurotoxic effects, and effects on second-messenger systems such as calcium movements (Harris and Elliott 2010). Effects attributed to PCBs include behavioral, physiological, and reproductive effects.

Most swallow studies assessing PCB effects on hatching success have shown no effect on reproductive endpoints probably because exposure was below an effect's threshold (Bishop et al. 1999; C. Custer et al. 1998; Neigh et al. 2006b; Papp et al. 2005). In only a few studies, where PCB concentrations were very high (>20 μg/g ww, C. Custer et al. 2003b), has an association with hatching success been convincingly established. Small, but statistically significant, effects on overwinter survival of adult tree swallows have also been documented (C. Custer et al. 2007b). Effects of PCBs on biomarkers, which have been extensively documented, will be covered in the biomarker discussions below.

Polychlorinated Dibenzo-*p*-Dioxins and Polychlorinated Dibenzofurans

Accumulating. This class of contaminants includes some of the most toxic chemicals present in the environment. They are relatively unstudied in tree swallows, however. Dioxins and furans are a group of 17 congeners that are produced by

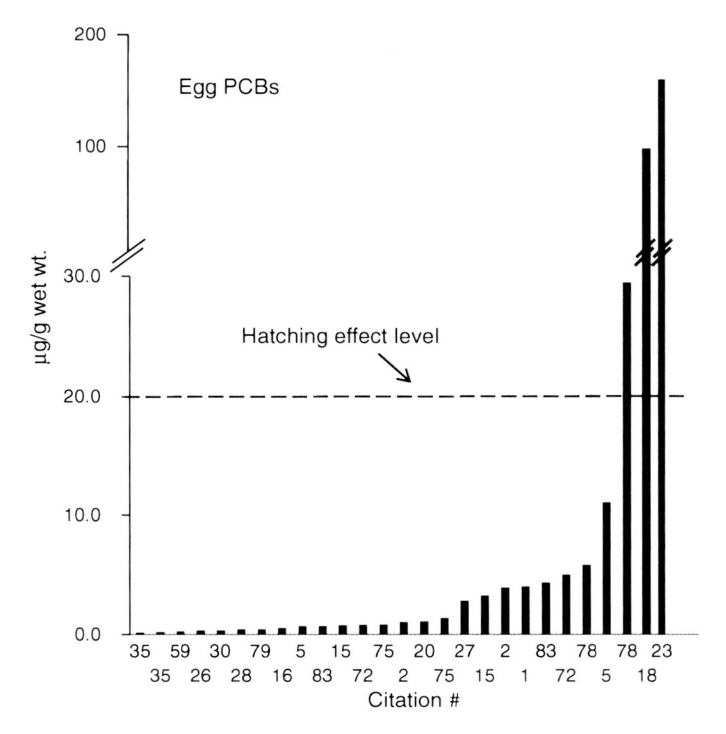

Fig. 3.8 Mean concentrations of total PCBs in swallow eggs from studies across North America. Only 1 or 2 means per study are presented. Citation numbers referred to are in brackets at the end of citations in the references section

incomplete combustion, and as a contaminant in herbicides, PCBs, and hexachlorophene formulations. Dioxins and furans were also distributed into the environment as a result of pulp and paper mill processes. Like PCB congeners, the polychlorinated dibenzo-*p*-dioxins and polychlorinated dibenzofurans (dioxins and furans or PCDD-Fs) congeners are variously distributed among bird species because of differences within their food webs. Insectivorous birds accumulated more PCDD-Fs than sympatrically nesting piscivorous birds (C. Custer 2010b), which is counter to the traditional food chain paradigm. Using multivariate statistical methods, the distribution of various congeners can sometimes be used to differentiate among possible sources. For example, some congeners (octachlorodibenzo-*p*-dioxin) are more associated with wood preservatives or as a contaminant in PCBs (mainly the furans; Harris and Elliott 2000).

Adverse reproductive effects of PCDD-Fs have been documented in swallows. Along the Woonasquatucket River, Rhode Island, TCDD exposure was four times higher in nests that failed compared to successful nests (C. Custer et al. 2005); dioxin concentrations in those swallow eggs and nestlings were some of the highest concentrations ever reported in birds and reached geometric mean concentrations of 1,013 pg/g ww. Total nest failures caused by abandonment of the clutch by the adult female seem to be the hallmark of dioxin and furan effects. This behavioral effect is similar to what was found in wood ducks (White and Seginak 1994) in the southeastern U.S and in herring gulls in the Great Lakes (Gilbertson 1983). Rivers below pulp and paper mills are variously contaminated with dioxins and furans (T. Custer et al. 2002; Harris and Elliott 2000), as are rivers that have been contaminated with certain Aroclor formulations of PCBs (C. Custer et al. 2000, 2003b). Along the Housatonic River in Massachusetts hatching success was negatively associated partially with PCDD-F concentrations (C. Custer et al. 2003b), which were probably introduced into the system as a contaminant of the PCBs used.

Polycyclic Aromatic Hydrocarbons

Nonaccumulating. This group (>40 individual compounds) of related chemicals can originate from petroleum, as a combustion byproduct, or from natural sources. They are similar to the ALHs except that they contain benzene rings. High proportions of methylated polycyclic aromatic hydrocarbons (PAHs) are indicative of a petroleum source and a lack thereof is indicative of a combustion source. These chemicals are rapidly metabolized by birds, so to quantify exposure, the GI tract and its contents are the preferred organ for analysis. PAHs can be found in whole carcasses, however, the concentration will always be low because they are so rapidly metabolized. In Wyoming, the maximum tree swallow carcass concentration was 0.06 μg/g ww (T. Custer et al. 2001). Concentrations in duck carcasses, from a highly petroleum-contaminated site, equaled 0.16 μg/g ww (T. Custer et al. 2000), which is probably the maximum concentration that might be expected in carcasses. In Louisiana, PAHs were detected in GI tract contents of barn swallows, but not in their whole carcasses (T. Custer et al. 2006).

p,p-Dichlorodiphenyldichloroethylene

Accumulating. p,p-Dichlorodiphenyldichloroethylene (*p,p*-DDE) is a metabolite of, or a co-contaminant in, the technical formulation of DDT (1,1,-trichloro-2,2-bis (*p*-chlorophenyl)ethane), a widely used pesticide that was banned in the USA in 1973. *p,p*-DDE is the isomer of DDT that was linked to the decline of numerous bird species including many raptor species, brown pelicans (*Pelecanus occidentalis*), double crested cormorants (*Phalacrocorax auritus*), and white-faced ibis (*Plegadis chihi*, Blus 1996). The most notable documented mode of action of DDE is the thinning of egg shells which then results in partial or total hatching failure.

In tree swallows in southern Ontario there was no association between mean concentrations of *p,p*-DDE in eggs and hatching success in various apple orchards (Bishop et al. 2000a) where the maximum mean concentrations of *p,p*-DDE were less than 2.56 μg/g. In British Columbia, there was no effect of *p,p*-DDE on either percent hatching or nesting success in successful nests (Elliott et al. 1994). Mean *p,p*-DDE concentrations in eggs ranged up to 11.2 μg/g ww in that study. Concentrations in other swallow studies were below levels, considered dangerous even to the most sensitive avian species (~3 μg/g ww, Blus et al. 1974; Blus 1995). The overall reduction in *p,p*-DDE exposure to biota across the USA and Canada since the ban of DDT, coupled with the lower trophic position of swallows, accounts for its lack of importance as a problematic contaminant in swallows.

Other Contaminants

Radiation (Gamma)

Accumulating. Gamma radiation is a type of ionizing radiation that is very penetrating and can cause chemical, as well as somatic and genetic changes in animals. Tissue damage is caused by the changed energy state of atoms and molecules.

In combination laboratory and field studies, Zach and Mayoh (1984, 1986) found that a single dose of gamma radiation, either of embryonated eggs or nestlings reduced the asymptotic weight of nestlings in one or two of the highest exposure groups. The nestlings in the highest exposure group in both studies also had shorter feathers and feet. There was no mortality associated with any of the exposures, however. The highest exposure in these two studies, 3.2 and 4.5 Gy, were at levels that have adverse effects on humans, including death. In two-field dosing experiments, Zach and Mayoh (1982) found that boxes with high-radiation exposure were avoided. There was no effect on asymptotic weight of nestlings or their growth rate in nest boxes that were used. The dosage, however, in the used boxes was reasonably low (3.9 μC/kg/day) compared to their other studies. There was a negative association between percent fledging and radiation exposure in 1 of 2 years, but the dosage was actually lower in the year with the significant negative association than in the year with higher exposures. This may indicate that other confounding factors were at work. In a second field-dosing experiment, where nesting swallows were continually treated with up to 45 times background exposure levels (approximately = 5.96 μGy/h), no effects of radiation on box selection, reproductive success, or on nestling growth were found (Zach et al. 1993). Barn swallows as a result of the nuclear accident at Chernobyl had mutation rates two to ten times higher than the rates at a control area (Ellegren et al. 1997) and had reduced adult survival and hatching success (Møller et al. 2005).

Measurement Endpoints from Cellular to Population Level

A major advantage of using swallows compared to some other avian species is that there are many endpoints already developed and available for use, and other endpoints in various stages of development and testing. One way to organize a discussion of endpoints is to categorize them by the level of biological organization at which they are measured. Effects at a lower level of biological organization may translate up to effects at higher levels, although this has been difficult to establish. That is a particular challenge in the entire field of ecotoxicology not solely an issue with swallows. Effects at lower levels, however, have been proffered as an early-warning system so that catastrophic population or higher level effects can be avoided. Three levels of organization will be used in this paper. Starting at the top and working down they are (1) population endpoints, (2) whole organism endpoints, and (3) suborganismal endpoints including effects on tissues and organs, cellular processes, and molecular effects. These suborganismal endpoints will be collectively termed biomarkers.

Population Level

Population endpoints are of particular interest to state and federal management agencies in their NRDAs, FCSAPs, and ERAs because of their potential for larger, more wide-spread, and longer term effects. There are only two endpoints that have been used so far to assess population level effects – births (reproduction) and deaths (survival). The other two terms in the population equation, immigration and emigration, have not been studied in swallows relative to contaminants. Additionally, nearly all of the work to date has concentrated at breeding areas. This leaves unstudied questions about nonbreeders and effects that may be occurring away from nesting sites. This is a shortcoming that is also common to many contaminant studies on birds.

Hatching Success

Hatching success is defined hereafter as the percent of eggs laid that hatch. Although simple in principle, this endpoint takes considerable field time and effort to accomplish well. Nest survival statistical techniques (e.g., Mayfield, Shafer, Kaplan–Meier) can be used to provide a more refined assessment of nest success especially if clutches are not all found when initiated. A benefit to these methods is that they can also include information on variation and uncertainty. Early hatching success studies statistically compared average concentrations (by location or before and after an event) with the hatching success among those units (Bishop et al. 1999; Brasso and Cristol 2008; C. Custer et al. 1998, 2005; Dods et al. 2005; Elliott et al. 1994; Gerrard and St. Louis 2001; Harris and Elliott 2000; Longcore et al. 2007a; McCarty and Secord 1999a; Neigh et al. 2006b; Nichols et al. 1995; Smits et al. 2000). This approach is not as useful as another, more powerful and more recently used option, which is the sample egg method (Blus et al. 1974). This method determines whether there is an association between the concentration of a contaminant in the clutch and the hatching success of that clutch. The experimental unit is the clutch rather than the location or time period. In some situations, rather than measuring exposure in the egg sample, exposure can be quantified by measuring contaminant concentration in a blood sample from the female which laid the eggs. This has the benefit of not reducing the number of eggs in the clutch, which could influence subsequent success measures, however, it requires that the female be captured and bled. See Brown and Brown (2009) for survival issues related to blood sampling in cliff swallows. Because of the small mass of a swallow blood sample, it may not be possible to analyze simultaneously for a large number of contaminants, which is possible when an egg is sampled. Regardless of whether an egg or blood sample is used, this method is more powerful because information is not lost which happens when averages across locations or time periods are used. The sample egg method removes aggregation bias, also called ecological fallacy. Ecological fallacy is the mistake of assuming that relationships found for groups (aggregated data) are also found among individuals of those groups.

Using the sample egg method is generally more expensive because added field time and effort are needed, and more contaminant analyses are usually required as well. As a rule of thumb, approximately 75 nests are needed to make this assessment, but that exact number depends on the magnitude of the effect. Examples of the use of the sample egg method with swallows include a negative association between total PCBs and PCDD-Fs and hatching success on the Housatonic River in western Massachusetts (C. Custer et al. 2003b). Along the Shenandoah River, Virginia, there was a significant but weak relationship between blood mercury concentrations and the proportion of the eggs that hatched (Brasso and Cristol 2008). Other studies that used a regression approach include Bishop et al. (2000a) in apple orchards, Zach and Mayoh (1982) with radiation gradients, and C. Custer et al. (2003a) with Pb on the upper Arkansas River. A logistic model is often preferable because an egg either hatches or it does not, a binary outcome. If the egg sample is collected at or near hatching then the number of eggs in the clutch is not reduced during the incubation period and the fate of every egg, whether it hatched or not, whether it contained a live or dead embryo, etc. can be determined. There is no consistent effect of laying order on organic contaminant concentrations (C. Custer et al. 2010a) so collecting one or two random eggs will be representative of contamination in the entire clutch.

The sample egg or sample nest method is preferable for a variety of reasons, but especially when variation is high within a location or time period. An important assumption when comparing among locations or time frames, is that those groups are relatively homogeneous. That often is not the case. For example, on the Housatonic River, Massachusetts, PCBs in eggs at one of the contaminated sites ranged between 3.0 and 187 μg/g ww (C. Custer et al. 2003b, 2007b and unpublished data). That is not a homogeneous group and one would not logically expect that a clutch with 3 μg/g PCBs would experience the same effect as a clutch with 187 μg/g total PCBs even though they were both from the same location. Another aspect of the sample egg method's better performance is in detecting more subtle effects. The importance of sample size and effect size are illustrated by data from the Woonasquatucket River, Rhode Island (C. Custer et al. 2005). In that study, there was no association between contaminant concentration in the clutch and hatching success of that clutch, even though there were significant differences in hatching success between the contaminated and reference locations. This probably resulted because the sample size was relatively small (<50 clutches, C. Custer et al. 2005), which reduced the statistical power of the logistic analysis. There were drastic effects, however, with only 50% of the eggs hatching at the contaminated sites which resulted in statistical differences between locations.

Adult Survival

Using mark-recapture methods on banded birds, the apparent survival rate can be compared between contaminated and reference locations, or before and after an event. Using these techniques, a 5% reduction in survival was documented on the highly PCB-contaminated Housatonic River (C. Custer et al. 2007b). This is one of only two studies where overwinter adult survival has been quantified in relation to contaminants in swallows. Møller et al. (2005) monitored overwinter survival in barn swallows inside and outside the exclusion zone at Chernobyl and found reduced survival inside the exclusion zone.

Organism Level

These are endpoints measured at the organism level and are metrics that are specific to individuals.

Accumulation Rate

This is a method to isolate and quantify local exposure to a contaminant. Because nestlings are only fed from the local area and this technique factors out what is contained in the egg, which may have come from elsewhere, it quantifies only local exposure. It has been used so far only with organic contaminants, but in principal it should work with at least some inorganic contaminants that bioaccumulate such as mercury. The mass (μg) of the contaminant is calculated for the egg and known-aged nestling, ideally from the same nest. The contaminant mass is simply the sample weight (g) multiplied by the contaminant concentration in that sample (μg/g). Twelve-day-old nestlings are most often used for tree swallows, but other ages can and have been used. The contaminant mass in the egg is subtracted from the contaminant mass in the nestling carcass and then divided by the age of the nestling. The results are expressed as microgram of contaminants accumulated per day by the nestling. As few as five pairs of sibling eggs and nestlings can be used to characterize a location.

Along the Housatonic River in western Massachusetts, 75.9 μg of total PCBs were accumulated per day (C. Custer et al. 2003b). At a more modestly contaminated location in Green Bay, Wisconsin, between 6.7 and 15 μg of total PCBs were accumulated per day (Ankley et al. 1993; C. Custer et al. 1998). Along the Woonasquatucket River in Rhode Island, 1,736 pg of TCDD were accumulated per day (C. Custer et al. 2003c) which compared to 49 pg/day PCDF at a modestly contaminated furan site on the Wisconsin River (T. Custer et al. 2002). If an area is uncontaminated the accumulation rate will be very small or even negative. For instance, at two reference locations near Green Bay, Wisconsin, between 0.06 and 0.42 μg of PCBs were accumulated per day (C. Custer et al. 1998) and along the Mississippi River the rate was 0.21 μg per day (C. Custer et al. 2000). Rarely is there a negative accumulation rate for PCBs because of their ubiquitous presence in the environment; the accumulation rate was negative at a reference area in Rhode Island (−0.04 μg per day, Jayaraman et al. 2009), however. Different PCB congeners are accumulated at different rates (Maul et al. 2010); this may need to be accounted for depending on the objective of the study.

Fluctuating Asymmetry

Fluctuating asymmetry (FA) is nonrandom deviation from perfect bilateral symmetry in an organism. It has been postulated that FA can occur more frequently, or be expressed more dramatically, in the face of environmental or chemical stresses that interrupt developmental stability. There are many tissues that can be used to quantify FA. Those used so far with swallows are differences in the length of the right and left wing, right and left tail feathers or tarsus, or differences between the two halves of the brain. See Teather (1996) and Hovorka and Robertson (2000) for a review of issues and concerns surrounding FA methods. For FA to be a useful endpoint, other stressors, such as weather, food shortages, etc. must be taken into account.

Tree swallow brains were not asymmetrical as a result of extreme dioxin exposure (C. Custer et al. 2005) even though dioxin effects on brain asymmetry have been hypothesized for other bird species. Møller and Mousseau (2003) measured asymmetry in a variety of barn swallow tissues, but there did not appear to be an effect of radiation on those endpoints.

Growth Rates

Growth rates are merely the change in body weight or a structural component over time. Growth has been measured mainly in nestlings either at a beginning and end point, or at multiple time steps in between the beginning and the end points. Gerrard and St. Louis (2001) found no effect of mercury exposure on growth of wing or tail feathers, bill length and width, or tarsus length, nor was there an effect on whole body weight. Longcore et al. (2007b) found only occasional differences among differentially mercury contaminated locations; in some cases the sites higher in mercury exhibited greater growth rates. Methyl mercury concentrations in feathers, however, were associated with lower body mass growth rate between 2 and 10 days of age in that study. At oil sand areas in Alberta, there was no effect on wing length at either 6 or 12 days of age, nor an effect on nestling body weight at 6 days of age in either of 2 years (Gentes et al. 2006). In 1 of 2 years, however, the reference nestlings weighed more than the PAH exposed nestlings at 12 days of age. Most of these studies concluded that food supply and parental care, among other factors, played a larger role in nestling growth rates than did contaminant exposure. Mainwaring et al. (2009) found that the order of hatching affected growth rates (weight and structural components) in barn swallow broods, as did the number of feathers in the nest (Stephenson et al. 2009). Growth rates may not be usable for contaminant assessments unless all these other factors can be accounted for appropriately. See McCarty (2001/2002) and McCarty and Winkler (1999) for a complete discussion on confounding factors and the use of growth rates in toxicological field studies.

Histology

For histological endpoints, tissues are removed from the organism, fixed, and stained. The tissue or organ is then sectioned and microscopically examined for abnormalities (Bishop et al. 1998a, b; Mayne et al. 2004). Examples of irregularities include degree of vacuolation, presence or absence of heterophils and presence of pathological changes (Bishop et al. 1998a, b; C. Custer et al. 2009; Mayne et al. 2004). Tree swallow tissues analyzed include adrenal glands, bursa of Fabricius, kidney, spleen, and thymus.

There were quantifiable seasonal effects on many histological endpoints in nestling swallows from orchards previously contaminated with *p,p*-DDE and sprayed with a variety of OP and carbamate pesticides, as well as fungicides and growth regulators, (Bishop et al. 1998a). After taking season into account, few of the histological endpoints were related to spray category except there was more inflammation of the bursa of Fabricius and less maturation of the thymus in sprayed versus nonsprayed orchards (Bishop et al. 1998a). There were no effects on numerous histological testis measurements (Bishop et al. 1998b). In a later study, adverse effects on thymus (increased cortical area and lymphocyte densities) and spleen (higher number of hyperplastic spleens and increased germinal centers) were documented, but no effects on the adrenal gland or bursa were seen (Mayne et al. 2004). These histological changes suggested an increase in immune organ activity. There was no effect of cadmium concentrations on kidney histology in Colorado, however, exposure to cadmium was relatively low (C. Custer et al. 2009).

Nest Structure

The size of the nest is weighed or measured, and the numbers of feathers used for insulation are counted. Nest volume and mass were greater at the reference area compared to a waste water treatment plant site (Dods et al. 2005). The associations between possible adverse effects of smaller nests and fewer insulating feathers and a longer incubation period and fewer number of fledglings were extremely weak (both $r^2s=0.05$), however. This may indicate that other factors were affecting these two endpoints to a greater degree than contaminant levels. In another study, the number of feathers in nests was less at more contaminated sites along the Hudson River than at the reference location (McCarty and Secord 1999b). Contrary to these two studies, there was no effect of high mercury exposure along the South River in Virginia on the total weight of nests, nor on the average mass of feathers or grasses in nests at the mercury contaminated versus reference location (D.Cristol, unpublished data). Furthermore, Lombardo (1994) found that bigger and more insulated nests in certain situations were not always better for nestling survival, casting doubt on the assumption that bigger is always better. Lombardo also found that female age and season had an effect on nest structure, effects that would have to be accounted for before a contaminant effect could be established.

In neither Dods et al. (2005) nor McCarty and Secord (1999b) was the availability of nest material or feathers taken into account. At Agassiz National Wildlife Refuge, Minnesota, an area managed for duck nesting, tree swallow nests contained an extraordinary number of feathers (C. Custer, personal observation). This may indicate that the availability of feathers, a resource often in short supply, or the supply of dried grasses especially late in the season, needs to be accounted for in any study attempting to use nest structure as a contaminant endpoint. There may have been no effect of contaminant exposure on the mercury contaminated stretch of the South River, Virginia, possibly because there was a poultry transportation route through that area and hence an abundant supply of feathers.

Plumage Coloration

Tree swallows are unique among passerines in that a large proportion of first year females exhibit delayed maturation (Hussell 1983); that is, first year females retain more brown plumage than is present in older females. It has been postulated that PCBs increase the proportion of "intermediate females" in the breeding population (McCarty and Secord 2000). The brighter females in the intermediate category bred earlier and had larger clutches than the less bright females, however, which could argue for a positive benefit from PCB exposure, not a negative effect. No other study on this endpoint has been done to validate these results, so further work is needed to test this hypothesis.

Somatic Indices

The formula to calculate a somatic indices (SI) is the fresh weight of the organ divided by the whole body weight minus that organ weight (Papp et al. 2005). These indices can be calculated for almost any organ, such as liver, bursa, or spleen to name just a few.

There was no association between somatic indices for tree swallow bursa or spleen, but there was for log PCB concentrations and liver SI (Papp et al. 2005).

Near a municipal outfall in greater Vancouver, there was no effect of alkylphenol ethoxylates on weights of bursa, spleen, thymus, testis, or kidney; however, nestling liver weights were heavier at the outfall compared to a nearby reference area (Dods et al. 2005). Along the St. Lawrence River there was no difference in the liver or adrenal SIs among sites that differed in organic chemical exposure (Bishop et al. 1999; Martinovic et al. 2003b), nor was there a difference in liver SI above and below pulp and paper mills in Alberta and Saskatchewan (Wayland et al. 1998). Organophosphate and carbamate pesticides also caused no change in various SIs measured, including liver SI, in orchards that were sprayed and not sprayed (Bishop et al. 1998a). DeWitt et al. (2006) found weak correlations (all $r^2s = 0.10$) between heart SI and various measures of PCB exposure. Similarly, Henshel and Sparks (2006) found relatively few significant effects for 16 different SI measurements from five passerine species at PCB-contaminated sites. The effects seen were often contradictory in their direction and differed among species and locations. In the latter two studies, these associations could have been spurious and due to chance alone because of the large number of statistical comparisons (100s) made. Mora et al. (2006) found a correlation between a variety of SIs (liver, gonads, and spleen) in adult cave and cliff swallows along the Rio Grande River in south Texas and some organic and inorganic contaminants, but the effects were not necessarily consistent across species, sex, or locations. The lack of strong and consistent results in many of these studies indicate that much more work is needed to understand which of these associations are related to contaminant exposure and what other environmental or biological factors may be affecting SI endpoints.

The most consistent SI endpoint to date is the increase in liver weight relative to body weight in some of the above studies. One function of the liver is detoxification; its enlargement could be interpreted as a successful response to contaminant exposure. Whether liver enlargement leads to any other type of adverse effect, e.g., reduced fitness, remains to be determined. This link needs to be established before this endpoint can be used for anything other than a measure of exposure and a physiological response to that exposure.

Biomarkers

Biomarkers have been used not only to quantify whether an organism has responded to the presence of a contaminant, but also have been suggested as a more cost-effective tool to document the exposure in lieu of more expensive chemical analyses. One disadvantage of biomarker assessments for this latter purpose, however, is that they are often highly variable, so a greater number of samples may be needed to achieve the same level of certainty when compared to chemical analyses. The magnitude of a biomarker response may determine whether adverse effects will occur in the organism and whether those adverse responses translate to effects at higher levels of biological organization. This linkage has generally not been established, except for some of the genetic endpoints where associations between genetic abnormalities and reduced fitness have been established (Ellegren et al. 1997). Biomarkers, however, can be used to further our understanding of mode of action for a particular contaminant. See van der Oost et al. (2003) for a more complete discourse on the usage of biomarkers.

Alkoxyresorufin-*O*-Dealkylase (alkEROD)

This group is the most well studied of the biomarkers and is comprised of several related enzymes, the most common one being ethoxyresorufin-*O*-dealkylase (EROD). For a current review see Harris and Elliott (2010). Others are benzyloxyresorufin-*O*-dealkylase (BROD), methylresorufin-*O*-dealkylase (MROD), and pentoxyresorufin-*O*-dealkylase (PROD). Because their activities are often highly correlated with one another (Papp et al. 2005), EROD is usually the one presented perhaps because it has higher activity levels than the others (Papp et al. 2005). This family of enzymes oxidizes, hydrolyzes, or reduces compounds through the insertion of an atom of atmospheric oxygen. This generally increases their water solubility, thereby enhancing elimination from the body. Induction of all of these related enzymes are indicative of exposure to a variety of chemicals including PCBs (Bishop et al. 1999; Papp et al. 2005), dioxins and furans (C. Custer et al. 2005), and PAHs (T. Custer et al. 2001; Gentes et al. 2006). Induction by such a broad array of contaminants may limit its usefulness unless paired with chemical analyses because the contaminant causing the induction must be verified.

EROD and BROD were induced ninefold as a result of PAH exposure in Wyoming when compared to a reference location (T. Custer et al. 2001), but only induced a maximum of 1.9 times in Alberta (Gentes et al. 2006). EROD was induced three to five times compared to the reference locations as a result of dioxin exposure on the Woonasquatucket River (C. Custer et al. 2005). EROD was positively correlated with total PCBs in 13-day old nestling carcass remainders, at least up to a concentration of 1 μg/g (Papp et al. 2005) and was induced at more highly PCB-contaminated sites along the St. Lawrence River than at less contaminated sites (Bishop et al. 1999). Melancon et al. (2006) found a dose–response between EROD and BROD in tree swallow nestlings injected with β-naphthoflavone, a known inducer. They also found similar results when using immunohistochemistry techniques on a 1-cm^2 skin sample. This immunohistochemistry technique might allow for nonlethal sampling to quantify enzyme induction rather than the traditional assay of liver tissue.

EROD induction, like many other biomarkers, indicates a physiological response to a contaminant and does not necessarily equate to a higher level effect. Additionally, EROD activity has been shown to drop off above a certain threshold and some individuals are nonresponsive, therefore the lack of induction does not necessarily mean lack of exposure (Harris and Elliott 2010). Finally, the variation of EROD is quite high (C. Custer et al. 2005; T. Custer et al. 2001; Gentes et al. 2006) which must be taken into account when determining sample size requirements.

δ-Aminolevulinic Acid Dehydratase

δ-Aminolevulinic acid dehydratase (ALAD) is an enzyme that is responsible for the second step in the production of heme, the iron-containing portion of hemoglobin. Heme molecules are incorporated into hemoglobin and packaged into red blood cells, or are used in the liver for the production of certain liver enzymes. This enzyme is particularly sensitive to Pb exposure, so it can be used as a biomarker of substantial exposure to Pb. It is a subclinical measure of effect, however. In the NRDA context, depression of ALAD enzyme activity >50% of normal is de facto indication of injury which is why it is still used even though other metrics are also available. Along the upper Arkansas River in central Colorado, exposure to Pb was sufficiently high to indicate injury (C. Custer et al. 2003a). Grue et al. (1984) found reductions in average ALAD activity of >30% in adult barn swallows and 14% in nestling barn swallows nesting along highway rights-of ways contaminated with lead from automotive exhaust.

Deoxyribonucleic Acid Mutation Rates or Damage

Deoxyribonucleic Acid (DNA) damage can result from direct damage to the DNA or by the formation of adducts. An adduct is a piece of DNA covalently bonded to a cancer-causing chemical and is used to quantify the amount of cancer in the subject. There are several methods that have been used to quantify DNA damage in tree swallow studies. Fingerprinting techniques have been used to examine multilocus profiles in bands (DNA fragments). Mutation rates are estimated from the number of DNA bands that do not match any bands from the putative parents. It is assumed that mutations have occurred if the number of novel fragments is ≥2 or if the rate exceeds the background mutation rates of one to three per 1,000 DNA bands (Stapleton et al. 2001). Somatic cell chromosome damage can also be measured by the variation in the amount of DNA in 10,000 red blood cells using a flow cytometer. Somatic chromosome damage is suggested when the half-peak coefficient of variation (HPCV) in the exposed population exceeds that of a reference population.

There was no difference in mutation rate between locations with high PCBs (Hudson River, NY) and background concentrations (Saukville, WI, Stapleton et al. 2001).

Chromosome damage has been associated with radiation exposure (Ellegren et al. 1997) and with cadmium exposure (C. Custer et al. 2003a), but not with dioxins and furans (C. Custer et al. 2005) or a complex mixture of organic and inorganic contaminants (cave swallows, Sitzlar et al. 2009). There was greater HPCV in cliff swallows between two sites in the lower Rio Grande River Valley, Texas, but whether the greater HPCV was associated with specific contaminants, or which contaminants, was unclear (Sitzlar et al. 2009). Other genetic damage assessment tools, such as single-cell gel assay (Comet assay), have not been attempted in swallows.

Glucocorticoid Stress Hormones

Corticosterone is one of the primary glucocorticoids in birds. Elevation of circulating corticosterone concentrations is an important mechanism to cope with short-term stresses. Chronic elevation of corticosterone, however, can promote severe protein loss, cause neuronal cell death, suppress growth, and dampen immune responsiveness (Mayne et al. 2004). The difference between baseline and poststress (10 or 30-min holding period) corticosterone levels is one metric that can be used. Additionally, the difference in the response to injected adrenocorticotropic hormone (ACTH), which should stimulate synthesis of corticosterone, can be used an as endpoint. More recently, corticosterone has been measured in feather extracts which may provide a more time-integrated assessment of corticosterone concentrations (Harms et al. 2010). There are time-of-day effects (Martinovic et al. 2003b), gender and body condition differences (Harms et al. 2010), and perhaps other environmental interactions that need to be accounted for when using this endpoint (Scanes and Griminger 1990).

There were no effects on corticosterone levels, either for baseline or 10-min poststress, in tree swallows nesting in orchards previously treated with DDE and currently sprayed with a variety of OP and carbamate pesticides (Mayne et al. 2004). There was, however, a heightened response to ACTH injection which may have indicated a sensitization from previous sprays, but many other possibilities were also postulated. No difference in basal or poststress corticosterone levels were found in swallows nesting along the St. Lawrence River among exposure groups (Martinovic et al. 2003b); there were occasional site differences, but these seemed unrelated to contaminant exposures. In that study there was no correlation with PCBs, but there was a negative correlation with furans. Along the Woonasquatucket River, Rhode Island, the baseline corticosterone levels were less at the highly TCDD-contaminated sites (Franceschini et al. 2008) compared to the references areas, which was consistent with the results of Martinovic et al. (2003b). The relationship between PCBs and either baseline or stress-induced corticosterone levels were not clear-cut in either nestling or adult tree swallows along the Housatonic River (Franceschini et al. 2008). In Virginia, there was a stress-induced increase in plasma corticosterone in relation to mercury in some but not all age groups; the direction of change also was not consistent among the age groups (Wada et al. 2009). Additionally, baseline corticosterone concentrations were elevated in 13- to 17-day-old nestlings, but blood mercury concentrations were less in that age group. Somewhat in contrast, Franceschini et al. (2009) found a negative association with blood mercury levels and baseline corticosterone concentrations in both adult and 11–13-day-old nestlings, but found no effect of mercury on stress-induced corticosterone concentrations. Additional work seems to be needed to better understand the relationship between individual contaminants, environmental and biological factors, and stress responses.

Eggshell Thinning

Eggshell thinning is correlated with DDE exposure above a specific threshold concentration in the egg. There are important differences among bird species, however, in whether thinning will occur and at what concentration. DeWeese et al. (1985) found no eggshell thinning in tree swallows in Colorado; however, mean concentrations were not particularly high (1.3–2.8 μg/g ww). This endpoint has not been used often with swallows for two possible reasons. Swallow eggshells are quite fragile and thin so this technique is more difficult to do well. Also, the minimum concentration in even the most sensitive bird species that causes eggshell thinning is often higher than concentrations typically found in tree swallows.

Immune Response

There are two types of immune responses that have been measured in swallows. T-lymphocyte-mediated immune response is a measure of the response to phytohaemagglutinin (PHA) injected into the wing patagium and B-cell response which quantifies an antibody response. Because PHA is a T-cell mutagen, T-lymphocytes migrate to, accumulate in, and proliferate at the injection site which results in swelling at that location compared to swelling with a sham injection of phosphate buffered saline. B-cell responses are the response to the injection of sheep red-blood cells. Pre-injection antibody titers are compared to postinjection titers. A differential response at contaminated versus reference areas is generally the metric used. See other measures of immune responses in the histological paragraphs above. Because the PHA response has been shown to be influenced by short-term fluctuations in energy balance in tree swallows, for instance, caused by ambient temperatures fluctuations or changes in food abundance (Lifjeld et al. 2002), by differences due to latitude or life history strategies (Ardia 2005), and by drought conditions (Fair and Whitaker 2008), great care must be taken not to confound possible contaminant effects with effects due to these other causes.

In nestling swallows from orchards previously contaminated with *p,p*-DDE and sprayed with a variety of OP and carbamate pesticides there were quantifiable seasonal effects on the immunological endpoints (Bishop et al. 1998a). After taking season into account, few of the immunological endpoints were related to spray category. There were, however, increased proliferation of lymphocytes when stimulated by pokeweed mitogen in nestlings from sprayed versus nonsprayed orchards (Bishop et al. 1998a). No difference in PHA response in swallows at previously DDE-contaminated and currently sprayed OP and carbamate apple orchards were found compared to reference areas (Mayne et al. 2004) or at reclamation ponds that differed in PAH contaminant (Smits et al. 2000). Along an extremely mercury contaminated river in Virginia, however, PHA response was significantly reduced at the contaminated versus the reference locations (Hawley et al. 2009). There was no effect, however, on the B-cell antibody responses in that study.

Metallothionein

This is a low molecular weight, cysteine-rich protein which is important in the metabolic processing of essential trace elements. Metallothionein (MT) also regulates, stores, and detoxifies toxic metals and can be correlated with some elements, such as Cd (Elliott et al. 1992). It is often quantified in liver or kidney tissues (St. Louis et al. 1993). In Ontario, MT was associated with the first principal component which represented Cu and Zn (St. Louis et al. 1993) but not with the principal component that represented Cd. Until further research is done, as with EROD induction or liver size, MT may be indicative of a positive physiological response to exposure rather than a metric to measure a negative effect on fitness.

Oxidative Stress Measures

This category includes a variety of endpoints including reduced glutathione (GSH), total sulfhydryl (TSH), protein bound thiol (PBSH), thiobarbituric acid reactive substances (TBARS), and oxidized glutathione (GSSG). Oxidative stress is a change in the cellular reduction potential or a change in the reducing capacity of cellular redox couples. The effects of oxidative stress depend upon the size of these changes. A cell may be able to overcome small perturbations, but not large changes. Particularly destructive to cells is the production of free radicals and peroxides.

Background activity levels of various oxidative stress measures are provided in C. Custer et al. (2006). In North Dakota there was a negative association between mercury and TSH, PBSH and GSSG, a positive association with GSH, and no association with TBARS (T. Custer et al. 2008). This was dissimilar to the lack of association of Hg with TSH, GSH, and TBARS, but a positive correlation with PBSH (C. Custer et al. 2006) in northwestern Minnesota even at similar concentrations of trace elements as found at the North Dakota sites. Work needs to be done, perhaps in controlled laboratory settings, to understand the associations among oxidative stress chemicals and the plethora of contaminants that can cause cellular disruptions.

P450 Aromatase Activity

This enzyme system converts C19 androgen steroids to phenol A ring estrogens. Endocrine disrupting chemicals can cause detrimental physiological and reproductive effects that are reflected in this enzyme system. In cave and cliff swallows, there were differences in aromatase activity levels in gonad but not in brain tissue, and there were differences between males and females. Females had higher activity levels than males (Sitzlar et al. 2009). There were occasional location differences found in male cliff swallows, but this did not occur for females or for cave swallows. Further study seems warranted to understand how this endpoint can be used in contaminant studies.

Porphyrins

There are several highly carboxylated porphyrins including uroporphyrin, heptaprophyrin, hexaporphyrin, and pentaporphyrin. There are mixed results in assessing the relation between porphyrins and PCBs (Bishop et al. 1999) and porphyrins and possible dioxin and furan exposure (Wayland et al. 1998). For example, both Hamilton and Wheatley Harbors had similar concentrations of total PCBs in nestling carcasses (0.66 and 0.64 μg/g ww), but vastly different total porphyrin levels (159 pmol/g liver vs. 44.8 pmol/g). A reference location (0.05 μg/g ww total PCBs) had 34.1 pmol/g liver. At only one of two locations were total porphyrins elevated immediately downstream of pulp and paper mills (Wayland et al. 1998). These mixed results along with high variation among individual swallows indicate that more work is needed to sort out whether porphyrins are a useful biomarker.

Sex Hormones

Androgens and 17β-oestradiol are measured in blood plasma samples at a standardized age or stage in the reproductive cycle. There were no differences in androgen concentrations in tree swallows above and below pulp and paper mills in west central Canada (Wayland et al. 1998). The levels of 17β-oestradiol were decreased in effluent-exposed birds at one of two study sites, but only in 1 of 2 years indicating that more work is needed to understand this relationship if it exists. Bishop et al. (1998b), in an exhaustive study at OP and carbamate-sprayed orchards, found no effects on oestradiol or testosterone in either nestling or adult tree swallows.

Thyroid Hormones

Triidothyronine (T_3) and thyroxine (T_4) are measured in blood plasma and the thyroid gland. Effects on these hormones and the pituitary–thyroid axis can affect metabolism, behavior, thermoregulation, feather development, and molt (Gentes et al. 2007). There are contradictory reports about the effect of nestling age on T_3 and T_4 levels. T_3, but not on T_4 concentrations, change with nestling age (Gentes et al. 2007), however in another study T_4 but not T_3 varied by age (Wada et al. 2009). The changes perhaps coincided with the onset of thermoregulation (Wada et al. 2009). There were also seasonal effects (Gentes et al. 2007), so all these factors must be accounted for when this biomarker is used. Additionally, food availability and dietary composition can influence thyroid function in birds (Scanes and Griminger 1990).

Gentes et al. (2007) found a dose–response in T_3 levels in plasma relative to PAH levels in reclaimed wetlands, but no effect on T_4 levels. In the thyroid gland itself, there seemed to be more of a threshold effect; reclaimed wetlands had greater levels of T_3 and T_4 than the reference location, but it was not a dose–response relationship. The larger effects in plasma rather than in the thyroid gland proper may have occurred because of various mechanisms including deiodinases operating outside the thyroid gland, and possibly enhanced hormone synthesis and storage within the thyroid gland. Along the Rio Grande River in Texas, T_4 levels in adult cave swallows did not vary among locations with different contaminant profiles and the reference locations, but there were some location effects in adult cliff swallows (Mora et al. 2006). T_4 levels were negatively correlated with DDE, copper, and selenium exposures in that study, when data from both swallow species were combined. No differences in T_3 levels were found in male or female nestling plasma between orchards sprayed and not sprayed with a variety of OP and carbamate pesticides, however, there was a positive correlation between the total number of mixed-spray episodes and T_3 in male, but not female nestlings (Bishop et al. 1998b).

Vitamin A

Vitamin A, which occurs as retinol, is an essential vitamin. Retinol can be converted to retinoic acid, which regulates growth and development, or converted to retinyl palmatate which is a storage form. Depressed levels of vitamin A can cause delayed growth, ocular discharge, and reduced reproduction in some avian species (Martinovic et al. 2003a). It is unclear, however, whether there is a link between environmental contaminants and vitamin A in tree swallows. Bishop et al. (1999), at sites along the St. Lawrence River that differed in PCB and PCDD-F exposure, found no difference in kidney concentrations of either form of vitamin A, but in the liver, retinol and retinyl palmitate were reduced at the two highest PCB sites (Bishop et al. 1999). Somewhat in contrast, Martinovic et al. (2003a) did not find significant differences in concentrations of either hepatic or renal retinol or retinyl palmitate between contaminant exposure groups (high vs. low), nor did they find a correlation between contaminant concentrations and those two forms of vitamin A. There were occasional differences in mean concentrations among the nine sites sampled and the variation within sites was quite high. Liver retinyl palmitate was reduced at some high PCB contaminated sites, but not at all highly contaminated sites (Martinovic et al. 2003a). Additional research is needed to determine whether vitamin A is a useful endpoint for contaminant assessments and especially research on what factors, such as vitamin A levels in food items, may affect our ability to use this biomarker (Bishop et al. 1999). See Harris and Elliott (2010) for a more complete discussion of vitamin A.

Conclusions

A common theme across many of the biomarkers discussed above is the need for more research to better understand the multiplicity of factors that influence these endpoints. These factors must be understood and then accounted for in the experimental design and statistical analyses before contaminant associations can be conclusively demonstrated. Additionally, whether biomarker responses lead to or can predict adverse effects at higher levels of biological organization remains an open question. The value of biomarkers may be in understanding the mode-of-action of contaminants.

Summary

Tree swallows are an important model species to study the effects of contaminants in wild bird populations. Tree swallows and their conspecifics have been used extensively across North America to study a wide variety of environmental contaminants. Exposure to most classes of contaminants can now be quickly put into context with other sites including an assessment of what constitutes background concentrations and what is elevated. Target tissues for exposure assessments have been identified and differ depending on the contaminant of concern. Endpoints range from population level assessments down to effects at the subcellular level. Many of the biomarkers need further work to more fully understand the role of environmental and biological variation. For some of the legacy contaminants, including mercury, PCBs, and dioxins, reproductive effect levels have been established. Research with tree swallows has proven very useful for a number of NRDA and ERA site assessments because of their close connection to and reflection of sediment contamination. Tree swallows can also be used to assess the success of and quantify the level of clean up achieved at remediated sites.

Acknowledgments I wish to thank Thomas W. Custer, Daniel A. Cristol, John E. Elliott, Kim J. Fernie, Megan L. Harris, and Kirk Lohman for review of earlier drafts of the document, and Paul M. Dummer for assistance with editorial aspects of this chapter. Use of trade, product, or firm names does not imply endorsement by the US Government.

References[1]

Ankley GT, Niemi GJ, Lodge KB, Harris HJ, Beaver DL, Tillitt DE, Schwartz TR, Giesy JP, Jones PD, Hagley C (1993) Uptake of planar polychlorinated biphenyls and 2,3,7,8-substituted polychlorinated dibenzofurans and dibenzo-*p*-dioxins by birds nesting in the lower Fox River and Green Bay, Wisconsin, USA. Arch Environ Contam Toxicol 24:332–344 [1]

Ardia DR (2005) Tree swallows trade off immune function and reproductive effort differently across their range. Ecology 86:2040–2046

Arenal CA, Halbrook RS, Woodruff M (2004) European starling (*Sturnus vulgaris*): avian model and monitor of polychlorinated biphenyl contamination at a superfund site in southern Illinois, USA. Environ Toxicol Chem 23:93–104

Bishop CA, Koster MD, Chek AA, Hussell DJT, Jock K (1995) Chlorinated hydrocarbons and mercury in sediments, red-winged blackbirds (*Agelaius phoeniceus*) and tree swallows (*Tachycineta bicolor*) from wetlands in the Great Lakes-St. Lawrence River basin. Environ Toxicol Chem 14:491–501 [2]

Bishop CA, Boermans HJ, Ng P, Campbell GD, Struger J (1998a) Health of tree swallows (*Tachycineta bicolor*) nesting in pesticide-sprayed apple orchards in Ontario, Canada. I. Immunological parameters. J Toxicol Environ Health A 55:531–559 [3]

Bishop CA, Van Der Kraak GJ, Ng P, Smits JEG, Hontela A (1998b) Health of tree swallows (*Tachycineta bicolor*) nesting in pesticide-sprayed apple orchards in Ontario, Canada. II.

[1]Numbers in brackets at the end of each citation below refer to numbers in tables and figures. Only papers on contaminants and swallows are numbered.

Sex and thyroid hormone concentrations and testes development. J Toxicol Environ Health A 55:561–581 [4]
Bishop CA, Mahony NA, Trudeau S, Pettit KE (1999) Reproductive success and biochemical effects in tree swallows (*Tachycineta bicolor*) exposed to chlorinated hydrocarbon contaminants in wetlands of the Great Lakes and St. Lawrence River basin, USA and Canada. Environ Toxicol Chem 18:263–271 [5]
Bishop CA, Collins B, Mineau P, Burgess NM, Read WF, Risley C (2000a) Reproduction of cavity-nesting birds in pesticide-sprayed apple orchards in southern Ontario, Canada, 1988–1994. Environ Toxicol Chem 19:588–599 [6]
Bishop CA, Ng P, Mineau P, Quinn JS, Struger J (2000b) Effects of pesticide spraying on chick growth, behavior, and parental care in tree swallows (*Tachycineta bicolor*) nesting in an apple orchard in Ontario, Canada. Environ Toxicol Chem 19:2286–2297 [7]
Blus LJ (1995) Organochlorine pesticides. In: Hoffman DJ, Rattner BA, Burton GA Jr, Cairns J Jr (eds) Handbook of ecotoxicology. Lewis Publishers, Boca Raton, pp 275–300
Blus LJ (1996) DDT, DDD, and DDE in birds. In: Beyer WN, Heinz GH, Redmon-Norwood AW (eds) Environmental contaminants in wildlife interpreting tissue concentrations. Lewis Publishers, Boca Raton, pp 49–71
Blus LJ, Neely BS Jr, Belisle AA, Prouty RM (1974) Organochlorine residues in brown pelican eggs: relation to reproductive success. Environ Pollut 7:81–91
Blus LJ, Henny CJ, Hoffman DJ, Grove RA (1995) Accumulation in and effects of lead and cadmium on waterfowl and passerines in northern Idaho. Environ Pollut 89:311–318 [8]
Brasso RL, Cristol DA (2008) Effects of mercury exposure on the reproductive success of tree swallows (*Tachycineta bicolor*). Ecotoxicology 17:133–141 [9]
Brasso RL, Abdel Latif MK, Cristol DA (2010) Relationship between laying sequence and mercury concentration in tree swallow eggs. Environ Toxcol Chem 29(5):1155–1159. doi:doi:10.1002/etc.144 [10]
Brown MB, Brown CR (2009) Blood sampling reduces annual survival in cliff swallows (*Petrochelidon pyrrhonota*). Auk 126:853–861
Bunck CM, Prouty RM, Krynitsky AJ (1987) Residues of organochlorine pesticides and polychloribiphenyls in starlings (*Sturnus vulgaris*), from the continental United States, 1982. Environ Monit Assess 8:59–75
Burgess NM, Hunt KA, Bishop C, Weseloh DV (1999) Cholinesterase inhibition in tree swallows (*Tachycineta bicolor*) and eastern bluebirds (*Sialia sialis*) exposed to organophosphorus insecticides in apple orchards in Ontario, Canada. Environ Toxicol Chem 18:708–716 [11]
Carson R (1962) Silent spring. Houghton Mifflin, Boston
Condon AM, Cristol DA (2009) Feather growth influences blood mercury level of young songbirds. Environ Toxicol Chem 28:395–401 [12]
Cristol DA, Brasso RL, Condon AM, Fovargue RE, Friedman SL, Hallinger KK, Monroe AP, White AE (2008) The movement of aquatic mercury through terrestrial food webs. Science 320:335 [13]
Custer CM, Read LB (2006) Polychlorinated biphenyl congener patterns in tree swallows (*Tachycineta bicolor*) nesting in the Housatonic River watershed, western Massachusetts, USA, using a novel statistical approach. Environ Pollut 142:235–245 [14]
Custer CM, Custer TW, Allen PD, Stromborg KL, Melancon MJ (1998) Reproduction and environmental contamination in tree swallows nesting in the Fox River drainage and Green Bay, Wisconsin, USA. Environ Toxicol Chem 17:1786–1798 [15]
Custer TW, Custer CM, Hines RK, Gutreuter S, Stromborg KL, Allen PD, Melancon MJ (1999) Organochlorine contaminants and reproductive success of double-crested cormorants from Green Bay, Wisconsin, USA. Environ Toxicol Chem 18:1209–1217
Custer CM, Custer TW, Coffey M (2000a) Organochlorine chemicals in tree swallows nesting in Pool 15 of the Upper Mississippi River. Bull Environ Contam Toxicol 64:341–346 [16]
Custer TW, Custer CM, Hines RK, Sparks DW (2000b) Trace elements, organochlorines, polycylic aromatic hydrocarbons, dioxins, and furans in lesser scaup wintering on the Indiana Harbor Canal. Environ Pollut 110:469–482

Custer TW, Custer CM, Dickerson K, Allen K, Melancon MJ, Schmidt LJ (2001) Polycyclic aromatic hydrocarbons, aliphatic hydrocarbons, trace elements and monooxygenase activity in birds nesting on the North Platte River, Casper, WY, USA. Environ Toxicol Chem 20:624–631 [25]

Custer TW, Custer CM, Hines RK (2002) Dioxins and congener-specific polychlorinated biphenyls in three avian species from the Wisconsin River, Wisconsin. Environ Pollut 119:323–332 [26]

Custer CM, Custer TW, Archuleta AS, Coppock LC, Swartz CD, Bickham JW (2003a) A mining impacted stream: exposure and effects of lead and other trace elements on tree swallows (*Tachycineta bicolor*) nesting in the upper Arkansas River basin, Colorado. In: Hoffman DJ, Rattner BA, Burton GA Jr, Cairns J Jr (eds) Handbook of ecotoxicology, 2nd edn. CRC Press, Boca Raton, pp 787–812 [17]

Custer CM, Custer TW, Dummer PM, Munney KL (2003b) Exposure and effects of chemical contaminants on tree swallows nesting along the Housatonic River, Berkshire County, Massachusetts, USA, 1998–2000. Environ Toxicol Chem 22:1605–1621 [18]

Custer CM, Custer TW, Rosiu CJ (2003c) Accumulation of dioxins and furans in tree swallows (*Tachycineta bicolor*) nesting near Centredale Manor Restoration Project Superfund site, Rhode Island. Organohalogen Compounds 62:391–394 [19]

Custer CM, Custer TW, Rosiu CJ, Melancon MJ, Bickham JW (2005) Exposure and effects of 2,3,7,8-tetrachlorodibenzo-*p*-dioxin in tree swallows (*Tachycineta bicolor*) nesting along the Woonasquatucket River, Rhode Island, USA. Environ Toxicol Chem 24:93–109 [20]

Custer CM, Custer TW, Warburton D, Hoffman DJ, Bickham JW, Matson CW (2006a) Trace element concentrations and bioindicator responses in tree swallows from northwestern Minnesota. Environ Monit Assess 118:247–266 [21]

Custer TW, Custer CM, Goatcher B, Melancon MJ, Matson CW, Bickham JW (2006b) Contaminant exposure of barn swallows nesting on Bayou D'Inde, Calcasieu Estuary, Louisiana, USA. Environ Monit Assess 121:543–560 [27]

Custer CM, Custer TW, Hill EF (2007a) Mercury exposure and effects on cavity-nesting birds from the Carson River, Nevada. Arch Environ Contam Toxicol 52:129–136 [22]

Custer CM, Custer TW, Hines JE, Nichols JD, Dummer PM (2007b) Adult tree swallow (*Tachycineta bicolor*) survival on the polychlorinated biphenyl-contaminated Housatonic River, Massachusetts, USA. Environ Toxicol Chem 26:1056–1065 [23]

Custer TW, Dummer PM, Custer CM, Li AU, Warburton D, Melancon MJ, Hoffman DJ, Matson CW, Bickham JW (2007c) Water level management and contaminant exposure to tree swallows nesting on the Upper Mississippi River. Environ Monit Assess 133:335–345 [28]

Custer TW, Custer CM, Johnson KM, Hoffman DJ (2008) Mercury and other element exposure to tree swallows (*Tachycineta bicolor*) nesting on Lostwood National Wildlife Refuge, North Dakota. Environ Pollut 155:217–226 [29]

Custer CM, Yang C, Crock JG, Shern-Bochsler V, Smith KS, Hageman PL (2009a) Exposure of insects and insectivorous birds to metals and other elements from abandoned mine tailings in three Summit County drainages, Colorado. Environ Monit Assess 153:161–177 [24]

Custer TW, Kannan K, Tao L, Saxena AR, Route B (2009b) Perfluorinated compounds and polybrominated diphenyl ethers in great blue heron eggs from Indiana Dunes National Lakeshore, Indiana. J Great Lakes Res 35:401–405. doi:10.1016/j.jglr.2009.02.003

Custer CM, Gray BR, Custer TM (2010a) Effects of egg order on organic and inorganic element concentrations and egg characteristics in tree swallows, *Tachycineta bicolor*. Environ Toxicol Chem 29:909–921 [24a]

Custer CM, Custer TM, Dummer PM (2010b) Patterns of organic contaminants in eggs of an insectivorous, omnivorous, and piscivorous bird nesting on the Hudson River, NY, USA. Environ Toxicol Chem 29:2286–2296 [24b]

DeWeese LR, Cohen RR, Stafford CJ (1985) Organochlorine residues and eggshell measurements for tree swallows *Tachycineta bicolor* in Colorado. Bull Environ Contam Toxicol 35:767–775 [30]

DeWeese LR, McEwen LC, Hensler GL, Petersen BE (1986) Organochlorine contaminants in Passeriformes and other avian prey of the Peregrine Falcon in the western United States. Environ Toxicol Chem 5:675–693 [31]

DeWitt JC, Millsap DS, Yaeger RL, Heise SS, Sparks DW, Henshel DS (2006) External heart deformities in passerine birds exposed to environmental mixtures of polychlorinated biphenyls during development. Environ Toxicol Chem 25:541–551 [32]

Dods PL, Birmingham EM, Williams TD, Ikonomou MG, Bennie DT, Elliott JE (2005) Reproductive success and contaminants in tree swallows (*Tachycineta bicolor*) breeding at a wastewater treatment plant. Environ Toxicol Chem 24:3106–3112 [33]

el Halawani ME, Burke WH, Dennison PT (1980) Effect of nest-deprivation on serum prolactin level in nesting female turkeys. Biol Reprod 23:118–123

Ellegren H, Lindgren G, Primmer CR, Møller AP (1997) Fitness loss and germline mutations in barn swallows breeding in Chernobyl. Nature 389:593–596 [34]

Elliott JE, Scheuhammer AM, Leighton FA, Pearce PA (1992) Heavy metal and metallothionein concentrations in Atlantic Canadian seabirds. Arch Environ Contam Toxicol 22:63–73

Elliott JE, Maratin PA, Arnold TW, Sinclair PH (1994) Organochlorines and reproductive success of birds in orchard and non-orchard areas of central British Columbia, Canada, 1990–91. Arch Environ Contam Toxicol 26:435–443 [35]

EPA and Army Corps of Engineers (2003) Ecological risk assessment for general electric (GE)/ Housatonic river site, rest of river, vols 1 and 2, 949 pp

Evers DC, Burgess NM, Champoux L, Hoskins B, Major A, Goodale WM, Taylor RJ, Poppenga R, Daigle T (2005) Patterns and interpretation of mercury exposure in freshwater avian communities in northeastern North America. Ecotoxicology 14:193–221 [36]

Fair JM, Whitaker SJ (2008) Avian cell-mediated immune response to drought. Wilson J Ornithol 120:813–819 [37]

Federal Register (2005) 70(210):65932–65933

Fernie KJ, Shutt JL, Letcher RJ, Ritchie JI, Sullivan K, Bird DM (2008) Changes in reproductive courtship behaviors of adult American kestrels (*Falco sparverius*) exposed to environmentally relevant levels of the polybrominated diphenyl ether mixture, DE-71. Toxicol Sci 102:171–178

Fernie KJ, Shutt JL, Letcher RJ, Ritchie IJ, Bird DM (2009) Environmentally relevant concentrations of DE-71 and HBCD alter eggshell thickness and reproductive success of American kestrels. Environ Sci Technol 43:2124–2130

Franceschini MD, Custer CM, Custer TW, Reed JM, Romero LM (2008) Corticosterone stress response in tree swallows nesting near polychlorinated biphenyl- and dioxin- contaminated rivers. Environ Toxicol Chem 27:2326–2331 [38]

Franceschini MD, Lane OP, Evers DC, Reed JM, Hoskins B, Romero LM (2009) The corticosterone stress response and mercury contamination in free-living tree swallows, *Tachycineta bicolor*. Ecotoxicology 18:514–521 [39]

Franson JC (1996) Interpretation of tissue lead residues in birds other than waterfowl. In: Beyer WN, Heinz GH, Redmon-Norwood AW (eds) Environmental contaminants in wildlife interpreting tissue concentrations. CRC Lewis Publishers, Boca Raton, pp 265–279

Fredericks TB, Giesy JP, Coefield SJ, Seston RM, Haswell MM, Tazelaar DL, Bradley PW, Moore JN, Roark SA, Zwiernik MJ (2010) Dietary exposure of three passerine species to PCDD/DFs from the Chippewa, Tittabawassee, and Saginaw River floodplains, Midland, Michigan, USA. Environ Monit Assess. doi:10.1007/s10661-010-1319-5 [40]

Froese KL, Verbrugge DA, Ankley GT, Niemi GJ, Larsen CP, Giesy JP (1998) Bioaccumulation of polychlorinated biphenyls from sediments to aquatic insects and tree swallow eggs and nestlings in Saginaw Bay, Michigan, USA. Environ Toxicol Chem 17:484–492 [41]

Furness RW (1996) Cadmium in birds. In: Beyer WN, Heinz GH, Redmon-Norwood AW (eds) Environmental contaminants in wildlife: interpreting tissue concentrations. CRC Lewis Publishers, Boca Raton, Florida pp 389–404

Gentes ML, Waldner C, Papp Z, Smits JEG (2006) Effects of oil sands tailings compounds and harsh weather on mortality rates, growth and detoxification efforts in nestling tree swallows (*Tachycineta bicolor*). Environ Pollut 142:24–33 [42]

Gentes ML, McNabb A, Waldner C, Smits JEG (2007) Increased thyroid hormone levels in tree swallows (*Tachycineta bicolor*) on reclaimed wetlands of the Athabasca oil sands. Arch Environ Contam Toxicol 53:287–292 [43]

Gerrard PM, St. Louis VL (2001) The effects of experimental reservoir creation on the bioaccumulation of methylmercury and reproductive success of tree swallows (*Tachycineta bicolor*). Environ Sci Technol 35:1329–1338 [44]

Gilbertson M (1983) Etiology of chick edema disease in herring gulls in the lower Great Lakes. Chemosphere 12:357–370

Golden NH, Rattner BA (2003) Ranking terrestrial vertebrate species for utility in biomonitoring and vulnerability to environmental contaminants. Rev Environ Contam Toxicol 176:67–136

Grue CE, Hunter CC (1984) Brain cholinesterase activity in fledgling starlings: implication for monitoring exposure of songbirds to ChE inhibitors. Bull Environ Contam Toxicol 32:282–289

Grue CE, Shipley BK (1984) Sensitivity of nestling and adult starlings to dicrotophos, an organophosphate pesticide. Environ Res 35:454–465

Grue CE, Powell GVN, McChesney MJ (1982) Care of nestlings by wild female starlings exposed to an organophosphate pesticide. J Appl Ecol 19:327–335

Grue CE, O'Shea TJ, Hoffman DJ (1984) Lead concentrations and reproduction in highway-nesting barn swallows. Condor 86:383–389 [45]

Harms NJ, Fairhurst GD, Bortolotti GR, Smits JEG (2010) Variation in immune function, body condition, and feather corticosterone in nestling tree swallows (*Tachycineta bicolor*) on reclaimed wetlands in the Athabasca oil sands, Alberta, Canada. Environ Pollut 158:841–848 [46]

Harris ML, Elliott JE (2000) Reproductive success and chlorinated hydrocarbon contamination in tree swallows (*Tachycineta bicolor*) nestling along rivers receiving pulp and paper mill effluent discharges. Environ Pollut 110:307–320 [47]

Harris ML, Elliott JE (2010) Effects of polychlorinated biphenyls, dibenzo-*p*-dioxins and dibenzofurans and polybrominated diphenyl ethers in wild birds. In: Beyer WN, Meadow J (eds) Environmental contaminants in wildlife interpreting tissue concentrations. CRC Press, New York, pp 471–522

Hawley DM, Hallinger KK, Cristol DA (2009) Compromised immune competence in free-living tree swallows exposed to mercury. Ecotoxicology 18:499–503 [48]

Heinz GH, Hoffman DJ, Klimstra JD, Stebbins KR, Kondrad SL, Erwin CA (2009) Species differences in the sensitivity of avian embryos to methylmercury. Arch Environ Contam Toxicol 56:129–138 [49]

Henny CJ, Blus LJ, Stafford CJ (1982) DDE not implicated in cliff swallow, *Petrochelidon pyrrhonota*, mortality during severe spring weather in Oregon. Can Field-Nat 96:210–211 [50]

Henny CJ, Hill EF, Hoffman DJ, Spalding MG, Grove RA (2002) Nineteenth century mercury: hazard to wading birds and cormorants of the Carson River, Nevada. Ecotoxicology 11:213–231

Henny CJ, Kaiser JL, Grove RA, Johnson BL, Letcher RJ (2009) Polybrominated diphenyl ether flame retardants in eggs may reduce reproductive success of ospreys in Oregon and Washington, USA. Ecotoxicology 18(7):802–813

Henshel DS, Sparks DW (2006) Site specific PCB-correlated interspecies differences in organ somatic indices. Ecotoxicology 15:9–18 [51]

Hothem RL, Trejo BS, Bauer ML, Crayon JJ (2008) Cliff Swallows *Petrochelidon pyrrhonota* as bioindicators of environmental mercury, Cache Creek Watershed, California. Arch Environ Contam Toxicol 55:111–121 [52]

Hovorka MD, Robertson RJ (2000) Food stress, nestling growth, and fluctuating asymmetry. Can J Zool 78:28–35

Hussell DJT (1983) Age and plumage color in female tree swallows. J Field Ornithol 54:312–318

Jayaraman S, Nacci DE, Champlin DM, Pruell RJ, Rocha KJ, Custer CM, Custer TW, Cantwell M (2009) PCBs and DDE in tree swallow (*Tachycineta bicolor*) eggs and nestlings from an estuarine PCB superfund site, New Bedford Harbor, MA, USA. Environ Sci Technol 43:8387–8392 [53]

Jones J (2003) Tree swallows (*Tachycineta bicolor*): a new model organism? Auk 120:591–599

Kay DP, Blankenship AL, Coady KK, Neigh AM, Zwiernik MJ, Millsap SD, Strause K, Park C, Bradley P, Newsted JL, Jones PD, Giesy JP (2005) Differential accumulation of polychlorinated biphenyl congeners in the aquatic food web at the Kalamazoo River superfund site, Michigan. Environ Sci Technol 39:5964–5974 [54]

King KA, Custer TW, Weaver DA (1994) Reproductive success of barn swallows nesting near a selenium-contaminated lake in east Texas, USA. Environ Pollut 84:53–58 [55]

Kraus ML (1989) Bioaccumulation of heavy metals in pre-fledgling tree swallows, *Tachycineta bicolor*. Bull Environ Contam Toxicol 43:407–414 [56]

Lifjeld JT, Dunn PO, Whittingham LA (2002) Short-term fluctuations in cellular immunity of tree swallows feeding nestlings. Oecologia 130:185–190

Lombardo MP (1994) Nest architecture and reproductive performance in tree swallows (*Tachycineta bicolor*). Auk 111:814–824

Longcore JR, Haines TA, Halteman WA (2007a) Mercury in tree swallow food, eggs, bodies, and feathers at Acadia National Park, Maine, and an EPA superfund site, Ayer, Massachusetts. Environ Monit Assess 126:129–143 [57]

Longcore JR, Dineli R, Haines TA (2007b) Mercury and growth of tree swallows at Acadia National Park, and at Orono, Maine, USA. Environ Monit Assess 126:117–127 [58]

Mainwaring MC, Rowe LV, Kelly DJ, Grey J, Bearhop S, Hartley IR (2009) Hatching asynchrony and growth trade-offs within barn swallow broods. Condor 111:668–674 [59]

Martin PA, Weseloh DV, Bishop CA, Legierse K, Braune B, Norstrom RJ (1995) Organochlorine contaminants in avian wildlife of Severn Sound. Water Qual Res J Canada 30:693–711

Martinovic B, Lean DRS, Bishop CA, Birmingham E, Secord A, Jock K (2003a) Health of tree swallow (*Tachycineta bicolor*) nestlings exposed to chlorinated hydrocarbons in the St. Lawrence River basin. Part I. Renal and hepatic vitamin A concentrations. J Toxicol Environ Health A 66:1053–1072 [60]

Martinovic B, Lean D, Bishop CA, Birmingham E, Secord A, Jock K (2003b) Health of tree swallow (*Tachycineta bicolor*) nestlings exposed to chlorinated hydrocarbons in the St. Lawrence River basin. Part II. Basal and stress plasma corticosterone concentrations. J Toxicol Environ Health A 66:2015–2029 [61]

Maul JD, Belden JB, Schwab BA, Whiles MR, Spears B, Farris JL, Lydy MJ (2006) Bioaccumulation and trophic transfer of polychlorinated biphenyls by aquatic and terrestrial insects to tree swallows (*Tachycineta bicolor*). Environ Toxicol Chem 25:1017–1025 [62]

Maul JD, Schuler LJ, Halbrook RS, Lydy MJ (2010) Congener-specific egg contributions of polychlorinated biphenyls to nestlings in two passerine species. Environ Pollut 158:2725–2732 [62a]

Mayne GJ, Martin PA, Bishop CA, Boermans HJ (2004) Stress and immune responses of nestling tree swallows (*Tachycineta bicolor*) and eastern bluebirds (*Sialia sialis*) exposed to nonpersistent pesticides and *p,p'*-dichlorodiphenyldichloroethylene in apple orchards of southern Ontario, Canada. Environ Toxicol Chem 23:2930–2940 [63]

McCarty JP (2001/2002) Use of tree swallows in studies of environmental stress. Rev Toxicol 4:61–104

McCarty JP, Secord AL (1999a) Reproductive ecology of tree swallows (*Tachycineta bicolor*) with high levels of polychlorinated biphenyl contamination. Environ Toxicol Chem 18:1433–1439 [64]

McCarty JP, Secord AL (1999b) Nest-building behavior in PCB-contaminated tree swallows. Auk 116:55–63 [65]

McCarty JP, Secord AL (2000) Possible effects of PCB contamination on female plumage color and reproductive success in Hudson River tree swallows. Auk 117:987–995 [66]

McCarty JP, Winkler DW (1999) Relative importance of environmental variables in determining the growth of nestling tree swallows *Tachycineta bicolor*. Ibis 141:286–296

McFarland VA, Clarke JU (1989) Environmental occurrence, abundance, and potential toxicity of polychlorinated biphenyl congeners: considerations for a congener-specific analysis. Environ Health Perspect 81:225–239

McKernan MA, Rattner BA, Hale RC, Ottinger MA (2009) Toxicity of polybrominated diphenyl ethers (DE-71) in chicken (*Gallus gallus*), mallard (*Anas platyrhynchos*), and American kestrel (*Falco sparverius*) embryos and hatchlings. Environ Toxicol Chem 28:1007–1017

Melancon MJ, Kutay AL, Woodin BR, Stegeman JJ (2006) Evaluating cytochrome P450 in lesser scaup (*Aythya affinis*) and tree swallow (*Tachycineta bicolor*) by monooxygenase activity and immunohistochemistry: possible nonlethal assessment by skin immunohistochemistry. Environ Toxicol Chem 25:2613–2617 [67]

Møller AP, Mousseau TA (2003) Mutation and sexual selection: a test using barn swallows from Chernobyl. Evolution 57:2139–2146 [68]

Møller AP, Mousseau TA, Milinevsky G, Peklo A, Pysanets E, Szép T (2005) Condition, reproduction and survival of barn swallows from Chernobyl. J Anim Ecol 74:1102–1111 [69]

Mora MA, Boutton TW, Musquiz D (2005) Regional variation and relationships between the contaminants DDE and selenium and stable isotopes in swallows nesting along the Rio Grande and one reference site, Texas, USA. Isotopes Environ Health Stud 41:69–85 [70]

Mora MA, Musquiz D, Bickham JW, Mackenzie DS, Hooper MJ, Szabo JK, Matson CW (2006) Biomarkers of exposure and effects of environmental contaminants on swallows nesting along the Rio Grande, Texas, USA. Environ Toxicol Chem 25:1574–1584 [71]

Neigh AM, Zwiernik MJ, Bradley PW, Kay DP, Park CS, Jones PD, Newsted JL, Blankenship AL, Giesy JP (2006a) Tree swallow (*Tachycineta bicolor*) exposure to polychlorinated biphenyls at the Kalamazoo River superfund site, Michigan, USA. Environ Toxicol Chem 25:428–437 [72]

Neigh AM, Zwiernik MJ, MacCarroll MA, Newsted JL, Blankenship AL, Jones PD, Kay DP, Giesy JP (2006b) Productivity of tree swallows (*Tachycineta bicolor*) exposed to PCBs at the Kalamazoo River superfund site. J Toxicol Environ Health A 69:395–415 [73]

Neigh AM, Zwiernik MJ, Blankenship AL, Bradley PW, Kay DP, MacCarroll MA, Park CS, Jones PD, Millsap SD, Newsted JW, Giesy JP (2006c) Exposure and multiple lines of evidence assessment of risk for PCBs found in the diets of passerine birds at the Kalamazoo River superfund site, Michigan. Hum Ecol Risk Assess 12:924–946 [74]

Nichols JW, Larsen CP, McDonald ME, Niemi GJ, Ankley GT (1995) Bioenergetics-based model for accumulation of polychlorinated biphenyls by nestling tree swallows, *Tachycineta bicolor*. Environ Sci Technol 29:604–612 [75]

Opperhuizen A, Sijm DTHM (1990) Bioaccumulation and biotransformation of polychlorinated dibenzo-*p*-dioxins and dibenzofurans in fish. Environ Toxicol Chem 9:175–186

Pain DJ (1996) Lead in waterfowl. In: Beyer WN, Heinz GH, Redmon-Norwood AW (eds) Environmental contaminants in wildlife: interpreting tissue concentrations. CRC Lewis Publishers, Boca Raton, Florida, pp 251–264

Papp Z, Bortolotti GR, Smits JEG (2005) Organochlorine contamination and physiological responses in nestling tree swallows in Point Pelee National Park, Canada. Arch Environ Contam Toxicol 49:563–568 [76]

Papp Z, Bortolotti GR, Sebastian M, Smits JEG (2007) PCB congener profiles in nestling tree swallows and their insect prey. Arch Environ Contam Toxicol 52:257–263 [77]

Peakall DB, Gilman AP (1979) Limitations of expressing organochlorine levels in eggs on a lipid-weight basis. Bull Environ Contam Toxicol 23:287–290

Reynolds KD, Skipper SL, Cobb GP, McCurry ST (2004) Relationship between DDE concentrations and laying sequence in eggs of two passerine species. Arch Environ Contam Toxicol 47:396–401

Robertson RJ, Stutchbury BJ, Cohen RR (1992) Tree swallow. In: Poole A, Stettenheim P, Gill F (eds) The birds of North America No. 11. Academy of Natural Sciences and AOU, Philadelphia and Washington, DC

Scanes CG, Griminger P (1990) Endocrine-nutrition interactions in birds. J Exp Zool 4:98–105

Scheuhammer AM (1987) The chronic toxicity of aluminum, cadmium, mercury, and lead in birds: a review. Environ Pollut 46:263–295

Secord AL, McCarty JP, Echols KR, Meadows JC, Gale RW, Tillitt DE (1999) Polychlorinated biphenyls and 2,3,7,8-tetrachlorodibenzo-*p*-dioxin equivalents in tree swallows from the upper Hudson River, New York state, USA. Environ Toxicol Chem 18:2519–2525 [78]

Shaw GG (1983) Organochlorine pesticide and PCB residues in eggs and nestlings of tree swallows, *Tachycineta bicolor*, in central Alberta. Can Field-Nat 98:258–260 [79]

Sitzlar MA, Mora MA, Fleming JGW, Bazer FW, Bickham JW, Matson CW (2009) Potential effects of environmental contaminants on P450 aromatase activity and DNA damage in swallows from the Rio Grande and Somerville, Texas. Ecotoxicology 18:15–21 [80]

Smits JE, Wayland ME, Miller MJ, Liber K, Trudeau S (2000) Reproductive, immune, and physiological end points in tree swallows on reclaimed oil sands mine sites. Environ Toxicol Chem 19:2951–2960 [81]

Smits JEG, Bortolotti GR, Sebastian M, Ciborowski JJH (2005) Spatial, temporal, and dietary determinants of organic contaminants in nestling tree swallows in Point Pelee National Park, Ontario, Canada. Environ Toxicol Chem 24:3159–3165 [82]

Spears BL, Brown MW, Hester CM (2008) Evaluation of polychlorinated biphenyl remediation at a superfund site using tree swallows (*Tachycineta bicolor*) as indicators. Environ Toxicol Chem 27:2512–2520 [83]

St. Louis VL, Breebaart L, Barlow JC, Klaverkamp JF (1993) Metal accumulation and metallothionein concentrations in tree swallow nestlings near acidified lakes. Environ Toxicol Chem 12:1203–1207 [85]

Stapleton M, Dunn PO, McCarty J, Secord A, Whittingham LA (2001) Polychlorinated biphenyl contamination and minisatellite DNA mutation rates of tree swallows. Environ Toxicol Chem 20:2263–2267 [84]

Stephenson S, Hannon S, Proctor H (2009) The function of feathers in tree swallow nests: insulation or ectoparasite barrier? Condor 111:479–487

Teather K (1996) Patterns of growth and asymmetry in nestling tree swallows. J Avian Biol 27:302–310

Thompson DR (1996) Mercury in birds and terrestrial mammals. In: Beyer WN, Heinz GH, Redmon-Norwood AW (eds) Environmental contaminants in wildlife, interpreting tissue concentrations. CRC Lewis Publishers, Boca Raton, Florida pp 341–356

Tsipoura N, Burger J, Feltes R, Yacabucci J, Mizrahi D, Jeitner C, Gochfeld M (2008) Metal concentrations in three species of passerine birds breeding in the Hackensack Meadowlands of New Jersey. Environ Res 107:218–228 [86]

Van den Steen E, Eens M, Covaci A, Dirtu AC, Jaspers VLB, Neels H, Pinxten R (2009) An exposure study with polybrominated diphenyl ethers (PBDEs) in female European starlings (*Sturnus vulgaris*): toxicokinetics and reproductive effects. Environ Pollut 157:430–436

Van der Oost R, Beyer J, Vermeulen NPE (2003) Fish bioaccumulation and biomarkers in environmental risk assessment: a review. Environ Toxicol Pharm 13:57–149

Wada H, Cristol DA, McNabb FMA, Hopkins WA (2009) Suppressed adrenocortical responses and thyroid hormone levels in birds near a mercury-contaminated river. Environ Sci Technol 43:6031–6038 [87]

Wayland M, Trudeau S, Marchant T, Parker D, Hobson KA (1998) The effect of pulp and paper mill effluent on an insectivorous bird, the tree swallow. Ecotoxicology 7:237–251 [88]

White DH, Seginak JT (1994) Dioxins and furans linked to reproductive impairment in wood ducks. J Wildl Manage 58:100–106

Youngren OM, el Halawani ME, Silsby JL, Phillips RE (1991) Intracranial prolactin perfusion induces incubation behavior in turkey hens. Biol Reprod 44:425–431

Zach R, Mayoh KR (1982) Breeding biology of tree swallows and house wrens in a gradient of gamma radiation. Ecology 63:1720–1728 [89]

Zach R, Mayoh KR (1984) Gamma radiation effects on nestling tree swallows. Ecology 65:1641–1647 [90]

Zach R, Mayoh KR (1986) Gamma irradiation of tree swallow embryos and subsequent growth and survival. Condor 88:1–10 [91]

Zach R, Hawkins JL, Sheppard SC (1993) Effects of ionizing radiation on breeding swallows at current radiation protection standards. Environ Toxicol Chem 12:779–786 [92]

Chapter 4
The Rocky Mountain Arsenal: From Environmental Catastrophe to Urban Wildlife Refuge

Jeffrey T. Edson, James V. Holmes, John E. Elliott, and Christine A. Bishop

An Army commander at the Arsenal reportedly called a one square mile (2.6 km^2) section located in the center to the Arsenal "the most contaminated square mile in the nation," or even "the most contaminated square mile on earth" (Hoffecker 2001). Although to our knowledge the quote has never been verified, it became part of the folklore of the Arsenal, regardless of its accuracy.

Abstract In 1942, the US Army purchased 70 km^2 near Denver, Colorado, to construct the Rocky Mountain Arsenal, where they manufactured chemical and incendiary weapons (mustard, lewisite, napalm) in support of the war effort. After World War II, Shell Oil and its predecessors manufactured pesticides, insecticides, and herbicides at the Arsenal, and the Army manufactured and then decommissioned Sarin (GB) nerve agent. The industrial manufacturing was concentrated near the center of the site, with many square kilometers of undeveloped land providing a buffer from urban Denver. Millions of liters of liquid wastes, including cyclodiene pesticides, such as dieldrin, aldrin, and endrin, were disposed of in open basins, pits, and trenches on the site. Cyclodiene pesticides impact the central nervous system of exposed biota, causing disorientation, emaciation, and eventually death. Thousands of wildlife mortalities were documented at the site, including an estimate of 20,000 duck deaths over a 10-year period in the 1950s, and over 1,800 waterfowl deaths in one basin alone between 1981 and 1987. After manufacturing ceased in the 1980s, the Army, Shell, the US Environmental Protection Agency, and the State of Colorado endeavored to address the contamination and the Arsenal's future land use. In 1992, the USA passed the Rocky Mountain Arsenal National Wildlife Refuge Act to create wildlife habitat from uncontaminated and remediated areas and reduce risk of exposure to humans. Many aspects of the cleanup were based on pesticide risks to wildlife, and the parties rarely reached agreement on cleanup thresholds and the amount of area requiring

J.V. Holmes (✉)
Stratus Consulting Inc., Boulder, CO, 80302, USA
e-mail: jholmes@stratusconsulting.com

J.E. Elliott et al. (eds.), *Wildlife Ecotoxicology: Forensic Approaches*,
Emerging Topics in Ecotoxicology 3, DOI 10.1007/978-0-387-89432-4_4,

remediation. Ultimately, in 2010, after nearly 30 years of investigation and cleanup efforts and over $2 billion, "significant environmental cleanup" was completed. Millions of metric tons of toxic sludges and soils had been placed in hazardous waste landfills, and millions of metric tons more were buried in place and capped. The refuge now comprises approximately 6,000 ha, roughly the size of Manhattan. The cleanup process was contentious, resulting in multiple lawsuits, which is not a model that the authors recommend. Although contamination at the Arsenal likely is still present and should continue to be evaluated, the refuge will be an important protected oasis of wildlife habitat in the midst of the Denver urban sprawl.

Introduction

The Rocky Mountain Arsenal ("the Arsenal") is located on the high plains ecosystem near Denver, Colorado. The Arsenal is a well-known example of a hazardous waste site in which wildlife toxicology played a central role in defining site remediation and future land use. Located about 18 km northeast of downtown Denver, the Arsenal evolved from cattle-grazed shortgrass prairie and irrigated fields to a munitions and pesticide manufacturing complex to a large urban wildlife refuge over a 70-year period (Fig. 4.1).

In 1942, the US Army purchased 70 km^2 from local ranchers and constructed a munitions manufacturing facility near the center of the plot. The Army selected the Arsenal location because it was far from the threat of enemy aircraft and close to major railroad lines. In addition, the Denver location provided adequate water and

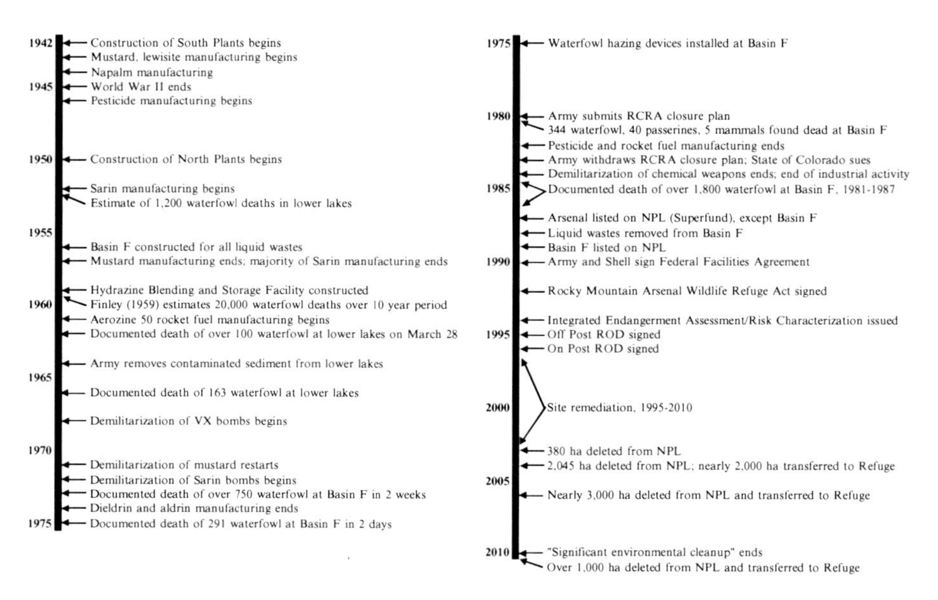

Fig. 4.1 Timeline of Arsenal activities and significant documented waterfowl deaths, 1942–2010

electrical power supplies, readily available skilled labor, and sufficient land to include several square kilometers of undeveloped habitat buffering the manufacturing facilities from the public (Army 2004). From 1942 until the late 1960s, the Army manufactured chemical warfare agents, including rockets and projectiles containing blister agents (e.g., mustard gas and lewisite), nerve agents (Sarin), and pulmonary agents (phosgene), along with incendiary bombs (napalm and phosphorous). The Army later used the Arsenal for the "demilitarization" of obsolete and off-specification chemical agents and bombs. As a result of these activities, the Army disposed of large volumes of manufacturing products and byproducts, unexploded ordnance, and other highly toxic hazardous wastes on-site.

After World War II, the Army leased portions of the site to private industry. Shell Oil Co. and its predecessors manufactured pesticides, insecticides, herbicides, and other chemicals from 1952 to 1982 (Fig. 4.1). Those manufacturing activities generated tons of toxic solid waste and millions of liters of hazardous liquid wastes that were discharged on-site into open natural depressions, trenches, and pits.

When manufacturing ceased, the Army was left with two massively contaminated industrial complexes, and numerous disposal pits, landfills, and open basins filled with a toxic stew of liquid waste from decades of chemical munitions and pesticide manufacturing.

The waste disposal sites at the Arsenal released contaminants into soils, groundwater, surface water, sediments, and air. At the same time, the buffer area between the manufacturing facilities and the property boundary was undisturbed by urban encroachment, providing a large swath of shortgrass prairie and wetland habitat, while at the same time attracting a wide diversity of wildlife that could be subsequently exposed to this contamination.

The on-site manufacturing and waste disposal activities at the Arsenal resulted in thousands of documented wildlife mortalities at the site, and undoubtedly many more wildlife mortalities that were not discovered or documented. Organochlorine pesticides, particularly dieldrin and aldrin, were responsible for the majority of the adverse effects. Starting in the 1980s, the Army, Shell, and federal and state government regulators set out to determine an appropriate remediation plan. They were tasked with determining the contaminant concentrations throughout the huge Arsenal complex and establishing "safe" (i.e., acceptable risk level) concentrations for organic and inorganic contaminants. Wildlife toxicology and risk analysis played a key role in determining the level of cleanup plan for the site.

The presence of bald eagles (*Haliaeetus leucocephalus*), ferruginous hawks (*Buteo regalis*), burrowing owls (*Athene cunicularia*), and other resident threatened species at the Arsenal, combined with high liability and remediation costs, prompted the US Congress to pass the Rocky Mountain Arsenal National Wildlife Refuge Act in 1992. This established a plan for a refuge comprising close to 6,000 ha of uncontaminated or remediated habitat, with the Army managing in perpetuity about 400 ha of unremediated or capped areas, constructed hazardous waste landfills, and groundwater treatment facilities. This refuge is now one of the largest urban wildlife refuges in the USA, providing a large swath of protected habitat within the Denver urban area.

In this chapter, we describe some of the manufacturing history of the Arsenal and the high volumes of hazardous wastes generated. We then discuss the environmental impacts that resulted from contaminant releases and waste disposal practices. Wildlife toxicology played a key role in cleanup decisions and ultimately in the current and future management of the wildlife refuge. While the establishment of such a large wildlife refuge within a major metropolitan area is a tremendous asset for wildlife, our hope is that the US Fish and Wildlife Service (USFWS) will monitor wildlife mortality after remediation is complete, and adequately address any potential ongoing continuing and ongoing impacts from past contaminant releases at the site.

Industrial History: Army Activity

The majority of the manufacturing at the Arsenal occurred in the South Plants Complex ("South Plants" – Fig. 4.2), a 200-ha area with buildings, roads, parking lots, railroad tracks, sewer lines, culverts, steam pipes, manholes, water mains, and some open space (Ebasco Services et al. 1989b). Army products manufactured in the South Plants included both chemical agents and incendiary munitions. Chemical agents manufactured at the site included vesicants (blistering agents) such as mustard and lewisite, and choking agents such as phosgene (Table 4.1). In 1942 and 1943, the Army manufactured over 3.2 million tons of mustard, including 302 tons determined to be off-specification and ultimately treated and disposed of on-site (DOJ 1986; Ebasco Services et al. 1988a). In total, the Army produced about 1.6 million nerve gas munitions, including cluster bombs, shells, bomblets, rockets, and warheads at the Arsenal (Goldstein 2001).

Incendiary munitions manufactured at the site included napalm (a powdered thickening agent added to gasoline to produce gelatinous fuel), white phosphorous, and a mixture of potassium chlorate, red phosphorous and glass known as "button bombs" and "sandwich button bombs" (Table 4.1). Napalm was produced from 1943 to 1945, with a total output of over 2.6 million bombs. White phosphorous-filled munitions, including white phosphorous cups, igniters, grenades, and 105 mm shells, followed from 1945 until 1970. By the end of World War II, the Arsenal had created more than 91,000 tons of incendiary munitions. By 1968, over 1.7 million button bombs and 7 million sandwich button bombs had been manufactured at the Arsenal (DOJ 1986; Ebasco Services et al. 1988a; Environmental Science & Engineering et al. 1988b; Army 2004).

The Army constructed the 36-ha North Plants Complex ("North Plants" – Fig. 4.2) from 1950 to 1952 (see timeline, Fig. 4.1). From 1952 to 1957, the Army used the North Plants to manufacture Sarin, an anticholinesterase nerve agent similar in structure and function to widely used organophosphorus insecticides (Table 4.1). They also used the North Plants to fill munitions with Sarin and assemble cluster bombs, as well as to store Sarin, feedstock chemicals, and munitions (Fig. 4.3). The facility was later used for the demilitarization of chemical warfare agents. The Army stored Sarin in 1-ton containers and in underground tanks (DOJ 1986; RVO 2004).

Thankfully, most of the chemical warfare agents manufactured at the Arsenal were not used during wartime. As a result, the Army retooled part of the Arsenal for demili-

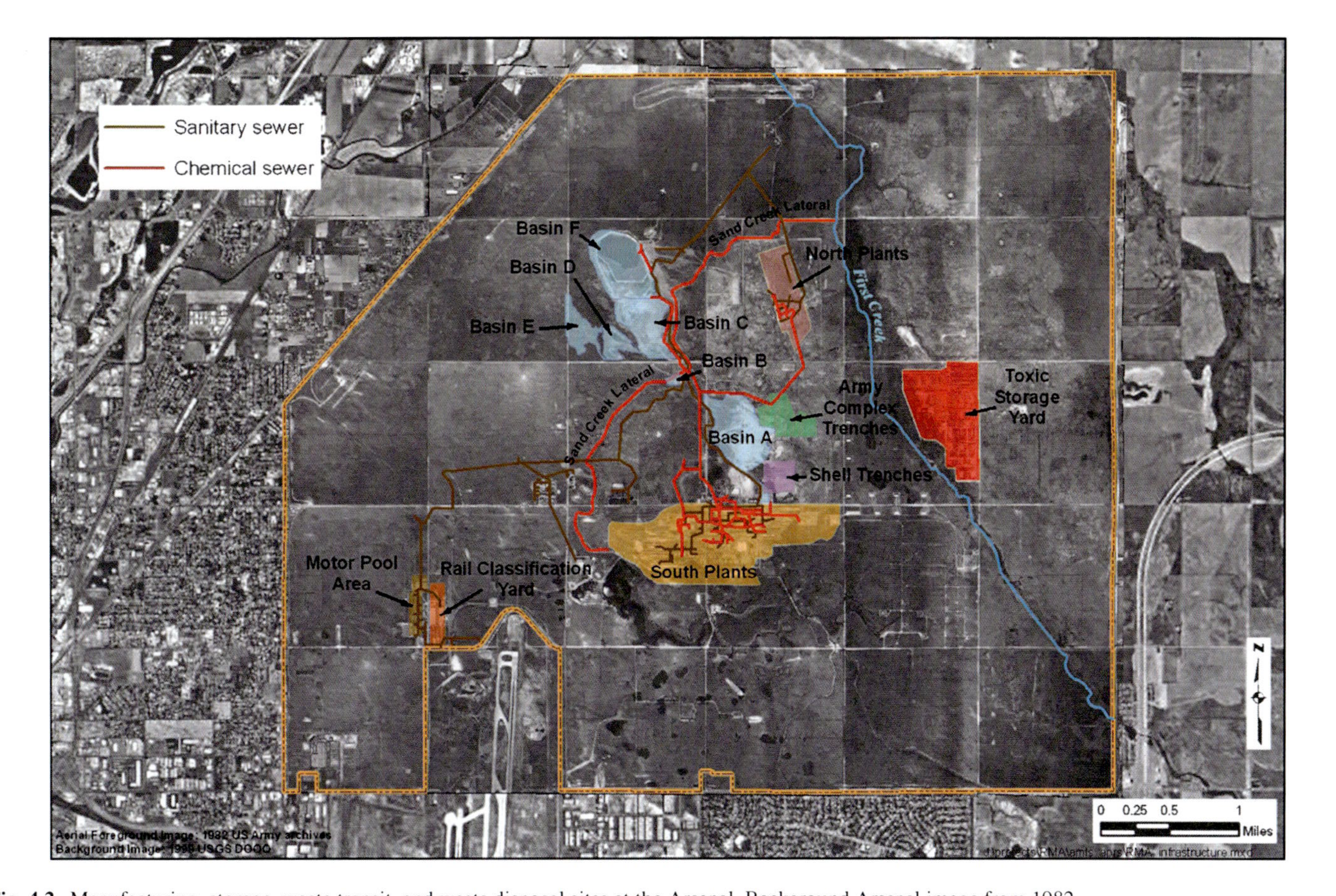

Fig. 4.2 Manufacturing, storage, waste transit, and waste disposal sites at the Arsenal. Background Arsenal image from 1982

Table 4.1 Chemical agents and munitions manufactured or handled at the Arsenal

Product	Description	Sources
Mustard [bis (2-chloroethyl) sulfide; H; HS; Levinstein mustard]	Mustard is a vesicant, or blistering agent. Raw materials used in the production of mustard were ethyl alcohol, Freon 114, sulfur monochloride, calcium chloride, bleaching powder, coke, kerosene, fuel oil, caustic soda, and hexamine. Mustard and mustard-filled munitions were manufactured at the Arsenal in 1943 and from 1950 to 1957. Between 1945 and 1946, the Army reprocessed mustard into distilled mustard.	Ebasco Services et al. (1988a)
Lewisite [2-chlorovinyl dichloro arsine]	Lewisite is a vesicant, sometimes used in combination with mustard. Raw materials associated with lewisite production included acetylene, arsenic trichloride, thionyl chloride, hydrochloric , and mercuric chloride, many of which were manufactured at the Arsenal. The Army produced lewisite from April through November 1943.	Kuznear and Trautmann (1980)
Phosgene [carbonyl chloride]	Phosgene is a choking agent. The phosgene bomb-filling plant at the Arsenal operated from January 1944 to December 1944.	Kuznear and Trautmann (1980), Army (2004)
Napalm	Napalm is a powdered thickening agent added to gasoline to produce the incendiary mixture called NP gel. This thickened fuel was used in incendiary bombs and flamethrowers. The incendiary oil plant, which mixed the napalm thickener and gasoline, operated from April 1943 to August 1945.	Ebasco Services et al. (1988a)
White and red phosphorus	White phosphorus was used for the production of incendiary bombs. Red phosphorus was used as a chemical constituent in button bombs and sandwich button bombs. The phosphorus plant operated in 1944–1945, 1952, 1953, and 1958–1960.	Ebasco Services et al. (1988a)
Sarin [GB; isopropyl methylphosphonofluoridate]	Sarin is an extremely toxic nerve agent. Raw materials included methylphosphonic dichloride, hydrofluoric acid, isopropyl, tributylamine, methanol, carbon tetrachloride, ethyl ether, methylene chloride, and calcium chloride. Sarin was manufactured at the North Plants from July 1952 through March 1957.	Ebasco Services et al. (1988a)
VX [*O*-ethyl-*S*-(2-diisopropylaminoethyl) methylphosphonothioate]	VX is an extremely toxic nerve agent. VX was not manufactured at the Arsenal, but VX bombs were decommissioned along with Sarin bombs at the site in the 1960s, and byproducts of the process were disposed of at the site.	Ebasco Services et al. (1988a)

Fig. 4.3 Rabbit used to check for Sarin leaks at the North Plants, c. 1955. Source: Library of Congress

tarization of chemical weapons. Beginning in 1947, and continuing sporadically through the 1970s, obsolete and deteriorating mustard-filled munitions were disassembled at the Arsenal. The mustard was incinerated in a furnace located in the South Plants (Fig. 4.2), and the casings were decontaminated in an acid bath and/or burned (DOJ 1986). Demilitarization of Sarin munitions began in the 1950s, with over 204,000 Sarin-filled munitions demilitarized between 1955 and 1970. In 1969, the Army initiated "Project Eagle" to demilitarize excess toxic agent at the Arsenal. The Army drained Sarin bomblets, mixed the Sarin with caustic in a reactor chamber, and spray-dried the brine. Approximately 1.7 million liters of Sarin from over 21,000 munitions were deactivated as part of Project Eagle, resulting in almost 2.8 million kilogram of contaminated spray-dried salt. The salt was stored in tens of thousands of steel and fiberboard drums on-site (Fig. 4.4) in the "Toxic Storage Yard" (Fig. 4.2; Army 1978; DOJ 1986).

In 1968, the Army began the demilitarization of the nerve agent VX, which was "the deadliest nerve agent ever created" (Council on Foreign Relations 2006). VX was not produced at the Arsenal, but from 1964 to 1969, one million kilograms were transported to the site and stored in steel containers, spray tanks, rockets, and mines. Approximately 1,630 kg of VX munitions were neutralized at the Arsenal (DOJ 1986).

In addition to chemical agents and incendiary munitions, the Army manufactured rocket fuel at the Arsenal. In 1959, in coordination with the US Air Force, the Army constructed a facility to blend anhydrous hydrazine with unsymmetrical dimethylhydrazine to produce Aerozine 50 (Ebasco Services et al. 1988b). Aerozine 50 is a hypergolic fuel (ignites spontaneously) when used with nitrogen oxidizers. It was used primarily in the Titan and Delta missile programs (NASA KSC 2006). Blending of Aerozine 50 began in 1961 and continued through 1982 (Fig. 4.1). During its operation, hundreds of liters of hydrazine and byproducts of hydrazine production were accidentally released into the environment (Ebasco Services et al. 1988b).

Fig. 4.4 Disintegrated drum of contaminated salt, a waste product from demilitarization of Sarin bombs, in the Toxic Storage Yard. Source: Jeff Edson

Industrial History: Pesticide Manufacturing

Pesticide manufacturing at the Arsenal occurred from 1947 to 1982 (Fig. 4.1). Colorado Fuel & Iron (CF&I) manufactured dichlorodiphenyltrichloroethane (DDT) in the South Plants in 1947 and 1948 (Ebasco Services et al. 1988c). Julius Hyman & Co. developed aldrin, dieldrin, and chlordane, and produced those pesticides, as well as endrin, at the Arsenal between 1947 and 1952 (Army 2004). Shell Oil Co. merged with Hyman and took over the Hyman pesticide manufacturing business at the Arsenal in 1952. From 1952 until 1982, Shell manufactured many products at the Arsenal, including chlorinated hydrocarbon insecticides, organophosphorus insecticides, carbamate insecticides, herbicides, and soil fumigants (Table 4.2).

Waste Management

Thousands of tons of contaminants were released at the Arsenal. Shell released an estimated 136,000 tons of contaminants into the environment, with the Army releasing an estimated 24,000 additional tons. Contaminants disposed of at the Arsenal included aldrin, arsenic, benzene, cadmium, carbon tetrachloride, chlordane, chlorobenzene, chloroform, chromium, cyanide, dichlorodiphenylethane (DDE), DDT, dieldrin, endrin, hydrochloric and hydrofluoric acids, isodrin, mercury, mustard, parathion, phosgene, phosphorous, Sarin (and its byproducts), trichloroethane, and vinyl chloride (U.S. DOJ 1986; Ebasco Services et al. 1988a).

Table 4.2 Herbicides and pesticides that Shell produced at the Arsenal

Substance	Years produced	Description
Aldrin/dieldrin	1947–1974	An organochlorine pesticide used on cotton and corn, among others. Use now banned or severely restricted.
Chlordane	1947–1952	A pesticide used on corn and citrus crops and on home lawns and gardens. Use now banned or severely restricted.
Endrin	1952–1965	A chlorinated hydrocarbon pesticide used to control insects, rodents, and birds. Use now banned or severely restricted.
Isodrin	1952–1965	A process intermediate of endrin with insecticidal properties similar to those of aldrin.
Methyl parathion	1957–1967	An insecticide for farm crops, especially cotton. Use now banned or severely restricted.
Ethyl parathion	1964–1966	An organophosphate insecticide used primarily on row crops and fruit.
Vapona	1960–1982	An insecticide used in Shell "No Pest Strips."
Supona	1963–1969	A pesticide used to kill parasites in sheep and cattle. No longer used in the USA.
Bidrin	1962–1979	A pesticide used to control aphids, mites, thrips, fleahoppers, grasshoppers, boll weevils, and other insects on cotton, ornamental trees, and fruit crops.
Dibrom	1962–1970	An organophosphate pesticide used for, among other things, mosquito control.
Ciodrin	1962–1976	An organophosphate insecticide.
Azodrin	1965–1977	An insecticide used to control a broad spectrum of pests. Use now banned or severely restricted.
Atrazine	1977–1987	An agricultural herbicide.
Gardona	1967–1968	An insecticide used to control flies.
Akton	1952–1974	An organophosphate insecticide.
Landrin	1969–?	An insecticide.
Nudrin	1973–1977	An insecticide included on EPA's Superfund Extremely Hazardous Substances list.
Nemagon[a]	1955–1977	A soil fumigant used to control soil-inhabiting nematodes. Use now banned or severely restricted.
Bladex	1970–1971, 1974–1975	A restricted-use herbicide used to control weeds and invasive grasses.
Planavin	1966–1975	An herbicide used to control weeds and grass.
Nemafere	Unknown	A nemacide used to control nematodes.
Phosdrin	1956–1973	An insecticide used on vegetables, alfalfa, fruits, and nuts. Use now banned or severely restricted.

Sources: DOJ (1986), Ebasco Services et al. (1988a), MK-Environmental Services (1993), ATSDR (2003), Scorecard.org (2005), U.S. EPA (2007)

[a] Nemagon is the trade name for dibromochloropropane (DBCP)

Waste from the South Plants were released to the environment through direct disposal into, among other locations, Basins A through F, Lime Settling Basins, Army Complex Trenches, Shell Trenches, sewers, ditches, and the Arsenal's lakes (Figs. 4.2 and 4.5), as well as directly to the ground from leakage, spills, and overflow of underground tanks and sumps. Uncontrolled contaminant releases, totaling

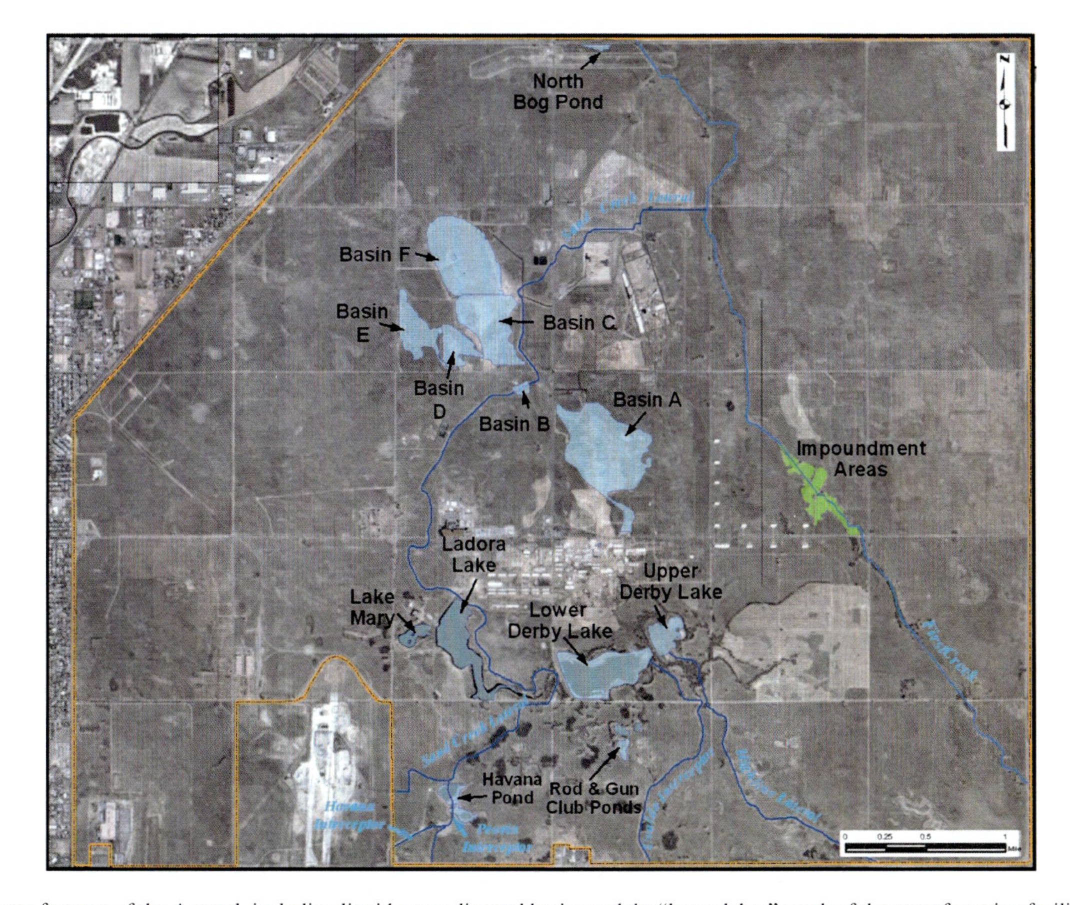

Fig. 4.5 Open water features of the Arsenal, including liquid waste disposal basins and the "lower lakes" south of the manufacturing facility

hundreds of thousands of liters, occurred regularly from the late 1940s to the end of manufacturing activities in the early 1980s.

Several natural and man-made lakes are located directly south and southwest of the South Plants (Fig. 4.5). The use of the lakes in conjunction with the South Plants cooling water process resulted in the release of numerous contaminants into the lakes, including aldrin, dieldrin, DBCP, methylene chloride, methylisobutyl ketone, arsenic, mercury, chromium, and lead (Ebasco Services et al. 1987, 1988g).The North Plants' manufacturing process resulted in the release of contaminants into, among other locations, Basins A and F, Army Complex Trenches, the Toxic Storage Yard, sewers, trenches, pits, and ditches, as well as directly to the ground from leakage and overflow of underground tanks and sumps within the site (Fig. 4.2). Contaminants released from the North Plants included benzene, 1,1 dichloroethane, 1,2 dichloroethane, carbon tetrachloride, 1,1,1 trichloroethane, trichloroethylene (TCE), chloroform, arsenic, mercury, cadmium, chromium, lead, chloroacetic acid, aldrin, and dieldrin (Ebasco Services et al. 1989a).

In 1988, the Army estimated the number of potentially contaminated sites on the Arsenal at 174, including 15 basins and lagoons; 14 ditches, lakes, ponds; 28 excavation and surface disturbance sites; 8 ordnance testing areas; 38 solid waste burial sites; 12 spill areas; 46 contaminated storage sites, buildings, equipment areas; and 13 sewers (Program Manager's Office 1988). Dozens of contaminants were found in hundreds of different samples around the manufacturing areas. Table 4.3 presents an example of some of the contaminant concentrations found in basins and lagoons at the site. Similar contaminants were found throughout manufacturing and waste disposal areas, and adjacent areas that received deposition of contaminants via aeolian transport.

Sources of Contamination

Liquid Wastes: Basins

For most of the industrial history of the Arsenal, liquid wastes were transported through chemical sewer systems to open basins, where they would either evaporate or infiltrate to underlying groundwater. The Army discharged the majority of its liquid wastes into six disposal basins (Basin A through F – see Figs. 4.2 and 4.5). In total, these basins covered almost 140 hectares. Basins A through E were used in the 1940s and 1950s; Basin F received nearly all liquid wastes from its construction in 1956 through the end of industrial operations in the early 1980s.

Basin A was an unlined natural depression covering 50 ha (Figs. 4.2 and 4.5). It was the primary disposal area for liquid chemical waste from the South Plants from the early 1940s until the mid-1950s, when construction of Basin F was completed (Environmental Science & Engineering 1987). Basin A received off-specification mustard, neutralized with caustic with varying degrees of success, as well as white phosphorus, waste pesticides and herbicides from South Plants, and wastes from Sarin production at North Plants (Ebasco Services et al. 1988a). Basin A was a major source of groundwater contamination on the Arsenal because of the quantity of liquids disposed in the basin and its shallow depth to groundwater.

Table 4.3 Range of concentrations of selected analytes in soil samples from basin and lagoon sites

Analyte	Detected	Samples	Detected concentration range (μg/g)	Median of detected samples (μg/g)
Volatile halogenated organics				
Chlorobenzene	9	508	0.34–5.0	1.7
Chloroform	22	538	0.12–70	1.8
Methylene chloride	58	422	0.27–6.7	1.0
Tetrachloroethene	19	537	0.20–40	2.8
Trichloroethene	4	538	0.14–1.0	0.51
Volatile hydrocarbons				
2,2-Oxybisethanol	36	36	0.6–8	2
2-Pentanone	9	9	1–20	10
Bicycloheptadiene	16	541	0.95–5,100	9.7
DCPD	58	1,079	0.35–22,000	24
Volatile aromatic organics				
Benzene	17	552	0.24–6.0	1.2
Ethylbenzene	11	552	0.14–9.3	1.6
m-Xylene	15	552	0.14–12	1.0
o- and *p*-Xylene	9	553	0.15–12	2.0
Toluene	18	550	0.13–1,000	0.95
Organosulfur compounds – mustard-agent related				
Chloroacetic acid	23	242	43–7,900	140
Dithiane	8	1,065	0.47–370	0.70
Thiodiglycol	11	242	6.0–120	25
Organosulfur compounds – herbicide related				
Chlorophenylmethyl sulfide	18	1,070	0.50–700	5.4
Chlorophenylmethyl sulfone	83	1,070	0.34–300	2.5
Chlorophenylmethyl sulfoxide	25	1,067	0.58–70	2.6
Dimethyldisulfide	3	760	2.0–70	10
Organophosphorous compounds – Sarin-agent related				
DIMP	119	968	0.12–10	1.3
Dimethylmethyl phosphonate	9	767	3.0–70	6.0
Isopropylmethyl phosphonic acid	34	234	4.6–3,700	39
Methylphosphonic acid	24	122	3.1–400	16
Phosphoric acid, triphenyl ester	33	33	1–20	10
DBCP	14	1,099	0.0061–20	0.019
Fluoroacetic acid	25	124	3.4–200	8.5
Polynuclear aromatic hydrocarbons (PAHs)				
Fluoranthene	4	4	4–8	5
Pyrene	3	3	1–100	4

(continued)

Table 4.3 (continued)

Analyte	Detected	Samples	Detected concentration range (μg/g)	Median of detected samples (μg/g)
Semivolatile halogenated organics				
Hexachlorocyclopentadiene	24	1,121	0.0040–2,600	0.018
Organochlorine pesticides				
Aldrin	169	1,249	0.0024–18,000	2.0
Chlordane	56	1,125	0.072–660	0.44
DDE	79	1,256	0.0014–28	0.038
DDT	56	1,251	0.0028–60	0.030
Dieldrin	401	1,247	0.0014–2,100	0.60
Endrin	217	1,246	0.0011–1,100	0.085
Isodrin	108	1,251	0.0014–11,000	1.5
Arsenic	531	1,253	2.4–110,000	12
Mercury	313	1,099	0.050–35,000	0.27
Metals				
Cadmium	163	1,047	0.63–3,900	1.3
Chromium	768	1,047	6.9–110	14
Copper	919	1,047	4.8–2,300	12
Lead	305	1,039	10–1,100	25
Zinc	969	1,047	11–910	44

Source: Ebasco Services et al. (1992)

Basins B through E were smaller reservoirs, primarily collecting overflow from Basin A. Basin B was an intermediate reservoir that temporarily held overflow wastes en route to Basin C. Basin C, colloquially known as the "straw-bottomed basin" because of the rapidity with which the liquid wastes infiltrated to groundwater, was an unlined depression covering 31 ha, with a capacity of 720 million liters. Basin D covered 8.5 ha and Basin E covered 12 ha (Fig. 4.5). Contaminants released to these basins included xylene, aldrin, dieldrin, DDE, chlorophenylmethyl sulfide, arsenic, mercury, lead, cadmium, and chromium (Environmental Science & Engineering et al. 1987a, b, c, d; Environmental Science & Engineering and Harding Lawson Associates 1988a, b, c, d).

In 1956, with Basin A regularly overflowing to the other basins and the Army's recognition that toxic wastes infiltrating to groundwater might cause problems for downgradient users, Basin F was constructed, complete with engineered dikes and an asphalt liner. Basin F covered 38 ha, with a capacity of 910 million liters. The toxic stew in Basin F included, among other wastes, aldrin, dieldrin, endrin, isodrin, toluene, trichloroethane, TCE, DBCP, diisopropyl methylphosphonate (DIMP), ethyl benzene, xylene, DDE, arsenic, mercury, lead, and chromium (Environmental Science & Engineering et al. 1988a). On at least three occasions in the 1960s and 1970s, Basin F was filled to capacity (Foster Wheeler 1996).

The Army's attempts to contain liquid waste with an asphalt liner in Basin F were unsuccessful. Many of the contaminants in Basin F, in particular DIMP, a byproduct of Sarin manufacturing, infiltrated to the underlying alluvial aquifer and migrated nearly 20 km down gradient.

In the early 1980s, the Army attempted to accelerate the evaporation of liquids in Basin F by aerating the pond, spraying contaminated liquids into the air. Unfortunately, the spray covered downwind soils with contaminants. In the late 1980s and early 1990s, the Army pumped the basin's liquids and surface sediments into tanks, surface impoundments, and waste piles. The Basin F response action became a contentious issue between the Army, the EPA, the Colorado Department of Health, and local residents, because of repulsive odors from the underlying sediments after the liquid was removed. Finally, after pressure from the federal and state environmental agencies, the Army covered the remaining Basin F sediments with soil. They incinerated over 40 million liters of Basin F liquids on-site between 1993 and 1995 (Foster Wheeler 1996).

Buried Waste: Trenches

The Army and Shell buried solid and liquid wastes, including unexploded ordnance, off-specification products, and other contaminated wastes, in numerous trenches. The Army Complex Trenches (Fig. 4.2) covers 40 ha and contains an assortment of wastes, including chemical warfare agents, chemical agent-filled unexploded ordnance, rejected incendiaries, contaminated tools, vehicles, and equipment. (Foster Wheeler 2001). Contaminants found in soils and groundwater near these trenches included aldrin, dieldrin, endrin, chlordane, chlorobenzene, dichloroethane, dichloroethene, tetrachloroethane, TCE, trichloroethane, DDT, DBCP, fluoroacetic acid, arsenic, mercury, cyanide, cadmium, chromium, lead, and xylene (Harding Lawson Associates 1993).

The Shell Trenches (Fig. 4.2) comprised a 3.2-ha area with 31 unlined trenches. Shell used the trenches to dispose of organic and inorganic compounds, process intermediates, and off-specification and unused pesticide and herbicide products produced between 1952 and 1965 (Table 4.2). Contaminants in the environment near the trenches include aldrin, dieldrin, endrin, isodrin, benzene, chlorobenzene, chloroform, DBCP, dicyclopropane, ethylbenzene, hexane, toluene, xylene, 1,1-dichloroethane, 1,2-dichloroethane, 1,2-dichloroethene, methylene chloride, tetrachloroethylene, and trichloroethene (Environmental Science & Engineering et al. 1987e, 1988c; Environmental Science & Engineering and Harding Lawson Associates 1988e, f; MK-Environmental Services 1993).

Toxic Storage Yard

The Toxic Storage Yard (Fig. 4.2) was originally constructed as a storage site for chemical agents and was subsequently used as a storage site for demilitarized chemical agent waste products. In the mid-1950s, the yard contained 625 large containers that likely contained Sarin. In the early 1960s, the Army discovered that containers and cluster bombs containing Sarin in the Toxic Storage Yard were leaking (Ebasco Services et al. 1988d, h).

In the early 1970s, the Army stored approximately 76,000 drums of demilitarized Sarin salts in the Toxic Storage Yard. Approximately 1.7 million liters of Sarin from over 21,000 munitions were deactivated in the North Plants, resulting in almost 2.8 million kilograms of contaminated spray-dried salt stored in the drums. Spills from leaking bombs and containers (e.g., Fig. 4.4) resulted in the release of chloroacetic acid, arsenic, chromium, and lead (Army 1978; DOJ 1986; Ebasco Services et al. 1988d, e, f).

Other Disposal Areas

The basins, trenches, and the Toxic Storage Yard are but a few examples of areas where hazardous wastes were disposed of on the Arsenal; both solid and liquid toxic wastes were placed in over a hundred different sites throughout the Arsenal, including areas where wildlife was in direct contact with these wastes. The result was a tremendous amount of mortality, most of which was likely undocumented.

Wildlife Exposure and Mortality

The Arsenal comprised 70 km^2, with an inner core of manufacturing and waste disposal areas surrounded by a buffer between Army and Shell manufacturing and disposal sites and publicly accessible land (Fig. 4.2). Over the years, the manufacturing and disposal sites, along with the undisturbed buffer areas, became home to a wide diversity of wildlife. In the 1980s and 1990s, Army contractors identified 26 species of mammals on the Arsenal, including all of the common mammals that inhabit shortgrass prairie in the Front Range of Colorado, as well as 176 species of birds and at least 17 species of reptiles and amphibians. There was a high species richness of birds at the Arsenal. Ground-nesting songbirds and other birds preferring open habitat were common in the primary Arsenal habitats of open grassland and weedy plains (Ebasco Services et al. 1994).

Unfortunately, many birds and other wildlife that frequented the Arsenal could not distinguish between the undeveloped buffer areas and the highly contaminated waste disposal areas. Basins A through F appeared to migrating waterfowl to be oases of open water in the semi-arid shortgrass prairie. Wildlife that came in contact with waste areas, particularly waterfowl, passerine birds, and prairie dogs, frequently died. Collection and analysis of carcasses was not routine until the 1980s, and even then, the collection and analysis was far from systematic or thorough. Here, we present the mortality data and older anecdotal evidence of mortality, focused primarily on avian impacts.

Starting in 1951 and continuing through the 1950s and 1960s (see Fig. 4.1), the Colorado Division of Wildlife and the USFWS Wildlife Research Laboratory visited the Arsenal, attempting to quantify some of the mortalities that were occurring in the

Arsenal's lower lakes (Fig. 4.5). Lake water became contaminated when used as process water in the South Plants. Returning the process water to the lake resulted in widespread contamination and subsequent exposure to waterfowl and other aquatic biota. In the spring of 1952, the USFWS estimated that 1,200 ducks died at the lakes, stating that the cause of death was toxic agents entering the lakes through the process-water drain from the chemical plant area (Finley 1959). Similarly, in April 1959, 119 dead birds and animals were counted on a single day around the shore of one of the lakes. An interview with a Shell employee revealed that he had gathered approximately 500 dead ducks for burial during the first 3 months of 1959 (Finley 1959). The USFWS stated that 2,000 ducks would be a conservative annual estimate of duck mortality in the lakes' area, with 20,000 or more ducks dying over a 10-year period (Finley 1959). The USFWS report noted that high wildlife mortality occurred at the lakes when extensive mud flats were exposed. In addition, they reported an absence of frog choruses, egg masses, and tadpoles at the lakes in 1960 (USFWS 1961).

Waterfowl mortality in the lakes continued in the 1960s, including more than 100 ducks found dead on March 28, 1962. In 1964, the Army removed contaminated sediments from the lakes and later reported that waterfowl mortality declined from previous years. However, this did not fully alleviate the problem, as an additional 163 waterfowl deaths were reported between January and May 1966 (Environmental Science and Engineering 1989).

Much of the avian mortality data from the Arsenal came from Basin F. Formal mortality reports from Basin F were sporadic prior to the onset of investigative activities in the 1980s (see timeline, Fig. 4.1). In a 2-week period in early April 1973, approximately 750 dead ducks and grebes were collected from the Basin F area over the course of three visits. Several hundred additional carcasses were observed on June 13 and 14, 1973 (Ward and Gauthier 1973). During 2 days in May 1975, investigators removed 291 bird carcasses from the shoreline of Basin F, including waterfowl, raptors, pheasants, and songbirds (Environmental Science and Engineering 1989). On May 7, 1980, government inspectors at Basin F found 389 wildlife carcasses, including 344 waterfowl, 40 birds other than waterfowl, and 5 small mammals (Seidel 1980). Another 49 waterfowl carcasses were collected at Basin F between October and December 1980 (Environmental Science and Engineering 1989).

Routine collection and recording of waterfowl mortality occurred at Basin F between 1981 and 1987. Between 139 and 444 dead birds were found each year. In total, over 1,800 dead waterfowl were found over this 7-year period (Table 4.4). In 1989, contaminated liquids were moved to storage tanks and lined holding basins, and the basin was covered with a soil cap, ending direct exposure of waterfowl to liquid waste in Basin F.

Transition to Remediation

By the end of 1980, manufacturing at the Arsenal was in decline, and the US Congress had passed the Resource Conservation and Recovery Act (RCRA) and the Comprehensive Environmental Response, Restoration Liability Act (CERCLA),

Table 4.4 Reported waterfowl mortalities at Basin F (1981–1987)

Year	Number of dead waterfowl
1981	202
1982	222
1983	444
1984	418
1985	140
1986	236
1987	139
Total	1,801

Source: Environmental Science and Engineering (1989), Table 4.1-1.

or "Superfund." The legacy of four decades of chemical weapon and pesticide manufacturing would require almost three decades of remediation to reduce the risk of harm to human health and the environment.

Arsenal manufacturing and disposal areas were to be closed under RCRA Army Closure Plans approved by the US Environmental Protection Agency (EPA) and later the Colorado Department of Health. On November 18, 1980, the Army submitted a RCRA Part A application, calling for the closure of five hazardous waste management units including the Toxic Storage Yard and Basin F, with the remainder of the facility addressed through CERCLA. In 1983, the Army withdrew its RCRA Closure Plan application, resulting in lengthy litigation between the State of Colorado and the USA. The State contended that through the Closure Plan, it had RCRA regulatory jurisdiction over the clean-up of Basin F. The US Court of Appeals for the 10th Circuit agreed with Colorado, concluding that states were not precluded from RCRA enforcement actions at federal facilities regardless of whether the site was listed on the CERCLA National Priorities List (NPL).

The EPA listed most of the Arsenal on the NPL on July 11, 1987. The remainder, Basin F in particular, was included on March 13, 1989. The listing made the Arsenal a priority for the federal cleanup program. The nature and scope of that cleanup was a matter of debate, hinging in part on the intended future use of the facility. Stapleton Airport was immediately south of the Arsenal (visible runways in Fig. 4.2) and was in need of expansion or replacement. Urban sprawl from the City of Denver had reached the Arsenal by the 1980s, with demand for additional growth. If the Arsenal was to be used for commercial or residential purposes, the amount of cleanup required to alleviate risk of adverse effects to humans would have been tremendous. At the same time, a large portion of the Arsenal served as a buffer to keep residential and commercial areas far away from the Arsenal industries, and a portion of that habitat was uncontaminated and supported a wide diversity of wildlife.

In 1989, the Army and Shell reached a Federal Facilities Agreement (FFA) with various federal agencies, although the State of Colorado declined to sign. The FFA established a framework and process for cleaning up the Arsenal. It also placed institutional controls on the entire Arsenal property, specifically stating that (1) the property could never be used for residential or agricultural purposes; (2) any hunting or fishing on the property would be restricted to nonconsumptive use; and (3) water on or under the site could never be used as a drinking water supply. The FFA also explicitly stated that it was a goal of the federal signatories to make significant portions of the site available for beneficial public use and require habitat preservation.

Three years later, in October 1992, the US Congress passed, and President George H.W. Bush signed into law, the Rocky Mountain Arsenal National Wildlife Refuge Act, establishing a goal of creating a large wildlife refuge in uncontaminated and adequately remediated areas (Foster Wheeler 1996). Creation of a wildlife refuge at the Arsenal on one hand can be viewed as a conservation victory, compensating for wildlife harm and preventing future habitat destruction. On the other hand, a wildlife refuge restricts human access and potential exposure to the site, which allowed the Army and Shell to leave the contamination in place and reduce the scope of site cleanup.

Risk Assessment

Once manufacturing drew to a close in the early 1980s and the primary site activity transitioned to remediation, the Army, Shell, the EPA, USFWS, and Colorado Department of Public Health and Environment (CDPHE) were faced with numerous fundamental questions: What methods can adequately characterize 70 km^2 of potential contamination? What are the appropriate remediation goals given the proposed future land use? What final remedial actions will be required to achieve a protective clean-up? Answers to these questions required knowledge of regulatory requirements, politics, wildlife toxicology, and public policy – a process staged over many years.

Throughout the 1980s, Army and Shell contractors conducted assessments of the Arsenal property to attempt to delineate the extent of the problem. They developed a list of over 600 contaminants potentially released over approximately 1,215 ha of habitat, plus 15 separate groundwater plumes and nearly 800 potentially contaminated structures (Ebasco Services et al. 1992). Contamination was detected in soil, ditches, stream and lakebed sediments, sewers, groundwater, surface water, air, and biota (Foster Wheeler 1996).

With the initial environmental investigation complete and the future use established, remedial managers were then assigned the task of determining the primary contaminants of concern and the appropriate cleanup targets for contaminants. This effort comprised many drafts, garnered volumes of public comments, cost millions of dollars, and required years of wrangling between Army and Shell contractors, EPA, and CDPHE. At the Arsenal, contractors combined the baseline risk assessment for both human health and ecological risks into an Integrated Endangerment Assessment/Risk Characterization (IEA/RC). The IEA/RC generally identified contaminants of concern, assessed exposure to and toxicity of those contaminants, and estimated human and ecological risk based on those assessments (Ebasco Services et al. 1994).

The ecological risk assessment focused on three categories of representative species: (1) avian predators, including bald eagle, great horned owl, American kestrel, and great blue heron (*Ardea herodias*); (2) species with special feeding niches, including shorebirds such as the mallard, blue-winged teal (*Anas discors*), and American coot (*Fulica Americana*), and small passerine birds such as the mourning dove (*Zenaida macroura*), vesper sparrow (*Pooecetes gramineus*), and western meadowlark (*Sturnella neglecta*); and (3) prey, including deer mouse (*Peromyscus maniculatus*), 13-lined ground squirrel (*Spermophilus tridecemlineatus*), black-tailed prairie dog (*Cynomys ludovicianus*), and desert cottontail (*Sylvilagus audubonii*). Food chain effects were estimated based on food webs that culminated in a single top predator. The IEA/RC evaluated potential ecological risk for 14 target compounds: aldrin, dieldrin, DDT, DDE, endrin, mercury, arsenic, cadmium, copper, chlordane, chlorophenyl methyl sulfide, chlorophenyl methyl sulfone, DCPD, and DBCP. These contaminants were selected based on toxicity, persistence, amount used or produced at the Arsenal, and areal extent of contamination (Ebasco Services et al. 1994). Of the 14 target compounds, pesticides were most prevalent in the lakes and surface soils, had been shown to be toxic in laboratory studies, and had caused considerable mortality at other sites.

Pesticide Toxicity

Mechanisms of Toxicity

The majority of organochlorine pesticides released at the Arsenal are cyclodienes, including aldrin, dieldrin, chlordane, endrin, and isodrin, and dichlorodiphenylethanes, including DDT and associated metabolites DDE and dichlorodiphenyldichloroethane (DDD). Cyclodienes are among the most acutely toxic of organochlorine pesticides (Hudson et al. 1984; Jorgenson 2001). Like DDT, cyclodiene pesticides are both highly soluble in fat and slowly degraded or metabolized, and thus tend to accumulate in fatty tissues such as liver, adipose, gall bladder, heart, muscle, bones, adrenals, brain, and spinal cord (Beyer and Gish 1980; Matsumura 1975).

Soil-dwelling earthworms or sediment-dwelling aquatic macroinvertebrates can have very high body burdens of toxic compounds because they have direct contact with the contamination and are often insensitive to pesticide toxicity. Consumers of those worms bioaccumulate and biomagnify the pesticides in fat and in fatty tissues such as the brain and liver, with organochlorine pesticide concentrations in these tissues increasing by three to ten times the ingested concentrations (Harris et al. 2000; Klaassen 2001). As a result, predatory birds and mammals at the top of the food web generally have the highest levels of persistent pesticides in their tissues. When these animals mobilize their fat stores during times of stress such as cold weather or migration, they can be poisoned when the accumulated pesticides are released into their circulatory system, even if the actual consumption of the pesticides happened days, weeks, or months previously (e.g., Henriksen et al. 1996).

Cyclodiene pesticides such as dieldrin target the central nervous system (Hayes and Laws 1991). A specific mechanism by which cyclodienes affect the brain is via attachment to the picrotoxin binding site on the gamma-aminobutyric acid (GABA) receptor-chloride channel complex, disrupting chloride ion flow. That leads to only partial repolarization of the neuron after each action potential, resulting in sustained excitation (Matsumura and Ghiasudding 1983; Bloomquist and Soderlund 1985; Eldefrawi et al. 1985; Coats 1990). Dieldrin can also inhibit neuronal Na+ – K+ – ATPase and Ca+ – Mg+ – ATPase, with the latter enzyme being crucial for control of free Ca+ at the synaptic membrane of the nerve terminus. Elevated [Ca+] at the terminus region induces release of neuro-transmitters from storage vesicles and the subsequent stimulation of adjacent neurons and increased central nervous system stimulation (Matsumura 1975; Wafford et al. 1989). Symptoms of cyclodiene poisoning via this mechanism include emaciation, rigid paralysis, convulsions, respiratory failure, and death (Ecobichon 1991; Nagata and Narahashi 1994).

Pesticide-Induced Mortality at Other Sites

Numerous studies since at least the 1950s have documented the poisoning of wildlife by acutely toxic cyclodiene pesticides, particularly dieldrin. In 1956 in Britain, the introduction of aldrin and dieldrin as seed treatments resulted in immediate poisonings of birds, particularly wood pigeons (*Columba livia*) and pheasants (*Phasianus colchicus*) (Peakall 1996). Those poisonings, involving hundreds of incidents a year and thousands of individual birds, continued unabated until the use of dieldrin as a wheat seed treatment was discontinued in 1975.

Blus (2003) described poisoning of pheasants and other wildlife from use of dieldrin to control Japanese beetle (*Popillia japonica*) during the 1960s in the USA. Aldrin-treated rice seed caused poisoning of numerous waterfowl and other birds during the 1960s and 1970s in Texas, and apparently caused the fulvous whistling duck (*Dendrocygna bicolor*) population to decline (Flickinger and King 1972), an effect which persisted into the 1980s (Flickinger et al. 1986).

Several researchers have postulated that dieldrin, rather than DDT, is likely to be responsible for the decline of peregrine falcons, although direct evidence such as tissue analysis of dead falcons is limited. Nisbet (1988) contended that the rapid crash of the peregrine falcon populations in the 1950s was more consistent with a mechanism of breeding bird mortality, rather than reproductive failure because of DDT/DDE effects on shell quality. Similarly, Ratcliffe (1973, 1988) and Newton (1988) have argued that adult poisoning caused by the cyclodienes, dieldrin, aldrin and also possibly heptachlor, was more responsible than DDT for the decline of European peregrines.

The organochlorine pesticides identified as the primary contaminants of concern at the Arsenal accumulated in organic matter within soils and sediments. Plant roots, and to a lesser extent plant green tissues, also accumulate pesticides (International Programme on Chemical Safety 1992). When higher trophic level organisms such as earthworms, insects, birds, or mammals ingest contaminated soils, sediments, or lower order biota, they bioaccumulate and frequently biomagnify the pesticides

(Schwarzenbach et al. 2003). The causative role of dieldrin in the decline of a variety of other British raptors, sparrow hawks, merlins, and barn owls, has been extensively investigated (Newton 1973; Newton and Bogan 1974; Newton and Bogan 1978; Newton 1986; Newton and Haas 1988; Newton et al. 1992; Walker and Newton 1999; Sibly et al. 2000). Despite evidence of relatively high concomitant exposure to DDT, and evidence of effects of DDT on shell quality and reproduction (Newton 1986), the reduction in juvenile and adult survival due to dieldrin poisoning has been determined to be the more significant factor influencing population stability (Sibly et al. 2000).

Linking Arsenal Mortality to Pesticides

To prioritize pollutants and assess ecological risk at the Arsenal, scientists from the Army, Shell, EPA, and CPDHE also reviewed data from other sites, and they evaluated oral dose and contaminant residue concentrations associated with adverse effects and lethality in wildlife. Several studies have investigated the link between pesticide dose, mortality, and tissue residue. Dose information is useful for evaluating when adverse effects are likely to happen in the future, while residue data helps with forensic evaluation of carcasses, determining whether pesticide exposure caused mortality in the past. Here, we review such studies and then compare data from the Arsenal to the thresholds published in the literature.

Hudson et al. (1984) compiled acute toxicity (LD_{50}) data for several organochlorine pesticides on several different birds (Table 4.5). Different pesticide compounds have different toxicities, and different bird species have different sensitivities to the same compounds. Generally, cyclodiene pesticides such as aldrin and dieldrin are among the most toxic organochlorine pesticides. However, aldrin is rapidly converted to dieldrin in environmental media such as soil or water, as well as in the tissues of biological organisms (WHO 1989). Therefore, aldrin is rarely detected at high concentrations. Endrin is about ten times more toxic than dieldrin for common bird test species, with acute (LD_{50}) oral doses ranging from 1 to 6 mg/kg (Table 4.5). Other cyclodiene pesticides (chlordane, oxychlordane, heptachlor, heptachlor epoxide, and isodrin) are somewhat less toxic than dieldrin to birds (Friend and Trainer 1974).

For each compound, there is also marked variation in toxicity among bird species. For example, mallards (*Anas platyrhyncos*) are less sensitive to pesticides than California quail (*Callipepla californica*), meaning that mallards can ingest a higher dose of pesticide without being killed. The variation in toxicity among species is likely a result of differences in the birds' ability to metabolize these compounds (Ronis and Walker 1989).

Subacute exposure to organochlorine pesticides has the potential to alter wildlife behavior and cause chronic health problems. Chronic low-level dosing results in a steady accumulation of dieldrin in an animal. Altered behavior occurring as a result of this low-level dosing includes reduced alertness to predators, altered courtship behavior, and altered aggression (Sharma et al. 1976). Other chronic toxicity effects

Table 4.5 Concentrations of different pesticides that cause mortality ("acute toxicity") in birds

Chemical	Test species	Number of samples	Sex	Age (months)	Acute oral dose (LD_{50} mg/kg)
Aldrin	Mallard	16	F	3–4	520
	Bobwhite	12	F	3–4	6.59
	Pheasant	12	F	3–4	16.8
Dieldrin	Mallard	12	F	6–7	381
	California quail	12	M	7	8.8
	Pheasant	9	M	10–23	79
	Rock dove	15	M, F	–	26.6
DDT	Mallard	8	F	3	>2,240
	California quail	12	M	6	595
	Pheasant	15	F	3–4	1,334
Chlordane	Mallard	12	F	4–5	1,200
	California quail	12	M	12	14.1
	Pheasant	4	F	3	24.0–72.0
Endrin	Mallard	12	F	12	5.6
	California quail	12	F	9–10	1.2
	Pheasant	12	M	3–4	1.8
	Rock dove	16	M, F	–	2.0–5.0

Source: Hudson et al. (1984)

include increased genetic mutations, higher cancer rates, and endocrine disruption (WHO 1989). Population effects of pesticide exposure for birds include delayed egg-laying, decreased egg production, reduced egg weights, and reduced eggshell thickness, all of which contribute to reduced hatchability and post-hatching mortality (Dahlgren et al. 1970; Sharma et al. 1976; Busbee 1977; Newton 1988; Walker and Newton 1999).

Because the mechanism of toxicity is similar for most organochlorine pesticides, the toxicity of multiple compounds may be additive. Stansley and Roscoe (1999) reported apparent additive toxicity in field collections of 425 birds killed by mixtures of dieldrin and chlordane at levels below the demonstrated lethality of either compound. Similarly, Ludke (1976) found additive toxicity in bobwhite quail that were fed a mixture of chlordane and endrin. Recently, Elliott and Bishop (2011) compiled examples supportive of the additive toxicity of both cyclodiene and DDT insecticides, and they have suggested a dieldrin toxic equivalence approach.

A wide range of organochlorine pesticide concentrations have been found in tissues of birds that have also shown adverse effects from exposure. Table 4.6 is a compilation from the literature of dieldrin brain and liver concentrations from dead birds collected in the field and suspected of dying from dieldrin exposure. The data generally show dieldrin residues ranging from about 2 to 50 μg/g in these specimens (Table 4.6).

Peakall (1996) reviewed data on toxicity of aldrin/dieldrin and other cyclodiene insecticides to wildlife with the specific goal of providing guidance for the interpretation of tissue concentrations. He found the range in lethality associated with dieldrin in brain ranged from about 4 to 10 μg/g (Table 4.7). Newton (1998) evaluated

Table 4.6 Dieldrin concentrations in tissues of dead birds found in the field, where mortality was attributed to dieldrin exposure

Species	Sample size	Organ	Residue (μg/g wet wt) Mean	Range	References
Sarus Crane (*Grus antigone*)	5	Brain	19.3	3.6–43.5	Muralidharan (1993)
Pink-footed goose (*Anser brachyrhynchus*)	6	Liver	31	15–48	Stanley and Bunyan (1979)
Snow goose (*Chen caerulescens*)	5/157[a]	Brain	17.3	13.0–24.5	Babcock and Flickinger (1977)
	8/112[a]	Brain	8.2	4.9–14	Flickinger (1979)
Lesser scaup (*Aythya affinis*)	4	Brain	11.9	7.7–16	Sheldon et al. (1963); cited in Stickel et al. (1969)
Buzzard (*Buteo buteo*)	14	Liver	19.7[b]	7.8–31.2	Fuchs (1967)
Lanner falcon (*Falco biarmicus*)	2	Brain	2.7	2.0–3.3	Jefferies and Prestt (1966)
Peregrine (*Falco peregrinus*)	2	Brain	11.6	6.8–16.4	Bogan and Mitchell (1973)
	2	Liver	35.6	17.3–53.9	Bogan and Mitchell (1973)
	1	Brain	5.4	5.4[c]	Reichel et al. (1974)
	2	Brain	5.7	3.5–7.8[d]	Jefferies and Prestt (1966)
	2	Liver	6.7	4.0–9.3[e]	Jefferies and Prestt (1966)
European kestrel (*Falco tinnunculus*)	84	Liver	NA	6–30[f]	Newton et al. (1992)
Sparrow hawk (*Accipter nisus*)	25	Liver	NA	5–21[g]	Newton et al. (1992)
Owls[h]	10	Brain	15	11–25	Jones et al. (1978)
Barn owl (*Tyto alba*)	51	Liver	14	6–44	Newton et al. (1991)
Sandwich tern (*Sterna sandvicensis*) chick	6	Liver	5.6	2.4–12[i]	Koeman et al. (1967)
Sandwich tern juvenile	8	Liver	4.6	1.9–6.6[j]	Koeman et al. (1967)
Sandwich tern adult	5	Liver	5.5+	4.7–7.2[k]	Koeman et al. (1967)
Meadowlark (*Sturnella magna*)	5	Brain	9.3	8.6–12.1	Stickel et al. (1969)
	5	Liver	13.1	7.9–15.9	Stickel et al. (1969)
American robin (*Turdus migratorius*)	7	Brain	9.6	5.0–17.0	Stickel et al. (1969)

[a] Number at each site presumed dead from dieldrin over total numbers
[b] Wet weight assumed
[c] Also 34 μg/g of DDE and 55 μg/g of PCBs
[d] Also 45 μg/g and 44 μg/g of DDE
[e] Also 70 μg/g and 60 μg/g of DDE
[f] Measurement given as μg/g; a few outliers as high as 99 μg/g
[g] Measurement given as μg/g; a few outliners as high as 85 μg/g
[h] A number of different species maintained at the London Zoo
[i] Also 2.3 μg/g of telodrin and 0.47 μg/g of endrin
[j] Also 0.86 μg/g of telodrin and 0.43 μg/g of endrin
[k] Also 1.0 μg/g of telodrin and 0.67 μg/g of endrin

Table 4.7 Benchmarks of toxicological effects of dieldrin on birds

Endpoint	Tissue	Concentration (μg/g)	References
Lethality	Brain	4.0	Stickel et al. (1969)
		4.0	Elliott and Bishop (2011)
		5.8	Linder et al. (1970)
		10	Robinson et al. (1967)
		10	Peakall (1996)
Anorexia	Brain	1.0	Heinz and Johnson (1981)
			Elliott and Bishop (2011)
Population decline	Liver	1.0	Newton (1998)
Population decline	Egg	0.7	Newton (1998)

occurrences of population declines and concluded that in birds, liver concentrations of dieldrin as low as 1 μg/g is a critical value associated with population decline. Similarly, he recommended 0.7 μg/g dieldrin in eggs as a benchmark indicator of potential population decline in birds (Table 4.7). Elliott and Bishop (2011) have extended Peakall's earlier review and recommended using the value of 4 mg/kg dieldrin or dieldrin equivalents in brain tissue as a protective criterion for lethality. Based on the work of Heinz and Johnson (1981) and a review of other studies which explored the relationship between dieldrin and suppression of feeding and body condition, they also proposed 1 mg/kg of dieldrin (or dieldrin equivalents) in brain, eggs, or plasma as the threshold concentration for potential anorexic effects, particularly during migration or the breeding season.

The symptoms of cyclodiene pesticide toxicity and subsequent body burdens noted by researchers at other sites both support the conclusion that these pesticides were causing much of the toxicity at the Arsenal. Although relatively few of the specimens collected at the Arsenal had necropsies to confirm that pesticide exposure played a role in the mortality, many of the bird carcasses at the site had symptoms indicative of organochlorine pesticide poisoning, such as emaciation, disorientation, and tremors. During his 22 years of working as CDPHE's Project Manager at the Arsenal, author Jeff Edson observed many birds sick and dying at the site, including a large woodpecker that nearly landed on his head after it expired and fell from a tree.

We compiled data from those specimens that were collected and analyzed for pesticide tissue residue at the Arsenal between 1981 and 2005 (Appendix). Over 700 specimens were analyzed, including nearly 170 dead birds analyzed for dieldrin in brain tissue. Over 70% of these specimens contained dieldrin concentrations exceeding the 1.0-μg/g adverse effect threshold (Table 4.7), and nearly 50% exceeded a 4.0-μg/g threshold for potential lethality (Table 4.7). One American Robin (*Turdus migratorius*) collected in 2001 contained 210 μg/g dieldrin in brain tissue, many times greater than the toxicity thresholds that have been proposed.

Other lines of evidence were also investigated to quantify potential avian pesticide poisoning at the Arsenal. Using radio-telemetry techniques, Frank and Lutz (1999) studied a population of great horned owls (*Bubo virginianus*) breeding at nests within the core contaminated area of the Arsenal and from nests where adult birds had been observed foraging in the more contaminated areas. Although they reported no differences in hatching or fledging success between owls nesting or foraging in contaminated versus uncontaminated locations, they observed a significant negative relationship between dieldrin concentrations in plasma of juvenile owls and postfledging survival. Interval survival of juvenile owls in a high dieldrin plasma concentration category of greater than 0.1 μg/g (100 μg/L) was significantly lower ($p<0.01$) than that of birds in a low dieldrin plasma concentration category of less than 0.05 μg/g (50 μg/L).

In a complementary laboratory study, Vander Lee and Lutz (2000) administered dieldrin to nestling black-billed magpies (*Pica pica*) and examined plasma and tissue distribution of dieldrin residues. They reported a strong correlation between dieldrin residue in plasma and in the brain ($y=0.882x+0.007$, $p<0.001$, $r^2=0.959$, $n=38$), indicating a strong likelihood that the great horned owls with high plasma dieldrin concentrations in the previous study also had correspondingly high dieldrin concentrations in the brain.

From 1993 to 1996, Roy (1997) investigated reproduction of American kestrels (*Falco sparverius*) in nest boxes placed at the Arsenal and at reference sites. He reported no significant differences in hatching success but did find a significant effect of dieldrin on the number of young fledged per nest attempt and on overall nest success. Typically, he observed entire nest failure during egg laying or early incubation periods. However, the kestrels were also exposed to DDE, preventing the author from drawing conclusions about the effects specific to dieldrin. He concluded that nest failure was the result of the disappearance and presumed mortality of breeding males. In support of that assertion, we note that dead or moribund male kestrels from the Arsenal were found during the breeding season, several with brain dieldrin concentrations indicative of lethal poisoning (Appendix).

Even after direct exposure to Basin F liquids ended, and with no systematic quantification of mortality, wildlife kills were reported in the late 1980s and 1990s. The USFWS collected "fortuitous specimens," or "samples of opportunity," which were dead or dying birds and mammals that they happened to find in the course of other activities at the site. From 1989 to 1993, USFWS collected 192 bird samples and 52 mammal samples. More than 30 different bird species were collected, including raptors, waterfowl, and passerines (CDPHE 1994).

The 1999 USFWS Annual Progress Report for the Rocky Mountain Arsenal (USFWS 2000) provides a cumulative list of the bird species for which mortalities were attributed to dieldrin or endrin poisoning between 1990 and 1998. A total of 102 bird mortalities from 19 species were attributed to dieldrin or endrin poisoning. The report also noted that several birds had pronounced keels (i.e., they were emaciated) and displayed other classic symptoms of dieldrin poisoning (USFWS 2000).

Mortality estimates at the Arsenal that are based on fortuitous specimen collection greatly underestimate the total amount of wildlife mortality at the site caused by pesticide poisoning. Although collected and identified, necropsies were never performed on many dead animals found on the Arsenal. For the specimens to have been seen, they likely died during normal work hours near a road or a trail or a building where humans happened to be working. Even if an attempt had been made to systematically search for carcasses, the search efficiency would likely have been low (Witmer et al. 1995; Mineau 2005), given the large area of contamination, difficulty searching in prairie habitat, and scavenger removal. However, the fortuitous collection program provided sufficient samples of dead birds and waterfowl to allow one to conclude that dieldrin and similar cyclodiene pesticides were driving the toxicity at this site.

Establishing Remediation Thresholds

With both qualitative and quantitative evidence that cyclodiene pesticides were the primary cause of adverse effects and mortality at the Arsenal, the risk assessors for the Army, Shell, EPA, and CPDHE needed to determine what soil concentration of each priority contaminant could be considered sufficient to ensure that no future adverse ecological effects would occur. In the IEA/RC, investigators evaluated the dietary exposure of a variety of species (or species groups) to different chemicals. The evaluation required estimating exposure area soil concentrations; species-specific feeding behavior, including identification of dietary items, fraction of items consumed, and feed rates; and trophic transfer, based on both chemical- and species-specific bioaccumulation and biomagnification factors (Ebasco Services et al. 1994).

Dietary exposure estimates were compared to toxicity reference values (TRVs), defined as the dose above which ecologically relevant effects might occur to wildlife species following chronic dietary exposure, and below which it is reasonably expected that such effects will not occur (U.S. EPA 2005). It is generally expressed in units of mass of contaminant per mass of ingesting animal per day (e.g., mg dieldrin/kg body weight/day). Comparisons of Arsenal dietary exposure estimates to TRVs were expressed as hazard quotients (HQs), where

$$\text{HQ} = \text{Exposure estimate/TRV}.$$

Perhaps the biggest challenge for toxicologists and regulators is accounting for uncertainty when establishing specific threshold doses. These thresholds should account for all target species and all target contaminants, including potential adverse or synergistic effects of multiple contaminants. At the Arsenal, most of the target contaminants are bioaccumulated and biomagnified, further complicating safe dose estimates. Finally, calculating a safe concentration in soil based on a proposed safe dietary dose for predators adds additional uncertainty.

Because of the extent of contamination at the Arsenal, changes in soil cleanup criteria could have had multi-million-dollar repercussions. Not surprisingly, the parties had trouble reaching agreement on modeling inputs and output, and the Army, Shell, and the EPA all reached different conclusions regarding concentrations of contaminants that represented risk to target species at the Arsenal (Fig. 4.6). Ultimately, dieldrin and aldrin were chosen as the contaminants most likely to cause adverse effects at the Arsenal, because of additive toxicity, symptoms in dying birds that matched symptoms of birds dosed with these pesticides in a laboratory setting,

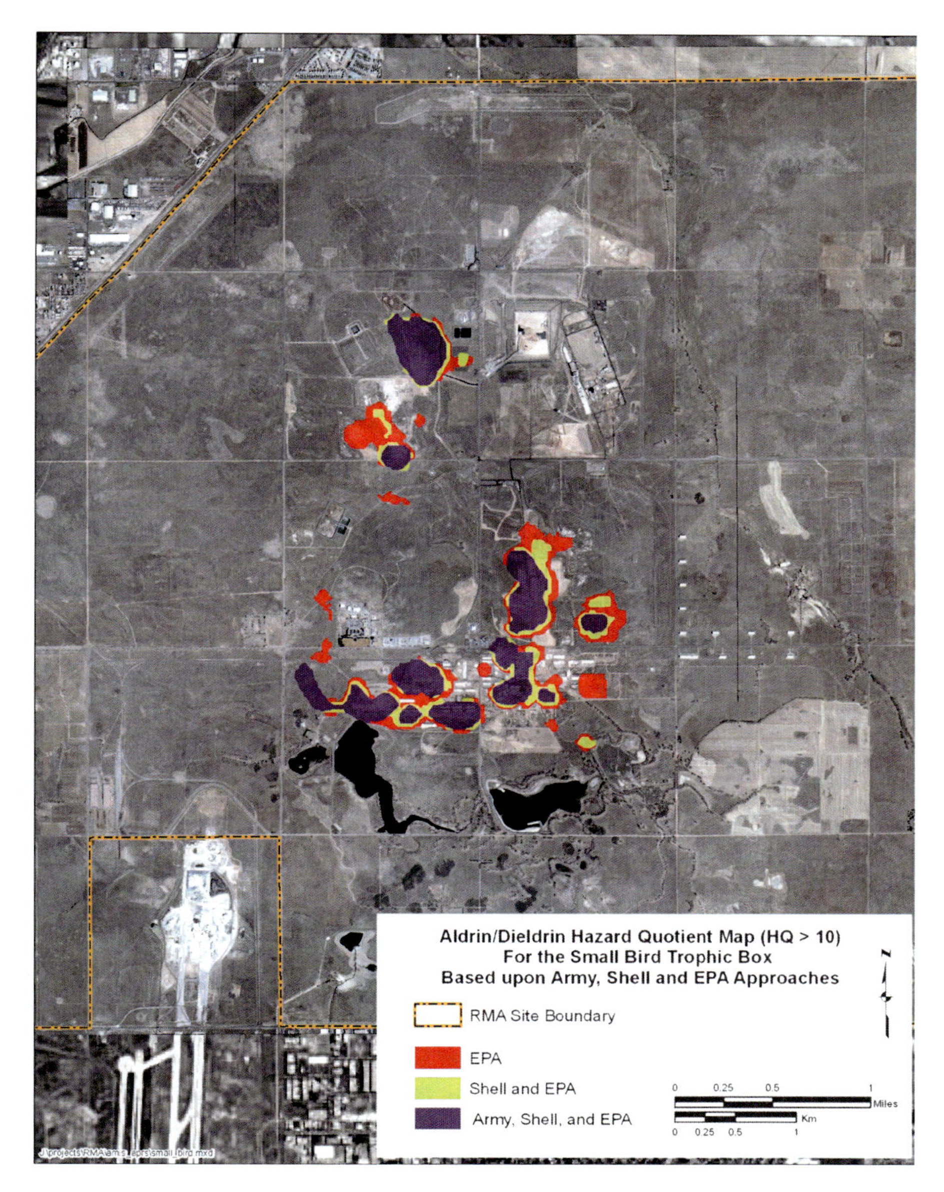

Fig. 4.6 Differing areal estimates of soils posing high dieldrin risk to small birds (HQ>10), as depicted in the IEA/RC report. The Army estimate covers 97 ha, while the EPA estimate covers 188 ha. Data source: Ebasco Services et al. (1994), Fig. C.3-33

and because previous biota studies identified dieldrin in 60% of nonmigrant biota samples and in all trophic levels (Stollar et al. 1992). Contractors integrated a database of surface soil concentrations into a geographic information system (GIS) and produced maps showing areas of ecological risk (i.e., HQ>1) similar to Fig. 4.6 (although we note that Fig. 4.6 shows only those areas with HQ>10). These areas of potential risk were also cross-referenced with observed effects in wildlife (Foster Wheeler 1996). However, the parties did not reach consensus on estimates of risk, safe dose, trophic transfer, and safe soil concentrations.

Final Remediation

Despite the lack of agreement among the parties, the EPA issued a Record of Decision (ROD) for the off-post remedy in December 1995 (Harding Lawson Associates 1995) and for the on-post remedy in July 1996 (Foster Wheeler 1996). These decisions spelled out the environmental remediation required by the Army to ensure protection of human health and the environment from exposure to the Arsenal's soils, groundwater, surface water, sediments, and air.

The RODs required the Army to continue (and expand) operation of the three RMA boundary groundwater containment and treatment systems, demolish structures with no planned future use, and excavate millions of cubic meters of contaminated soils over thousands of hectares (Fig. 4.7). The Army and Shell constructed two on-site landfills, including one with a triple lined cell for the soils with the highest contaminant concentrations (Fig. 4.8). The second landfill, unlined and located within the former Basin A, was used for the consolidation of less contaminated soils and other waste such as demolition and sanitary landfill debris. Generally, areas with contaminated soils were required to have the upper 1.5–3 m of soil removed and placed in one of the landfills. All contaminated soils were scraped and replaced with uncontaminated soils from other areas of the property ("borrow" soils) and replanted with native vegetation (Foster Wheeler 1996). The Army estimates that over 2.3 million cubic meters of contaminated soils were excavated and landfilled, covering thousands of hectares (RVO 2009).

Areas such as the trenches that were considered too dangerous to excavate were surrounded with slurry walls to prevent contaminant migration. These areas, as well as areas that contained contamination that was deeper than 1.5–3 m, were then covered with engineered caps that included "biota barriers" comprising crushed pieces of runway from the former Stapleton Airport that are intended to prevent burrowing animals from reaching contaminated soils at depth.

The remediation that the RODs required were based on EPA's conclusions of areas where the HQ > 1 (e.g., Foster Wheeler 1996). However, the parties continued to debate cleanup criteria for several more years after the RODs were released. Wildlife experts from the Army, Shell, USFWS, EPA, and CDPHE formed the Biological Advisory Subcommittee (BAS) to resolve the differences. The BAS was also tasked to work with USFWS to ensure that the Arsenal's biomonitoring program considered actual exposures for the individual species sampled. It was not until 2002, over 6 years after the RODs were issued, that the BAS reached an agreement and soil remediation targets were settled.

The BAS (2002) concluded that small insectivorous birds exposed to dieldrin were most at risk at the Arsenal, and they established a TRV of 0.028 mg dieldrin/kg

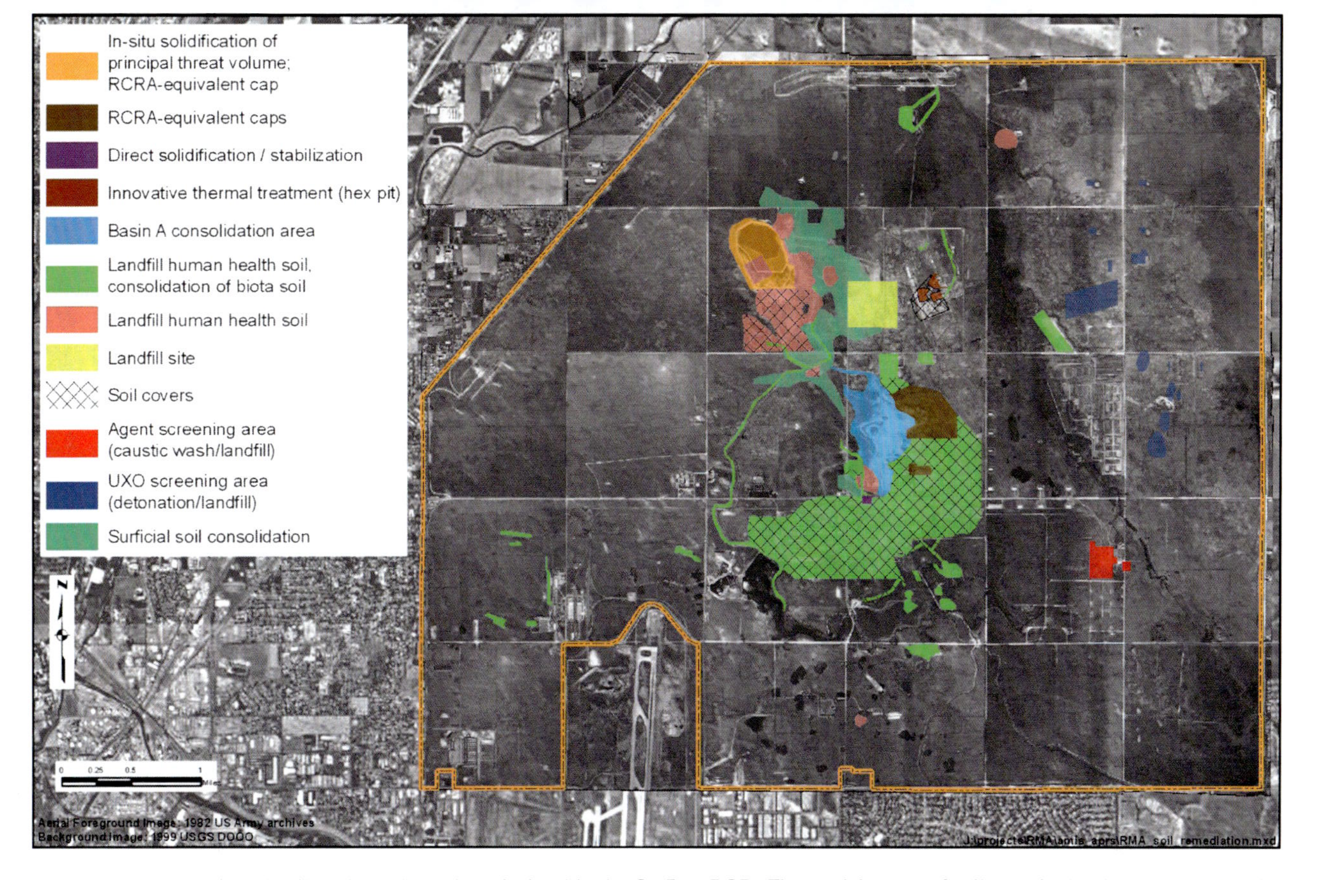

Fig. 4.7 Areas with contaminated soils to be addressed, as depicted in the On-Post ROD. The spatial extent of soils required to be removed, landfilled, and capped is extensive. Highly contaminated soils were placed in the hazardous waste landfill site (*yellow*), with less contaminated soils and other debris consolidated in Basin A (*light blue*). Original data source: Foster Wheeler (1996)

Fig. 4.8 Construction of the lining on-post hazardous waste landfill for contaminated soils. Source: US Army archives

body weight/day as an "estimated safe level" for these birds. The BAS TRV was based on an estimated critical dose of 0.28 mg dieldrin/kg body weight/day from a study with homing pigeons (*C. livia*) (Robinson and Crabtree 1969). They divided this critical dose by an uncertainty factor of 5 to consider potential interspecies differences in sensitivity, and by an additional uncertainty factor of 2 to account for differences between laboratory and field conditions and the potential for variability within a species (BAS 2002). The BAS TRV is lower than the 0.0709 mg dieldrin/kg body weight/day TRV that U.S. EPA (2005) subsequently established based on adverse reproductive effects, after a comprehensive dieldrin literature review for establishing ecological soil screening criteria. The BAS (2002) food web model determined that the concentration of dieldrin/aldrin in soils should not exceed 0.065 mg/kg, to protect insectivorous birds from receiving a dose greater than 0.028 mg/kg body weight/day.

Surface soils with aldrin/dieldrin concentrations exceeding this threshold covered hundreds of hectares of the Arsenal. From the mid 1990s when the RODs were released until 2010, Army contractors implemented the remedy, removing hundreds of buildings, constructing the waste repository in Basin A and the triple-lined hazardous waste repository (Fig. 4.8), and filling the repositories with millions of cubic meters of contaminated surface soils and sediment. Biota barriers and borrow soils were placed over hundreds of hectares of remediated area and replanted with native shortgrass prairie seeds. Despite the multibillion dollar effort to remove surface soils and restore native habitat, disagreements among the parties continued, including some who disagreed with a remedy that left in place a large volume of deep (below 3 m) subsurface contamination at the site, and some who were not convinced that the Army and Shell had adequately characterized the spatial extent of contamination. Thus, particular attention has been paid to the plans to monitor potential postremediation toxicity.

Potential for Ongoing Impacts

While the 30-year effort to remediate the Arsenal addressed the vast majority of the on-site surface contamination, there is no guarantee that the Army and Shell identified and removed all soils containing elevated concentrations of dieldrin and other contaminants. The Army will continue to be responsible for long-term monitoring of the site. In addition to monitoring and maintaining the caps and covers of landfills and consolidation areas, as well as operating the groundwater treatment systems in perpetuity, the Army, the USFWS, and the BAS are monitoring biota to evaluate the effectiveness of the selected remedy.

The BAS (2006) developed a long-term monitoring program for terrestrial ecological receptors at the Arsenal. Based on their previous work (e.g., BAS 2002), the BAS targeted aldrin and dieldrin as the contaminants most likely to affect terrestrial receptors. Because aldrin rapidly converts to dieldrin in environmental media (WHO 1989), the BAS selected dieldrin as the target contaminant for long-term monitoring.

The BAS (2006) biomonitoring plan called for collection and tissue analysis of adult starlings (*Sturnus vulgaris*) and eggs of American kestrels. Starling and kestrel nest boxes were to be placed throughout both the central manufacturing area and the periphery of the Arsenal, in a grid pattern based primarily on spatial representation and habitat type and quality. In total, there would be 24 starling nest box arrays and 22 kestrel nest box arrays, a subset of the 38 kestrel nest box arrays already established at the Arsenal from previous studies. Additional or different kestrel nest boxes may be used depending on occupancy rate and other field conditions (BAS 2006).

The BAS (2006) studies focus on dieldrin concentrations in the brain. The BAS concluded that the maximum allowable tissue concentration (MATC) for dieldrin in the brain is 1.0 μg/g for both passerines and raptors. Specifically, they concluded that brain concentrations of dieldrin between 5.0 and 20.0 μg/g would likely result in the death of adult birds, and concentrations of dieldrin as low as 1.0 μg/g may result in irreversible starvation in some sensitive individuals (Table 4.7).

Because the European starling is very common and an invasive species, the BAS concluded that adult birds could be captured and sacrificed for monitoring. The kestrel, on the other hand, is a native raptor, which the BAS (2006) considered a "valuable" species that should not be sacrificed. The plan for evaluating kestrel exposure to dieldrin was first to harvest and evaluate dieldrin concentrations in eggs. The BAS estimated that 0.05 μg/g dieldrin in eggs, or five percent of the brain MATC, would be an appropriate no observed adverse effect concentration

(NOAEC). If dieldrin data from kestrel eggs were deemed insufficient for drawing conclusions about exposure, kestrel chicks would be harvested just before fledging, and their brain dieldrin concentrations would be compared to the 1.0-μg/g MATC (BAS 2006).

The BAS (2006) plan also called for collection of at least 30 starlings, a minimum of ten individuals per year over a 3-year period starting in 2007, from each of the 24 designated exposure areas at the Arsenal. For kestrels, the BAS recommended sampling one egg per nest box per year. For the cleanup to be considered effective, the BAS expected dieldrin concentrations in adult starling brains and, if collected, kestrel chick brains to be below the MATC of 1.0 μg/g, and concentrations of dieldrin in kestrel eggs to be below the NOAEC of 0.05 μg/g. If dieldrin concentrations exceeded either threshold prior to the completion of 3 years of sampling, the BAS (2006) plan suggested that biomonitoring would be discontinued until the source of exposure had been identified and addressed. However, the plan had flexibility, allowing the BAS to recommend changes or revisions over time, including changes in sampling frequency, duration, or location, as indicated by the data based on new information (BAS 2006).

In May 2009, the USWFS published a draft report of the 2007 and 2008 starling and kestrel sampling (USFWS 2009). Over 250 starling brain samples were collected. One sample contained 3.7 μg/g dieldrin, exceeding the 1.0 μg/g MATC threshold, and the remainder contained dieldrin concentrations below the MATC threshold (USFWS 2009). However, the sampling was not consistent with the BAS (2006) biomonitoring plan. After performing habitat suitability analyses, the USFWS (2009) concluded that 18 (rather than 24) exposure areas were suitable for starling nest box arrays, and no sites were suitable for kestrel studies. Thus, no kestrel eggs or chicks were harvested and analyzed for residual dieldrin in 2007 or 2008. In addition, the BAS (2006) plan called for a minimum of ten starling brain samples per exposure area; in 2007, USFWS collected only 72 samples total from the 18 exposure areas. Moreover, rather than analyzing brain dieldrin residues of adult birds as specified in the biomonitoring plan, the USFWS (2009) analyzed fledglings that had been alive for 3 weeks or less, which may not have been the sufficient time for dieldrin to accumulate. The federal and state agencies analyzing these data are not in agreement as to whether the sampling that has occurred is adequate to draw conclusions about current avian exposure to dieldrin at the site.

Transition to National Wildlife Refuge

The decision to make the Arsenal a National Wildlife Refuge was driven by a combination of political, economic, and scientific factors. The large areas of undeveloped and relatively unimpacted habitat in the buffer areas, protecting shortgrass prairie and wetland habitat from urban encroachment, were certainly worth preserving. The budget and logistics required to clean up the Arsenal to standards suitable for commercial human use would likely have been extraordinary. And the stigma associated with the Arsenal's history and the "most contaminated square mile on earth" likely contributed to the decision to prevent commercial and industrial development on the site.

The final cleanup plan for the Arsenal required the Army to maintain, possibly in perpetuity, approximately 400 ha of on-site landfills, consolidation areas, and groundwater treatment facilities. In total, approximately 6,000 ha, an area roughly the size of Manhattan, has been transferred to the USFWS for management as a National Wildlife Refuge. By 2006, nearly 5,000 ha had already been transferred to the USFWS (RVO 2009), and the remaining 1,000 ha were transferred in October 2010, when "significant environmental cleanup" had concluded (RVO 2010). Hundreds of hectares of the refuge are shortgrass prairie habitat that Army contractors created, after removing and emplacing in hazardous waste landfills the underlying 1.5–3 m of contaminated soils, emplacing the "biota barriers" of crushed runway concrete, and covering with borrow soils and seeds from native plants. The Rocky Mountain Arsenal National Wildlife Refuge is now one of the largest urban wildlife refuges in the USA.

In addition to their remediation of the property and re-establishment of native habitat, the Army and Shell paid $35 million in natural resource damages to the State of Colorado in 2008 to compensate for the interim past harm to biota and groundwater over several decades, as well as for ongoing harm to habitat and resources that cannot be restored, such as the hazardous waste landfills and groundwater treatment plants. The authors of this report helped the State to reach this settlement. Colorado intends to use the settlement money to protect and restore additional shortgrass prairie, aquatic, and riparian habitat in areas in close proximity to the Arsenal.

Conclusions

After almost 40 years of manufacturing astronomical quantities of chemical weapons and pesticides, with toxic waste products sluiced into open disposal basins, pits, trenches, and unlined shallow landfills, followed by nearly 30 years to design and implement a strategy to remediate the problem, the Arsenal has become one of the largest urban wildlife refuges in the USA. The Army and Shell spent $2.1 billion on-site investigation and remediation (RVO 2010) and an additional $35 million to settle Colorado's natural resource damages litigation. Many millions of additional dollars will be spent in the future for treating contaminated groundwater and monitoring and maintaining the long-term effectiveness of the landfill and capped areas.

Today, large portions of the Arsenal are open to the public as part of the Rocky Mountain Arsenal Wildlife Refuge, where visitors can see a wide diversity of wildlife and even a herd a bison, which the USFWS introduced in March 2007. Wildlife toxicology played a key role in the site assessment, remedial goals, and the wildlife refuge designation at this site. However, one should be cautious about concluding that wildlife toxicology and a desire to compensate for the atrocities of yesteryear were the driving forces behind these conservation gains. The quantity of subsurface contamination, the cost to remove that contamination (particularly if the contamination needed to be transported offsite), and the public stigma associated with the Arsenal also played pivotal roles in the land use decisions. The Army, Shell, EPA, and the State rarely agreed on the adequacy of the site assessment, appropriate toxicity thresholds, appropriate quantity of contaminants to leave in place, and many other decisions that were required to clean up this site. Multiple lawsuits were filed when the parties could not reach agreement on remediation and restoration of the site. Thus, the process of transforming this site from a highly contaminated industrial area to a national wildlife refuge is perhaps not a model that we would recommend for other sites.

The conservation gains at the Arsenal are impressive, where hundreds of hectares of industrial area have been returned to shortgrass prairie, and thousands of more hectares of undeveloped habitat have been preserved in perpetuity, which is clearly a positive outcome for wildlife. On the other hand, many thousands of cubic meters of highly contaminated soils remain, serving as a source of groundwater contamination in perpetuity, and two very large hazardous waste landfills are in the middle of the refuge, which will forever require monitoring and maintenance.

The Army has now entered the Arsenal's long-term operations and maintenance program. This process ensures the final remedy is protective through monitoring the existing landfill covers and treatment systems while maintaining the institutional controls set out in the Army's FFA. Verifying the success of the remediation effort should be a high priority for the Army, BAS, and the USFWS. Although large areas of surface soils have been remediated, there is certainly no guarantee that all contaminated areas were identified and addressed across this very large site. The BAS (2006) biomonitoring plan, relying on starling and kestrel tissue data, is important, and it is the hope of the authors that the USFWS will more closely follow the plan in the future (e.g., include kestrels, and measure dieldrin residue in adult starlings).

In addition, we hope that the USFWS will investigate other lines of evidence to evaluate the efficacy of the remedy. For example, we hope the USFWS will continue their "fortuitous specimen" program at a minimum, with implementation of a systematic carcass search program preferable to relying on "samples of opportunity." Dead birds with elevated dieldrin/aldrin concentrations were still being found at the Arsenal at least through 2005 (Appendix), when much of the most highly contaminated soil had already been removed and landfilled. Should bird or other wildlife mortality continue now that remediation is complete, it should be incumbent on the USFWS to implement additional biomonitoring studies in the areas where carcasses are discovered. If unusual mortality continues or elevated dieldrin concentrations are found in starlings and kestrels, the Army and Shell will need to implement additional sampling to identify and address possible hot spots of contamination still present in the surface soils.

Appendix

Contaminant concentrations in tissues of selected specimens collected at the arsenal between 1981 and 2005

Species	Sample date	Substrate analyzed	Contaminant	Concentration (μg/g)
Avocet, American (*Recurvirostra americana*)	1999	Egg	Dieldrin	2.34
			Endrin	0.189
			DDE	0.478
Blackbird, Brewers (*Euphagus cyanocephalus*)	1982	Brain	Dieldrin	7.9
			DDE	4.1
			DDD	0.2
			DDT	0.1
			Oxychlordane	0.28
			Nonachlor	0.18
	1982	Brain	Dieldrin	12
			DDE	2.5
			Mercury	0.18
			Oxychlordane	0.27
			Nonachlor	0.4
			PCB	0.52
	1988	Dressed carcass	DDE	1.1
	1999	Brain	Dieldrin	12
			DDE	0.393
		Liver	Dieldrin	9.3
			Endrin	0.0243
			DDE	0.336
Blackbird, Red-winged (*Agelaius phoeniceus*)	2003	Brain	Dieldrin	7.0468
			Endrin	0.221
			DDE	1.3272
Coot, American (*Fulica Americana*)	1988	Dressed carcass	Aldrin	0.0658
	1988	Dressed carcass	Mercury	0.137
	1988	Dressed carcass	Dieldrin	0.101
	1988	Dressed carcass	Dieldrin	0.142
	1988	Dressed carcass	Endrin	0.135
	1988	Dressed carcass	DDE	0.359
	1988	Dressed carcass	Mercury	0.18
	1988	Dressed carcass	DDT	0.569
	c. 1989	Liver ($n=9$)	Mercury	0.3–1.77
			Dieldrin	<0.124–0.693
	c. 1989	Muscle ($n=9$)	Mercury	<0.05–0.339
Cormorant, Double-crested (*Phalacrocorax auritus*)	2001	Kidney	Arsenic	0.0362
			Mercury	0.42

(continued)

Appendix (continued)

Species	Sample date	Substrate analyzed	Contaminant	Concentration (μg/g)
Dove, Mourning (*Zenaida macroura*)	1981	Breast	*p*-Chlorophenylmethyl sulfone	0.22
			Dieldrin	0.24
			Endrin	0.03
	1981	Breast	Dieldrin	0.22
			Endrin	0.03
	1981	Breast	DDE	0.03
	1981	Breast	*p*-Chlorophenylmethyl sulfone	0.22
	1981	Breast	*p*-Chlorophenylmethyl sulfone	0.53
			Dieldrin	3.4
			Endrin	0.3
	1981	Breast	*p*-Chlorophenylmethyl sulfone	2.53
			Dieldrin	9.96
			Endrin	1.21
	1988	Dressed carcass	DDE	0.0766
	1988	Dressed carcass	Isodrin	0.151
	1988	Dressed carcass	Endrin	0.352
	1988	Dressed carcass	DDT	0.308
	1988	Dressed carcass	Mercury	0.4
	1988	Dressed carcass	DDE	0.455
	1988	Dressed carcass	Endrin	1.3
	1988	Dressed carcass	Aldrin	1.3
	1988	Dressed carcass	Dieldrin	32
	1988	Dressed carcass	Dieldrin	14
	1988	Dressed carcass	Arsenic	2.63
	c. 1989	Carcass	Aldrin	1.83
			Dieldrin	56.3
			Endrin	3.44
	c. 1989	Liver	Dieldrin	7.37
			Endrin	3.74
	1994	Brain	Endrin	2.4
		Liver	Dieldrin	1.61
			Endrin	3.1
	1999	Brain	Dieldrin	0.187
			Endrin	4.4
			Isodrin	1.6
		Liver	Dieldrin	0.71
			Endrin	12
			Isodrin	4
	1999	Brain	Dieldrin	0.626
			Endrin	0.91
		Liver	Dieldrin	3.7
			Endrin	4.9

(continued)

Appendix (continued)

Species	Sample date	Substrate analyzed	Contaminant	Concentration (μg/g)
	1999	Liver	Dieldrin	0.12
	2000	Liver	Dieldrin	1.2
			Endrin	0.0882
			Heptachlor epoxide	0.0559
			trans-Chlordane	0.0882
	2001	Liver	Dieldrin	2.17
	2001	Brain	Dieldrin	0.939
		Liver	Dieldrin	1.9
			Endrin	0.185
Dove, Rock (*Columba livia*)	1999	Brain	Dieldrin	17
			Endrin	0.182
Eagle, Bald (*Haliaeetus leucocephalus*)	1988	Muscle	Mercury	0.0542
	1988	Brain	Dieldrin	0.112
	1988	Liver	DDT	0.135
	1988	Liver	Mercury	0.153
	1988	Brain	DDE	0.4
	1988	Liver	DDE	0.404
	1988	Muscle	DDE	1.7
Eagle, Golden (*Aquila chrysaetos*)	1988	Brain	Mercury	0.0969
	1988	Liver	DDE	0.124
	1988	Muscle	Mercury	0.14
	1988	Liver	Mercury	0.304
	1988	Muscle	DDE	0.639
	c. 1989	Liver	Mercury	0.216
			Dieldrin	0.221
	c. 1989	Brain	Mercury	0.257
Finch, House (*Carpodacus mexicanus*)	1995	Brain	Dieldrin	15
Goose, Canada (*Branta canadensis*)	1999	Liver	Dieldrin	0.205
	2001	Liver	Dieldrin	0.116
Grackle, Common (*Quiscalus quiscula*)	1995	Brain	Dieldrin	9.83
	2001	Brain	Dieldrin	15
			DDE	3.11
		Liver	Dieldrin	8.1
			DDE	1.4
	2001	Liver	Dieldrin	0.225
			DDE	0.372
	2001	Brain	Dieldrin	1.57
			DDE	3.83
		Liver	Dieldrin	1.04
			DDE	2.7
	2003	Brain	Dieldrin	0.0607
			DDE	0.244

(continued)

Appendix (continued)

Species	Sample date	Substrate analyzed	Contaminant	Concentration (μg/g)
Grebe, Eared (*Podiceps nigricollis*)	1982	Brain	Dieldrin	0.18
			Oxychlordane	0.36
			Nonachlor	0.19
			PCB	5.9
Harrier, Northern (*Circus cyaneus*)	c. 1989	Egg	Dieldrin	0.303
	c. 1989	Egg	Dieldrin	0.676
Hawk, Cooper's (*Accipiter cooperii*)	2004	Brain	*cis*-Chlordane	0.894
			Dieldrin	0.2312
			Heptachlor epoxide	0.0854
			DDE	2.6124
Hawk, Ferruginous (*Buteo regalis*)	1987	Brain	Dieldrin	7.7
			DDE	1.8
			DDT	5.5
	1987	Brain	Dieldrin	10
			DDE	2.3
	1987	Brain	DDE	0.92
			DDT	2.8
	1988	Whole body	Mercury	0.076
	1988	Whole body	DDE	0.169
	1988	Whole body	Endrin	0.233
	1988	Whole body	Dieldrin	11
	c. 1989	Liver	Mercury ($n=1$)	0.293
			Dieldrin ($n=5$)	0.263–4.79
Hawk, Ferruginous (cont.)	c. 1989	Brain	Mercury ($n=1$)	0.152
			Dieldrin ($n=5$)	<0.238–9.98
Hawk, Red-tailed (*Buteo jamaicensis*)	1982	Brain	Dieldrin	3.7
			Endrin	0.37
	1982	Brain	Dieldrin	4.2
	1987	Brain	Dieldrin	9.5
			DDE	1.1
			DDT	3.5
	1987	Brain	DDT	6.9
			DDE	2.3
			Arsenic	1.2
			Endrin	0.96
	1988	Brain	Dieldrin	9.4
	1988	Liver	Mercury	0.0489
	1988	Liver	Endrin	0.125
	1988	Liver	DDE	0.145
	1988	Liver	Isodrin	0.206
	c. 1989	Liver	Mercury ($n=1$)	0.345
			Dieldrin ($n=3$)	0.520–6.59
			DDE ($n=3$)	<0.313–0.759
	c. 1989	Brain	Mercury ($n=1$)	0.093
			Dieldrin ($n=3$)	<0.751–9.44

(continued)

Appendix (continued)

Species	Sample date	Substrate analyzed	Contaminant	Concentration (μg/g)
Hawk, Sharp-shinned (*Accipiter striatus*)	1994	Brain	Dieldrin	0.888
			DDE	4.92
Heron, Great Blue (*Ardea Herodias*)	1982	Brain	Dieldrin	11
			Endrin	0.22
			DDE	15
			DDD	0.42
			Oxychlordane	0.66
			cis-Chlordane	0.57
			PCB	15
	2003	Brain	*cis*-Chlordane	0.863
			Dieldrin	7.035
			Endrin	0.124
			DDE	1.99
		Liver	*cis*-Chlordane	0.109
			Dieldrin	6.98
			Endrin	0.0903
			DDE	2.673
		Kidney	Dieldrin	4.51
			Endrin	0.0682
			DDE	1.75
Kestrel, American (*Falco sparverius*)	1988	Dressed carcass	DDT	0.14
	1988	Egg	DDE	0.351
	1988	Dressed carcass	Dieldrin	1.7
	1988	Dressed carcass	DDE	0.401
	1988	Dressed carcass	DDT	0.446
	1988	Dressed carcass	DDE	0.811
	1988	Egg	DDT	0.62
	1988	Egg	Dieldrin	1.7
	1988	Dressed carcass	Dieldrin	2.2
	1988	Egg	DDE	5.3
	1988	Egg	Dieldrin	5.6
	1988	Dressed carcass	Dieldrin	3.7
	c. 1989	Egg	Mercury ($n=34$)	<0.05–0.405
			Dieldrin ($n=33$)	<0.031–3.63
			DDE ($n=1$)	1.25
	1993	Brain	Dieldrin	1.5
	1993	Brain	Dieldrin	12
	1993	Brain	Dieldrin	4
	1993	Brain	Dieldrin	12
	1994	Brain	Dieldrin	0.36
			DDE	0.76
	1995	Dressed carcass	Dieldrin	2.32
		Liver	Dieldrin	6.37
		Viscera	Dieldrin	0.92

(continued)

Appendix (continued)

Species	Sample date	Substrate analyzed	Contaminant	Concentration (μg/g)
	1995	Dressed carcass	Dieldrin	2.46
		Liver	Dieldrin	3.05
		Viscera	Dieldrin	0.94
	1995	Dressed carcass	Dieldrin	2.26
		Liver	Dieldrin	1.54
		Viscera	Dieldrin	0.77
	1995	Dressed carcass	Dieldrin	0.021
		Liver	Dieldrin	0.279
		Brain	Dieldrin	<0.065
		Viscera	Dieldrin	0.03
	1995	Dressed carcass	Dieldrin	0.127
		Liver	Dieldrin	0.155
		Brain	Dieldrin	<0.143
		Viscera	Dieldrin	0.049
	1995	Liver	Dieldrin	1.7
	1995	Brain	Dieldrin	4.4
	1999	Egg	DDE	0.0579
	1999	Brain	Dieldrin	2.6
			DDE	0.374
		Liver	Dieldrin	7.9
			DDE	0.575
	2001	Liver	Dieldrin	0.68
	2002	Brain	Dieldrin	0.9191
	2003	Brain	Dieldrin	0.12
	2003	Brain	Dieldrin	0.11
	2003	Brain	Dieldrin	0.12
	2005	Brain	Dieldrin	0.073
		Liver	Dieldrin	0.6686
			Dieldrin	0.1243
	2005	Brain	Dieldrin	4.3127
			DDE	4.2608
			DDT	0.7279
		Liver	Dieldrin	5.1005
			DDE	4.0295
	2005	Brain	Dieldrin	0.1908
			DDE	0.293
		Liver	Dieldrin	1.3534
			DDE	0.9042
			DDT	0.1036

(continued)

Appendix (continued)

Species	Sample date	Substrate analyzed	Contaminant	Concentration (μg/g)
Killdeer (*Charadrius vociferous*)	1988	Dressed carcass	Endrin	0.0714
	1988	Dressed carcass	Endrin	0.0975
	1988	Dressed carcass	Mercury	0.109
	1988	Dressed carcass	Mercury	0.305
	1988	Dressed carcass	DDT	1
	1988	Dressed carcass	DDT	0.955
	1988	Dressed carcass	Dieldrin	1.1
	1988	Dressed carcass	DDE	7
	1988	Dressed carcass	DDE	4.7
	1988	Dressed carcass	Dieldrin	6.6
	1994	Liver	Dieldrin	0.649
			DDE	0.167
	2001	Brain	α-Chlordane	0.29
			Dieldrin	3.6
			Endrin	1.3
			DDE	0.61
			DDT	0.151
		Kidney	Arsenic	0.13
			Mercury	0.41
		Liver	Endrin	0.387
			γ-Chlordane	0.128
			Isodrin	0.172
			DDT	0.31
	2004	Brain	*cis*-Chlordane	0.1153
			Dieldrin	7.6053
			Endrin	0.0836
			Heptachlor epoxide	0.0706
	2004	Brain	Dieldrin	5.1611
			DDE	1.4652
Kingbird, Eastern (*Tyrannus tyrannus*)	1995	Brain	Dieldrin	7.34
	1999	Brain	DDE	0.184
		Liver	Dieldrin	0.609
			DDE	0.262
	2005	Liver	Dieldrin	0.0646
Kingbird, Western (*Tyrannus verticalis*)	1994	Brain	Dieldrin	1.78
			DDE	1.25
	1994	Liver	Dieldrin	6.43
			DDE	0.25
	1995	Liver	Dieldrin	15.1
	1995	Brain	Dieldrin	14.1
	1995	Brain	Dieldrin	4.32
	1995	Brain	Dieldrin	10.1
	1995	Brain	Dieldrin	8.83
	2001	Liver	Dieldrin	0.25

(continued)

Appendix (continued)

Species	Sample date	Substrate analyzed	Contaminant	Concentration (μg/g)
Magpie, Black-billed (*Pica pica*)	1982	Brain	Dieldrin	5.3
			DDE	5.5
			DDT	0.28
			PCB	0.55
	1988	Dressed carcass	DDT	0.235
	1988	Dressed carcass	DDE	2.7
	1988	Dressed carcass	Dieldrin	5.9
	1995	Brain	Dieldrin	0.47
	1996	Brain	Dieldrin	9.6
	1996	Brain	Dieldrin	1.1
	1996	Brain	Dieldrin	3.4
	1996	Brain	Dieldrin	6.1
	1997	Brain	Dieldrin	13
	1997	Brain	Dieldrin	1.8
	1997	Brain	Dieldrin	3.5
	1997	Brain	Dieldrin	2.3
	1997	Brain	Dieldrin	1.6
	1997	Brain	Dieldrin	2.2
	1997	Brain	Dieldrin	5.3
	1997	Brain	Dieldrin	2.7
	1997	Brain	Dieldrin	3.4
	1997	Brain	Dieldrin	2.5
	1997	Brain	Dieldrin	14
	1999	Brain	DDE	0.393
		Liver	Dieldrin	0.66
			DDE	1.3
	1999	Liver	Dieldrin	0.263
			DDE	0.109
	2000	Brain	Dieldrin	0.252
			DDE	0.137
		Liver	Dieldrin	3.3
			DDE	0.308
	2000	Brain	Dieldrin	1.1
			DDE	0.36
		Liver	Dieldrin	2.2
			DDE	0.57
	2001	Liver	Dieldrin	0.391
			DDE	0.365
	2001	Brain	Dieldrin	0.5313
			DDE	4.52
			DDT	0.0877
	2002	Brain	Dieldrin	8.08
			Endrin	0.1354
			Heptachlor epoxide	0.3696
			DDE	1.1568

(continued)

Appendix (continued)

Species	Sample date	Substrate analyzed	Contaminant	Concentration (μg/g)
	2003	Brain	Dieldrin	0.0646
	2003	Brain	Dieldrin	0.216
	2003	Brain	DDE	0.0743
	2003	Brain	DDE	0.326
	2003	Brain	Dieldrin	0.28
			DDE	2.632
	2003	Brain	Dieldrin	0.0809
			Dieldrin	0.128
			DDE	0.448
	2003	Brain	Dieldrin	0.214
			DDE	1.1111
	2003	Brain	Dieldrin	0.0847
			DDE	1.144
	2003	Brain	Dieldrin	4.179
			DDE	0.154
	2003	Liver	DDE	0.0855
	2003	Brain	Dieldrin	0.199
			DDE	0.336
Mallard (*Anas platyrhyncos*)	1981	Breast/leg	DDE	0.21
	1981	Breast/leg	Dieldrin	0.08
			Endrin	0.03
			DDE	0.1
			DDT	0.03
	1981	Breast/leg	Dieldrin	0.97
			Endrin	0.03
			DDT	0.03
	1981	Breast/leg	Dieldrin	0.06
			DDE	0.09
			DDT	0.02
	1981	Breast/leg	Dieldrin	0.03
	1981	Breast/leg	Dieldrin	0.04
			Endrin	0.03
			DDE	0.03
			DDT	0.02
	1981	Breast/leg	Aldrin	0.03
			Dieldrin	0.88
			Endrin	0.04
			DDT	0.02
	1981	Breast/leg	Dieldrin	0.68
			DDT	0.02
	1981	Breast/leg	Aldrin	0.02
			Dieldrin	0.42
			Endrin	0.02
			DDT	0.02

(continued)

Appendix (continued)

Species	Sample date	Substrate analyzed	Contaminant	Concentration (μg/g)
	1981	Breast/leg	Dieldrin	0.28
			Endrin	0.02
			DDT	0.02
	1982	Brain	Endrin	1.3
	1988	Dressed carcass	Mercury	0.0681
	1988	Dressed carcass	Mercury	0.143
	1988	Dressed carcass	Mercury	0.242
	1988	Dressed carcass	Aldrin	0.0934
	1988	Dressed carcass	DDE	0.146
	1988	Dressed carcass	Endrin	0.104
	1988	Liver	Mercury	0.353
	1988	Dressed carcass	DDE	0.183
	1988	Liver	Mercury	0.679
	1988	Dressed carcass	DDE	0.404
	1988	Dressed carcass	Dieldrin	1.1
	1988	Liver	Dieldrin	0.68
	1988	Liver	DDE	0.633
	1988	Liver	Dieldrin	1.5
	1988	Dressed carcass	Dieldrin	3.8
	1988	Dressed carcass	Dieldrin	4.2
	c. 1989	Egg ($n=2$)	Mercury	0.173–0.185
			Dieldrin	3.0–4.89
			DDE	0.606–0.919
	2005	Brain	Dieldrin	0.2191
		Liver	Dieldrin	0.8463
Meadowlark, Western (*Sturnella neglecta*)	1988	Dressed carcass	Endrin	0.0569
	1988	Dressed carcass	Endrin	0.103
	1988	Dressed carcass	DDE	0.122
	1988	Dressed carcass	Endrin	0.13
	1988	Dressed carcass	DDE	0.149
	1988	Dressed carcass	Isodrin	0.177
	1988	Dressed carcass	Dieldrin	1.6
	1988	Dressed carcass	Dieldrin	2.1
	1988	Dressed carcass	Dieldrin	4.4
	1995	Brain	Dieldrin	7.15
	1995	Brain	Dieldrin	8.35
	1999	Brain	Dieldrin	0.36
		Liver	Dieldrin	0.481
	2004	Brain	Dieldrin	0.6746
			DDE	0.1455
Oriole, Bullock's (*Icterus bullockii*)	1982	Brain	Dieldrin	3.4
	1995	Brain	Dieldrin	10.4
Owl, Barn (*Tyto alba*)	1994	Liver	Dieldrin	0.17
			DDE	0.43

(continued)

Appendix (continued)

Species	Sample date	Substrate analyzed	Contaminant	Concentration (μg/g)
Owl, Burrowing (*Athene cunicularia*)	1988	Dressed carcass	Endrin	0.0583
	1988	Dressed carcass	DDE	0.128
	1988	Dressed carcass	Dieldrin	0.281
	1988	Dressed carcass	Dieldrin	1
	1988	Dressed carcass	DDE	0.195
Owl, Great Horned (*Bubo virginianus*)	1987	Brain	Dieldrin	16
			Aldrin	1.7
			DDE	10
			DDT	1.7
	1987	Brain	Dieldrin	10
			DDE	2.2
			DDT	0.81
	1987	Brain	Dieldrin	9
			DDE	1.4
	1988	Liver	Mercury	0.0581
	1988	Muscle	Mercury	0.0664
	1988	Egg	Mercury	0.106
	1988	Egg	Endrin	0.171
	1988	Muscle	Isodrin	0.174
	1988	Egg	DDE	0.951
	1988	Muscle	DDE	0.667
	1988	Egg	Dieldrin	6.6
	c. 1989	Liver	Mercury ($n=2$)	<0.05–0.086
			Dieldrin ($n=4$)	0.143–27.7
			DDE ($n=3$)	<0.094–15.5
	c. 1989	Brain	Dieldrin ($n=3$)	<0.175–15.6
			DDE ($n=3$)	<0.529–10.3
	1994	Brain	Dieldrin	3.2
	1994	Brain	Dieldrin	0.09
			DDE	0.18
		Liver	Dieldrin	0.1
			DDE	0.18
	1994	Brain	Dieldrin	0.37
			DDE	0.047
	1995	Brain	Dieldrin	2.06
	1995	Brain	Dieldrin	0.12
	1995	Brain	Dieldrin	0.04
	1995	Brain	Dieldrin	0.19
	1995	Brain	Dieldrin	10.55
	1995	Brain	Dieldrin	8.1
	1995	Brain	Dieldrin	0.28
	2000	Brain	Dieldrin	14
			Endrin	0.06
			DDE	0.68
		Liver	Dieldrin	20
			Endrin	0.165
			DDE	1.2

(continued)

Appendix (continued)

Species	Sample date	Substrate analyzed	Contaminant	Concentration (μg/g)
	2001	Brain	Dieldrin	1.9
			DDE	0.159
		Liver	Dieldrin	3.4
			DDE	0.49
Pheasant, Ring-necked (*Phasianus colchicus*)	1981	Breast/leg	Dieldrin	0.04
			Mercury	1.2
	1981	Breast/leg	Dieldrin	0.16
			Mercury	0.23
	1981	Breast/leg	Dieldrin	0.04
	1981	Breast/leg	Dieldrin	0.02
			Mercury	0.32
	1981	Breast/leg	Mercury	0.26
	1981	Breast/leg	DIMP	0.1
			Dieldrin	0.06
	1981	Breast/leg	DIMP	0.1
			Dieldrin	0.02
	1981	Breast/leg	*p*-Chlorophenylmethyl sulfone	0.36
			DIMP	0.34
			Dieldrin	2.83
			Endrin	0.03
	1988	Liver	DDE	0.812
			Dieldrin	1.2
	1988	Dressed carcass	DDE	0.0701
	1988	Liver	Dieldrin	0.165
	1988	Dressed carcass	Dieldrin	0.108
	1988	Dressed carcass	Mercury	0.122
	1988	Dressed carcass	DDE	0.214
	1988	Dressed carcass	DDE	0.43
	1988	Liver	DDE	0.464
	1988	Liver	DDE	0.81
	1988	Liver	Dieldrin	1.2
	1988	Dressed carcass	Dieldrin	4.7
	1988	Liver	Dieldrin	2.2
	1988	Liver	Dieldrin	5.9
	1988	Dressed carcass	Dieldrin	5.9
	1988	Liver	DDE	0.368
	c. 1989	Egg	Dieldrin ($n=11$)	<0.031–5.38
			Endrin ($n=1$)	0.143
	c. 1989	Liver	Dieldrin ($n=6$)	<0.018–2.3
			Endrin ($n=1$)	0.091
			DDE ($n=1$)	0.44
Redhead (*Aythya americana*)	c. 1989	Liver	Mercury ($n=5$)	0.08–0.368

(continued)

Appendix (continued)

Species	Sample date	Substrate analyzed	Contaminant	Concentration (μg/g)
			Aldrin ($n=1$)	0.088
			Dieldrin ($n=5$)	0.307–0.747
			Endrin ($n=1$)	0.074
			DDE ($n=1$)	0.156
	c. 1989	Muscle	Mercury ($n=5$)	<0.05–0.071
			Dieldrin ($n=5$)	0.117–0.320
Robin, American (*Turdus migratorius*)	1988	Dressed carcass	Mercury	0.0623
	1988	Dressed carcass	DDT	0.339
	1988	Dressed carcass	Endrin	0.98
	1988	Dressed carcass	DDT	0.95
	1988	Dressed carcass	Endrin	0.97
	1988	Dressed carcass	DDE	8.3
	1988	Dressed carcass	DDE	6.5
	1994	Brain	Dieldrin	1.5
			DDE	2.42
	1994	Brain	DDE	4.77
	1994	Brain	DDE	0.441
		Liver	Dieldrin	0.24
			DDE	0.438
	1999	Brain	DDE	14
			DDT	1.3
		Liver	Dieldrin	1.4
			DDE	19
			DDT	0.53
	1999	Egg	Dieldrin	2.3
			DDE	1.8
			DDT	0.227
	1999	Egg	Dieldrin	2.9
			DDE	2.2
			DDT	0.274
	1999	Brain	Dieldrin	7.5
			Endrin	0.386
			DDE	1.8
			DDT	0.158
		Liver	Dieldrin	12
			Endrin	0.493
			DDE	2
	1999	Brain	α-Chlordane	0.0872
			Endrin	0.0955
			γ-Chlordane	0.0251
			DDE	0.196
		Liver	Dieldrin	12
			Endrin	0.0792
			DDE	0.301

(continued)

Appendix (continued)

Species	Sample date	Substrate analyzed	Contaminant	Concentration (μg/g)
	1999	Liver	Dieldrin	0.549
			γ-Chlordane	0.0476
			DDE	0.179
	1999	Brain	Dieldrin	0.96
			Endrin	0.0935
			Heptachlor epoxide	0.0998
			DDE	1.3
			DDT	0.295
		Liver	Dieldrin	1.2
			Endrin	0.104
			Heptachlor epoxide	0.116
			Isodrin	0.0869
			DDE	1.2
	1999	Brain	Dieldrin	0.0732
		Liver	Dieldrin	0.365
	2001	Brain	Dieldrin	210
		Liver	Dieldrin	34
			Endrin	0.873
			DDE	1.3
			DDT	0.163
	2003	Liver	Dieldrin	0.332
			Dieldrin	0.302
	2003	Liver	Dieldrin	0.17
			DDE	0.6525
	2004	Brain	Dieldrin	0.1333
			DDE	0.964
			DDT	0.1717
	2005	Brain	Dieldrin	0.3817
			DDE	13.688
			DDT	0.6728
		Liver	Dieldrin	0.3373
			DDE	16.291
Sora (*Porzana carolina*)	2001	Kidney	Arsenic	0.849
			Mercury	4
Sparrow, House (*Passer domesticus*)	1995	Brain	Dieldrin	9.66
Sparrow, Vesper (*Pooecetes gramineus*)	1988	Dressed carcass	Dieldrin	1.9
	2000	Brain	Dieldrin	29
		Liver	Dieldrin	33
Starling, European (*Sturnus vulgaris*)	1982	Brain	Dieldrin	7.9
			Endrin	0.18
			DDE	6.1
	1982	Brain	Dieldrin	3.3
	1982	Brain	Dieldrin	5.7
			DDE	0.51

(continued)

Appendix (continued)

Species	Sample date	Substrate analyzed	Contaminant	Concentration (μg/g)
	1988	Dressed carcass	Mercury	0.0477
	1988	Dressed carcass	Endrin	0.394
	1988	Dressed carcass	DDE	4.3
	1988	Dressed carcass	Dieldrin	5.9
	1994	Brain	Dieldrin	8.34
	1994	Brain	Dieldrin	4.9
			DDE	0.842
	1995	Gastrointestinal (GI) tract	DDE	0.0529
			Dieldrin	0.705
		Kidney	Mercury	0.453
		Liver	DDE	0.663
			Dieldrin	13
	1995	GI tract	Dieldrin	0.501
			Endrin	0.0208
	1995	GI tract	DDE	0.0209
			Dieldrin	0.96
			Endrin	0.0323
	1995	GI tract	DDE	0.0274
			Dieldrin	0.718
			Endrin	0.0294
	1995	GI tract	DDE	0.0224
			Dieldrin	0.821
			Endrin	0.037
	1995	GI tract	Dieldrin	0.0166
	1995	Liver	Dieldrin	0.0675
	1995	GI tract	Dieldrin	0.0227
	1995	GI tract	Dieldrin	0.0152
	1995	GI tract	DDE	0.0405
			Dieldrin	0.0724
	1995	Liver	Dieldrin	0.0803
	1995	Brain	Dieldrin	7.38
	1995	Brain	Dieldrin	4.1
	1995	Brain	Dieldrin	1.66
	1995	Brain	Dieldrin	3.77
	1995	Brain	Dieldrin	3.72
	1995	Brain	Dieldrin	4.58
	1995	Brain	Dieldrin	7.6
	1995	Brain	Dieldrin	1.86
	1995	Brain	Dieldrin	3.01
	1995	Brain	Dieldrin	5
	1995	Brain	Dieldrin	9.48
	1995	Brain	Dieldrin	7
	1995	Brain	Dieldrin	5.5
	1995	Brain	Dieldrin	2.41
	1995	Brain	Dieldrin	5.37

(continued)

Appendix (continued)

Species	Sample date	Substrate analyzed	Contaminant	Concentration (μg/g)
	1995	Brain	DDE	3.02
			Dieldrin	24
	1995	Brain	Dieldrin	2.17
	1995	Brain	Dieldrin	2.57
	1995	Brain	Dieldrin	2.57
	1995	Brain	DDE	1.16
			Dieldrin	4.7
	1995	Brain	Dieldrin	6.99
	1995	Brain	Dieldrin	2
	1995	Brain	Dieldrin	2.31
	1995	Brain	Dieldrin	5.99
	1995	Brain	Dieldrin	6.9
	1995	Brain	Dieldrin	5.24
	1995	Brain	Dieldrin	4.77
	1995	Brain	Dieldrin	5.28
	1995	Brain	DDE	1.08
	1995	Brain	Dieldrin	3.03
			DDE	1.08
	1995	Brain	Dieldrin	4.31
	1995	Brain	Dieldrin	1.6
	1995	Brain	Dieldrin	9.2
	1996	Brain	Dieldrin	1.6
	1996	Brain	Aldrin	0.886
			α-Chlordane	0.832
	1999	Brain	Dieldrin	3.4
			DDE	0.201
		Liver	Dieldrin	4.2
			DDE	0.249
	1999	Brain	Dieldrin	4.1
		Liver	Dieldrin	5.4
			Heptachlor epoxide	0.074
			Isodrin	0.124
			trans-Chlordane	0.0653
	1999	Brain	Dieldrin	6.2
			DDE	2.4
		Liver	Dieldrin	4.8
			DDE	1.7
	2000	Brain	Dieldrin	50
		Liver	Dieldrin	12
	2000	Brain	Dieldrin	16
			Endrin	0.0646
			DDE	0.97
		Liver	Dieldrin	18
			Endrin	0.106
			DDE	1.3

(continued)

Appendix (continued)

Species	Sample date	Substrate analyzed	Contaminant	Concentration (μg/g)
	2000	Brain	Dieldrin	15.25
			DDE	0.94
		Liver	Dieldrin	17.6
			Endrin	0.0993
	2000	Brain	Dieldrin	13.35
	2000	Blood	Dieldrin	67.5
	2000	Brain	Dieldrin	0.143
		Liver	Dieldrin	1.5
		Liver	Dieldrin	0.438
		Liver	Dieldrin	0.75
		Liver	Dieldrin	0.508
	2000	Liver	Dieldrin	0.84
	2000	Liver	Dieldrin	0.285
	2000	Brain	Dieldrin	0.246
		Liver	Dieldrin	1.4
	2001	Brain	Dieldrin	0.167
		Liver	Dieldrin	4.5
	2001	Brain	Dieldrin	0.239
		Liver	Dieldrin	3.4
	2001	Liver	Dieldrin	0.215
	2001	Brain	Dieldrin	0.167
		Liver	Dieldrin	1.1
	2001	Brain	Dieldrin	0.91
			DDE	1.5
		Liver	Dieldrin	5.4
			DDE	6.9
	2001	Liver	Dieldrin	0.16
	2001	Liver	Dieldrin	0.28
	2001	Liver	Dieldrin	0.161
	2003	Liver	DDE	1.19
	2003	Brain	Dieldrin	6.69
			Endrin	0.0776
			Heptachlor epoxide	0.0503
			DDE	1.875
	2003	Liver	Dieldrin	0.1
	2003	Brain	DDE	0.062
	2003	Brain	Dieldrin	1.304
	2005	Brain	Dieldrin	0.0881
			DDE	0.1966
		Liver	Dieldrin	0.2838
			DDE	0.2708
	2005	Liver	Dieldrin	0.7534
			DDE	0.0599
	2005	Liver	Dieldrin	1.3837
			DDE	0.9832

(continued)

Appendix (continued)

Species	Sample date	Substrate analyzed	Contaminant	Concentration (µg/g)
	2005	Brain	Dieldrin	13.736
			DDE	2.5064
		Liver	Dieldrin	17.978
			DDE	2.8245
Swallow, Barn (*Hirundo rustica*)	2000	Brain	DDE	1.5
		Liver	DDE	2.5
Swallow, Cliff (*Petrochelidon pyrrhonota*)	2000	Liver	Dieldrin	0.149
	2001	Liver	Dieldrin	0.68
	2001	Liver	Dieldrin	0.61
	2001	Kidney	Arsenic	0.084
			Mercury	0.13
		Liver	Dieldrin	0.126
	2004	Brain	DDE	2.3714
Swallow, Tree (*Tachycineta bicolor*)	1999	Egg	Dieldrin	0.232
			DDE	1.71
	1999	Egg	Dieldrin	0.0984
			DDE	1
	2000	Liver	Dieldrin	0.301
			DDE	0.201
Teal, Blue-winged (*Anas discors*)	c. 1989	Liver ($n=3$)	Mercury	0.371–1.64
			Dieldrin	0.183–0.281
	c. 1989	Muscle ($n=3$)	Mercury	0.259–0.559
			Dieldrin	0.09–0.164
	1988	Dressed carcass	DDE	0.127
	1988	Dressed carcass	Mercury	0.338
	1988	Dressed carcass	Dieldrin	0.57
	1999	Liver	Dieldrin	0.125

Data sources: Thorne (1982), McEwen (1983), McEwen and DeWeese (1984), Environmental Science and Engineering (1989), R.L. Stollar & Associates (1990a, b), USFWS (1996, 2000), USFWS (unpublished data), US Army, Rocky Mountain Arsenal Environmental Database (unpublished data)

References

Army (1978) Final Report, Project Eagle – Phase II, Demilitarization and Disposal of the M-34 Cluster at Rocky Mountain Arsenal

Army (2004) Rocky Mountain Arsenal Site History. http://www.rma.army.mil/site/sitefrm.html. Accessed 28 Dec 2009

ATSDR (2003) Toxicological Profile for Atrazine. http://www.atsdr.cdc.gov/toxprofiles/tp153.html. Accessed 28 Apr 2010

Babcock KM, Flickinger EL (1977) Dieldrin mortality of lesser snow geese in Missouri. J Wildl Manage 41:100–103

BAS (2002) Assessment of residual ecological risk and risk management recommendations at the Rocky Mountain Arsenal. Part 1: Terrestrial pathways and receptors. Prepared by the Rocky Mountain Arsenal Biological Advisory Sub committee (BAS). Final Report

BAS (2006) Long-term contaminant biomonitoring program for terrestrial ecological receptors at the Rocky Mountain Arsenal. Revision 0. Prepared by the Rocky Mountain Arsenal Biological Advisory Subcommittee (BAS), with technical assistance from Syracuse Research Corporation. November 3

Beyer WN, Gish CD (1980) Persistence in earthworms and potential hazards to birds of soil applied DDT, dieldrin and heptachlor. J Appl Ecol 17:295–307

Bloomquist JR, Soderlund DM (1985) Neurotoxic insecticides inhibit GABA-dependent chloride uptake by mouse brain vesicles. Biochem Biophys Res Commun 133:37–43

Blus LJ (2003) Organochlorine pesticides. In: Hoffman DJ, Rattner BA, Burton GA, Cairns J (eds) Handbook of ecotoxicology. CRC, Boca Raton, FL, pp 313–339

Bogan J, Mitchell J (1973) Continuing dangers of peregrines from dieldrin. Br Birds 66:437–439

Busbee EL (1977) The effects of dieldrin on the behaviour of young loggerhead shrikes. Auk 94:28–35

CDPHE (1994) Response to comments of Shell Oil Company on the Arsenal Fortuitous Specimen Collection and Necropsies Dated February 1994. Colorado Department of Public Health and Environment. August 11

International Programme on Chemical Safety (1992) Endrin. Environmental Health Criteria 130. World Health Organization, Geneva

Coats JR (1990) Mechanisms of toxic action and structure-activity relationships for organochlorine and synthetic pyrethroid insecticides. Environ Health Perspect 87:255–262

Council on Foreign Relations (2006) What is VX? www.cfr.org/publication/9556. Accessed 27 Aug 2007

Dahlgren RB, Linder RL, Ortman KK (1970) Dieldrin effects on susceptibility of penned pheasants to hand capture. J Wildl Manage 34(4):957–959

DOJ (1986) Assessment of CERCLA hazardous substances released by Shell Oil Company and the United States Army at the Rocky Mountain Arsenal, Volumes I & II. December 30

Ebasco Services, R.L. Stollar & Associates, California Analytical Laboratories, Datachem, Technos Inc., and Geraghty & Miller (1987) Litigation technical support and services, Rocky Mountain Arsenal: Final Phase I Contamination Assessment Report, Site 1–2: Upper and Lowry Derby Lakes. Prepared for U.S. Army Program Manager's Office for Rocky Mountain Arsenal Contamination Cleanup. June

Ebasco Services, R.L. Stollar & Associates, California Analytical Laboratories, Datachem, Technos Inc., and Geraghty & Miller (1988a) Litigation Technical Support and Services, Rocky Mountain Arsenal. Rocky Mountain Arsenal Chemical Index. Prepared for U.S. Army Program Manager's Office for Rocky Mountain Arsenal Contamination Cleanup. August

Ebasco Services, R.L. Stollar & Associates, California Analytical Laboratories, Datachem, and Geraghty & Miller (1988b) Litigation Technical support and services, Rocky Mountain Arsenal: Final Phase I Contamination Assessment Report: Site 1–7 Hydrazine Blending and Storage Facility, Version 3.2. Prepared for U.S. Army Program Manager's Office for Rocky Mountain Arsenal Contamination Cleanup. September

Ebasco Services, R.L. Stollar & Associates, California Analytical Laboratories, Datachem, and Geraghty & Miller. 1988c. Litigation Technical Support and Services, Rocky Mountain Arsenal: Final Phase I Contamination Assessment Report, Sites 1–13 and 2–18, South Plants Manufacturing Complex, Shell Chemical Company Spill Sites, Version 3.1. Task No. 2 – South Plants. Prepared for U.S. Army Program Manager's Office for Rocky Mountain Arsenal Contamination Cleanup. July

Ebasco Services, R.L. Stollar & Associates, California Analytical Laboratories, Datachem, and Geraghty and Miller (1988d) Litigation technical support and services, Rocky Mountain Arsenal: Final Phase I Contamination Assessment Report, Site 31–4: Toxic Storage Yard (Version 3.1) – Task No. 15. Prepared for U.S. Army Program Manager's Office for Rocky Mountain Arsenal Contamination Cleanup. June

Ebasco Services, R.L. Stollar & Associates, California Analytical Laboratories, Datachem, and Geraghty & Miller (1988e) Litigation technical support and services, Rocky Mountain Arsenal: Final Contamination Assessment, Site 3–4. Nemagon Spill Area (Version 3.2) – Task No. 7 –

Lower Lakes. Prepared for U.S. Army Program Manager's Office for Rocky Mountain Arsenal Contamination Cleanup. March

Ebasco Services, R.L. Stollar & Associates, California Analytical Laboratories, Datachem, and Geraghty & Miller (1988f) Litigation Technical Support and Services, Rocky Mountain Arsenal: Final Phase II Data ADDEndum, Site 31–4: Toxic Storage Yard (Version 3.1) – Task No. 22 – Army Sites – South. Prepared for U.S. Army Program Manager's Office for Rocky Mountain Arsenal Contamination Cleanup. October

Ebasco Services, R.L. Stollar & Associates, California Analytical Laboratories, Datachem, and Geraghty & Miller (1988g) Litigation technical support and services, Rocky Mountain Arsenal: Final Phase II Data ADDEndum, Site 2–17: Lake Ladora and Lake Mary. Prepared for U.S. Army Program Manager's Office for Rocky Mountain Arsenal Contamination Cleanup. October

Ebasco Services, R.L. Stollar & Associates, California Analytical Laboratories, Datachem, and Geraghty & Miller (1988h) Litigation technical support and services, Rocky Mountain Arsenal: Final Phase II Data Addendum, Site 31–4. Toxic Storage Yard (Version 3.1) – Task No. 22 – Army Sites – South. Prepared for U.S. Army Program Manager's Office for Rocky Mountain Arsenal Contamination Cleanup. October

Ebasco Services, Applied Environmental, CH2M Hill, Datachem, and R.L. Stollar & Associates (1989a) Technical support for Rocky Mountain Arsenal: Final Remedial Investigation Report Volume IX, North Plants Study Area, Text, Version 3.3. Prepared for U.S. Army Program Manager's Office for Rocky Mountain Arsenal Contamination Cleanup. July

Ebasco Services, Applied Environmental, CH2M Hill, Datachem, and R.L. Stollar & Associates (1989b) Technical support for Rocky Mountain Arsenal: Final Remedial Investigation Report Volume XIII, South Plants Study Area, Text, Version 3.3. Prepared for U.S. Army Program Manager's Office for Rocky Mountain Arsenal Contamination Cleanup. July

Ebasco Services, Applied Environmental, CH2M Hill, Datachem, and R.L. Stollar & Associates (1992) Final remediation summary report, Version 3.2. Prepared for U.S. Army Program Manager's Office for Rocky Mountain Arsenal Contamination Cleanup. January

Ebasco Services, James M. Montgomery, International Dismantling & Machinery, Greystone Environmental, Hazen Research, Data Chem, BC Analytical, and Terra Technologies (1994) Technical support for Rocky Mountain Arsenal. Final Integrated Endangerment Assessment/ Risk Characterization. Version 4.2. Prepared for U.S. Army Program Manager's Office for the Rocky Mountain Arsenal Contamination Cleanup. July

Ecobichon DJ (1991) Toxic effects of pesticides. In: Amdur MO, Doull J, Klaassen CD (eds) Casarett and Doull's toxicology. Pergamon, New York, pp 565–622

Eldefrawi MES, Sherby SM, Abalis IM, Eldefrawi AT (1985) Interactions of pyrethroid and cyclodiene insecticides with nicotinic acetylcholine and GABA receptors. Neurotoxicology 6:47–62

Elliott JE, Bishop CA (2011) Cyclodienes and other organochlorine pesticides in birds. In: Beyer WN, Meador J (eds) Environmental contaminants in wildlife – interpreting tissue concentrations. CRC, Boca Raton, FL

Environmental Science & Engineering (1987) Litigation technical support and services, Rocky Mountain Arsenal: Final Phase I Contamination Assessment Report, Site 36–1: Basin A (Version 3.2). Prepared for U.S. Army Program Manager Office for Rocky Mountain Arsenal. July

Environmental Science & Engineering (1989) Biota remedial investigation final report. Prepared for Office of the Program Manager, Rocky Mountain Arsenal Contamination Cleanup. May

Environmental Science & Engineering, Harding Lawson Associates, and Midwest Research Institute (1987a) Litigation technical support and services, Rocky Mountain Arsenal: Final Phase I Contamination Assessment Report, Site 26–3: Basin C (Version 3.3) – Task No. 6 (Section 26 and 35). Prepared for U.S. Army Program Manager's Office for Rocky Mountain Arsenal. December

Environmental Science & Engineering, Harding Lawson Associates, and Midwest Research Institute (1987b) Litigation technical support and services, Rocky Mountain Arsenal: Final Phase I Contamination Assessment Report, Site 26–4. Basin D (Version 3.3) – Task No. 6

(Section 26 and 35). Prepared for U.S. Army Program Manager's Office for Rocky Mountain Arsenal. October

Environmental Science & Engineering, Harding Lawson Associates, and Midwest Research Institute (1987c) Litigation technical support and services, Rocky Mountain Arsenal: Final Phase I Contamination Assessment Report, Site 26–5: Basin E (Version 3.2). Prepared for U.S. Army Program Manager's Office for Rocky Mountain Arsenal. July

Environmental Science & Engineering, Harding Lawson Associates, and Midwest Research Institute (1987d) Litigation technical support and services, Rocky Mountain Arsenal: Final Phase I Contamination Assessment Report, Site 35–3. Basin B (Version 3.3). Prepared for U.S. Army Program Manager's Office for Rocky Mountain Arsenal. July

Environmental Science & Engineering, Harding Lawson Associates, and Midwest Research Institute (1987e) Litigation technical support and services, Rocky Mountain Arsenal: Draft Final Phase I Contamination Assessment Report, Site 36–17: Complex Disposal Activity (Version 2.2). Prepared for U.S. Army Program Manager's Office for Rocky Mountain Arsenal. October

Environmental Science & Engineering and Harding Lawson Associates (1988a) Litigation technical support and services, Rocky Mountain Arsenal: Final Phase II Data Addendum, Site 26–3: Basin C (Version 3.1). Prepared for U.S. Army Program Manager's Office for Rocky Mountain Arsenal. September

Environmental Science & Engineering and Harding Lawson Associates (1988b) Litigation technical support and services, Rocky Mountain Arsenal: Final Phase II Data Addendum, Site 26–4: Basin D (Version 3.1). Prepared for U.S. Army Program Manager's Office for Rocky Mountain Arsenal. September

Environmental Science & Engineering and Harding Lawson Associates (1988c) Litigation Technical support and services, Rocky Mountain Arsenal: Final Phase II Data Addendum, Site 26–5: Basin E (Version 3.1). Prepared for U.S. Army Program Manager's Office for Rocky Mountain Arsenal. September

Environmental Science & Engineering, Harding Lawson Associates, and Midwest Research Institute (1988a) Litigation technical support and services, Rocky Mountain Arsenal: Final Phase I Contamination Assessment Report, Site 26–6: Basin F (Version 3.2) – Task No. 6 (Section 26 and 35). Prepared for U.S. Army Program Manager's Office for Rocky Mountain Arsenal. May

Environmental Science & Engineering, Harding Lawson Associates, and Midwest Research Institute (1988b) Litigation Technical Support and Services, Rocky Mountain Arsenal: Draft Final Phase I Contamination Assessment Report, Site 36–9: Incendiary or Munition Test Area (Version 2.2) – Task No. 14 (Army Sites North). Prepared for U.S. Army Program Manager's Office for Rocky Mountain Arsenal. January

Environmental Science & Engineering and Harding Lawson Associates (1988d) Litigation technical support and services, Rocky Mountain Arsenal: Final Phase II Data Addendum, Site 35–3: Basin B (Version 3.1). Prepared for U.S. Army Program Manager's Office for Rocky Mountain Arsenal. September

Environmental Science & Engineering and Harding Lawson Associates (1988e) Litigation technical support and services, Rocky Mountain Arsenal: Final Phase II Data Addendum, Site 36-3: Insecticide Pit (Version 3.1). Prepared for U.S. Army Program Manager's Office for Rocky Mountain Arsenal. September

Environmental Science & Engineering and Harding Lawson Associates (1988f) Litigation technical support and services, Rocky Mountain Arsenal: Final Phase II Data Addendum, Site 36-17: Complex Disposal Activity (Version 3.1). Prepared for U.S. Army Program Manager's Office for Rocky Mountain Arsenal. September

Finley RB (1959) Investigations of waterfowl mortality at the Rocky Mountain Arsenal. U.S. Fish and Wildlife Service, Wildlife Research Laboratory, Denver, CO

Flickinger EL (1979) Effects of aldrin exposure on snow geese in Texas rice fields. J Wildl Manage 43:94–101

Flickinger EL, King KA (1972) Some effects of aldrin-treated rice on Gulf Coast wildlife. J Wildl Manage 36:706–777

Flickinger EL, Mitchell CA, Krynitsky AJ (1986) Dieldrin and endrin residues in fulvous whistling ducks in Texas in 1983. J Field Ornithol 57:85

Foster Wheeler (1996) Record of decision for the on-post operable unit. Version 3.1. Foster Wheeler Environmental. July 2

Foster Wheeler (2001) Rocky Mountain Arsenal Complex Army Trenches Groundwater Barrier Project Construction Completion Report. Prepared for Rocky Mountain Arsenal Remediation Venture Office, Department of the Army, Shell Oil Company, U.S. Fish and Wildlife Service. April 30

Frank RA, Lutz RS (1999) Productivity and survival of Great Horned Owls exposed to dieldrin. Condor 101(2):331–339

Friend M, Trainer DO (1974) Experimental dieldrin – duck hepatitis virus interaction studies. J Wildl Manage 38(4):896–902

Fuchs P (1967) Death of birds caused by application of seed dressings in The Netherlands. Med Rijksf Landb Gent 32:855–859

Goldstein S (2001) I Volunteered for the Front Lines of Chemical Warfare. Philadelphia Inquirer. January 14. http://www.jtfcs.northcom.mil/News/2001/001.html. Accessed 7 Jan 2010

Harding Lawson Associates (1993) Second year reevaluation report for the complex disposal trenches interim response action

Harding Lawson Associates (1995) Rocky Mountain Arsenal Off-post Operable Unit: Final Record of Decision. Rocky Mountain Arsenal, Commerce City, Colorado. Prepared for the Program Manager of the Rocky Mountain Arsenal. December 19

Harris ML, Wilson LK, Elliott JE, Bishop CA, Tomlin AD, Henning KV (2000) Transfer of DDT and metabolites from fruit orchard soils to American robins (*Turdus migratorius*) twenty years after agricultural use of DDT in Canada. Arch Environ Contam Toxicol 39:205–220

Hayes WJ, Laws ER (1991) Handbook of pesticide toxicology: classes of pesticides, vol 2. Academic, San Diego, CA

Heinz GH, Johnson RW (1981) Diagnostic brain residues of dieldrin: some new insights. In: Lamb DW, Kenaga EE (eds) Avian and mammalian wildlife toxicology: second conference. ASTM STP 757, pp 72–92

Henriksen EO, Gabrielsen GW, Skaare JU (1996) Levels and congeners patterns of polychlorinated biphenyls in kittiwakes (*Rissa tridactyla*) in relation to mobilization of body lipids associated with reproduction. Environ Pollut 92:27–37

Hoffecker JF (2001) Twenty-seven square miles: landscape and history at Rocky Mountain Arsenal National Wildlife Refuge. U.S. Fish and Wildlife Service, Rocky Mountain Arsenal National Wildlife Refuge

Hudson RH, Tucker RK, and Haegele MA (1984) Handbook of toxicity of pesticides to wildlife. U.S. Fish and Wildlife Service, Washington DC. Resource Publication, p 153

Jefferies DJ, Prestt I (1966) Post-mortems of Peregrines and Lanners with particular reference to organochlorine residues. Br Birds 59:49–64

Jones DM, Bennett D, Elgar KE (1978) Deaths of owls traced to insecticide-treated timber. Nature 272:52

Jorgenson JL (2001) Aldrin and dieldrin: a review of research on their production, environmental deposition and fate, bioaccumulation, toxicology, and epidemiology in the United States. Environ Health Perspect 109(suppl 1):113–139

Klaassen CD (ed) (2001) Casarett and Doull's toxicology: the basic science of poisons. McGraw-Hill, New York

Koeman JH, Oskamp G, Veen J, Brouwer E, Rooth J, Zwart P, van den Brock E, van Genderen H (1967) Insecticides as a factor in the mortality of the Sandwich Tern (*Sterna sandvicensis*). Med Rijksf Landb Gent 32:841–854

Kuznear C, Trautmann WL (1980) History of pollution sources and hazards at Rocky Mountain Arsenal. September

Linder RL, Dahlgren RB, Greichus YA (1970) Residues in the brain of adult pheasants given dieldrin. J Wildl Manage 34:954–956

Ludke JL (1976) Organochlorine pesticide residues associated with mortality: additivity of chlordane and endrin. Bull Environ Contam Tox 16:253–260

Matsumura F (1975) Toxicology of insecticides. Plenum, New York
Matsumura F, Ghiasudding SM (1983) Evidence for similarities between cyclodiene type insecticides and picrotoxin in their action mechanisms. J Environ Sci Health B 18:1–14
McEwen LC (1983) Letter re: Summary table showing organochlorine residues detected in brain tissues of some of the animals found dead at the RMA in 1982. To William McNeill, Rocky Mountain Arsenal. March 4
McEwen LC, DeWeese LR (1984) U.S. Fish and Wildlife Service Investigations of Chemical Contaminants in Animals and Habitats of the Rocky Mountain Arsenal, Denver, Colorado. U.S. Fish & Wildlife Service, Fort Collins, Colorado. May
Mineau P (2005) Direct losses of birds to pesticides – beginnings of a quantification. USDA Forest Service General Technical Report PSW-GTR-191, pp 1065–1070
MK-Environmental Services (1993) Data compilation and Interpretation, Shell Section 36 Trenches, Location and Content, August 1993
Muralidharan S (1993) Aldrin poisoning of Sarus cranes (*Grus antigone*) and a few granivorous birds in Keoladeo National Park, Bharatpur, India. Ecotoxicology 2:196–202
Nagata K, Narahashi T (1994) Dual action of the cyclodiene insecticide dieldrin on the gamma-aminobutyric acid receptor-chloride channel complex of rat dorsal root ganglion neurons. J Pharmacol Exp Ther 269:164–171
NASA KSC (2006) Aerozine-50 specifications & DOT shipping information. Updated October 6. http://propellants.ksc.nasa.gov/commodities/Aerzone50.pdf. Accessed 27 Aug 2007
Newton I (1973) Success of sparrowhawks in an area of pesticide usage. Bird Study 20:1–8
Newton I (1986) The sparrowhawk. T&AD Poyser, Calton, UK, 396 pp
Newton I (1988) Determination of critical pollutant levels in wild populations, with examples from organochlorine insecticides in birds of prey. Environ Pollut 55:29–40
Newton I (1998) Population limitation in birds. Academic, London, UK
Newton I, Bogan J (1974) Organochlorine residues, eggshell thinning and hatching success in British sparrowhawks. Nature 249:582–583
Newton I, Bogan J (1978) The role of different organochlorine compounds in the breeding of British sparrowhawks. J Appl Ecol 15:105–116
Newton I, Haas MB (1988) Pollutants in merlin eggs and their effects on breeding. Br Birds 81:258–269
Newton I, Wylie I, Asher A (1991) Mortality causes in British barn owls (*Tyto alba*) with a discussion of aldrin and dieldrin poisoning. Ibis 133:1629
Newton I, Wylie I, Asher A (1992) Mortality from the pesticides aldrin and dieldrin in British sparrowhawks and kestrels. Ecotoxicology 1:31–44
Nisbet ICT (1988) The relative importance of DDE and dieldrin in the decline of peregrine falcon populations. In: Cade TJ, Enderson JH, Thelander CG, White CM (eds) Peregrine falcon populations: their management and recovery. Peregrine Fund, Boise, ID, pp 351–375
Peakall DB (1996) Dieldrin and other cyclodiene pesticides in wildlife. In: Beyer WN, Heinz GH, Redmon-Norwood AW (eds) Environmental contaminants in wildlife interpreting tissue concentrations. Lewis, Boca Raton, FL, pp 73–97
Program Manager's Office (1988) Rocky Mountain Arsenal Contamination Cleanup: Draft Final Technical Program Plan FY88-FY92 (Remedial Investigation/Feasibility Study/Interim Response Actions). Prepared by the Program Manager's Office for the Rocky Mountain Arsenal Contamination Cleanup. February 22
R.L. Stollar & Associates, Harding Lawson Associates, Ebasco Services, DataChem, Enseco-Cal Lab, and Midwest Research Institute (1992) Comprehensive monitoring program: Biota Annual Report for 1990 and Summary Report for 1988 to 1990. Prepared for U.S. Army Program Manager for Rocky Mountain Arsenal. July
R.L. Stollar & Associates, Harding Lawson, Ebasco Services, Datachem, and Midwest Research Institute (1990a) Comprehensive monitoring program, Final Biota Annual Report for 1988 (Version 2.1) Volume I. Prepared for U.S. Army Program Manager for Rocky Mountain Arsenal. May
R.L. Stollar & Associates, Harding Lawson, Ebasco Services, Environmental Science & Engineering, Datachem, and Midwest Research Institute (1990b) Comprehensive monitoring

program, Final Biota Annual Report for 1989 (Version 2.0) Volume I. Prepared for U.S. Army Program Manager for Rocky Mountain Arsenal. June
Ratcliffe DA (1973) Studies of the recent breeding success of the peregrine, Falco peregrinus. J Reprod Fertil Suppl 19:377–389
Ratcliffe DA (1988) The peregrine falcon. Buteo, Vermillion, SD, 416 pp
Reichel WL, Locke LN, Prouty RM (1974) Case report: Peregrine falcon suspected of pesticide poisoning. Avian Dis 18:487–489
Robinson J, Crabtree AJ (1969) The effect of dieldrin on homing pigeons. Med Rijksf Landb Gent 34(3):413–427
Robinson J, Brown VKH, Richardson A, Roberts M (1967) Residues of dieldrin (HEOD) in the tissues of experimentally poisoned birds. Life Sci 6:1207–1220
Ronis MJJ, Walker CH (1989) The microsomal monooxygenases of birds. Rev Biochem Toxicol 10:301–384
Roy R (1997) Results from the American Kestrel (Falco sparvenus) biomonitoring study at the Rocky Mountain Arsenal National Wildlife Refuge. 1993–1996. USFWS. Dept. of Interior. p 54
RVO (2004) North Plants Area Fact Sheet. Rocky Mountain Arsenal Remediation Venture Office. http://www.rma.army.mil/site/n-plants.html. Accessed 28 Dec 2009
RVO (2009) Quick Facts About Rocky Mountain Arsenal. http://www.rma.army.mil/cleanup/facts/Quick%20Facts.pdf. Accessed 22 Jan 2010
RVO (2010) Ceremony to mark fulfillment of vision for RMA: End of fieldwork, refuge expansion signal transformation complete. http://www.rma.army.mil/involve/Newsletters/Milstones Fall2010.pdf. Accessed 15 Oct 2010
Schwarzenbach RP, Gschwend PM, Imboden DM (2003) Environmental organic chemistry. Wiley, New York
Scorecard.org. (2005) Chemical profiles. http://www.scorecard.org/chemical-profiles. Accessed 28 Apr 2010
Seidel J (1980) Letter re: Tour of Rocky Mountain Arsenal with the intent of checking Reservoir "F" for duck mortality. To Darryl Todd, Colorado Division of Wildlife. May 8
Sharma RP, Winn DS, Low JB (1976) Toxic, neurochemical and behavioral effects of dieldrin exposure in mallard ducks. Arch Environ Contam Toxicol 5(1):43–53
Sibly RM, Newton I, Walker CH (2000) Effects of dieldrin on population growth rates of sparrowhawks 1963–1986. J Appl Ecol 37:540–546
Stanley PI, Bunyan PJ (1979) Hazards to wintering geese and other wildlife from the use of dieldrin, chlorfenvinphos and carbophenothion as wheat seed treatments. Proc R Soc Lond B 205:31–45
Stansley W, Roscoe DE (1999) Chlordane poisoning of birds in New Jersey, USA. Environ Toxicol Chem 18(9):2095–2099
Stickel WH, Stickel LF, Spann JW (1969) Tissue residues of dieldrin in relation to mortality in birds and mammals. In: Miller MW, Berg GG (eds) Chemical fallout: current research on persistent pesticides. Thomas, Springfield, IL, pp 174–203
Thorne DS (1982) Contaminants in fish and game animals on Rocky Mountain Arsenal, 1977–1982. Results of the Rocky Mountain Arsenal biological monitoring program
U.S. EPA (2005) Ecological soil screening levels for dieldrin. Interim Final
U.S. EPA (2007) Types of pesticides. http://www.epa.gov/pesticides/about/types.htm. Accessed 28 Apr 2010
USFWS (1961) Special pesticide problems. Unpublished report, Denver Wildlife Research Center
USFWS (1996) Rocky Mountain Arsenal National Wildlife Refuge Fiscal Year 1995 Annual Progress Report. January
USFWS (2000) Rocky Mountain Arsenal National Wildlife Refuge Fiscal Year 1999 Annual Progress Report. February
USFWS (2009) Rocky Mountain Arsenal 2007 and 2008 annual biomonitoring report. May. Draft
Vander Lee B, Lutz RS (2000) Dose-tissue relationships for dieldrin in nestling black-billed magpies. Bull Environ Contam Toxicol 65:427–434

Wafford KA, Lummis SCR, Sattelle DB (1989) Block of an insect central nervous system GABA receptor by cyclodiene and cyclohexane insecticides. Proc R Soc Lond B 237:53–61

Walker CH, Newton I (1999) Effects of cyclodiene insecticides on raptors in Britain – correction and updating of an earlier paper by Walker and Newton in Ecotoxicology 7:185–189 (1998). Ecotoxicology 8:425–429

Ward FP, Gauthier DA (1973) Research prospectus: studies to elucidate the cause of waterfowl mortalities at Basin F and Vicinity, Rocky Mountain Arsenal, Colorado. Edgewood Arsenal/Dugway Proving Ground Ecological Research Team. August

WHO (1989) Aldrin and Dieldrin Health and Safety Guide. IPCS International Programme on Chemical Safety. Health and Safety Guide No. 21. World Health Organization. http://www.inchem.org/documents/hsg/hsg/hsg021.htm. Accessed 10 May 2004

Witmer GW, Pipas MJ, Campbell DL (1995) Effectiveness of search patterns for recovery of animal carcasses in relation to pocket gopher infestation control. Int Biodeter Biodegrad 36:177–187

Chapter 5
Abnormal Alligators and Organochlorine Pesticides in Lake Apopka, Florida

Allan R. Woodward, H. Franklin Percival, R. Heath Rauschenberger, Timothy S. Gross, Kenneth G. Rice, and Roxanne Conrow

Mike Jennings and Ab Abercrombie paddled their way along the shoreline of Lake Apopka searching for the twinkling red eye reflections of recently hatched alligators. They spotted an opening in the marsh leading to a trail, perhaps made by a mother alligator, which might lead them to hatchlings and possibly a nest. Freshly disturbed mud in the trail confirmed the recent passage of a large alligator, so they grabbed a paddle and snake tongs and carefully climbed out of the canoe onto the unstable floating tussocks (mats of peat and vegetation), typical of certain areas along the Lake Apopka shoreline. The tongs were for catching elusive hatchlings and the paddle was the first line of defense against a protective mother. Mike and Ab worked their way along the trail, occasionally breaking through tussocks and sinking up to their waists in putrid, reddish brown muck. They were able to wallow their way through the mire and managed to arrive at a 60-cm high mound of peat and duck potato. Mike and Ab had been hired as biologists on a project to determine the sustainable harvest rate of hatchling alligators for supplementing alligator farming operations. Lake Apopka was a particularly difficult lake to work because much of the nesting habitat was covered with an intermediate to heavy tree canopy, and nests were difficult to spot from a helicopter, as was typically done on other lakes. Most nests on Lake Apopka had to be found from the ground by probing into likely-looking areas. The nest they had found that night had not been opened by the mother, which was unusual for that time in mid-September. So, they decided to dig into the nest to see if the eggs were close to hatching. The eggs at the top layer were discolored by peat but they also had a yellow cast that usually indicated dead or undeveloped eggs. Ab opened the first egg and was greeted with the typical rotten egg odor. The next egg popped as he opened it, spraying him with a foul gray fluid, which would cling to him for several days. Mike and Ab went through the entire clutch of eggs and found no viable eggs. A completely non-viable clutch of eggs was unusual for alligators but it soon became commonplace on Lake Apopka. It was also the first indication that alligators on Lake Apopka had a serious problem with reproductive success. That September 1981 observation set the stage for three decades of alligator investigation on the lake.

A.R. Woodward (✉)
Fish and Wildlife Research Institute, Florida Fish and Wildlife Conservation Commission, 1105 SW Williston Road, Gainesville, FL, 32601, USA
e-mail: allan.woodward@myfwc.com

J.E. Elliott et al. (eds.), *Wildlife Ecotoxicology: Forensic Approaches*, Emerging Topics in Ecotoxicology 3, DOI 10.1007/978-0-387-89432-4_5,

Abstract Lake Apopka is a 12,400-ha hypereutrophic lake in central Florida that was the recipient of nutrient and pesticide pollution from adjacent agricultural operations for 50 years. The abnormal American alligator (*Alligator mississsippiensis*) population in Lake Apopka has been the object of a number of studies including investigations of a population crash, the epidemiology of egg failure, and anomalous endocrine function. Several hypotheses of the causes of these abnormalities have been proposed and examined by multiple research organizations over the past three decades. Initially, organochlorine pesticide (OCP) contamination was considered the most likely factor causing poor reproductive success. DDE concentrations in alligator eggs sampled in 1984–1985 were approximately 4 mg/kg and toxaphene concentrations were approximately 2.5 mg/kg. These levels were known to cause reproductive failure in certain birds. However, transmissible diseases, population age and density, cyanotoxins, nutritional deficiencies, and combinations thereof, were also investigated for their contribution to poor alligator reproductive success. Investigations of an alligator mortality and reproductive failure event on Lake Griffin, a lake similar to Lake Apopka but with lower OCP levels, revealed analogous reproductive abnormalities that were associated with a dietary thiamine deficiency. Thiamine deficiency appeared to be associated with a diet of almost exclusively gizzard shad, which contain thiaminase, an enzyme that breaks down thiamine. OCP contaminants may contribute to these maladies, perhaps through endocrine disruption and increased stress. The findings of the past 30 years of work at Lake Apopka have affected local management decisions as well as policy at the national level.

Introduction

Lake Apopka was once the second largest lake in Florida and renowned for its sport fishery. Agricultural, industrial, and urban development around the lake during the last century combined to contribute to a general degradation of lake's water quality, sport fish habitat, and wildlife habitat (see Chap. 6). The decline of Lake Apopka's sport fishery (Fig. 5.1) and water quality has elicited concerns from recreational fishers, the public, and natural resource managers (see Canfield et al. 2000; Schelske et al. 2000 for reviews). However, the decline of Lake Apopka's American alligator (*Alligator mississippiensis*) population during the 1980s and associated anomalies have also captured widespread attention, at the scientific and popular levels. In this chapter, we will review the 30-year effort to diagnose the causes of abnormal alligators on Lake Apopka and summarize lessons learned from our experiences.

Lake Apopka's Alligator Population

Historically, Lake Apopka was a large (12,400-ha), shallow (mean depth of 1.3 m) lake with an adjacent 7,300-ha marsh along the north shore. During the 1940s, the north marsh was impounded, drained, and converted to vegetable farms (Fig. 5.2).

Fig. 5.1 Ghostly remains of one of the 15 fish camps that prospered around Lake Apopka, Florida, before the largemouth bass fishery collapsed during the 1960s

The St. Johns River Water Management District (SJRWMD) purchased the vegetable farms during the 1990s, and is now restoring both the north marsh and the improving water quality on the lake. Scant documentation is available of the American alligator population on Lake Apopka prior to 1979. Anecdotal reports from alligator hunters indicated that a substantial population probably existed during the 1950s (C. White, pers. comm.) but no empirical data were available until the late 1970s. The deep, peat soils of the marsh (Hortenstine and Forbes 1972) and nutrient-rich bottom substrate provided a fertile base for primary production (Clugston 1963; Huffstutler et al. 1965). Abundant vegetation leads to prolific secondary production of insects, crustaceans, fish, and turtles, which are the predominant diet of alligators on large Florida lakes (Delany and Abercrombie 1986; Delany et al. 1999). As primary and secondary production increases, there is typically a concomitant increase in populations of vertebrates, such as alligators, higher up the food chain, (Wood et al. 1985; Evert 1999). Lake Apopka showed signs of anthropogenic eutrophication as early as the 1950s (Clugston 1963; Huffstutler et al. 1965), and probably before, and was considered hypereutrophic as early as 1970 (Brezonik and Shannon 1971). The north shore marsh (see Chap. 6) and perimeter marsh/swamp undoubtedly provided ideal habitat for nesting and juvenile alligators. This combination of high food production and desirable nesting habitat would certainly have contributed to a high carrying capacity for the alligator population. Although there is little doubt that Lake Apopka was subjected to the cyclic nature of alligator harvest pressure before effective protection in 1970, as described by Allen and Neill (1949) and Hines (1979), it is reasonable to assume that Lake Apopka historically had a relatively dense alligator population.

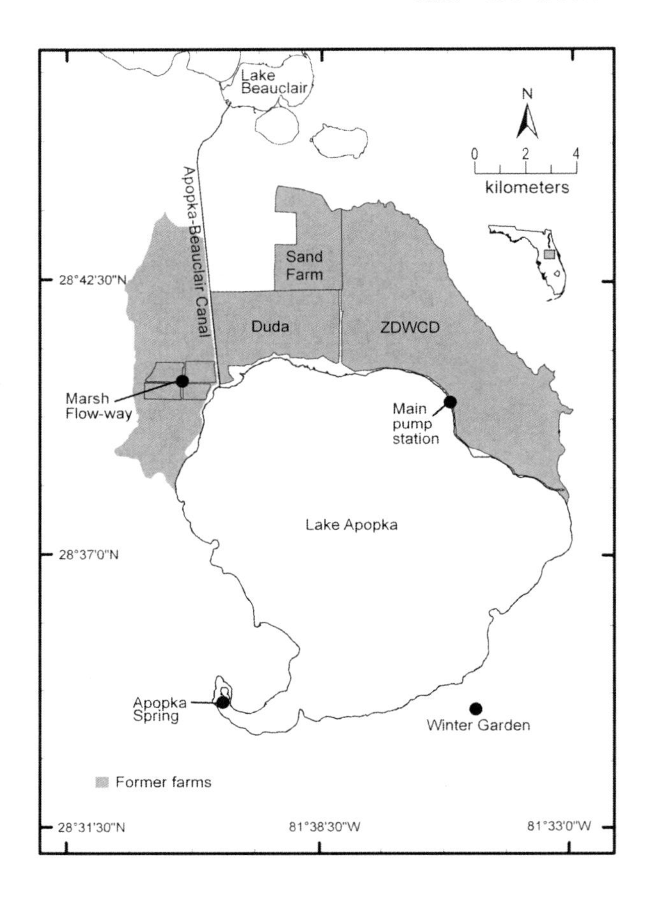

Fig. 5.2 Area map of Lake Apopka, Florida, with the former farms (presently the St. Johns River Water Management District North Shore Restoration Area) shaded in *gray*

The first published reference to Lake Apopka alligators was a fish and reptile mortality event during May 1971–January 1972, following a lake drawdown to consolidate bottom sediments. A team of scientists investigating the die-off concluded that stress, resulting from the weakened condition of reptiles due to overwintering and low water levels, may have led to bacterial (*Aeromonas* sp.) infections and subsequent deaths of Florida softshell turtles (*Apalone ferox*) and of at least several dozen adult alligators (Shotts et al. 1972). Unfortunately, the investigation was limited by a shortage of moribund and freshly dead specimens on which to conduct tests and postmortem examinations (Shotts et al. 1972).

Alligator Egg Viability Problems

In 1979, Lake Apopka's reputation as having a productive alligator population led to it being considered as a candidate for collection of eggs and hatchlings with which to supplement alligator farms (Jennings et al. 1988). Alligator farming was a fledgling industry in Florida in the 1970s, and farms were having difficulty producing economically feasible quantities of viable eggs for commercial rearing. To supplement captive production on alligator farms, the Florida Alligator Farmers Association (FAFA) requested permission to harvest alligator eggs and hatchlings on Lake Apopka and nearby Lakes Griffin and Jesup. In 1981, the Florida Game and Fresh Water Fish Commission (GFC) approved a pilot study to determine the effects of a 50% harvest of the annual production of alligator eggs or hatchlings on those lakes (Jennings et al. 1988; Rice et al. 1999). The study was conducted by the Florida Cooperative Fish and Wildlife Research Unit (FCFWRU), and funded by the GFC (2/3) and the FAFA (1/3). The FCFWRU is a natural resource research consortium, which in 1981 was comprised of the GFC (now the Florida Fish and Wildlife Conservation Commission), the US Fish and Wildlife Service (the research component is now a part of the US Geological Survey), the University of Florida, and the Wildlife Management Institute. Initial investigations included limited population monitoring and preliminary assessment of the effects of egg and hatchling harvests on the alligator population.

Hatchling collections on Lake Apopka began in 1981. However, fewer hatchlings than expected were found, and pod (sibling group) size was smaller than expected (Jennings et al. 1988). Small pod size was first attributed to limited accessibility to pods due to floating peat mats with dense stands of small- to medium-sized trees along the shoreline. Further examination of nests in 1982 revealed a high percentage of nonviable eggs (C. Abercrombie 1982, unpublished data). This failure rate was considered abnormally high, compared to rates seen in studies in Louisiana (Joanen and McNease 1977), but few studies had closely evaluated hatch rates of wild alligator eggs.

There were no immediate explanations for the apparently poor hatching success. Then, in October 1982, a newspaper article reported high levels of the pesticide, toxaphene, in catfish on Lake Apopka (Churchville 1982). Toxaphene was suspected of contributing to reproductive failure of alligators because it had been shown to cause increased reproductive failure in pheasants, rats, and mice (Eisler and Jacknow 1985). Other organochlorine pesticides (OCP), including DDT and chlordane (see Chap. 6), had been used extensively on the muck farms adjacent to Lake Apopka since the 1940s. DDT was a suspect in fish kills on Lake Apopka during the early 1960s but scientists from the Florida State Board of Health concluded DDT was unlikely to be the direct cause of the fish kill. They reported that catfish contained 4.4 mg/kg of DDT and its metabolites in edible tissue (all contaminant concentrations in this chapter reflect wet weight values) and 14.4 mg/kg in fat, liver, and roe (Huffstutler et al. 1965).

In 1983, the FCFWRU began collecting alligator eggs (Fig. 5.3), in addition to hatchlings, from Lakes Apopka, Griffin, Jesup, and Okeechobee (Jennings et al. 1988). Eggs were incubated in artificial incubators at alligator farms during 1983–1986, and monitored for survival by FCFWRU staff. During this period, "clutch or egg viability rate" was the term used to represent the proportion of eggs from a clutch hatching with newly emerged hatchlings surviving at least 1 day. This term was used to avoid potential confusion with other terms, such as "fertility rate,"

Fig. 5.3 Florida Game and Fresh Water Fish Commission biologist collecting alligator eggs from a nest on Lake Apopka during the 1980s

which has been widely used to refer to fertilization rate in poultry. Preliminary evaluations of egg viability provided critical insight into the variation in potential hatching success of wild alligator eggs in Florida.

Clutches from Lakes Apopka, Griffin, Jesup, and Okeechobee were evaluated for viability as part of other studies looking at the effects of time of collection on hatching success (Woodward et al. 1989). During 1983–1986, the FCFWRU found egg viability rates from each of Lakes Apopka, Griffin, Jesup, and Okeechobee to be well below what was considered normal for alligators. At that time, the only published hatch rate information for wild alligator eggs was for Rockefeller Wildlife Refuge, Louisiana, which reported an 86% hatch rate (Joanen and McNease 1987). The FCFWRU found substantially lower mean egg viability rates (13– 65%) on the four Florida lakes, depending on area and year (Woodward et al. 1993). Lake Apopka had the lowest hatching success of all areas studied, showed signs of a declining egg viability rate, and had a high percentage of clutches with complete failure (Woodward et al. 1993) (Fig. 5.4). In 1983, we became aware of a chemical spill at a nearby pesticide manufacturing plant in 1980 that included DDT. However, the extent to which this spill contributed to the observed levels of DDT in alligators was confounded by the long-term agricultural application of DDT on farms located along the north shore (see Chap. 6).

From 1980 to 1987, the FCFWRU documented a significant decline in the Lake Apopka alligator population (Fig. 5.5), with most of the decline in the juvenile (0.3–1.2 m TL) size class (Jennings et al. 1988; Woodward et al. 1993). Conversely, juvenile alligator populations on Lakes Griffin and Jesup subjected to similar harvest levels were stable or increasing during that period. This indicated a lack of

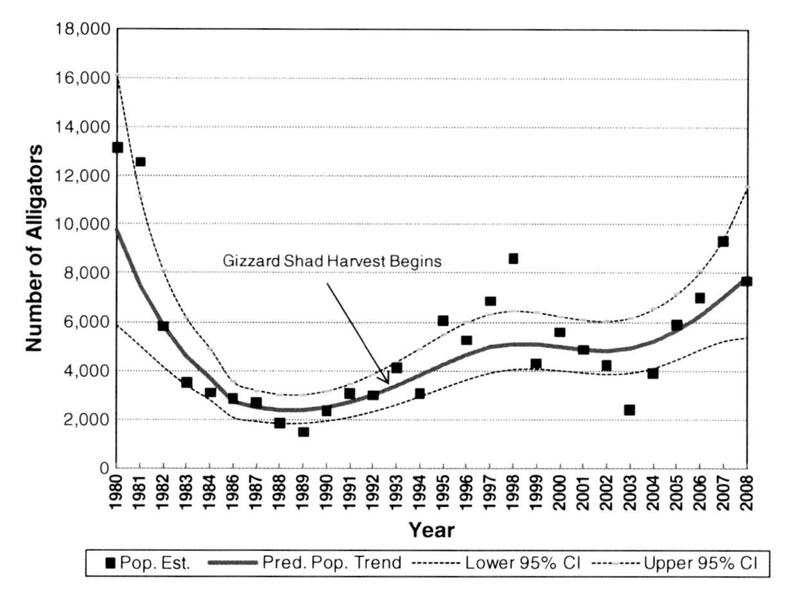

Fig. 5.4 Population trend of non-hatchling American alligators on Lake Apopka from 1980 to 2008

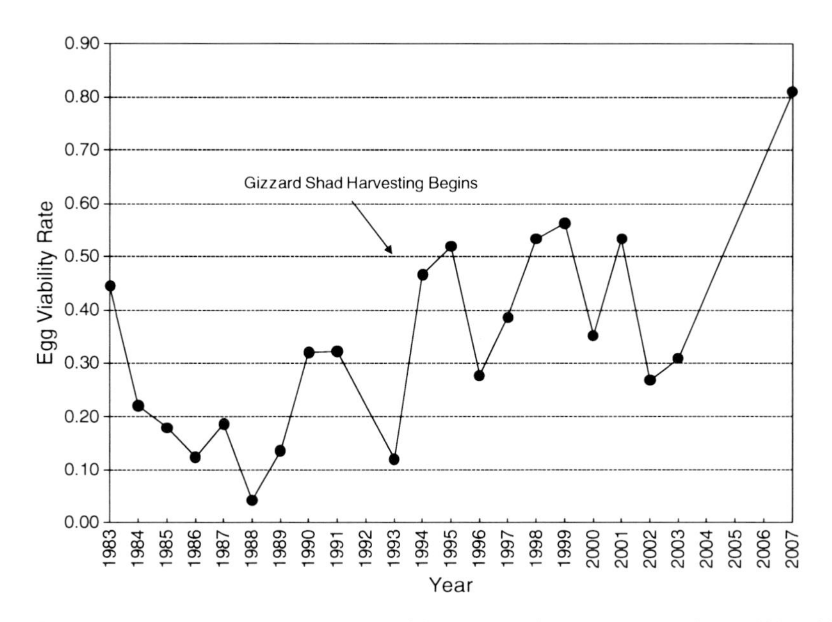

Fig. 5.5 Mean annual alligator clutch viability (*black circles*) for Lake Apopka from 1983 to 2007

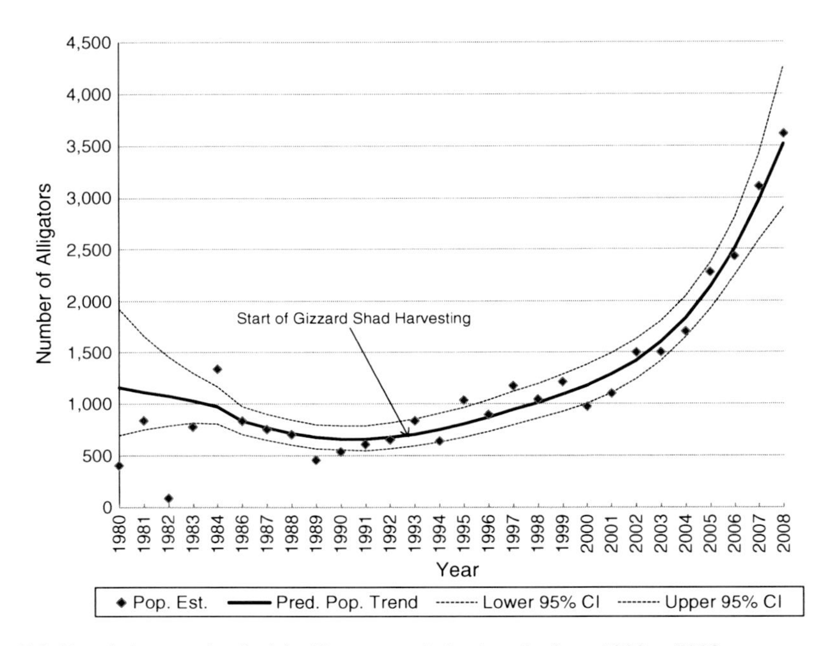

Fig. 5.6 Population trends of adult alligators on Lake Apopka from 1980 to 2008

recruitment into the juvenile population on Lake Apopka, suggesting something unique was likely affecting alligator recruitment (Jennings et al. 1988; Rice et al. 1999). We also observed a greater incidence of dead alligators, softshell turtles, and water snakes on Lake Apopka than on other lakes, but systematic surveys for dead reptiles were not performed (Woodward et al. 1993). The adult (≥1.8 m TL) alligator population declined during the early 1980s, suggesting elevated mortality or reduced recruitment, but rose steadily after 1988 (Fig. 5.6). We saw some similarities to the mortality event described by Shotts et al. (1972), but it was difficult to compare these two events because of differing levels of monitoring.

Hypothesis Development

By the mid-1980s, several hypotheses had emerged to explain the depressed egg viability. These included the following.

Depressed Egg Viability Was Caused by Toxicosis from Environmental Contaminants

The FCFWRU team was aware that Lake Apopka had been the recipient of decades of OCP pollution from adjacent agricultural operations (Woodward et al. 1993, see

Chap. 6). Muck farms typically pumped water out of the lake to irrigate crops, and flushed water into the lake when fields became saturated. Runoff from citrus groves adjacent to 50% of the lake shore prior to 1985 also contributed pesticides to the lake. There also were reports that Lake Apopka catfish, bass, and pan fish had elevated levels of DDT, DDD, DDE during the early 1960s (Huffstutler et al. 1965), as well as reports of elevated levels of toxaphene in bullhead catfish in the early 1980s (Churchville 1982). In the spring of 1980, a chemical spill occurred at a nearby pesticide manufacturing plant, Tower Chemical Company. At the time of the spill, the chemical identified was a miticide, dicofol [(brand name Kelthane) (*Use of trade, product, or firm names is for descriptive purposes only and does not imply endorsement by the federal government.*)] (US EPA 1980), which was composed of approximately 15% DDT (Clark 1990). A later assessment indicated that the spill included chlorobenzilate and perhaps other compounds, which also used DDT as an active ingredient (see Chap. 6) The spill contaminated a stream that drains into the Gourd Neck area of Lake Apopka (EPA 1980). Dicofol has been associated with adverse effects on reproduction in fish, birds, and mammals (Clark 1990). Similarly, DDT has been associated with acute mortality in reptiles (Herald 1949; George and Stickel 1949), and DDT and its metabolites have been linked to poor reproductive success in birds (see Fry 1995 for review). Whether toxins from the chemical spill entered the lake was uncertain but the decline in the juvenile alligator population was well documented and appeared to be coincident with the Tower Chemical spill (Woodward et al. 1993) (Fig. 5.4). At the time, our opinion was that the levels of DDT, DDD, and DDE did not appear to be sufficient to cause outright death of adult alligators but may have been high enough to cause mortality or developmental problems in embryos. As a result, pesticide toxicity was high on the list of suspects as a primary cause of poor hatching success.

Depressed Egg Viability Was Caused by Mechanical Damage During Collections

Eggs were collected for egg viability investigations when eggs were 0–30 days post-oviposition, transported to artificial incubators, then regularly inspected for viability during incubation. Because alligator embryos are susceptible to damage from trauma (Bock et al. 2004) and changing orientation during incubation (Chabreck 1978), we considered the possibility that handling associated with these activities contributed to increased embryo mortality.

Depressed Egg Viability Resulted from Environmental Causes Before Collection

Alligator embryos are susceptible to asphyxia due to flooding (Joanen et al. 1977) and to lethal cell damage from overheating (Ferguson and Joanen 1983). We speculated that flooding before collection or overheating before collection or during transport might have contributed to mortality. Elsey et al. (1994) reported mortality of embryos in nests constructed with *Sagittaria* spp. (arrowhead, duck potato) due to lethally high (>37°C) temperatures during the first few days of incubation. *Sagittaria* spp. have a greater water content and tend to degrade more quickly than other species of vegetation alligators used for nesting. We hypothesized that certain nesting media found on Lake Apopka, commonly known as flags [e.g. *Sagittaria* spp., *Arum* spp. (wild taro), and *Peltandra virginica* (arrow arum)], might either generate lethally high temperatures as part of an accelerated early composting process or provide less insulation from mid-day heat due to the tendency of their leaves to wilt and compress when dry.

Depressed Egg Viability Was Caused by Senescence of an Aging Adult Female Population

Ferguson (1985) suggested that older female alligators might be expected to produce smaller clutches with larger eggs, which would have lower fertility and ovum quality than those produced by middle-aged females. We observed unusually large females attending nests and pods on Lake Apopka, and speculated that they might be older animals with an increased probability of reproductive senescence.

Depressed Egg Viability Was Caused by Density-Related Stress

Stress hormones have been hypothesized to have an adverse effect on the female sex hormone, estradiol, which is essential for normal reproductive function in alligators (see Lance et al. 2001b; Rooney and Guillette 2001 for reviews). Adult female alligators kept in high-density situations on farms have been found to have elevated levels of the stress hormone, corticosterone (Elsey et al. 1990), which can alter reproductive hormone production and function. Prior to 1982, adult alligator densities on Lake Apopka were among the highest recorded in Florida (Wood et al. 1985). We thought that because of high adult alligator densities and relatively limited nesting habitat, adult female alligators might be exhibiting stress-related reproductive dysfunction.

Depressed Egg Viability Was Caused by Nutritional Diseases

Experiments on alligator farms during the 1970s and 1980s by Joanen and McNease (1987) suggested that vitamin supplements improved the reproductive success of alligators, particularly those maintained on fish diets. Investigations by Lance et al. (1983) found a positive relationship between vitamin E and hatch rates of alligator eggs on alligator farms, and suggested that lower vitamin E levels were related to a diet of vitamin-deficient fish. In related studies, certain fatty acids appeared to be associated with reproductive success and failure of alligators (Noble et al. 1993). Although we considered it was unlikely that alligators on a lake with a prolific prey base, such as Lake Apopka, would have nutritional problems, we thought it warranted inquiry.

Forensics: Phase 1 (1984–1992)

In 1984, we began several ancillary investigations to examine toxic chemicals in alligators. Investigations were expanded in 1988 to examine patterns in egg viability in other alligator populations, seek operational or ecological explanations for poor egg viability, and expand the expertise of the group of cooperating scientists. Clutches were collected from six study areas, Lakes Apopka, Griffin, Jesup, George, and Okeechobee, and Conservation Areas 2 and 3 (two components of the Greater Everglades Ecosystem) (Percival et al. 1992). Eggs were handled and transported carefully, temperatures during transportation were maintained within tolerance limits for alligator embryos, and all eggs were incubated under standard conditions (temperature, humidity, and nest material), so that any differences in hatch rates could be attributed to conditions at the study area or to alligator populations at those sites. We initially found that egg viability varied among areas, with Lake Apopka having the least viability and Conservation Areas 2 and 3 having the greatest (Percival et al. 1992; Masson 1995). Egg viability also varied by year within areas, and, interestingly, the viability rate improved on Lake Apopka , increasing from 3.9% in 1988 to 28.6% in 1991 (Masson 1995). The early hypotheses outlined above were, therefore, evaluated in light of these new findings.

Depressed Egg Viability Was Caused by Toxicosis from Environmental Contaminants

In 1984, we conducted a preliminary study to investigate whether concentrations of OCPs, polychlorinated biphenyls (PCB), and metals were greater in Lake Apopka than other areas. We collected eggs from a sample of clutches, and provided them to colleagues from the Patuxent Wildlife Research Center of the US Fish and Wildlife Service (now part of the US Geological Survey) for contaminant analyses. Relatively low levels of PCB (<0.17 mg/kg) were detected, and the more toxic metals were either not detected or found at relatively low levels (Heinz et al. 1991). We therefore, concluded that PCBs and metals were unlikely to be associated with poor egg viability. However, OCP concentrations, particularly those of DDE and toxaphene, in eggs from Lake Apopka were uniformly greater than eggs from Lakes

Griffin and Okeechobee (Heinz et al. 1991). We sampled two randomly selected eggs from each clutch to assess variation in OCPs within clutches. We found close agreement in OCP concentrations between the eggs (Heinz et al. 1991), which is consistent with the crocodilian pattern of developing and ovulating the entire clutch simultaneously (Palmer and Guillette (1992). Thereafter, we assumed low variation of contaminant concentrations among eggs within a clutch. We collected single eggs from a larger sample of clutches in 1985 to focus on assessing OCP concentrations. Mean level of DDE in Lake Apopka eggs was approximately 5 mg/kg, which was an order of magnitude higher than levels in Lakes Griffin and Okeechobee. How could this affect alligator reproductive success? Female alligators have shell glands in the posterior uterus that deposit a calcareous layer of shell, similar to that of birds (Palmer and Guillette 1992). DDE was associated with eggshell thinning and failure of brown pelican eggs to hatch (Blus et al. 1974), but Heinz et al. (1991) found no difference in eggshell thickness between eggs from Lakes Apopka and Griffin. It was likely that these pesticides originated from multiple sources, including muck farms on the north shore of the lake, citrus groves on the elevated lands surrounding the lake, and the Tower Chemical spill. The proportional contribution of OCPs to Lake Apopka from the various sources remains uncertain. Conrow et al. (see Chap. 6) provide further discussion on potential sources of OCPs in Lake Apopka. We, therefore, determined that this hypothesis warranted further evaluation.

Depressed Egg Viability Was Caused by Mechanical Damage During Collections

We employed careful handling procedures during collection, transportation, inspection, and incubation of alligator eggs to reduce the risk of embryo damage. When eggs were carefully handled, experiments indicated that mechanical damage was minimal (Woodward et al. 1989, 1993). We also found that much of the embryo mortality on Florida lakes was occurring prior to collection and artificial incubation (i.e., days 0–30 days of development) (Fig. 5.7). Early embryonic mortality was confirmed by more intensive investigation by Masson (1995), who found that 53% of mortality occurred before day 10 of incubation.

Alligator eggs with live embryos develop an opaque zone shortly after they are laid, which eventually appears as a contrasting band around the widest part of the egg (Ferguson 1982). Lack of a band was, at one time, considered an indication of an unfertilized egg. However, Masson (1995) found dead zygotes in eggs with no evidence of opaque banding on eggshells in a substantial proportion of eggs, indicating embryos probably died before oviposition. We also observed variability in egg viability among lakes, even though all clutches were collected and handled consistently, indicating a lake effect (Woodward et al. 1993; Masson 1995). Although causes of early embryo mortality required further evaluations, the issue of mechanical damage during collection did not appear to be a significant problem.

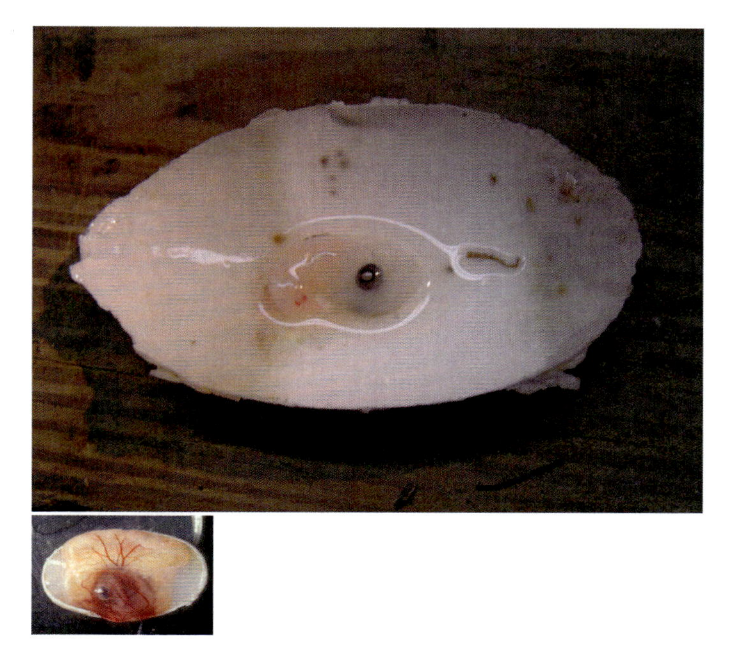

Fig. 5.7 Live (*bottom*) and dead (*top*) alligator embryos

Depressed Egg Viability Was Caused by Environmental Damage Prior to Collection

We used only clutches that had clearly not been flooded at any time prior to collection to eliminate flooding as a factor. We did this by monitoring water level changes on the lake and inspecting eggs for stains resulting from water contact. We measured nest temperatures by inserting a thermometer probe into the clutch cavity prior to uncovering the nest, but did not see any discernible patterns. Further, we found no association of egg viability with either nest temperatures or nest medium, suggesting that these environmental factors were not a significant problem for embryos (Percival et al. 1992; Masson 1995).

Depressed Egg Viability Was Caused by Senescence of the Adult Female Population

Several researchers have found evidence of a positive relationship between female body size and clutch size (Wilkinson 1983; Hall 1991; Rice 1996), suggesting that larger (and presumably older) females tend to produce more eggs. We observed one of the highest mean clutch sizes reported for alligators (45–46 eggs) on Lake Apopka

(Woodward et al. 1993; Masson 1995). The consistently elevated clutch size on Lake Apopka over multiple years (Woodward et al. 1993; Masson 1995; Rice 1996) suggests that the majority of females there tended to be larger than females on other lakes but that they were not senescent (Ferguson 1985). Large clutch sizes on Lake Apopka and several other lakes may also be the result of nutritional, physiological, or genetic variation. Also, older female alligators are more likely to have accumulated greater concentrations of OCPs, which might increase the risk of reproductive impairment. Thus, at the time, we could only speculate on what factors caused female size-related differences in egg viability.

Depressed Egg Viability Was Caused by Density-Related Stress

The average density of adult alligators observed along survey routes on Lake Apopka during 1980–1989 was 3.2 individuals/km, and showed some evidence of a decline during the mid-1980s. This density was less than that on most mesotrophic or eutrophic lakes (Woodward and Moore 1990; Woodward et al. 1993), and similar to or lower than that of Florida lakes that are now known to have high egg viability (Orange, Lochloosa, and Woodruff). Similarly, nesting densities were lower on Lake Apopka relative to Lakes Griffin and Jesup, both of which had higher egg viability during the 1980s (Jennings et al. 1988). Therefore, there was little evidence to suggest that high adult alligator densities and associated density-related stress were responsible for the low egg viability. Unfortunately, we did not monitor corticosterone and estrogen levels at that time to determine whether altered hormone levels were associated with depressed egg viability (Lance et al. 2001b).

Depressed Egg Viability Was Caused by Poor Nutrition

Evaluation of alligator nutrition during this phase of our forensics investigation was minimal. Adult female alligators appeared to be in good to excellent physical condition (A. Woodward, unpubl. data), perhaps to the point of obesity. Alligators on Lakes Apopka, George, Griffin, Jesup, and Okeechobee had greater clutch sizes and lower egg viability than did alligators on the Water Conservation Areas (Masson 1995). Given the lack of data, it appeared that additional evaluation of nutritional status and issues was warranted, and that other areas with high egg viability needed to be sampled for comparison with low egg viability areas.

Addition of Collaborators and Critical Contributions in Understanding Lake Apopka's Abnormal Alligators

A major step in our understanding of mechanisms that might contribute to depressed alligator egg viability on Lake Apopka was the collaboration with Dr. Louis Guillette and his lab at the Department of Zoology, University of Florida. In 1988, the FWC contracted with Dr. Guillette to lead an adult female alligator reproductive biology study. As part of that study, his lab described the morphology and function of the female alligator reproductive tract (Palmer and Guillette 1992) and the reproductive cycle of alligators in Florida (Guillette et al. 1997). Through this work, the Guillette lab began to look at reproductive hormones in alligators.

In 1992, the FCFWRU began to collaborate with Dr. Timothy Gross, also of the University of Florida. Dr. Gross had been working with researchers from the Archie Carr Sea Turtle Center to evaluate whether the gender of newly hatched sea turtles could be determined by measuring the ratio of estradiol:testosterone in the chorioallantoic fluid collected at the time of hatching. Because sea turtles were endangered, surrogate species were used to develop and validate the methods. The surrogate species included both alligators and the Florida redbelly turtle, *Pseudemys nelsoni,* from Lake Apopka and other central Florida lakes. It was through these method development studies that irregular sex hormone patterns, ratios, and gonadal tissue anomalies were first observed in neonate turtles and alligators from Lake Apopka.

Collectively, these efforts led to investigations by Gross et al. (1994) and Guillette et al. (1994) on reproductive system development of neonate and juvenile alligators on Lake Apopka and a relatively non-contaminated aquatic habitat in Florida, Lake Woodruff National Wildlife Refuge (Lake Woodruff). Field inspections of alligator clutches during previous work on Lake Woodruff indicated higher egg viability (Woodward et al. 1992) relative to Lake Apopka and other previously studied lakes (Woodward et al. 1993; Masson 1995). Initial comparative studies of clutch viability during 1994–1995 found a mean egg viability rate of 77% (A. Woodward, unpublished data). We then decided to use Lake Woodruff as a reference site for future comparisons. Studies by the Guillette and Gross labs provided intriguing findings, which led to additional assessments and continued interest in the role of contaminants in the alligator population decline on Lake Apopka. These scientists compared sex hormone levels, sex organ development, and phalli dimensions and found Lake Apopka alligators had poorer differentiation in these characteristics between the sexes than did alligators on Lake Woodruff. Further work by Guillette et al. (1999) on seven Florida lakes revealed that juvenile alligators on Lake Apopka had less differentiated sex steroid concentrations and smaller male phallus size than most other areas, but not all. It was apparent that we did not clearly understand what "normal" sex steroid concentrations were for juvenile alligators in Florida, but concentrations for Lake Apopka alligators were usually anomalous.

Forensics: Phase 2 (1993–2008)

During the 1990s, our consortium continued to examine the hypothesis that OCPs were primarily responsible for reproductive failure in Lake Apopka alligators. From 1994–2000, the Guillette lab, with support from National Institute of Environmental Health and Science (NIEHS) and US EPA grants, continued to examine endocrine disruption effects of OCPs (Matter et al. 1998; Guillette et al. 2000). From 1995–2005, Dr. Gross was able to secure two consecutive grants from the NIEHS to evaluate alligator egg viability and female alligator OCP levels and conducted field studies in collaboration with the FCFWRU (Giroux 1998; Rotstein et al. 2002).

Lake Apopka's alligator population increased steadily during the early 1990s (Rice 1996) (Fig. 5.4) and coincided with an increase in egg viability (Fig. 5.5) and a decrease in completely failed clutches (Rice 1996; Fujisaki et al. 2007). This prompted us to question whether this population recovery was the result of a new generation of less reproductively impaired female alligators gradually replacing older, impaired females (Fujisaki et al. 2007).

Until 1997, Lake Apopka had been the focal point for alligator anomalies. Then, an alligator die-off (Schoeb et al. 2002) and an associated drop in egg viability were documented on Lake Griffin (Cardeilhac et al. 1998). The event provided further insight into the possible causes of poor reproductive success of alligators. Lake Griffin is a large eutrophic lake approximately 35 km downstream from Lake Apopka. Like Lake Apopka, Lake Griffin historically had a large adjacent marsh that was converted to muck farms during the 1950s, with similar fertilizer and pesticide application practices as the Lake Apopka farms. However, it was not farmed for as many years as the Lake Apopka farms, and pesticide concentrations in alligator eggs sampled from Lake Griffin were an order of magnitude lower than those found in Lake Apopka alligator eggs during 1984–1985 (Heinz et al. 1991). During 1997–2004, at least 450 large (>1.2 m) dead alligators were documented on Lake Griffin (Honeyfield et al. 2008) (Fig. 5.8). During that same period and egg viability declined to 4% in 1997 (Cardeilhac et al. 1998) from approximately 60% during

Fig. 5.8 Dead, adult alligator found on Lake Griffin, Florida in 2001

1983–1986 (Woodward et al. 1993) and 43% during 1988–1991 (Masson 1995). This event had several similarities to the one we observed on Lake Apopka in the early 1980s, and this time, we were better able to document changes in alligator mortality and egg viability. We were unable to conduct a complete epidemiological investigation of the Lake Griffin alligator die-off, but we were able to assemble a team of scientists, with veterinary pathologists assisting in the investigation. Throughout the mortality event, we conducted weekly surveys on Lake Griffin (Honeyfield et al. 2008) and monthly surveys on Lakes Woodruff and Apopka to document mortality rates and trends. We collected moribund alligators from Lake Griffin with signs of lethargy and discoordination, and we assumed these alligators were afflicted with a disease that was at least partially responsible for the observed mortality. Moribund alligators from Lake Griffin and reference alligators from Lake Woodruff were evaluated using a series of neurological tests and postmortem examinations. The pathology team concluded Lake Griffin alligators suffered from peripheral neuropathy, and some showed a lesion in the mid-brain that was thought to be responsible for their lethargy and discoordination (Schoeb et al. 2002).

At about the same time that mortality increased and alligator egg viability plummeted on Lake Griffin, reports surfaced about a newly appearing species of cyanobacterium (blue-green algae), *Cylindrospermopsis raciborskii,* which had become the dominant bloom on Lake Griffin (Chapman and Schelske 1997). *C. raciborskii* is a species of nitrogen-fixing cyanobacterium that can produce neurotoxic and hepatotoxic metabolites, and is known to cause sickness and death of large vertebrates (Carmichael 1994). The coincidence of the appearance of this blue-green algae and the mortality event was very compelling, and led to further investigations.

As investigations progressed, a colleague of ours learned, through a serendipitous discussion with Dr. Dale Honeyfield (USGS) at a meeting, that Honeyfield and colleagues were studying a disease in young salmonines with somewhat similar characteristics to the embryonic mortality in Lake Apopka alligators. Their working hypothesis was salmon in the Great Lakes were eating prey fish with high thiaminase levels, which depleted thiamine levels in salmonines, resulting in developmental failure and death of embryos and fry (see Fitzsimons et al. 1999 for review; Czesny et al. 2009). Subsequent studies indicated consumption of a clupeid, the alewife (*Laos pseudoharengus*), by adult salmonines in the Great Lakes caused thiamine deficiencies, which resulted in adult mortality (Brown et al. 2005), mortality at early stages of embryo and fry development (Czesny et al. 2009), and poor developmental success of young fish (Fisher et al. 1996). Under experimental conditions, thiamine therapy reversed these conditions in salmon (Fisher et al. 1996; Ketola et al. 2000). Gizzard shad (*Dorosoma cepedianum*) have been a predominant component of the fish populations on eutrophic lakes in Florida and are also in the clupeid family. Honeyfield suggested that this might be a productive course of investigation for the Lake Griffin alligator team.

By comparing the mortality events on Lake Griffin and Lake Apopka and assessing the findings from previous investigations, we were able to reassess previous hypotheses and develop more focused hypotheses about depressed alligator egg viability and alligator mortality on eutrophic lakes in Florida.

Depressed Egg Viability Was Caused by Senescence of the Adult Female Population

Rice (1996) revisited the hypothesis that egg viability problems on Lake Apopka were linked to an aging female alligator population. He did not find a difference in egg viability between maternal females of similar sizes on Lakes Apopka and Woodruff during 1994–1995. However, he found some evidence of a difference in size distribution between Lake Apopka and Lake Woodruff, suggesting that size might influence the overall egg viability on these lakes. Predicted size distributions of maternal alligators on Lake Apopka, based on clutch characteristics, indicated a possible change in size distribution between 1986–1987 (very low egg viability) and 1994–1995 (moderately low egg viability) (Rice 1996). Although egg viability had increased on Lake Apopka from approximately 7% during 1988–1989 (Masson 1995) to 48% in 1994–1995, egg viability was still depressed relative to the 80% found in the reference area, Lake Woodruff. Rice (1996) also found nesting female alligators were larger in total length and weight on Lake Apopka than on Lake Woodruff, and he found egg viability was greater for medium-sized females than for the largest females. Rice (1996) was able to use a predictive model to reconstruct a size distribution of nesting female alligators from their clutch characteristics, and found that there was little difference in the mean size of nesting female alligators on Lake Apopka between the 1988–1989 and 1994–1995 periods. However, he noted that there were differences in the size distribution, with nesting females on Lake Apopka being more likely to be in the prime egg viability size range during 1994–1995. He also found that Lake Apopka had a greater percentage of larger females than Lake Woodruff in 1994–1995, which confounded the interpretation of size distribution data.

Mortality and Reproductive Dysfunction of Adult Alligators on Hypereutrophic Florida Lakes Was Caused by Transmissible Diseases

Initial investigations by pathologists, found adult alligators on Lake Griffin were generally in healthy condition and showed no remarkable evidence of infectious diseases (Schoeb et al. 2002; Honeyfield et al. 2008). Subsequent postmortem examinations of alligators revealed no association of morbidity with parasitic, bacterial, fungal, or viral diseases. No mycoplasmas, an organism attributed to causing a major mortality event of adult alligators at an alligator exhibit in Florida (Brown et al. 2001), were isolated from moribund alligator tissues. *Aeromononas hydrophila* was cultured from several individuals from each of Lake Griffin and Lake Woodruff but this potential pathogen was considered to be at normal background levels and not associated with alligator mortality or morbidity. Although *Aeromonas* sp. had been implicated as a secondary cause of mortality in the 1971 Lake Apopka mortality event (Shotts et al. 1972), *Aeromonas sp.* are common bacteria in freshwater lakes in

Florida and would be expected to occur in cultures. Neither reproductive tracts nor other organs associated with reproduction (brain and liver) showed signs of transmissible diseases (Schoeb et al. 2002; Honeyfield et al. 2008). Therefore, we concluded transmissible diseases were not causing the observed elevated mortality and depressed egg viability on Lake Griffin, and probably not on Lake Apopka.

Toxicity from OCPs Caused Direct Mortality of Alligators and Alligator Embryos

Egg dosing experiments by the Guillette lab found that topically applied *p,p*'-DDE on eggs from a reference area (Lake Woodruff), produced sex reversal in a significant proportion of embryos (Matter et al. 1998; Milnes et al. 2005). However, Milnes et al. (2005) did not find elevated embryo mortality on DDE-dosed eggs. This suggested that direct toxicity of DDE might not be causing mortality of embryos.

From 1994 through 2002, we trapped maternal alligators at nests to examine the link between contaminant levels of female alligators and egg viability. We took biopsies of abdominal adipose tissue from maternal alligators and yolk samples from one egg from each clutch to assess concentrations of DDT and its metabolites. Egg clutches were collected shortly after oviposition (usually 1–20 days) to reduce exposure of eggs to predation, flooding, and overheating. Eggs were then incubated in commercially obtained sphagnum moss under standard temperature and humidity conditions to standardize those sources of variation. We found high levels of DDE in adipose tissue of maternal alligators on Lake Apopka (48.6 mg/kg) relative to maternal alligators on Lake Woodruff (4.6 mg/kg) (Giroux 1998). We also found a positive correlation between concentrations of DDE in maternal alligators and their eggs on Lake Apopka. Giroux (1998) noted that eggs from nests on the north side of the lake, adjacent to the north shore farms, had lower hatch rates than eggs from the southern part of the lake but he did not find a significant difference in DDE levels between the two areas. Although this investigation confirmed maternal alligators carried high concentrations of DDE, the association of egg viability with DDE was unclear.

Adult Lake Griffin alligators had elevated levels of DDE (Schoeb et al. 2002). However, the mean level of DDE in adipose tissue of Lake Griffin alligators was 0.9 mg/kg (Honeyfield et al. 2008), which was substantially less than the 39.3–61.7 mg/kg found in female alligators on Lake Apopka during 1994–1995 (Giroux 1998) and 29.8 mg/kg found during 2001–2002 (Rauschenberger et al. 2004a). Toxaphene was elevated in adipose tissue of Lake Griffin alligators (2.1 mg/kg) as it was in Lake Apopka alligators during 2001–2002 (13.4 mg/kg). Our pathology teams concluded it was unlikely the relatively low concentrations of OCPs in tissues of Lake Griffin alligators were responsible for observed neuropathy and encephalopathy (Schoeb et al. 2002; Honeyfield et al. 2008).

Pesticide levels in Lake Griffin alligator eggs were an order of magnitude less than concentrations found in Lake Apopka eggs, in both 1984–1985 (Heinz et al.

Table 5.1 Mean concentrations (mg/kg) of major organochlorine pesticides in alligator eggs from Lake Apopka, Florida during 1985–2002

Years	*n*	*p,p'*-DDE	*p,p'*-DDD	*p,p'*-DDT	Toxaphene	Tissue	References
1985	23	3.5	0.4	<0.1	2.4	Whole eggs	Heinz et al. (1991)
1994–1995	29	4.1	n.a.	0.1		Egg yolks	Giroux (1998)
2000–2002	23	5.8	0.04	<0.1	2.7	Egg yolks	Rauschenberger et al. (2007)

1991) and in 2000 (Sepúlveda et al. 2004). However, Lake Griffin eggs showed a high rate of early embryonic mortality, similar to Lake Apopka eggs (A. Woodward, unpublished data), some of which died prior to oviposition (Rotstein et al. 2002). Sepúlveda et al. (2004) did not find a significant relationship between overall hatch rates of alligator eggs and pesticide levels on several Florida aquatic habitats, including Lake Apopka, but their sample size was relatively small ($n=20$ for all areas). However, this was the first published work that provided evidence that pesticides might not be the driving factor of depressed egg viability. An unpublished study done by Richey and Woodward in 2000 found weak evidence of a threshold effect, with egg viability declining as concentrations of DDE in egg yolk exceeded 5 mg/kg. Sepúlveda et al. (2006) also found an increased incidence in lesions in neonate alligators from clutches on Lake Griffin with high OCP levels. From these findings, OCP involvement remained a plausible causative agent in poor reproductive success, but it appeared that other drivers were likely involved.

It is particularly interesting to compare OCP pesticide loads in alligator eggs over time. Concentrations of *p,p'*-DDE, a metabolite comprising the greatest proportion of DDT metabolites, has remained relatively high in alligator eggs from when we first measured them in 1984–1985 to the most recent survey in 2000–2002 (Table 5.1). Yet, egg viability rates have fluctuated significantly over that time period and have shown signs of a "recovery" since 1988 (Fig. 5.5).

To complement field investigations, the Gross lab (Muller et al. 2007a, b) evaluated exposure and effects of topical egg treatments with a OCP-vehicle solution and effects of direct injection of OCPs via hypodermic needle to compare effectiveness of exposure methods, examine effects of direct exposure, and determine whether direct egg exposure to OCPs could elicit effects. The injection controls caused excessive mortality to the point that no comparisons could be made and suggested injection was not a viable technique. Topical application of OCPs in a vehicle did not result in vehicle-related mortality but did show that only a very small percentage of the applied OCPs actually crossed into the egg. Indeed, these efforts verified that transfer of pesticides from nest material into eggs was an unlikely route of exposure of alligator embryos to OCPs. Rather, embryonic exposure to OCPs is most likely from maternal transfer, as suggested by Rauschenberger et al. (2004a).

The Gross lab also examined the morphometry and histopathology of embryos from wild clutches with good clutch viability and low OCP egg concentrations, and compared them to embryos from wild clutches with poor clutch viability and high OCP egg concentrations (Rauschenberger 2004). Results indicated that embryo mortality occurring in alligator populations inhabiting reference and OCP-contaminated sites was characterized by developmental retardation, without gross deformities or overt presence of lesions to vital organs. The lack of lesions suggested that direct toxicity to vital organs was an unlikely cause.

However, variation in embryo morphology appeared to be associated with variation in OCP burdens of eggs, and the percentage composition of OCP analytes was equally as important as their concentration, suggesting the importance of mixtures. For example, head lengths of live embryos sampled on incubation Day 14, 33, and 43 were negatively correlated with the proportion of chlordane in the OCP mix measured in corresponding egg yolks (Rauschenberger 2004).

Indirect Toxicity from OCPs Causes Reproductive Dysfunction in Maternal Alligators, Which Results in Poor Egg Quality and Embryo Survival

Guillette and Edwards (2008), and Milnes and Guillette (2008) have proposed that OCP burdens in maternal alligators may affect reproductive success by disrupting normal endocrine function. Guillette et al. (2000) also hypothesized that exposure of alligators to contaminants at an early age may disrupt endocrine function, and thereby alter reproductive development and function. This concept has been supported by recent research, which has found OCPs can disrupt endocrine function by altering gene function (Kohno et al. 2008). We suspect that endocrine disruption, either at the early stages of development, or in adult alligators, may result in depressed egg viability.

To further examine the indirect OCP hypothesis, the Gross lab conducted field studies and a novel laboratory experiment with a captive population of adult alligators. Rauschenberger et al. (2004a) presented experimental evidence of maternal transfer of OCPs from maternal alligators to embryos and concluded that OCP concentrations in eggs were reliable predictors of levels in females. In field studies, Rauschenberger et al. (2007) examined 115 clutches from four areas and found that OCP burdens explained 39% of variation in clutch viability for Lake Apopka, 21% for Lake Griffin clutches, 9% for Emeralda Marsh clutches, and 0% for the reference site (Orange Lake) clutches.

Because there are so many uncontrollable factors in field studies, an OCP dosing experiment using a captive adult alligator population was also conducted (Rauschenberger et al. 2004b; 2007). Successful reproduction is difficult with captive alligators, and the resulting sample of clutches was relatively small (controls $n=9$, treated $n=7$). Nonetheless, results showed that breeding pairs of male and female alligators dosed with an ecologically relevant mixture of OCPs yielded eggs with significantly depressed hatch rates compared to nontreated cohorts. In addition, OCP concentrations in captive eggs were similar to those of wild eggs from Lake Apopka. Specifically, clutch viability for the control group was 35% higher than the treated group, and the treated group had a 42% higher incidence of nonbanded eggs.

Total OCP burdens in yolks from the control group (0.005 ± 0.003 mg/kg) were less than those of the treated group (13.3 ± 2.7 mg/kg), and less than those of wild alligators from the reference site (102.0 ± 15.5 mg/kg). Lipid content of eggs of alligators from the treated group (22% ± 0.7%) was significantly greater than those of eggs of the control group (19% ± 0.7%). Other reports suggest that differences in lipid and fatty acid content of alligator eggs may be associated with higher embryo mortality (Noble et al. 1993; Lance et al. 2001a), and altered lipid metabolism and transport have been shown in catfish (Lal and Singh 1987) and mink (Kakela et al. 2001) exposed to chlorinated hydrocarbons. Because both treated and control groups received similar diets, differences in yolk lipid content suggested that OCP exposure may also have altered lipid metabolism and/or follicular deposition.

Although the results of this lab experiment appear to be quite consistent with observations in wild populations, some key differences should be noted. Differences in effects between the captive and free-ranging alligator studies were that depressed clutch viability in populations from OCP-contaminated sites was primarily due to greater early (post-oviposition) and late embryo mortality, whereas higher rates of nonbanded eggs were the primary reason for lower clutch viability in the captive study (Rauschenberger et al. 2007)

Whether fertilization failure or very early embryo mortality caused the higher incidence of nonbanded eggs in the OCP-treated group was unknown because genetic testing was not conducted on the blastodiscs (Rotstein et al. 2002), and histological examination of the blastodiscs for the presence of zygotes was inconclusive. A factor to consider with possible fertilization failure is that male alligators from the OCP-treated group were also dosed, which suggested the possibility that the incidence of nonbanded eggs may have been related to alteration of male reproductive function (Rauschenberger et al. 2007).

Toxins from Blue-Green Algal Blooms Are Responsible for Adult Alligator Mortality and Poor Egg Viability on Eutrophic Florida Lakes

Lake Griffin experienced extensive blooms of *C. raciborskii* during the mortality event from 1997–2003. However, liver tissue and liver enzyme levels of alligators on Lake Griffin showed no evidence of liver disease, which might be caused by the hepatotoxin, cylindrospermopsin (Schoeb et al. 2002). Further, Schoeb et al. (2002) observed that neuropathy observed on Lake Griffin alligators was not consistent with pathology from algal neurotoxins. The validity of the *C. raciborskii* toxin hypothesis was further weakened by Ross (2000), who was not able to find a correlation between alligator deaths and *C. raciborskii* densities during the early part of the mortality event on Lake Griffin.

Lake Apopka has been beleaguered with algal blooms since the mid-1940s (Clugston 1963, Huffstutler et al. 1965) (Fig. 5.9). Blooms have been attributed to a green alga, *Botryococcus brauni* (Huffstutler et al. 1965), and more recently to

Fig. 5.9 Juvenile alligator swimming in algal bloom on Lake Apopka during the mid-1990s (photo by H. Suzuki)

cyanobacteria *Synechococcus sp.*, *Microcystis incera*, and *Lyngbya contorta* (Carrick et al. 1993). *Microcystis* can produce a hepatotoxin (Carmichael 1994) but we have not observed a level of alligator mortality on Lake Apopka since 1983 that would suggest widespread mortality due to algal blooms. Examination of alligator livers and liver enzymes has not been conducted on Lake Apopka alligators, and would be worthwhile, especially considering that the liver produces vitellogenin, which is a precursor protein in egg yolk production (Guillette et al. 1997; Guillette and Milnes 2000). A possible scenario is that algal hepatotoxins could subtly alter liver function and the quality of yolk, which may lead to egg failure. Although this hypothesis remains untested, our best available information does not support that algal toxicosis is directly involved in depressed alligator egg viability.

Adult Mortality and Poor Egg Viability Are Caused by Nutritional Deficiencies

Because nutritional deficiencies were still a plausible cause of the problems with Lake Apopka alligators, the Gross lab conducted preliminary evaluations of egg nutritional status and embryo mortality in alligators (Sepúlveda et al. 2004). The results showed a positive association between thiamine levels and hatching success of alligator eggs for four study areas, Lake Apopka, Lake Griffin, Emeralda Marsh, and Orange Lake. Specifically, thiamine concentrations in egg yolks explained approximately 40% of the variation in clutch viability, suggesting thiamine may play a role in depressed clutch viability. Although only five clutches were examined per site, the results were not unlike findings documented in quail, where mortality was characterized by developmental retardation caused by nutrient deficiencies in eggs, with the deficiencies caused by altering maternal diet (Donaldson and Fites 1970).

As a follow-up to the Sepúlveda study, four complementary studies were conducted by the Gross lab to further examine the association between clutch viability, thiamine levels, and OCPs, as well as other important nutrients and contaminants (Rauschenberger et al. 2009). The first of the studies, a case–control cohort study grouped clutches ($n=20$) by OCP egg yolk concentrations and clutch viability and compared the concentrations of polycyclic aromatic hydrocarbons (PAHs), PCBs, selenium, zinc, and vitamins E and A. Results indicated the non-thiamine, non-OCP contaminants were unlikely causes of subnormal clutch viability, as their levels did not show large differences across sites, nor were they significantly associated with differences among clutch survival parameters. Furthermore, total PCB and PAH burdens were less than those known to elicit adverse effects on avian development (Summer et al. 1996). However, thiamine levels in eggs accounted for approximately 38% of the variation in clutch survival parameters, and lower levels of thiamine were associated with lower clutch viability (Rauschenberger et al. 2009), which is consistent with similar studies involving fish (Fitzsimons et al. 1999).

The second study, an expanded field study, examined 72 clutches. Thiamine levels in this study accounted for only 27% of the variation in clutch survival variables, which is less than the 38% noted in the case–control cohort study. Another finding of the second study was that lipid concentrations in eggs and certain OCPs were negatively associated with thiamine levels, suggesting a potential relationship (Rauschenberger et al. 2009).

The third (thiamine amelioration) and fourth (thiamine deactivation) studies were laboratory egg treatment experiments that attempted to alter the thiamine levels and bioactivity to determine if alterations caused changes in hatch rates. In regard to transfer of thiamine into the egg yolk, thiamine concentrations were increased in the albumin, but not in the yolk compartment. Results showed no differences in hatch rates among the treatment groups of either study. The lack of effects may be attributed to a number of factors from insufficient dose to changes in susceptibility related to development stage. In summation, thiamine egg concentrations explained some of the variation in clutch viability, about 27%, for free-ranging alligators. Cause–effect relationships are yet unclear, but thiamine deficiency remains one of the leading hypotheses, along with effects of OCPs, as both accounted for a comparable amount of variation in clutch viability (Rauschenberger et al. 2007; 2009).

For the dying, adult alligators that were examined from Lake Griffin, the clinical signs were similar to those described for thiamine deficiency in crocodilians (Wallach 1978; Jubb 1992, see Huchzermeyer 2003 for review), and similar neurological lesions were reportedly observed in thiamine-deficient farm-raised alligators [Jubb and Huxtable 1993 (cited by Schoeb et al. 2002)]. Subsequent investigations found low thiamine levels in moribund adult alligators in Lake Griffin compared with Lake Woodruff alligators and seemingly healthy Lake Griffin alligators (Honeyfield et al. 2008). Ross et al. (2009) were able to induce depressed thiamine levels and brain lesions characteristic of thiamine deficiency in a small sample of captive alligators by feeding them a diet of gizzard shad. They were also able to reverse the thiamine deficiency in two alligators through thiamine therapy. These findings added support for the hypothesis that thiamine deficiency was a primary cause for elevated adult mortality on Lake Griffin. Further induction experiments of this kind would be enlightening.

But what could be causing thiamine deficiencies? Thiamine is important for certain metabolic processes in vertebrates that are critical to the development and maintenance of certain organs (McCandless and Schenker 1968; Akerman et al. 1998). Clupeid (herring family) fishes are typically high in thiaminase, an enzyme that breaks down thiamine (Tillitt et al. 2005). Gizzard shad can have much higher levels of thiaminase than alewife (Tillitt et al. 2005), and Ross et al. (2009) confirmed high levels of thiaminase in gizzard shad on both Lakes Apopka and Griffin. Gizzard shad densities have tended to increase as nutrient levels and resulting algal blooms increase. At times over the past 40 years, gizzard shad have comprised 65–90% of the fish biomass in Lake Apopka (Clugston 1963; Huffstutler et al. 1965; Holcomb et al. 1974; Benton et al. 1991; Crumpton and Godwin 1997).

On Lake Griffin, the emergence of gizzard shad as a principal forage fish appears to be more recent (Holcomb et al. 1974). Gizzard shad have comprised a significant proportion of the diet of alligators on many large Florida lakes (Delany and Abercrombie 1986, Delany et al. 1999), including Lake Apopka (Rice et al. 2007). On both Lakes Apopka and Griffin, gizzard shad densities have fluctuated over time, and periodic alligator mortality and hatching failure events may have been related to these fluctuations. One plausible hypothesis is that under certain conditions and with high gizzard shad densities, alligators consumed a critical level of gizzard shad, which resulted in increased mortality and reproductive failure (Ross et al. 2009).

The SJRWMD began a rough fish removal program on Lake Apopka in 1993 (Crumpton and Godwin 1997) and Lake Griffin in 2003 (W. Godwin, pers. comm.), designed to remove phosphorus from these aquatic systems. Between 1993 and 2008, over 6,800 mt of fish were harvested from Lake Apopka (W. Godwin, pers. comm.). This systematic shad harvesting can be expected to reduce shad populations and their occurrence in alligator diets (Rice et al. 2007). Although difficult to test and circumstantial in some respects, the recent decline in alligator mortality, increase in alligator abundance, and improvement in alligator egg viability on Lake Apopka may, in part, be related to shad reduction efforts.

Lower Egg Viability Is a Product of Natural Genetic Variation Among Populations

We have speculated whether the Lake Apopka alligator population has, over time, developed a reproductive strategy that selects for genes that produce numerous, poor quality eggs. However, Lake Apopka is connected to the Ocklawaha River basin by a stream that can be easily traversed to other alligator populations. So, it is likely that regular genetic exchange has occurred between Lake Apopka and other alligator populations. This is supported by genetics work, which found little genetic differentiation among Lakes Apopka, Griffin, Orange, and Woodruff (Davis et al. 2002).

Also, fluctuation in egg viability over time does not support inherited genetic variation as a major factor contributing to the persistent lower egg viability on Lake Apopka. These findings indicate that it is unlikely that major genetic differences at the population level are affecting alligator reproduction.

Summary and Challenges

Despite advances made by the many research groups, agencies and organizations involved over the 30-year investigation, a definitive cause-and-effect relationship for alligator anomalies on Lake Apopka has yet to be determined. Investigations have been able to rule out several prime suspects, leaving OCP toxicity and thiamine deficiency as the primary explanation for poor reproductive success.

Indeed, several studies have demonstrated that OCPs can cause developmental abnormalities and endocrine disruption in alligators (Guillette and Edwards 2008; Milnes and Guillette 2008). In field studies, Rauschenberger et al. (2007) estimated OCP levels in eggs accounted for 40 and 21% variation in clutch viability for Lake Apopka and Lake Griffin, suggesting OCPs are a contributing factor to embryo mortality.

Like OCPs, thiamine deficiency can impair the health of alligators. In field studies examining clutches from Lake Apopka and Lake Griffin (pooled), thiamine levels in eggs accounted for about 27% of the variation in clutch viability (Rauschenberger et al. 2009).

Given our inability to deduct thiamine or OCPs from the equation, there is the possibility that both are necessary to produce severe adult mortality events and depress clutch viability. OCP levels in eggs have not changed appreciably over the past 25 years, and our work has demonstrated a strong association between OCP levels in eggs and in maternal alligators. Hypothetically, elevated OCP levels may cause chronic stress, and when this chronic stress is combined with thiamine deficiency and other environmental stressors, the resiliency of alligators is exceeded such that they die and/or their clutch viability is depressed. One illustration of this potential relationship comes from field studies on central Florida lakes (including Lake Apopka and Lake Griffin) where OCP levels in alligator eggs accounted for 15% of the variation in thiamine levels in eggs (Rauschenberger et al. 2009).

In addition to OCPs, other chemicals, such as nitrates, may disrupt endocrine systems and cause developmental problems in vertebrates. Most watersheds in Florida have been subjected to nutrient pollution in the form of nitrogen and phosphorus compounds. Several studies have found reduced early survival, developmental anomalies, and depressed androgen levels in vertebrates associated with elevated nitrate levels (see Guillette and Edwards 2005 for a review). Therefore, researchers should not ignore the potential effects of these compounds when assessing reproductive and developmental anomalies in alligators.

Further studies to examine the association of thiamine, thiaminase, shad population, and OCP levels with alligator health and egg viability on multiple areas over multiple years would provide valuable insight into the plausibility of these factors as causes of alligator morbidity and depressed egg viability on Lake Apopka. Further, complementing field investigations such as these with controlled laboratory thiaminase dosing experiments to examine its effects on alligator development and health would increase our understanding of cause and effect.

Effects on Policy and the Human Element

Effects on Commercial Alligator Egg Collections and Recreational Adult Harvests

Commercial use of the alligator resource on Lake Apopka has been limited because of the uncertain population viability. Commercial harvest of hatchlings and eggs was allowed during the early research phase in 1981–1986. However, with declining egg viability and alligator population levels, all hatchlings produced from eggs collected on Lake Apopka were returned to the lake after 1986. Since 2008, the population has been considered sufficiently recovered and has a growth rate that would allow egg harvests. However, alligator ranchers are apprehensive about Lake Apopka eggs because of their history of poor hatching success. Because of the lack of interest by farmers and on-going research, commercial egg collections have not resumed as of 2010.

The adult population of alligators has recovered sufficiently to allow a limited recreational harvest. Although hunters and FWC management staff are interested in opening Lake Apopka to adult harvests, FWC has not opened up an adult harvest because of uncertainty about OCP levels in meat and resulting human health concerns. Indeed, a fish consumption advisory recommended against consuming bullhead catfish from Lake Apopka as recently as 2009 (FDOH 2009). The FWC likely will conduct future testing to provide a basis for determining if alligator meat is safe for human consumption.

Training and Developing a Generation of Scientists

The research effort on Lake Apopka's abnormal alligators has been ongoing since 1979 and involved several laboratories, producing a number of graduate theses devoted to the lake's infamous residents. In addition, many more students have been involved by assisting in the vast number of studies. To try to derive a number of graduate and undergraduate students that have benefited from the combined research efforts over the years would be difficult. However, one can gain an appreciation by scanning over the number of publications, their authors, and those listed in the respective acknowledgement sections. Lake Apopka's abnormal alligators have been used as case examples to inspire pre-collegiate students and teachers in environmental education endeavors.

Effects of Lake Apopka's Alligator Investigations on the Public Perception and Policy of Endocrine Disruption

Endocrine disruption became a more widely known public concern during the early 1990s. Because of the abnormalities in the endocrine function and the poor clutch viability of Lake Apopka's alligators, the lake emerged as an ecological illustration of how environmental contaminants may be causing adverse changes to the alligator's endocrine system at levels well below those considered to be acutely toxic. Public awareness grew and, in 1993, Dr. Guillette testified as an expert witness for the Hearing on Pesticides as Environmental Estrogens held by the US Congress Subcommittee on Health and the Environment (H. Waxman (CA), Chairman). Additional findings from Lake Apopka alligators and other studies from around the world were generated, and a book entitled "Our Stolen Future" (Colburn et al. 1996) was published, which popularized endocrine disruption. Soon thereafter, Congress passed the 1996 Food Quality and Protection Act (FQPA) and the Safe Drinking Water Act (SDWA), which mandated that the US Environmental Protection Agency (EPA) develop a screening and testing strategy for endocrine disruptors by August 1998 and implement the plan by August 1999. Although a difficult and complicated task, screening for endocrine disruptors is moving forward in three ways: (1) The Tier 1 battery composed of the validated assays has been finalized and Tier 2 tests are being validated; (2) EPA has issued the first test orders for pesticides to be screened for their potential effects on the endocrine system; and (3) Implementing the policies and procedures used to require testing (US EPA 2009). As an icon for endocrine disruption, Lake Apopka's alligators still have mass media appeal. For example, they were recently mentioned in a New York Times editorial that warned the public to avoid complacency regarding the potential dangers of endocrine disruptors (Kristof 2009).

Conservation Gains

A challenge in trying to relate the present findings to other OCP exposure studies involving birds, mammals, and fish is that the basic metabolic function of an alligator is vastly different from typical mammalian laboratory models. For example, blood flow of a 70 kg alligator (0.26 L/min) is less than 8% of that of a 70 kg human (Coulson and Hernandez 1983). These differences mean that xenobiotics and endogenous compounds circulate throughout the alligator at a decreased rate, which can affect excretion and elimination as well as the amount of time target organs are exposed to these substances. Another factor that may affect OCP toxicity is the temperature of the alligator, as low temperatures have been associated with increased DDT toxicity in exposed fish (Rattner and Heath 2003). Although alligators can live with relatively high tissue concentrations of OCPs, their physiology may render them more susceptible to sublethal effects. Speculatively, low blood flow, seasonally lower body temperatures, and seasonal fasting (resulting in mobilization of lipids and hydrophobic contaminants) may contribute to this susceptibility of alligators to reproductive modulation via OCP exposure.

From the species perspective, the last three decades of alligator research at Lake Apopka and other central Florida lakes have contributed greatly to our understanding of the effects of contaminants on alligators and alligator biology in general. This information is useful in providing a basis with which to prioritize contaminant risks among the other factors affecting recovery efforts for endangered crocodilians and other reptiles worldwide.

From a local restoration perspective, data from research efforts at Lake Apopka provide one context in which to evaluate the health of the lake's ecosystem and the effectiveness of restoration efforts. For example, researchers with the FWC, FCFWRU, and USFWS are currently examining alligator movements and reproductive ecology within the Lake Apopka North Shore Restoration Area. If it is found that alligators have home ranges that are limited to specific flooding blocks, and their OCP concentrations are correlated with OCP sediment concentrations, then it may be possible to use alligators as worst-plausible exposure models for more sensitive species such as endangered wood storks and other piscivorous birds.

What Did We Learn?

First, recognizing when a disease event occurs is critical and requires some understanding of normal conditions. For alligators and other ecological receptors, this means knowledge of normal or baseline population levels, mortality rates, reproductive rates, egg/embryo quality, and physiological parameters with which to identify a departure from normal conditions. Under ideal conditions, ongoing monitoring of basic population and biological indicators of population and community health should be conducted. This would require an elevated level of ongoing financial commitment to monitoring with no guarantee the information would eventually be used.

Second, in addition to having good baseline ecological data, it is important to be able to rapidly mobilize personnel and funding to support coordinated field and laboratory investigations when an event is identified. Expertise should include not only wildlife ecologists but specialists in veterinary pathology, toxicology, and biomedical sciences. Ideally, a dedicated project coordinator should be available to provide a broad perspective on the issue, coordinate and assist with management tasks, and facilitate communication internally among specialists and externally among stakeholders.

Third and specific to determining causes, we underscore the importance of captive animal studies, such as the adult alligator exposure studies summarized above, because field studies may be confounded by unknown and/or uncontrolled factors. However, captive studies are expensive and difficult to execute because of space and maintenance requirements. Further, captive-related stress and disease issues may confound inferences about alligator reproductive performance and general health. In addition, alligators require about 8–10 years to reach reproductive maturity, and may not ovulate every year. Despite such difficulties, captive studies are critical for establishing a cause–effect link and evaluating effects of contaminant exposure.

Fourth, we learned that scientists tend to pursue their areas of interest, and are often limited by the availability of funding and the need to respond to competitive funding sources. So, it is imperative to assemble a team of scientists and veterinary experts with a broad range of interests and expertise to cover as many bases as possible. This may have an added advantage of broadening the range of funding resources.

Final Thoughts

Although scientists and philosophers may, on occasion, rightly state the intrinsic value of our natural resources, it is policy makers and resource managers that make the difficult decisions about funding priorities. Although the average person might have a passing interest in alligators, it is difficult to garner support for research unless there is some tangible human-related interest at stake. In the present case, initial funding was obtained because the interest was an economic one; alligator farmers needed an inexpensive and reliable source of alligator hatchlings for their farms, and poor hatchling success threatened a program aimed to enhance the economic value of alligators to Floridians. As the potential link to pesticide toxicity, endocrine disruption, and nutritional deficiency grew, several other funding sources appeared because of implications to global wildlife conservation and human health. Most recently, studies to monitor OCP accumulation in alligators – as surrogates for other more sensitive or rare species – have been funded by the regional water management district, which is in the process of restoring former wetlands. Lastly, maybe most importantly, the professional connections, collaborations, and in-kind services among researchers and management personnel in participating agencies have been critical in sustaining investigations.

Acknowledgments We wish to thank the many students and collaborators who have participated in the last 30 years of research at Lake Apopka. Although trying to list their names would take more space than allotted here; and undoubtedly, result in leaving someone's name unmentioned, the reader may gain an appreciation of the number of dedicated individuals by looking at the coauthors on the many publications devoted to Lake Apopka's abnormal alligators. Similarly, we wish to recognize the many federal, state, and local agencies and other organizations that have provided support for work during the last three decades through grants and in-kind services. We thank T. O'Meara, J. Berish, B. Crowder, and J. Colvo for providing a critical review of earlier drafts of this chapter. The findings and conclusions in this article do not necessarily represent the views of the US Fish and Wildlife Service, the St. Johns River Water Management District, or the University of Florida.

References

Akerman G, Tjarnlund U, Noaksson E, Balk L (1998) Studies with oxythiamine to mimic reproduction disorders among fish early life stages. Mar Environ Res 46:493–497

Allen RE, Neill WT (1949) Increasing abundance of the alligator in the eastern portion of its range. Herpetologica 5:109–112

Benton J, Douglas D, Prevatt L (1991) Lake Apopka fisheries studies. Wallop-Breaux Project F-30-18 Completion Report, Tallahassee

Blus LJ, Neely BS Jr, Belisle AA, Prouty RM (1974) Organochlorine residues in brown pelican eggs: relation to reproductive success. Environ Pollut 7:81–91

Bock JL, Woodward AR, Linda SB, Percival HF, Carbonneau DA (2004) Hatching success of American alligator eggs when subjected to simulated collection trauma. Proc Annu Conf Southeast Assoc Fish Wildl Agencies 58:323–335

Brezonik PL, Shannon EE (1971) Trophic state of lakes in north central Florida. Publication No 13. Water Resources Research Center, University of Florida, Gainesville

Brown DR, Nogueira MF, Schoeb TR, Vliet KA, Bennett RA, Pye GW, Jacobson ER (2001) Pathology of experimental mycoplasmosis in American alligators. J Wildl Dis 37:671–679

Brown SB, Honeyfield DC, Hnath JG, Wolgamood M, Marcquenski SV, Fitzsimons JD, Tillitt DE (2005) Thiamine status in adult salmonines in the Great Lakes. J Aquat Anim Health 17:59–64

Canfield DE, Bachmann RW, Hoyer MV (2000) A management alternative for Lake Apopka. Lake Reserv Manage 16:205–221

Cardeilhac PT, Winternitz DL, Barnett JD, Foster KO, Froehlich E, Ashley JD (1998) Declining reproductive potential of the alligator population on Lake Griffin in central Florida. Proc Int Assoc Aquat Anim Med 29:30–37

Carmichael WW (1994) The toxins of cyanobacteria. Sci Am 270:78–86

Carrick JJ, Aldridge FJ, Schelske CL (1993) Wind influences phytoplankton biomass and composition in a shallow, productive lake. Limnol Oceanogr 38:1179–1192

Chabreck RH (1978) Collection of American alligator eggs for artificial incubation. Wildl Soc Bull 6:253–256

Chapman AD, Schelske CL (1997) Recent appearance of Cylindrospermopsis (Cyanobacteria) in five hypereutrophic Florida lakes. J Phycol 33:191–195

Churchville V (1982) Toxaphene traces show up in Lake Apopka catfish. Orlando Sentinel 29 October, Section A1

Clark DR Jr (1990) Docofol (Kelthane) as an environmental contaminant. US Fish and Wildlife Service, Washington, Technical Report 29

Clugston JP (1963) Lake Apopka, Florida, a changing lake and its vegetation. Q J Fla Acad Sci 26:168–174

Colburn T, Dumanoski D, Myers JP (1996) Our stolen future: are we threatening our own fertility, intelligence, and survival? – a scientific detective story. Dutton Penguin Books, New York

Coulson RA, Hernandez T (1983) Alligator metabolism: studies on chemical reactions in vivo. Pergamon, Oxford

Crumpton JE, Godwin WF (1997) Rough fish harvesting in Lake Apopka: summary report, 1993–97. Special Publication SJ97-SP23. St Johns River Water Management District, Palatka

Czesny S, Dettmers JM, Rinchard J, Dabrowski K (2009) Linking egg thiamine and fatty acid concentrations of Lake Michigan lake trout with early life stage mortality. J Aquat Anim Health 21:262–271

Davis LM, Glenn TC, Strickland DC, Guillette LJ Jr, Elsey RM, Rhodes WE, Dessauer HC, Sawyer RH (2002) Microsatellite DNA analyses support an east-west phylogeographic split of American alligator populations. J Exp Zool 294:352–372

Del Sepúlveda MS, Peiro F, Wiebe JJ, Rauschenberger HR, Gross TS (2006) Necropsy findings in American alligator late-stage embryos hatchlings from northcentral Florida lakes contaminated with organochlorine pesticides. J Wildl Dis 42:56–73

Delany MF, Abercrombie CL (1986) American alligator food habits in northcentral Florida. J Wildl Manage 50:348–353

Delany MF, Linda SB, Moore CT (1999) Diet and condition of American alligators in 4 Florida lakes. Proc Annu Conf Southeast Assoc Fish Wildl Agencies 53:375–389

Donaldson WE, Fites BL (1970) Embryo mortality in quail induced by cyclopropene fatty acids: reduction by maternal diets high in unsaturated fatty acids. J Nutr 100:605–610

Eisler R, Jacknow J (1985) Toxaphene hazards to fish, wildlife, and invertebrates: a synoptic review. US Fish and Wildlife Service, Washington, Biological Report 85 (1.4)

Elsey RM, Joanen T, McNease L, Lance V (1990) Stress and plasma corticosterone levels in the American alligator – relationships with stocking density and nesting success. Comp Biochem Physiol 95A:55–63

Elsey RM, Joanen T, McNease L (1994) Louisiana's alligator research and management program: an update. In: Proceedings of the 12th working meeting Crocodile Specialist Group, IUCN – The World Conservation Union, vol 1. Gland, Switzerland

Evert JD (1999) Relationships of alligator (*Alligator mississippiensis*) population density to environmental factors in Florida lakes. Thesis, University of Florida, Gainesville

FDOH (Florida Department of Health) (2009) Special fish consumption advisories. http://www.doh.state.fl.us/Environment/community/fishconsumptionadvisories/FWDPS.htm. Accessed 20 Oct 2009

Ferguson MWJ (1982) The structure and composition of the eggshell and embryonic membranes of *Alligator mississippiensis*. Trans Zool Soc Lond 36:99–152

Ferguson MWJ (1985) Reproductive biology and embryology of the crocodilians. In: Gans C, Billet F, Maderson P (eds) Biology of the reptilia, vol 14. Wiley, New York

Ferguson MWJ, Joanen T (1983) Temperature-dependent sex determination in *Alligator mississippiensis*. J Zool Soc Lond 200:143–177

Fisher JP, Fitzsimons JD, Combs GF Jr, Spitsbergen JM (1996) Naturally occurring thiamine deficiency causing reproductive failure in Finger Lakes Atlantic salmon and Great Lakes lake trout. Trans Am Fish Soc 125:167–178

Fitzsimons JD, Brown SB, Honeyfield DC, Hnath JG (1999) A review of early mortality syndrome (EMS) in great lakes salmonids: relationship with thiamine deficiency. Ambio 28:9–14

Fry DM (1995) Reproductive effects in birds exposed to pesticides and industrial chemicals. Environ Health Perspect 103(suppl 7):165–171

Fujisaki I, Rice KG, Woodward AR, Percival HF (2007) Generational effects of habitat degradation on alligator reproduction. J Wildl Manage 71:2284–2289

George JL, Stickel WH (1949) Wildlife effects of DDT dust used for tick control on a Texas prairie. Am Midl Nat 42:228–237

Giroux DJ (1998) Lake Apopka revisited: a correlational analysis of nesting anomalies and DDT contaminants. Thesis, University of Florida, Gainesville

Gross TS, Guillette LJ, Percival HF, Masson GR, Matter JM, Woodward AR (1994) Contaminant-induced reproductive anomalies in Florida. Comp Pathol Bull 26:1–8

Guillette LJ Jr, Edwards TM (2005) Is nitrate an ecologically relevant endocrine disrupter in vertebrates? Integr Comp Biol 45:19–27

Guillette LJ Jr, Edwards TM (2008) Environmental influences on fertility: can we learn lessons from studies of wildlife? Fertil Steril 89:e21–e24

Guillette LJ Jr, Milnes MR (2000) Recent observations on the reproductive physiology and toxicology of crocodilians. In: Grigg GC, Seebacher F, Franklin CE (eds) Crocodilian biology evolution. Surrey Beatty & Sons, Chipping Norton, NSW

Guillette LJ Jr, Gross TS, Masson GR, Matter JM, Percival HF, Woodward AR (1994) Developmental abnormalities of the gonad and abnormal sex hormone concentrations in juvenile alligators from contaminated and control lakes in Florida. Environ Health Perspect 102:680–688

Guillette LJ Jr, Woodward AR, Crain DA, Masson GR, Palmer BD, Cox MC, You-Xiang Q, Orlando EF (1997) The reproductive cycle of the female American alligator (*Alligator mississippiensis*). Gen Comp Endocrinol 108:87–101

Guillette LJ Jr, Woodward AR, Crain DA, Pickford DB, Rooney AA, Percival HF (1999) Plasma steroid concentrations and male phallus size in juvenile alligators from seven Florida lakes. Gen Comp Endocrinol 116:356–372

Guillette LJ Jr, Crain DA, Gunderson MP, Kools SAE, Milnes MR, Orlando EE, Rooney AA, Woodward AR (2000) Alligators and endocrine disrupting contaminants: a current perspective. Am Zool 40:438–452

Hall PM (1991) Estimation of nesting female crocodilian size from clutch characteristics: correlates of reproductive mode, and harvest implications. J Herpetol 25:133–141

Heinz GH, Percival HF, Jennings ML (1991) Contaminants in American alligator eggs from lakes Apopka, Griffin, and Okeechobee, Florida. Environ Monit Assess 16:277–285

Herald ES (1949) Effects of DDT-oil solution upon amphibians and reptiles. Herpetologica 5:117–120

Hines TC (1979) Past present status of the alligator in Florida. Proc Annu Conf Southeast Assoc Fish Wildl Agencies 33:224–232

Holcomb DE, Barwick DH, Jenkins J, Young N, Prevatt L (1974) Oklawaha Basin fisheries investigations. First Annual Performance Report, Florida Game Fresh Water Fish Commission, Tallahassee

Honeyfield DC, Ross JP, Carbonneau DA, Terell SP, Woodward AR, Schoeb TR, Percival HF, Hinterkopf JP (2008) Pathology, physiologic parameters, tissue contaminants, tissue thiamine in morbid healthy central Florida adult American alligators (*Alligator mississippiensis*). J Wildl Dis 44:280–294

Hortenstine CC, Forbes RB (1972) Concentrations of nitrogen, phosphorus, potassium, total soluble salts in soil solution samples from fertilized unfertilized soils. J Environ Qual 1:446–449

Huchzermeyer FW (2003) Crocodiles – biology, husbandry, diseases. CABO Publishing, Cambridge, MA

Huffstutler KK, Burgess JE, Glenn BB (1965) Biological, physical, chemical study of Lake Apopka. Report. Florida State Board Health, Jacksonville

Jennings ML, Percival HF, Woodward AR (1988) Evaluation of alligator hatchling egg removal from three Florida lakes. Proc Annu Conf Southeast Assoc Fish Wildl Agencies 42:283–294

Joanen T, McNease L (1977) Artificial incubation of alligator eggs post hatching culture in controlled environmental chambers. Proc Annu Meet World Mariculture Soc 8:483–490

Joanen T, McNease L (1987) Alligator farming research in Louisiana. In: Webb GJW, Manolis SC, Whitehead PJ (eds) Wildlife management: crocodiles alligators. Surrey Beatty & Sons, Chipping Norton, NSW

Joanen T, McNease L, Perry G (1977) Effects of simulated flooding on alligator eggs. Proc Annu Conf Southeast Assoc Fish Wildl Agencies 31:33–35

Jubb TF (1992) A thiamine responsive nervous disease in saltwater crocodiles (*Crocodylus porosus*). Vet Rec 131:347–348

Jubb KVF, Huxtable CR (1993) Thiamine deficiency. In: Jubb KVF, Kennedy PC, Palmer N (eds) Pathology of domestic animals, 4th edn. Academic, San Diego

Kakela R, Kinnunen S, Kakela A, Hyvarinen H, Asikainen J (2001) Fatty acids, lipids, cytochrome p-450 monooxygenase in hepatic microsomes of minks fed fish-based diets exposed to Aroclor 1242. J Toxicol Environ Health A 64:427–446

Ketola HG, Bowser PR, Wooster GA, Wedge LR, Hurst SS (2000) Effects of thiamine on reproduction of Atlantic salmon a new hypothesis for their extirpation in Lake Ontario. Trans Am Fish Soc 129:607–612

Kohno S, Bermudez DS, Katsu Y, Iguchi T, Guillette LJ Jr (2008) Gene expression patterns in juvenile American alligators (*Alligator mississippiensis*) exposed to environmental contaminants. Aquat Toxicol 88:95–101

Kristof ND (2009) It's time to learn from the frogs. The New York Times 27 June 2009. http://www.nytimes.com/2009/06/28/opinion/28kristof.html. Accessed 2 Nov 2009

Lal B, Singh TP (1987) Impact of pesticides on lipid metabolism in the freshwater catfish, *Clarias batrachus*, during the vitellogenic phase of its annual reproductive cycle. Ecotoxicol Environ Saf 13:13–23

Lance V, Joanen T, McNease L (1983) Selenium, vitamin E, trace elements in the plasma of wild farm-reared alligators during the reproductive cycle. Can J Zool 61:1744–1751

Lance VA, Morici LA, Elsey RM (2001a) Physiology endocrinology of stress in crocodilians. In: Grigg GG, Seebacher F, Franklin CE (eds) Crocodilian biology evolution. Surrey Beatty & Sons, Chipping Norton, NSW

Lance VA, Morici LA, Elsey RM, Lund ED, Place AR (2001b) Hyperlipidemia reproductive failure in captive-reared alligators: vitamin E, vitamin A, plasma lipids, fatty acids, steroid hormones. Comp Biochem Physiol Biochem Mol Biol 128:285–294

Masson GR (1995) Environmental influences on reproductive potential, clutch viability, embryonic mortality of the American alligator in Florida. Dissertation, University of Florida, Gainesville

Matter JM et al (1998) Effects of endocrine-disrupting contaminants in reptiles: alligators. In: Kendal R, Dickerson R, Giesy J, Suk W (eds) Principles and processes for evaluating endocrine disruption in wildlife. SETAC Press, Pensacola

McCandless DW, Schenker S (1968) Encephalopathy of thiamine deficiency: studies of intracerebral mechanisms. J Clin Investig 47:2268–2280

Milnes MR, Guillette LJ Jr (2008) Alligator tales: new lessons about environmental contaminants from a sentinel species. Bioscience 58:1027–1036

Milnes MR, Bryan TA, Medina JG, Gunderson MP, Guillette LJ Jr (2005) Developmental alterations as a result of in ovo exposure to the pesticide metabolite p,p′-DDE in Alligator mississippiensis. Gen Comp Endocrinol 144:257–263

Muller JK, Gross TS, Borgert CJ (2007a) Topical dose delivery in the reptilian egg treatment model. Environ Toxicol Chem 26:914–919

Muller JK, Scarborough JE, Sepúlveda MS, Casella G, Gross TS, Borgert CJ (2007b) Dose verification after topical treatment of alligator (*Alligator mississippiensis*) eggs. Environ Toxicol Chem 26:908–913

Noble RC, McCartney R, Ferguson MWJ (1993) Lipid fatty acid composition differences between eggs of wild captive-breeding alligators (*Alligator mississippiensis*): an association with reduced hatchability? J Zool Soc Lond 230:639–649

Palmer BD, Guillette LJ Jr (1992) Alligators provide evidence for the evolution of an archosaurian mode of oviparity. Biol Reprod 46:39–47

Percival HF, Masson GR, Woodward AR, Rice KG (1992) Variation in clutch viability among seven American alligator populations in Florida. Final Report. Florida Cooperative Fish Wildlife Research Unit, University of Florida, Gainesville

Rattner BA, Heath AG (2003) Environmental factors affecting contaminant toxicity in aquatic terrestrial vertebrates. In: Hoffman DJ, Rattner BA, Burton GA Jr, Cairns J Jr (eds) Handbook of ecotoxicology. Lewis Publishers, Boca Raton

Rauschenberger RH (2004) Developmental mortality in American alligators (*Alligator mississippiensis*) exposed to organochlorine pesticides. Dissertation, University of Florida, Gainesville

Rauschenberger RH, Sepúlveda MS, Wiebe JJ, Szabo NJ, Gross TS (2004a) Predicting maternal body burdens of organochlorine pesticides from eggs evidence of maternal transfer in *Alligator mississippiensis*. Environ Toxicol Chem 23:2906–2915

Rauschenberger RH, Wiebe JJ, Buck JE, Smith JT, Sepúlveda MS, Gross TS (2004b) Achieving environmentally relevant organochlorine pesticide concentrations in eggs through maternal exposure in *Alligator mississippiensis*. Mar Environ Res 58:851–856

Rauschenberger RH, Wiebe JJ, Sepúlveda MS, Scarborough JE, Gross TS (2007) Parental exposure to pesticides and poor clutch viability in American alligators. Environ Sci Technol 41:5559–5563

Rauschenberger RH, Sepúlveda MS, Wiebe JJ, Wiebe JE, Honeyfield DC, Gross TS (2009) Nutrient and organochlorine pesticide concentrations in American alligator eggs their associations with clutch viability. J Aquat Anim Health 21:249–261

Rice KG (1996) Dynamics of exploitation on the American alligator: environmental contaminants harvest. Dissertation, University of Florida, Gainesville

Rice KG, Percival HF, Woodward AR, Jennings ML (1999) Effects of egg hatchling harvest on American alligators in Florida. J Wildl Manage 63:1193–1200

Rice AN, Ross JP, Woodward AR (2007) Alligator diet in relation to alligator mortality on Lake Griffin, FL. Southeast Nat 6:97–110

Rooney AA, Guillette LJ Jr (2001) Biotic abiotic factors in crocodilian stress: the challenge of a modern environment. In: Grigg GG, Seebacher F, Franklin CE (eds) Crocodilian biology evolution. Surrey Beatty & Sons, Chipping Norton, NSW

Ross JP (2000) Effects of toxic algae on alligators alligator egg development. Final Report. University of Florida Water Resources Research Center, Gainesville

Ross JP, Honeyfield DC, Brown SB, Brown LR, Waddle AR, Welker ME, Schoeb TR (2009) Gizzard shad thiaminase activity its effect on the thiamine status of captive American alligators (*Alligator mississippiensis*). J Aquat Anim Health 21:239–248

Rotstein DS, Schoeb TR, Davis LM, Glenn TC, Arnold BS, Gross TS (2002) Detection by microsatellite analysis of early embryonic mortality in an alligator population in Florida. J Wildl Dis 38:160–165

Schelske CL, Coveney MF, Aldridge FJ, Kenney WF, Cable JE (2000) Wind or nutrients: historic development of hypereutrophy in Lake Apopka, Florida. Arch Hydrobiol Spec Issues Adv Limnol 55:543–563

Schoeb TR, Heaton-Jones TG, Clemmons RM, Carbonneau DA, Woodward AR, Shelton D, Poppenga RH (2002) Clinical necropsy findings associated with increased mortality among American alligators of Lake Griffin, Florida. J Wildl Dis 38:320–337

Sepúlveda MS, Wiebe JJ, Honeyfield DC, Rauschenberger HR, Hinterkopf JP, Johnson WE, Gross TS (2004) Organochlorine pesticides and thiamine in eggs of largemouth bass and American alligators and their relationship with early life-stage mortality. J Wildl Dis 40:782–786

Shotts EB Jr, Gaines JL Jr, Martin L, Prestwood AK (1972) Aeromonas-induced deaths among fish reptiles in an eutrophic inland lake. J Am Vet Med Assoc 161:603–607

Summer CL, Giesy JP, Bursian SJ, Render JA, Kubiak TJ, Jones PD, Verbrugge DA, Aulerich RJ (1996) Effects induced by feeding organochlorine-contaminated carp from Saginaw Bay, Lake Huron, to laying White Leghorn hens. II. Embryotoxic teratogenic effects. J Toxicol Environ Health 49:409–438

Tillitt DE, Zajicek JL, Brown SB, Brown LR, Fitzsimons JD, Honeyfield DC, Holey ME, Wright GM (2005) Thiamine thiaminase status in forage fish of salmonines from Lake Michigan. J Aquat Anim Health 17:13–25

US EPA (2009) Endocrine disruptor screening program. http://www.epa.gov/endo/. Accessed 12 Oct 2010

Wallach JD (1978) Feeding nutritional diseases. In: Fowler ME (ed) Zoo animal medicine. WB Saunders Company, Philadelphia, pp 123–129

Wilkinson PM (1983) Nesting ecology of the American alligator in coastal South Carolina. Study Completion Report. South Carolina Wildlife Marine Resources Department, Columbia

Wood JM, Woodward AR, Humphrey SR, Hines TC (1985) Night counts as an index of American alligator population trends. Wildl Soc Bull 13:262–273

Woodward AR, Moore CT (1990) Statewide alligator surveys. Florida Game Fresh Water Fish Commission, Final Report, Tallahassee

Woodward AR, Jennings ML, Percival HF (1989) Egg collecting hatch rates of American alligator eggs in Florida. Wildl Soc Bull 17:124–130

Woodward AR, Moore CT, Delany MF (1992) Experimental alligator harvest. Final Report. Florida Game and Fresh Water Fish Commission, Tallahassee

Woodward AR, Percival HF, Jennings ML, Moore CT (1993) Low clutch viability in American alligators on Lake Apopka. Fla Sci 56:52–64

Chapter 6
Restoration of Lake Apopka's North Shore Marsh: High Hopes, Tough Times, and Persistent Progress

Roxanne Conrow, Edgar F. Lowe, Michael F. Coveney, R. Heath Rauschenberger, and Greg Masson

The early morning bicycle ride along the dirt levee threaded between a wide expanse of open-water lake on one side and shallow marshes on the other. The humid Florida air carried the organic smells of soil and plants coupled with the beating of dragonfly wings, thousands and thousands of them hovering over the pickerelweed. Birds were everywhere—blackbirds and cardinals singing, northern harriers and red-shouldered hawks circling, herons of all kinds wading. A softshell turtle, bigger than my bike's tire, lifted its extra terrestrial head to watch us approach, and then eased back into the lake. An alligator crossed our path, walking high

R. Conrow (✉)
St. Johns River Water Management District, P.O. Box 1429, Palatka, FL, 32178, USA
e-mail: rconrow@sjrwmd.com

J.E. Elliott et al. (eds.), *Wildlife Ecotoxicology: Forensic Approaches*, Emerging Topics in Ecotoxicology 3, DOI 10.1007/978-0-387-89432-4_6,

on legs like a sumo wrestler, its ponderous jowl almost brushing the ground. Eleven years ago, we were on that same levee using binoculars to pan the black expanse of exposed muck soil, looking for the telltale flash of an American white pelican… dead, or dying, from poisons. But today was beautiful and we had come a long, long way.

Abstract The story of Lake Apopka is a familiar one to many Floridians and has gained international notoriety. The 12,500-ha lake was once a world-class bass fishery. Then, a century-long decline occurred, traced to the loss of over 8,000 ha of wetlands to farming operations, agricultural discharges laden with phosphorus to the lake, treated wastewater discharges, and input from citrus processing plants. The state of Florida and the Federal Government purchased the property with the goal of restoring the aquatic habitat. Shortly after flooding in the winter of 1998–1999, a bird mortality event occurred, resulting in the deaths of 676 birds, primarily American white pelicans (*Pelecanus erythrorhynchos*), and also including 43 endangered wood storks (*Mycteria americana*), 58 great blue herons (*Ardea herodias*), and 34 great egrets (*Casmerodius albus*). The deaths of the birds, attributed to pesticide toxicosis, resulted in years of research and remediation to ensure the future safety of wildlife on the property. Presently, about 3,000 ha of wetlands have been rehydrated since resuming restoration activities, with no adverse effects to wildlife. The following chapter presents the history of Lake Apopka, the efforts to restore it, and what we have learned along the way.

History and Background

As things go in Florida, Lake Apopka is old, dating back to the Miocene Epoch. It lies within view of the highest point in peninsular Florida, the 95-m Sugarloaf Mountain. At night from Sugarloaf, you can see lights from the city of Orlando, 25 km to the southeast. Lake Apopka is situated between two of the state's oldest sand dune ridge systems and has numerous recorded archeological sites along its shores. Its name is said to derive either from the Creek work *Ahapopka*, meaning potato-eating place, or from *Tsala Apopka*, meaning trout [bass]-eating place (Morris 1974). Maps dating to the early 1800s (Sime 1995) show a somewhat potato-shaped lake with marsh areas on the north that extended as a natural swale to connect at extreme high stages to the neighboring Lake Beauclair (Connell 1954). Those ancient sawgrass (*Cladium jamaicense*) marshes, which comprised roughly a third of the then 20,235-ha lake, resulted in a buildup of organic peat and muck deposits reaching a depth of up to 6 m thick along the northeastern shore (Davis 1946).

The natural drainage pattern to the north and the rich Everglades peat soils of the marsh inspired early developers interested in agriculture to dig a drainage canal (Apopka-Beauclair Canal) in the late 1800s. It lowered the water surface of Lake Apopka by about 1.2 m. Even so, farming attempts soon were aborted, and by 1910, the canal was filled with vegetation and the lake was back to its original depth of 2.4–2.7 m (Connell 1954). Five years later, another attempt to drain the marshes failed. In 1941, the Florida legislature, by a special act, created the Zellwood Drainage and Water Control District (Zellwood). By World War II, a levee was constructed to isolate the marshes from the lake. Aerial photography taken in 1947 shows approximately 1,162 ha in row crops with an additional 1,974 ha in preparation for what is locally called muck farming (Hoge et al. 2003). In 1958, the Apopka-Beauclair Canal was deepened.

Meanwhile, Lake Apopka had a reputation as one of the best fishing holes in the nation. Newspaper accounts reported that, in 1951, 10,000 bass were caught from just one dock of the 13 fish camps located along the south rim of the lake (Harper 1952). However, even as the good fishing was publicized and more fish camps were built, there were ominous indications that the lake's health was breaking down. The previous decade saw the first drainage water discharge into the lake from the muck farms. Increasing population in Winter Garden, the largest town on the shores of Lake Apopka, resulted in more wastewater discharge to the lake. Citrus grove production was booming in the area and wastewater discharges from facilities producing fruit concentrate entered the lake. Water hyacinth (*Eichhornia crassipes*), a non-native plant, was very thick around the shoreline and, in 1948, private citizen groups began spray applications of the herbicide 2,4-D, which resulted in a large number of decaying plants in the lake (Burgess 1964). By 1947, an extensive algal bloom appeared in the historically clear waters. Sport fishing was extremely poor during most of 1948 and 1949 (Dequine 1950).

By 1950, almost all of the rooted native vegetation – primarily eelgrass (*Vallisneria americana*), pondweed (*Potamogeton illinoensis*), and southern naiad (*Najas guadalupensis*) – had disappeared (Clugston 1963). The resurgence of record catches of black crappie (*Pomoxis nigromaculatus*), sunfish (*Lepomis* spp.), and largemouth bass (*Micropterus salmoides*) in the early 1950s was the last gasp of a lake once called trout-eating place. A control structure in the Apopka-Beauclair Canal stabilized the lake level beginning in 1952, effectively turning it into an impoundment. By 1957, rough fish, primarily gizzard shad (*Dorosoma cepedianum*), comprised 82% of the fish population. A series of selective shad treatments destroyed more than 1,600 metric tons (t) of shad in 1957, 4,500 t in 1958, and 2,900 t in 1959. A photograph of the floating carpet of dead Lake Apopka fish appeared in the December 9, 1957 issue of Life Magazine, along with a picture of the squadron of boats that had delivered the 19,000 liters of rotenone used to kill the shad. Most of the dead fish sank to the lake bottom, followed after each treatment by dense phytoplankton blooms.

The continuous algal blooms left extensive loose, unconsolidated deposits of organic material. Another constant source of nutrients came from the sewage treatment plant of Winter Garden, the largest municipality bordering the lake. However, the calculated amount of nitrogen and phosphates from the combined fish poisoning and hyacinth control spraying during 1957–1959 was about 150% and 78%, respectively, of the nutrients introduced from the sewage effluent in the previous 37 years (Burgess 1964).

Lake Apopka experiences an average 127 cm of rain each year (Jenab et al. 1986) and loses all but about 18 cm of that to evapotranspiration (Fernald and Purdum 1998). Its only other major water source is from Apopka Spring, located in the southwest arm of the lake called Gourd Neck. The spring typically discharges about 0.85 m^3/s (Hoge et al. 2003). Thus the lake, whose average depth is only 1.6 m, is particularly vulnerable to drought.

Based on National Oceanic and Atmospheric Administration (NOAA) records, Florida had low rain years in 1960–1963. Water levels dropped to an average depth of around 1.2 m (Burgess 1964) and a record temperature of 36°C was set in Orlando on June 17, 1963 (NOAA records). An estimated 4,500–9,000 t of fish died in May and June of that year. Both state and federal agencies investigated the event. Fish camp owners blamed spray planes from the farms overshooting the land and dumping pesticides in the water. Indeed, a report from the U.S. Fish and Wildlife Service (USFWS) biologist investigating the fish kill commented, "At the time the lake was being inspected the crops along the north and northeast shore were being sprayed with insecticides by airplanes ... The pilots apparently paid no attention to the fact that our boat was several hundred yards off-shore, since all three mornings they flew over our boat at a very low altitude with the insecticide still being released from their spray rigs." (Wellborn 1963).

The biologist also noted in his report that, according to the crop reporting service of the Department of Agriculture, 9 of the approximately 27 farms bordering Lake Apopka expended $389,400.00 annually for spray materials. Because of the subtropical climate, two or even three crop rotations were possible each year. How much chemicals did the farmers use? The best information is somewhat anecdotal, but according to a county agricultural agent, "... *this typical spray and dust (and bait) insecticidal program would reflect the following for 1 acre of sweet corn.*

0.4 lbs actual Chlordane per acre.
13.3 lbs actual Toxaphene per acre.
68 lbs actual DDT per acre.
11 lbs of actual Parathion per acre" (Swanson 1967).

Newspaper headlines indicated that "DDT Caused Fish Kill" (Rider 1963), citing USFWS and U.S. Public Health Service findings of 12 ppm of DDT and its breakdown products in the flesh of Lake Apopka bass. However, the Florida State Board of Health concluded that the fish kill was the result of an algae explosion brought on by high water temperature and too much nutrients in the water (Huffstutler et al. 1965).

Over the next decade (1970–1980) a number of governor-appointed advisory committees, and citizen and agency studies worked to improve water quality in Lake Apopka, with little success. An experimental gravity drawdown in 1971 resulted in alligator, turtle, bird, snake, and fish mortality, but did not improve water quality. Wastewater discharges from citrus processors were curtailed in 1969 and completely ceased in 1977. Direct sewage discharges from Winter Garden ended in 1980. However, by the 1980s, 7,290 ha along the north shore were in farm production. Stormwater discharges from the farms since the late 1940s, estimated at 76 billion liters of water annually (Hoge et al. 2003), increased areal phosphorus loading to the lake sevenfold (Battoe et al. 1999; Lowe et al. 1999; Schelske et al. 2000).

While most public and agency attention focused on the lake's nutrient pollution issues, an event in 1980 revived interest in other contaminants. An investigation report and risk assessment (U.S. EPA 1994, 2001) provides the details. Beginning in 1957, Tower Chemical Company (TCC) produced various pesticides, including chlorobenzilate, at their 6-ha plant located near the Gourd Neck area of Lake Apopka. When an intermediate product in the manufacture of chlorobenzilate became too difficult to obtain, TCC modified their manufacturing process without informing regulators. The modification involved the in-house production of an intermediate chemical from technical grade DDT (dichlorodiphenyltrichloroethane). In 1980, following heavy rains, the wastewater pond at the main facility overflowed into a stream and wetlands bordering Gourd Neck, killing vegetation and fish, and farm fowl drinking from the stream. The site was placed on the EPA Superfund list and, through the 1980s, Superfund spent about $6 million for remediation. In 1981, there was a major decline in Lake Apopka's juvenile alligators (Woodward et al. 1993) while other lakes in the region showed numerically stable populations (Jennings et al. 1988). An analysis of alligator eggs collected in 1985 indicated significantly elevated levels of DDT and its breakdown products DDD and DDE (Rice and Percival 1996).

In the mid-1980s, Florida passed several legislative acts directing that an "environmentally sound and economically feasible" means be found to restore Lake Apopka to a condition suitable for recreational use and for the propagation of fish and wildlife (Battoe et al. 1999). They identified the St. Johns River Water Management District (SJRWMD) as the lead agency and provided funding for Lake Apopka as one of several priority water bodies in the state requiring restoration work. Ultimately, SJRWMD determined that major reductions in phosphorus loads to Lake Apopka were needed to meet restoration goals.

In 1988, SJRWMD received authority to regulate agricultural discharges. That same month they entered into a consent order with a corporate farming entity, A. Duda & Sons (Duda), which owned 2,633 ha of muck farm fields bordering both sides of the Apopka-Beauclair Canal (Fig. 6.1). The consent order required Duda to construct a water recycling reservoir for their acreage east of the canal, and to sell their acreage on the west side of the canal to SJRWMD. In 1989, the farmers belonging to the Zellwood Drainage District signed a consent order with phosphorus reduction targets; however, a legal challenge delayed implementation for 2 years. This consent order was modified to allow the farmers to construct an alum injection system to help reduce phosphorus loads from their pump discharge, beginning in 1995.

Meanwhile, SJRWMD constructed a demonstration-scale wetland filtration system on a portion of the newly purchased Duda property. Beginning November 1990,

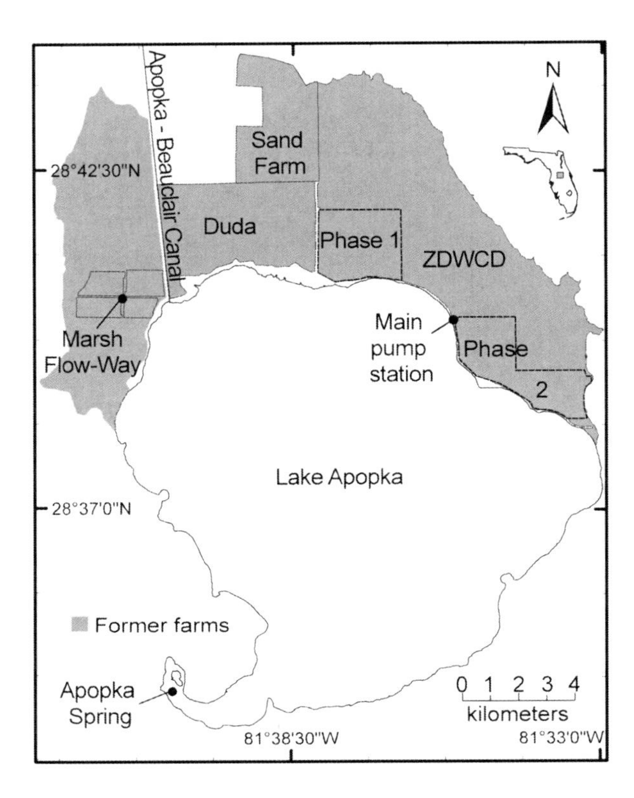

Fig. 6.1 Lake Apopka and its former farm area

lake water was circulated through the created wetland (flow-way) for 29 months. The flow-way met its target efficiency and removed 85% of suspended solids, and 30% of total phosphorus (Coveney et al. 2001). Using the demonstration project area and additional former farm acreage to the north of it, SJRWMD constructed a full-scale 300 ha wetland filtration system which became operational in 2003. Through 2009, the flow-way removed approximately 17 t of phosphorus and 28,000 t of total suspended solids from lake water (SJRWMD data). All other farm properties north of the flow-way and west of the Apopka-Beauclair Canal were purchased by SJRWMD and flooded for wetland restoration by 1994.

Beginning in 1993, and subsidized by SJRWMD, commercial fishers began harvesting gizzard shad from Lake Apopka during the winter/spring months when the fish are schooling (Fig. 6.2). Unlike earlier times when dead shad were allowed to sink to the lake bottom, the harvested fish were frozen for use as bait in the crab fishery, or as bait for crayfish farms in the Southeast. Between 1993 and 2008, over 7,600 t of fish were harvested from the lake, corresponding to a removal of approximately 53 t of phosphorus.

In 1994, Zellwood farmers challenged the regulatory authority of SJRWMD to set a phosphorus discharge limitation for Lake Apopka, which resulted in legal wrangling for several more years. In 1996, the state legislature specifically authorized

Fig. 6.2 Gizzard shad fishermen on Lake Apopka, Florida. Credit SJRWMD

SJRWMD to set a phosphorus concentration target for the lake. The agency established a phosphorus-loading target that represents a 76% reduction from baseline (1968–1992) conditions (Coveney et al. 2005). The legislature also determined it was in the public's interest for SJRWMD to pursue a buyout of the farms as the long-term solution to the nutrient loading problem.

By August 1998, SJRWMD and the Natural Resources Conservation Service (NRCS) Wetland Reserve Program had purchased most of the farms east of the Apopka-Beauclair Canal for $102 million, with the intent of restoring them to wetlands. As part of the due diligence process during the acquisition of these properties, Phase I Environmental Site Assessments were conducted on all 32 properties in accordance with the current American Society for Testing and Materials 1527 standard. As a result, Phase II Environmental Site Assessments were completed on 26 properties to investigate possible contamination. Eighteen of the Phase II assessments confirmed contamination and the subsequent remediation activities removed approximately 22,000 t of contaminated soil from former petroleum storage areas and mix/load sites to a class III landfill in south Florida.

SJRWMD also conducted an environmental risk assessment (ERA) of the residual pesticide levels expected to remain in the farm soils after remediation was completed. Approximately 400 soil samples were analyzed for pesticide residues in the ERA process. The risk assessment, performed on the first property acquired and a subsequent follow-up evaluation indicated no likelihood that residual organochlorine pesticide (OCP) levels would present an acute toxicity risk to aquatic or wetland animals, although long-term, sublethal effects on growth or reproduction of top predator piscivores were possible (ATRA, Inc. 1997, 1998). ATRA accurately assessed the available data, although their predictions clearly underestimated the real risk. In later investigations, SJRWMD found several factors that, in the aggregate, likely explain this underestimate. The more important factors included that early soil sampling focused on mix-load sites, storage areas, and other areas where contamination was known to occur, with fewer samples taken in the fields. Later extensive sampling in the fields showed high variability and some elevated OCP concentrations in individual fields despite identical soil and crop conditions. Later analyses were done by a single laboratory with special expertise in toxaphene quantitation. Also, experiments done subsequently showed greater bioaccumulation of OCPs in open-water habitats such as newly flooded fields than in the vegetated habitats that provided the earlier bioaccumulation data. Finally, in the early work geostatistical methods (kriging) were used to calculate spatially weighted averages across multiple fields for risk assessment. With our later understanding of field-to-field variability, we concluded that large-scale spatial averaging was not appropriate and we quantified risk for each individual field.

In 1998, after up to 50 years of tilling, compaction, and soil oxidation (mean organic carbon content of these peat soils 39% dry weight), the farms were 1.2–1.8 m below the surface elevation of the lake. Thus, a restoration that simply called for breaching the outer levee would create more lake, not the desired wetland habitat. Over the many decades of operation the farmers shallowly flooded their fields for several weeks after the final summer crops were harvested (Fig. 6.3). The flooding

Fig. 6.3 Aerial photograph of Zellwood farms after summer harvest in August 1996. One portion (*middle left* of photo) is flooded for erosion and nematode control. Credit SJRWMD

helped to control weeds, nematodes, and limit soil erosion, and oxidation. More pertinent to this account, the flooding coincided with the annual fall migration through Florida of shorebirds. As natural wetlands have disappeared in the state, the flooded Zellwood farms became increasingly useful to shorebirds as feeding and staging areas. They also became important to birdwatchers. Thus, when the state and federal government purchased the farms, citizen petitions and letters began pouring in asking that portions of the property be set aside for waterfowl management and for shorebirds. SJRWMD proposed that, once monitoring efforts showed that there were no adverse effects of contaminants upon wildlife using the restored wetland area, it would consider how best to manage sections for shorebird habitat. Ornithologists responded that even a 1-year disruption to a migration stopover was ill-advised (Williams 1999). Hence, it seemed like a double benefit when, in the summer of 1998, the farmers harvested their final crops, and SJRWMD requested that they leave the fields shallowly flooded, just as they always did. The flooding would inhibit weed growth and make it easier for the initial restoration activities scheduled for the winter, and it would allow the annual shorebird migration patterns to continue uninterrupted.

By fall, as more and more of the farms were flooded, there was a mosaic of wet conditions. What resulted was an unprecedented response from fish-eating birds, shorebirds, and water fowl (Fig. 6.4). Daily counts of individual birds rose progressively from 10,000 in September to 41,000 in December. Whereas typically a dozen or so American white pelicans (*Pelecanus erythrorhynchos*) would visit the lake each winter, 4,370 were counted on a single day at the flooded farm

Fig. 6.4 Birds feeding in flooded field ditches, January 1999. Credit SJRWMD

properties. In mid-November, dewatering of the north fields began, with the goal of moving water from the north section to other field blocks to sequentially dry areas in order to spread alum residual, a byproduct of the process used to clarify drinking water. An earlier pilot project had shown that the residual would reduce the flux of phosphorus from the organic soils into the water column, thus minimizing the phosphorus level in water discharged to the lake. The dewatering served to concentrate fish in field ditches and feeder canals providing a virtual smorgasbord for the birds. Also in November, a few carcasses of American white pelicans were visible in the fields. However, with tens of thousands of birds present, their deaths were not considered unusual. By December, an additional 35 American white pelicans had died; some moribund birds were observed exhibiting extreme neurological symptoms, and several were collected for necropsy. Meanwhile, a Christmas Bird Count (CBC) was held on December 20, 1998, organized by local birdwatchers as part of the annual national effort. The count tally was 174 species, the highest total ever obtained on an inland CBC anywhere in North America (NAS 1999).

Because of the dying birds, pumping to dewater the entire North Shore Restoration Area (NSRA) was accelerated and by mid-February 1999, the fields of the NSRA had little standing water. By the end of the mortality event, a total of 676 birds had died onsite including 441 American white pelicans, 58 great blue herons (*Ardea herodias*), 43 wood storks (*Mycteria americana*), 34 great egrets (*Casmerodius albus*), and smaller numbers of birds from 20 other species (USFWS 2004). In the spring of 1999, SJRWMD and NRCS initiated large-scale resampling and analysis of NSRA soils and funded the tissues analyses and necropsies of dead birds collected by Florida Audubon as part of an effort to determine the cause of the bird mortality (Table 6.1).

Table 6.1 Brain levels (mg/kg wet weight) of OCPs in 64 birds collected during the mortality event, December 1998 to April 1999

Contaminant	*N*	Mean	Median	Range
4,4'-DDE	64	37.0	12.4	0.94–234
DDT equivalents[a]	59	4.13	2.15	0.05–26.7
Dieldrin	71	4.55	2.10	0.01–35.7
Toxaphene	67	32.2	17.0	0.05–288
Endrin	12	0.43	0.42	0.03–0.86
Endrin (nd)[b]	39	–	–	0.01–9.90
Oxychlordane	43	0.78	0.26	0.01–5.95
Heptachlor epoxide	17	0.13	0.12	0.04–0.24
H. epoxide (nd)[a]	35	–	–	0.01–4.90

Most birds were collected on-site or in the near vicinity; however, five were collected in other parts of the state. Analyses were performed by nine laboratories. Splits of five brains were analyzed by two labs and one brain was analyzed by three labs

[a]*Sensu* Stickel et al. (1970), DDT equivalents = DDD/5 + DDE/15 + DDT

[b]nd are laboratory values reported at detection limit

Forensic Investigations

While SJRWMD, with assistance from the Florida Audubon Society and the Florida Fish and Wildlife Conservation Commission (FFWCC), was in the process of trying to determine the cause of the bird mortalities (Conrow et al. 2003), the U. S. Department of Justice (DOJ) and USFWS began a joint investigation of the bird mortality event on January 12, 1999 (USFWS 2001a). The federal investigation fell under the authority of the Migratory Bird Treaty Act (MBTA), the Bald and Golden Eagle Protection Act (BGEPA), and the Endangered Species Act (ESA), and was aimed at determining the cause and identifying the persons or entities that were potentially legally responsible. During the investigation, dead and dying birds were collected by USFWS contaminant specialists and law enforcement agents. The USFWS worked with FFWCC, NRCS, the Audubon Society, SJRWMD, wildlife disease labs of several universities, and the U.S. Geological Survey's (USGS) National Wildlife Health Laboratory in Madison, Wisconsin to determine the cause of death. Bird carcasses underwent necropsies and were screened for wildlife diseases. The U.S. Environmental Protection Agency, the Florida Department of Environmental Protection, University of Florida, and the local Orange County also assisted in the investigation analyzing bird, fish, and soil samples for contaminants of concern.

On February 9, 1999, the USFWS announced that initial studies of the dead birds had ruled out known wildlife diseases, such as botulism, as the cause of the mortalities (USFWS 1999). A week later, the USFWS announced that organochlorine toxicosis was determined as the preliminary cause of the bird mortalities. The USFWS used OCP levels measured in the brain as the measure for diagnosing OCP toxicosis (Table 6.2) because dosing experiments showed that lethal concentrations of OCPs in the brain do not significantly differ among species, age, sex, or dosing regimen (Wiemeyer 1996; Stickel et al. 1979b; Peakall 1996). In cases where a range of lethal brain concentrations were provided, the USFWS used the lower concentration of that range reported in experimental studies since wild birds may encounter periods of inadequate food and extreme environmental conditions. In some cases, the USFWS reported that dieldrin levels alone were high enough to cause death in some of the dead birds, including a wood stork (USFWS 2001a).

Table 6.2 Concentrations (mg/kg wet weight) of OCPs in brain tissue used to diagnose lethal exposure in birds that died on the NSRA during winter 1998 and 1999

Contaminant	Brain concentration	References
Dieldrin	>4	Stickel et al. (1969)
Endrin	>0.7	Stickel et al. (1979a)
DDT equivalents	>18	Blus (1996)
Oxychlordane	>5.8	Stickel et al. (1979b)
Heptachlor epoxide	>8	Wiemeyer (1996) and Stickel et al. (1979b)

Fortunately, well before the USFWS announcements, SJRWMD had already started pumping to dewater the site to discourage more birds from using the area; by mid-February there was little standing water and by March 1999 the entire area had been dewatered and the mortality ceased.

SJRWMD also contracted with E^{x}ponent, Inc., a toxicological consulting firm to conduct a forensic analysis using all available information: (1) levels of pesticides in soils, fish, and bird tissues; (2) necropsies of dead birds; (3) information on the number of birds that died by species; (4) flooding and drying patterns; and (5) literature information on the toxicities of the major pesticide contaminants in the NSRA. E^{x}ponent concluded that no definitive causal agent for the deaths could be identified with the available data (E^{x}ponent 2003). SJRWMD later concluded that organochlorine toxicosis was a significant factor in the bird mortality event (Lowe et al. 2003). Consequently, they proceeded with additional studies to determine safe concentrations of pesticides in the soils.

Regulatory Ramifications

SJRWMD was perplexed and distressed by the incident and fully cooperated with the federal investigation. In March 1999, SJRWMD and NRCS launched a $1.5 million project with help from a 13-agency Technical Advisory Group to investigate the cause of the bird mortality event and determine how restoration might be safely achieved.

During the course of the federal investigation, SJRWMD became the subject of a criminal investigation. In addition, the U.S. Department of Agriculture (USDA)-NRCS actions were investigated because it was the Federal agency that had funded the project without initiating interagency consultation with the USFWS as required by the ESA.

Because the USFWS had not been formally consulted prior to the flooding, USDA-NRCS and SJRWMD were potentially liable if it was found they were responsible for the death of any federally endangered species, such as wood storks. Indeed, DOJ considered filing criminal charges against SJRWMD for violation of the ESA, MBTA, and BGEPA. In addition, SJRWMD was later advised that a potential civil claim for natural resources damages could be filed by the USFWS under the authority of the Comprehensive Environmental Response, Compensation and Liability Act (CERCLA).

In response, the SJRWMD denied criminal or civil liability and provided a detailed factual and legal response to these potential charges. At this point, the issue was elevated, and attorneys and senior officials from DOJ, USFWS, and SJRWMD began lengthy discussions to find a mutually agreeable resolution.

As a result of the criminal investigation, prolonged negotiations among agency attorneys were required before scientists could exchange information or discuss the mortality event. Thus, what had been a cooperative effort of information sharing, became a bifurcated effort to independently understand the scientific facts (Angelo 2009).

Restarting Restoration

While the federal investigation and senior level discussions continued, the restoration area remained dry, and no further OCP-related bird mortalities were observed. To maintain dry conditions, pumping was necessary because the elevation of the NSRA averaged over a meter below Lake Apopka's surface level, and seepage from the lake and surrounding upland areas, as well as rain accumulation, required active management.

Operation and maintenance of the pumps and infrastructure, including mowing and roller chopping for vegetation control, cost over $1 million per year. More important than the cost was the fact that it was necessary to pump phosphorus-laden water into the lake. The continued discharge of water into Lake Apopka defeated the original reason why more than 100 million taxpayer dollars had been used to purchase the NSRA. Operating the pumps in perpetuity would not reduce phosphorus input to the lake and achieve the long-term restoration of Lake Apopka. However, running the pumps to keep the fields dry was the best intermediate solution to protect birds while research and a long-term plan for restoration was developed.

In 2000, local staff with the USFWS, SJRWMD, and NRCS began to formulate a plan to safely reflood a small section of the NSRA. After a review of contaminant levels in the soils across the NSRA provided by the 1999 resampling, a 276-ha site, known as Duda sub-East, was selected for potential rehydration. The Duda sub-East site was one of the last areas of the historic sawgrass marsh to be converted to farmland and had relatively low OCP soil levels. The Biota Sediment Accumulation Factor (BSAF) was used to predict OCP levels in fish on a lipid basis, using carbon normalized soil OCP levels. A dietary exposure for the great blue heron was calculated based on the predicted fish concentrations. The potential for adverse effects for each OCP – known as the hazard quotient (HQ) – was calculated as the ratio of the estimated dose to a Toxicity Reference Value, determined from literature values. The HQs for DDTx (DDD + DDE + DDT), chlordane, dieldrin, and toxaphene were summed to estimate a hazard index (HI) to provide a measure of the cumulative effect of the OCPs. An HI less than 1.0 indicates that a group of contaminants is unlikely to cause adverse ecological effects (U.S. EPA 1986).

Throughout 2001 and into the early spring of 2002, local staff of the three agencies refined the restoration plan, which included a follow-up monitoring plan to validate models and verify assumptions. The NRCS formally initiated ESA consultation with the USFWS in 2001 and on April 1, 2002, the USFWS issued a Biological Opinion (BO) on the proposed restoration plan (USFWS 2002). In the BO, the USFWS reviewed the proposal and concurred that restoration of the area would not jeopardize the recovery of any federally endangered or threatened species. However, since the HI (1.6) was above 1.0, the USFWS determined that the action might result in the incidental take of one bald eagle and one wood stork.

USFWS authorized the incidental take of one bald eagle and one wood stork, and required that NRCS and SJRWMD implement several reasonable and prudent measures in order to minimize the risk of take occurring from exposure to OCPs. These measures included visually surveying transects to detect mortality, monitoring OCP concentrations in bird eggs and prey, managing water levels to promote dense herbaceous vegetation coverage, and establishing contaminant trigger values that required reinitiation of consultation if exceeded (USFWS 2002). For example, SJRWMD and NRCS agreed to implement a posthydration monitoring plan with established trigger levels for OCPs in fish and wading bird eggs. The OCP trigger levels for fish and bird eggs were developed by USFWS contaminant specialists to provide a means of comparing samples collected during monitoring efforts. Exceedences of the established OCP trigger levels would initiate a meeting between NRCS, USFWS, and SJRWMD to discuss concerns and appropriate actions.

The trigger values established for the COCs in fish were DDTx – 3.6 mg/kg; toxaphene – 4.3 mg/kg; dieldrin – 0.69 mg/kg; total chlordane – 0.285 mg/kg. Trigger values for COCs in bird eggs were DDTx – 3 mg/kg, dieldrin – 3 mg/kg, total chlordane – 1.5 mg/kg; and avian eggshell thinning of greater than 10% for the surrogate species within a reproductive year when compared to reference data for the same species. SJRWMD developed a contingency plan for a worst-case scenario to provide a way for fields to be drained rapidly if monitoring efforts indicated trigger levels were greatly exceeded and birds were in imminent risk.

The posthydration monitoring plan and associated reports also provided data that could be used to verify the assumptions used in the risk assessment and to validate, refine, and reduce the uncertainty of future risk assessments, with an adaptive management approach.

By summer 2002, SJRWMD and NRCS had the reasonable and prudent measures developed and implemented, and the rehydration of Duda sub-East began. For over a year, vigorous monitoring of bird use and fish OCP concentrations was conducted. No bird mortalities occurred during this period and fish OCP levels were below the trigger levels stated in the BO. The USFWS met with NRCS and SJRWMD to discuss results and the three agencies agreed that the remaining Duda fields (over 1,214 ha) could be rehydrated, with the exception of a 15-ha section on the north edge of the property.

The Lake Apopka Agreement

Parallel with the progress made at the local level, attorneys and senior managers with DOJ, USFWS, and SJRWMD had continued working and making progress toward resolving the significant legal issues. On October 8, 2003, DOJ announced a memorandum of understanding (MOU) with SJRWMD that resolved the criminal investigation and the natural resource damage claims (DOJ 2003). Because of the MOU, both sides were able to focus collective efforts on the potential of the future rather than engage in expensive, protracted litigation that would focus on the past.

In the MOU agreement (DOJ and SJRWMD 2003), the USA agreed not to file criminal charges against the SJRWMD for wildlife violations, provided that it fulfilled the terms of the MOU. Furthermore, the MOU did not affect whether or not restoration in any form may proceed at the site. Indeed, by the time the MOU was signed, the USFWS, SJRWMD, and NRCS had safely rehydrated the entire Duda property. SJRWMD agreed to the following provisions found in Sections I–IV and VIII of the MOU, which resolved the criminal investigation. SJRWMD agreed to:

1. Bring all of its properties into compliance with the ESA and other wildlife statutes
2. Reimburse the wildlife rehabilitators that worked on affected birds approximately $90,000.00, collectively, that they expended in this effort
3. Run a conference to educate other Florida Water Management Districts and other state agencies and stakeholders regarding the avian mortality event on the NSRA
4. Make certain acknowledgments regarding the cause of death of the birds and its role therein
5. Provide training to its employees regarding legal obligations of the ESA, the MBTA, and the BGEPA
6. Conduct a 5-year program to monitor pesticide levels in fish from Lake Apopka
7. Monitor wood stork populations on its properties for 5 years
8. Develop an active management plan for threatened or endangered species on at least 200 acres of its properties

In regard to the natural resource damages, a Damage Assessment and Restoration Plan (USFWS 2004) was prepared to resolve the CERCLA claim. However, during its preparation a time-sensitive opportunity arose for the SJRWMD to purchase a 3,426-ha tract of land in St. Johns County, Florida (the Matanzas Marsh property) that contained one of the two largest wood stork colonies in Northeast Florida. Because immediate action was required to obtain the property, SJRWMD coordinated the purchase of the land prior to completion of the Damage Assessment and Restoration Plan. They undertook this action with the understanding that the USA was not obligated to determine that purchase of the Matanzas Marsh property and protection of the wood stork colony would satisfy SJRWMD's obligation for natural resource damages.

In the completed Damage Assessment and Restoration Plan, USFWS concluded that purchase of the Matanzas Marsh property, with protection of its wood stork colony, was the preferred alternative. Along with the $10 million SJRWMD applied toward the purchase, SJRWMD also agreed to conduct monitoring and management of the wood stork colony.

The MOU also provided for SJRWMD to pay past costs incurred by the Department of the Interior in connection with the damage assessment and future costs up to $1,500 per year to fund the Interior's participation in the monitoring of the Matanzas Marsh wood stork colony. In addition, SJRWMD paid $14,776 to USFWS to fund an update of the existing habitat management guidelines for the wood stork and an additional $10,450 to fund a study of eggshell thinning.

The MOU contained a covenant not to sue, with re-openers. If SJRWMD breached the agreement during the first 5 years after execution of the MOU, the USA would be able to bring an action against it for natural resource damages. The MOU further provided that the Damage Assessment and Restoration Plan would be subject to public comment. If public comments disclosed facts or considerations, which indicated that the actions set forth in the MOU were not appropriate compensation for the injuries to natural resources, USFWS might withdraw from the natural resource damage provisions of the MOU (USFWS 2004).

Implementing the MOU and Moving Forward

The signing of the MOU and the restoration of the Duda property (Fig 6.1) have been very important developments in the restoration of Lake Apopka's north shore, as well as providing a framework that has solidified the cooperative relationship between SJRWMD, NRCS, and USFWS. The MOU provided a resolution of all the legal issues and a framework for moving forward, with each agency afforded certain assurances and mutual benefits. The restoration of the Duda property provided the three agencies with a common goal toward which they could work together. Its successful approach became the model for future restoration efforts.

Following the restoration of the Duda property (Table 6.3), a biological assessment for the 300-ha Marsh Flow-Way Project was completed with a conclusion of not likely to adversely affect. With USFWS concurrence, operation began in 2003. In November 2004, SJRWMD completed a biological assessment and restoration plan for the 611-ha Sand Farm South property. The USFWS reviewed and accepted the plan and SJRWMD implemented it. Two consecutive years of fish monitoring resulted in fish OCP levels that were less than, or equal to, one-half the trigger levels provided in the biological opinion (USFWS 2002) for the Duda property, indicating risks to piscivorous birds were negligible.

In July 2006, SJRWMD and NRCS submitted a biological assessment and restoration plan for the 486-ha Unit 2 West site (also known as Phase I). The USFWS reviewed it and concurred with the NRCS that the project "may affect, but is not likely to adversely affect Federally listed species." However, despite approval to commence restoration flooding, drought conditions prevented restoration efforts until March 2008, when SJRWMD decided to use water from Lake Apopka to flood the fields. Because of extensive bird activity in the delivery canals after initial flooding, in April 2008, staff collected fish from the canals and recently flooded fields for OCP analysis. All data were well below established trigger values. The first planned quarterly fish sampling occurred in July 2008 and OCP results were, again, well below established trigger values.

Table 6.3 Select OCP levels (μg/kg wet weight) in composite samples of mosquitofish (*Gambusia holbrooki*) collected from Duda August 2002 through November 2007; $N = 133$

OCP	Mean	Median	Range	Trigger
4,4'-DDTx	52	31	0.5–366	3,600
Dieldrin	6.0	3.2	0.5–48	690
Toxaphene	193	130	23–860	4,300

DTTx is the sum of 4,4'-DDE, 4,4'-DDD, and 4,4'-DDT

Fig. 6.5 A tractor pulls a special plow that places potentially contaminated soil deep below the surface as part of a soil inversion project. Credit SJRWMD

In addition to the restoration of the properties described above, the SJRWMD has been investigating novel methods for large-scale remediation of contaminated organic soil areas since 1999. Depending upon the remediation technique employed, costs could range from \$6.7 to \$62 million (MACTEC 2005). One of the methods developed over the last decade for the NSRA is the deep soil plowing-inversion technique (Fig. 6.5). This technique uses 565–600 horsepower tractors to pull four-bottom Baker reversible disc plows, with 1.3-m disc blades. The plows were custom built to plow up to a meter deep and essentially flip the soil, bringing up uncontaminated soil to cover the top 30 cm of contaminated soil. Because the OCPs bind tightly to the highly organic peat soils and have not been shown to move vertically in these soil types, this technique seems to provide a safe and feasible method of large-scale remediation for the NSRA.

After years of validation studies, which showed that a 50% reduction of OCPs in the biologically active layer (surface to 0.3 m) was very feasible, the deep soil plowing and inversion technique were implemented at the field level beginning in 2007. By May 2009, 1,619 ha of highly contaminated land was remediated at a cost of \$9.6 million, with a median 67% reduction of DDE, down from a dry weight average of 2,330 μg/kg to 773 μg/kg ($n=339$) (Brown and Bartol 2009).

In 2008, SJRWMD and NRCS submitted a biological assessment and restoration plan for the 526-ha Phase 2 site, which was the first area to undergo deep soil plowing-inversion. Importantly, the Phase 2 restoration plan incorporated new information gained from the previous restoration activities, new trigger levels, and

Table 6.4 Select OCP levels (μg/kg wet weight) in composite fish samples collected from Phase 2 August 2009 through March 2010; $N=63$

OCP	Mean	Median	Range	Trigger
4,4'-DDE	583	476	19–2,520	1,500
Dieldrin	41	20	4–201	140
Toxaphene	344	232	30–1,450	5,000

information gained from a number of research studies conducted by SJRWMD and others. The USFWS reviewed the plan and concurred with NRCS that the project may affect, but was unlikely to adversely affect Federally listed species. Currently, monitoring is ongoing at this site and early results indicate that mean fish levels are below trigger levels (Table 6.4).

As monitoring results accrue, indicating whether the remediation is successful, biological assessments and restoration plans for rehydrating the remaining dry areas are being developed.

Conservation and Scientific Gains

A great number of conservation benefits have been realized since the purchase of the NSRA in 1998, even when considering the avian mortality event. The bird deaths brought attention to the potential risks of restoring agricultural lands back to wetlands, particularly on organic soils in the South where multiple crop rotations each year have left a legacy of high residual pesticide loads. Across the nation, organizations involved in wetland restoration took notice, including those involved in Everglades restoration. Indeed, risk assessment and risk management strategies have been developed and implemented to prevent a similar event from occurring in the Everglades. And, although no one can know whether another mortality event would have otherwise occurred in the Everglades, it is clear that the risk management strategies provided significant conservation benefits to the many species of piscivorous birds that inhabit the Everglades.

Another conservation benefit was the amount of research conducted in response to the mortality event. Studies funded by SJRWMD and NRCS have increased the understanding of how OCPs move from flooded soils into aquatic biota and the critical role of total organic carbon in soil and lipid levels in fish (Coveney et al. 2008; Sepúlveda et al. 2005). This increased understanding has led to improved models that more accurately predict OCP levels in fish based on preflood concentrations of OCPs in dry soil. Other studies funded by SJRWMD and NRCS have increased knowledge of how OCPs in fish are accumulated and stored in the fat, brain, and liver of fish-eating birds (Gross et al. 2009). These studies are completed and manuscripts are being prepared for publication in scientific journals. Other research that is presently underway is examining species abundance of amphibians, as well as OCP levels, movements, and nesting ecology of alligators in the NSRA (Rauschenberger et al. 2010).

In addition to novel research, the restoration of each individual area has a monitoring component to verify assumptions made during the planning phase and to validate bioaccumulation models at the field scale. This monitoring has not only verified assumptions and models, but has also provided important information regarding bird usage, vegetative changes, and reductions in OCP levels.

Nearly 50% of wetland habitat in Florida has been destroyed by drainage and development (Fretwell et al. 1996). As a consequence, wading bird populations have declined by almost 90% since the early 1900s (Robertson and Kushlan 1974). Watershed restoration at Lake Apopka will recreate approximately 8,000 ha of herbaceous wetland with substantial regional benefits to populations of wetland-dependent birds and other species. Restoration of shallow wetland habitat at Lake Apopka, along with the protected upland areas surrounding them, has already created an important bird-use area that now rivals other top birding spots in the state. Evidence for the regional importance of the area is partly provided by observations of high species diversity; for instance, 173 species of birds were observed in Audubon's 2008–2009 Christmas Bird Count (LeBaron 2009).

Eutrophication is one of the major causes of water quality degradation in Florida (FDEP 2008). Restoration of wetland habitat on the former farms will play a major role in correcting nutrient pollution to Lake Apopka; the fourth largest lake in Florida, and a headwater lake to the major tributary (Ocklawaha River) to the St. Johns River. Water quality benefits have already been achieved in Lake Apopka, including a 54% reduction of total phosphorus, a 37% decline in phytoplankton density, and a 47% improvement in water transparency (Coveney et al. 2005). Once the restoration goals for the lake have been achieved, approximately 20,500 ha of primarily wetland and aquatic habitat will have been restored for wetland-dependent and aquatic species.

Conclusions

The salient conclusion that can be drawn from this work is that ecological restoration is not without risks. At Lake Apopka it is clear that the pre-existing science and approaches for risk assessment underestimated the risks posed by organochlorine pesticides. In other locations, different risks could be underestimated. Nevertheless, given the scale of habitat loss nationally and globally, it is essential that restoration efforts proceed. It is important that, when unexpected, deleterious results occur, all parties react cooperatively rather than combatively, and that adjustments rapidly be made.

We also observed that restoration requires adaptive management. For adaptive management to work well, projects should proceed in a stepwise fashion, with the size of each step dependent on the situation and the potential risks at hand. By taking small steps that allow good control, and rapid reversal, deleterious effects can be minimized in time and space. Adaptive management cannot work without sufficient monitoring. In addition to ground-truthing models and assumptions, the monitoring provides information on habitats (e.g., acres of wetlands restored) and species (e.g., number of birds using the site).

Also, the success, to date, of the step-wise project at Lake Apopka demonstrates that simple site-specific bioaccumulation models such as the Biota Sediment Accumulation Factor, combined with a posthydration monitoring plan to validate assumptions, can effectively guide a restoration project. The benefit of using simple models is that model parameters are relatively easy to verify through monitoring, which closes the adaptive management loop.

Finally, ecological degradation occurring over decades – requires decades to reverse. Restoration of ecological forcing factors can take years of effort. In our case, we needed to deal with issues of legacy pesticides and nutrients, and hydrology. Once the forcing factors are where they need to be, it can take decades to reach project goals through ecological succession. Remaining risk posed by pesticides, although mitigated, will decline slowly over a period of many years. Moreover, initial plant communities may differ quite markedly from those that constitute the ultimate goal. Managers must have a long-term view and must communicate that perspective to policy makers and the public.

Acknowledgments We appreciate and thank the many people, both volunteers and agency personnel, who assisted with the bird mortality event. For those who have dedicated their careers to protecting and improving habitat for wildlife it was a trying time. In particular, we thank Resee Collins and the Audubon of Florida Center for Birds of Prey, the Suncoast Seabird Sanctuary, Gian Basili, Julie Hovis, Marilyn Spalding, Jay Herrington, Karen Benjamin, Ken Murray, Elizabeth Mace, Steven Richter, Paul Ek, James Peterson, Bruce Corley, Frank Kuncir, and Harold Weatherman. The findings and conclusions in this chapter are those of the authors and do not necessarily represent the views of the U.S. Fish and Wildlife Service or the St. Johns River Water Management District.

References

Angelo MJ (2009) Stumbling toward success: a story of adaptive law and ecological resilience. Neb Law Rev 87:701

ATRA, Inc (1997) Environmental risk assessment of a Lake Apopka muck farm wetlands restoration. Publication No. SJ98-SP7. St. Johns River Water Management District, Palatka, FL

ATRA, Inc (1998) Comparison of muck farm restoration levels to concentrations in Lake Apopka North Shore Restoration Area. Prepared for St. Johns River Water Management District, Palatka, FL

Battoe LE, Coveney MF, Lowe EF, Stites DL (1999) The role of phosphorus reduction and export in the restoration of Lake Apopka, Florida. In: Reddy KR, O'Connor GA, Schelske CL (eds) Phosphorus biogeochemistry of subtropical ecosystems. Lewis, Boca Raton, FL, pp 511–526

Blus LJ (1996) DDT, DDD, and DDE in Birds. In: Beyer W (ed) Environmental contaminants in wildlife. Lewis, Boca Raton, pp 49–71

Brown C, Bartol T (2009) Large scale pesticide remediation using deep soil inversion (abstract). In: Florida Remediation Conference; October 14–15 2009; Orlando, Florida

Burgess JE (1964) Summary report of Lake Apopka, February 18, 1964. Florida State Board of Health Bureau of Sanitary Engineering, Jacksonville, FL

Clugston JP (1963) Lake Apopka, Florida, a changing lake and its vegetation. Q J Fl Acad Sci 26(2):168–174.

Connell, M. H. & Associates, Inc (1954) Report on water control and navigation, Lake Apopka, Florida, and connecting waterways for Lake Apopka Recreation, Water Conservation and Control Authority. Orange County, Florida, p 27
Conrow R, Coveney M, Lowe E, Marzolf E, Spalding M (2003) Mortality of fish-eating birds on former farms at Lake Apopka, Florida. Platform Presentation. Abstract: SETAC North America 24th Annual Meeting, 9–13 Nov 2003, Austin, TX
Coveney MF, Lowe EF, Battoe LE (2001) Performance of a recirculating wetland filter designed to remove particulate phosphorus for restoration of Lake Apopka (Florida, USA). Water Sci Technol 44(11–12):131–136
Coveney MF, Lowe EF, Battoe LE, Marzolf ER, Conrow R (2005) Response of a eutrophic, shallow subtropical lake to reduced nutrient loading. Freshw Biol 50:1718–1730
Coveney MF, Conrow R, Lowe EF, Marzolf ER (2008) Bioaccumulation of organochlorine pesticide residues from organic soils to fish on flooded agricultural lands at Lake Apopka, Florida, USA. Poster Presentation. Abstract: SETAC North America 29th Annual Meeting, 16–20 Nov 2008, Tampa, FL
Davis JH (1946) The peat deposits of Florida, their occurrence, development, and uses. Florida Geological Survey Bulletin 30. Tallahassee, FL
Dequine JF (1950) Results of rough fish control operations in lake Apopka during December 1949 and January 1959. Mimeographed report, Florida Game and Fresh Water Fish Comm., p 7
DOJ (2003) Memorandum of understanding resolves criminal investigation and natural resources damages claims against the St. Johns River Water Management District. October 3, 2003. U.S. Dept. of Justice, Washington D.C. http://www.justice.gov/opa/pr/2003/October/03_enrd_556.htm. Accessed 14 Dec 2009
DOJ, SJRWMD (2003) Memorandum of understanding between the St. Johns River Water Management District and the United States of America. http://sjrwmd.state.fl.us/lakeapopka/MOU.html. Accessed 17 Dec 2009
Exponent (2003) Analysis of avian mortality at the North Shore Restoration Area of Lake Apopka in 1998–1999. Special Publication SJ2004-SP1. St. Johns River Water Management District, Palatka, FL
Fernald EA, Purdum ED (eds) (1998) Water resources atlas of Florida. Institute of Public Affairs, Florida State University, Tallahassee, FL
Florida Department of Environmental Protection (2008) Integrated water quality assessment for Florida: 2008 305(b) Report and 303(d) List Update. FDEP, Tallahassee, Florida, p 142
Fretwell JD, Williams JS, Redman PJ (1996) National water summary on wetland resources. USGS Water-supply Paper 2425
Gross TS, Sepúlveda MS, Grosso J, Fazio A (2009) Data report: an evaluation of the bioaccumulation of organochlorine pesticides in great egrets (*Casmerodius albus*): laboratory model for the North Shore Restoration Area at Lake Apopka. Special Publication SJ2009-SP5. St. Johns River Water Management District, Palatka, FL
Harper J (1952) Florida's best fishing hole. Orlando Sunday Sentinel-Star 23 [Orlando, FL] 23 March 1952:35
Hoge VR, Conrow R, Stites DL, Coveney MF, Marzolf ER, Lowe EF, Battoe LE (2003) SWIM (Surface Water Improvement and Management) Plan for Lake Apopka, Florida. St. Johns River Water Management District, Palatka, FL
Huffstutler KK, Burgess JE, Glenn BB (1965) Biological, physical and chemical study of Lake Apopka 1962–1964. Florida State Board of Health, Bureau of Sanitary Engineering, Jacksonville, FL
Jenab SA, Rao DV, Clapp D (1986) Rainfall analysis for northeast Florida. Part: Summary of monthly and annual rainfall data. Technical Publication SJ86-4. St. Johns River Water Management District, Palatka, FL
Jennings ML, Percival HF, Woodward AR (1988) Evaluation of alligator hatchling and egg removal from three Florida lakes. Proc. Annu. Conf. SEAFWA. pp 283–294
LeBaron GS (2009) The 109th Christmas Bird Count. Am Birds 63:2–8

Lowe EF, Battoe LE, Coveney MF, Stites DL (1999) Setting water quality goals for restoration of Lake Apopka: inferring past conditions. Lake Reserv Manage 15:103–120
Lowe E, Conrow R, Coveney M, Marzolf E, Richter S, Schell J, Gross T, Sepúlveda M (2003) Pesticide toxicosis in birds at Lake Apopka, Florida: remaining uncertainties. Platform Presentation. Abstract: SETAC North America 24th Annual Meeting, 9–13 Nov 2003, Austin, TX
MACTEC (2005) Lake Apopka North Shore Restoration Area feasibility study Orange and Lake counties, Florida. Final report Project No. 609604004 for the St. Johns River Water Management District, Palatka, FL, p 200
Morris A (1974) Florida place names. University of Miami Press, Coral Gables, FL
NAS (National Audubon Society) (1999) American birds. The 99th Christmas Bird Count, pp 245–246
Peakall DB (1996) Dieldrin and other cyclodiene pesticides in wildlife. In: Beyer WN, Heinz GH, Redmon-Norwood QW (eds) Environmental contaminants in wildlife: interpreting tissue concentrations (SETAC Special Publication). Lewis/CRC, Boca Raton, FL
Rauschenberger RH, Carter CB, Throm RWD, Woodward AR (2010) Alligator and amphibian monitoring on the Lake Apopka North Shore Restoration Area – assessing organochlorine pesticide levels and potential biomonitors. Final Report for the St. Johns River Water Management District. Palatka, FL
Rice KG, Percival HF (eds) (1996) Effects of environmental contaminants on the demographics and reproduction of Lake Apopka's alligators and other taxa. Florida Cooperative Fish and Wildlife Research Unit, U.S. Biol. Serv. Tech. Rep. 53, p 85
Rider D (1963) DDT Caused Fish Kill. Orlando Sentinel [Orlando, FL] 9 August 1963: A1+
Robertson WB, Kushlan JA (1974) The southern Florida avifauna. In: Gleason PJ (ed) Environments of South Florida: present and past. Miami Geological Society, Miami, FL, p 452
Schelske CL, Coveney MF, Aldridge FJ, Kenney WF, Cable JE (2000) Wind or nutrients: Historic development of hypereutrophy in Lake Apopka, Florida. Arch Hydrobiol Spec Issues Advanc Limnol 55:543–563
Sepúlveda MS, Gross TS, Ruessler SD, Grosso JA (2005) Microcosm evaluation of the bioaccumulation of organochlorine pesticides fromsoils in the North Shore Restoration Area at Lake Apopka. Special Publication SJ2005-SP11. St. Johns River Water Management District. Palatka, Fl
Sime W (1995) About some lakes and more in Lake County. Lake County Historical Museum, Tavares, FL
Stickel WH, Stickel LR, Spann JW (1969) Tissue residues of dieldrin in relation to mortality in birds and mammals. In: Miller MW, Gerg GG (eds) Chemical fallout: current research on persistent pesticides. Charles C. Thomas, Springfield, IL, pp 174–204
Stickel WH, Stickel LF, Coon FB (1970) DDE and DDD residues correlated with mortality of experimental birds. In: Deichmann WB (ed) Pesticides symposia, Seventh Int. Am. Conf. Toxicol. Occup. Med. Helios, Miami, FL, pp 287–294
Stickel WH, Reichel WL, Hughes DL (1979a) Endrin in birds: lethal residues and secondary poisoning. In: Deichman WB (Organizer) Toxicology and occupational medicince. Elsevier/North Holland, New York, pp 397–406
Stickel LR, Stickel WH, McArthur RA, Hughes DL (1979b) Chlordane in birds: a study of lethal residues and loss rates. In: Deichmann WB (Organizer) Toxicology and occupational medicine. Elsevier/North Holland, New York, pp 387–396
Swanson HF (1967) Letter to C.W. Sheffield, Chairman, Lake Apopka Technical Committee from H.F. Swanson, County Agent, Orange County Agricultural Center. July 24, 1967
U.S. EPA (1986) Guidelines for the health risk assessment of chemical mixtures. EPA/630/R-98/002. U.S. Environmental Protection Agency, Risk Assessment Forum, Office of Research and Development, Washington, DC
U.S. EPA (1994) Tower Chemical Company Superfund site biological assessment, March 1994. United States Environmental Protection Agency, Region IV, Athens, Georgia, pp 21
U.S. EPA (2001) Screening level biological risk assessment, Tower Chemical Site, Clermont, FL. United States Environmental Protection Agency, Region IV, Athens, Georgia, pp 26

USFWS (1999) U. S. Fish and Wildlife Service investigating wildlife deaths at Lake Apopka, Florida. U.S. Fish and Wildlife Service, Southeast Region. Press Release: R99-016, 9 February 1999. http://www.fws.gov/southeast/news/1999/r99-016.html. Accessed 12 Dec 2009

USFWS (2001a) Lab results released from Lake Apopka wildlife death investigation. U.S. Fish and Wildlife Service, Southeast Region. Press Release: 11 June 2001. http://www.fws.gov/southeast/news/2001/ga01-007.html, Accessed 12 Dec 2009

USFWS (2002) Biological opinion for the interim restoration plan for the Duda sub-East property. U.S. Fish and Wildlife Service, Jacksonville, Florida

USFWS (2004) Lake Apopka Natural Resource Damage Assessment and Restoration Plan. Produced by Industrial Economics, Incorporated, Cambridge Massachusetts, with assistance from St. Johns River Water Management District Palatka, Florida. http://restoration.doi.gov/Case_Document_Table_files/Restoraton_Docs/plans/FL_Lake_Apopka_RP_06-04.pdf. Accessed 13 Oct 2010

Wellborn TL (1963) Inspection of fish kills at Lake Apopka, Florida. U.S. Department of the Interior, Fish and Wildlife Service Bureau of Sport Fisheries and Wildlife. Atlanta, GA

Wiemeyer SN (1996) Other organochlorine pesticidies in birds. In: Beyer W (ed) Environmental contaminants in wildlife. Lewis, Boca Raton, FL, p 101

Williams T (1999) Lessons from Lake Apopka. Audubon Magazine July–August: 64–72

Woodward AR, Percival HF, Jennings ML, Moore CT (1993) Low clutch viability of American alligators on Lake Apopka. Fl Sci 56(1):52–64

Chapter 7
Controlling Wireworms Without Killing Wildlife in the Fraser River Delta

John E. Elliott, Laurie K. Wilson, and Robert Vernon

The Fraser River drains much of the Canadian province of British Columbia, flowing through a landscape dominated by mountain and forest, much of it in a relatively wild state. Emptying into the ocean in the southwestern corner of the province, it forms an alluvial delta of flat lands almost European in appearance, where intensive agriculture intersperses the urban development of Metro-Vancouver. The estuarine habitat is highly productive and still supports substantial populations of waterfowl and other wildlife. Reaching the Pacific Wildlife Research Centre on one of the estuarine islands involves a commute through areas of intensive farming. As with many roads, it is not uncommon to find dead or dying waterfowl and raptors en route, normally victims of vehicle strikes. Many of the raptors and some waterfowl that we found while commuting were later tested and a surprising number were poisoned by pesticides. One of us even found two pesticide-poisoned raptors during a noon hour jog on the dykes separating the Fraser River from adjacent farmland. It raises the question as to whether this was purely the coincidence of wildlife habitat, intensive agriculture and interested biologists? Alternately, was it evidence that pesticide poisoning of wildlife was, and possibly still is, a larger problem than is commonly recognized or appreciated?

Abstract We studied the poisoning of birds of prey and waterfowl by anti-cholinesterase insecticides from 1989 to the present in the lower Fraser Valley of British Columbia, Canada. It began as an investigation of causes of death of the bald eagle (*Haliaeetus leucocephalus*) with a focus on the role of lead shot. During the first year, however, a number of eagles and other birds of prey were discovered dead and debilitated from unknown causes. A forensic investigation revealed acute poisoning mainly by the carbamate compound, carbofuran. We subsequently showed that

J.E. Elliott (✉)
Environment Canada, Science and Technology Branch,
Pacific Wildlife Research Centre, Delta, BC, V4K 3N2, Canada
e-mail: John.elliott@ec.gc.ca

J.E. Elliott et al. (eds.), *Wildlife Ecotoxicology: Forensic Approaches*,
Emerging Topics in Ecotoxicology 3, DOI 10.1007/978-0-387-89432-4_7,

these non-persistent non-bioaccumulative pesticides did indeed persist in local soils from spring application well into the following winter. They could then be ingested by waterfowl as they intensely foraged across the delta farmlands. Seasonal and long-term trends in eagle populations and their winter foraging behaviour contributed to the high rates of poisoning. Carbofuran had been introduced as a replacement for the organochlorine insecticides such as aldrin and heptachlor primarily to control the introduced soil pests known as wireworm, larvae of the *Agriotes* click beetles. From 1990 to 1999, three organophosphorus (OP) insecticide wireworm control alternatives, fensulfothion, phorate and fonofos, were shown to persist and poison raptors and were each removed in turn from the market. In the early 2000s under the guidance of multi-stakeholder committee, the British Columbia Wireworm Committee, a fourth OP compound, chlorpyrifos, was introduced and has been used for 10 years under an integrated pest management framework to effectively control wireworm, and while it likely has killed some waterfowl, has not been linked to poisoning of birds of prey. Thus, with focused effort and cooperation among agricultural, wildlife and regulatory communities, effective pest control can be achieved without unacceptable poisoning of non-target wildlife.

Introduction

There is an extensive literature documenting the exposure and effects of agricultural pesticides and environmental contaminants on wildlife (for a brief history, see Rattner 2009; Rattner et al. 2011). In the 1930s biologists began to investigate the impact of vertebrate pest control products, such as thallium and strychnine, on non-target wildlife. By the 1960s, there were efforts in many countries to assess the environmental impact of organochlorine (OC) insecticides, particularly DDT. With restrictions on the use of those compounds, emphasis shifted in the 1970s to their replacements, the cholinesterase-inhibiting group of organophosphorus and carbamate insecticides. This chapter focuses on that latter group of chemicals and their effects, particularly on birds of prey. We have tried to put the science in context of the broader ecological as well as socio-economic factors behind the agricultural pest problems and regulatory decision making. We have also woven some human element into the narrative.

History and Background

With the nineteenth century expansion of European settlement across the southern half of North America, populations of large predators, such as the bald eagle (*Haliaeetus leucocephalus*), went into widespread decline from persecution and habitat destruction. Some eagles survived and reproduced until the early 1950s, even in regions with significant human presence. However, post WW II, the explosion in new technologies included development and use of many new synthetic organic chemicals. Associated with the increase in commercial chemical use, millions of kilograms of toxic materials were released to environments all over the world, intentionally as part of agricultural pest control and unintentionally from industrial wastes. Chemical stability was an integral trait of many of those chemicals, and if released to the ambient environment caused them to persist for years in sinks such as soils and sediments. That feature combined with a tendency to be highly soluble in fat resulted in ready uptake by biota and transfer from prey to predator. It combined for a new phenomenon, bioaccumulation and biomagnification or movement up food chains of xenobiotic chemicals, some of which were highly toxic. Populations of eagles, peregrine falcons (*Falco peregrinus*), ospreys (*Pandion haliaetus*) and some hawk species which had managed to survive persecution and habitat loss, now disappeared completely from much of the continental USA and parts of southern Canada. In the case of highly migratory species such as the peregrine falcon, they even declined at breeding sites in the arctic. The impact of DDT on reproduction and other organochlorine insecticides such as aldrin/dieldrin and heptachlor on eagles and other raptors is a well-known story and needs no further recounting here (see, e.g., Newton 1979, 1986, Stalmaster 1987; Poole 1989; Beans 1996; Elliott and Harris 2001). The western eagle populations of British Columbia, the Yukon and Alaska largely escaped such problems, given the relative lack of intensive agriculture and industrial development. Fortunately, for example, forest pests in this region were not controlled by widespread spraying with DDT. However, in the Fraser River Delta, eagles did largely disappear. In the early 1970s when regular Audubon Christmas Bird Counts began in the area, only a couple of bald eagles were observed.

As the bald eagle disappeared from the Delta landscape, a number of other much smaller, lower profile species were proliferating. That included wireworms, which are the larval, subterranean stage of several species of beetles known as "click beetles."[1] Wireworms, which can take between 3 and 5 years to develop in the soil depending on the species, can be major pests of many key agricultural crops including cereals such as wheat, forage crops especially corn, canola, pulse crops, potatoes, vegetables, berry fruit and many more (Fig. 7.1).

About 30 species of wireworms are known to be pests of agricultural land in British Columbia, but in the Fraser River Delta the lined click beetle, *Agriotes lineatus*, and the dusky wireworm, *A. obscurus*, are by far the most important (Wilkinson 1963). Sometime around 1900 or earlier, those wireworms were introduced from Western Europe into southwestern British Columbia and have gradually spread throughout the lower Fraser Valley and Vancouver Island. It is thought those exotic wireworms arrived in British Columbia via ship ballast (soil) or in soil accompanying imported crops such as hops. Where exotic wireworms have become

[1]Click beetles get their name from their ability to right themselves when on their backs by flexing and snapping a spine on the prosternum, which sommersaults the beetle into the air with a loud "click." These beetles have been known to keep entomologists (and their children and guests) amused for hours.

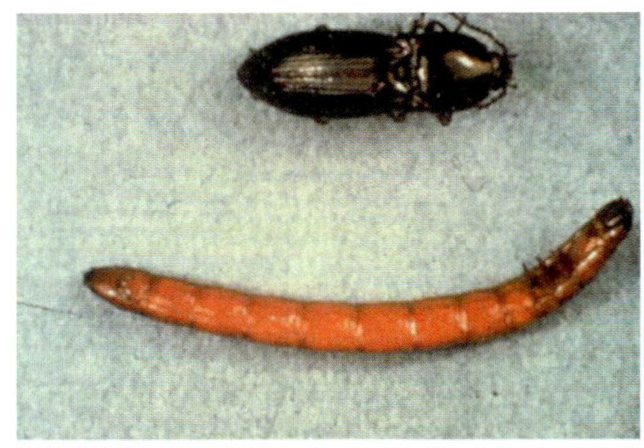

Fig. 7.1 The upper image is of a dead adult click beetle with its larval form, the wireworm, shown below (photo by B. Vernon)

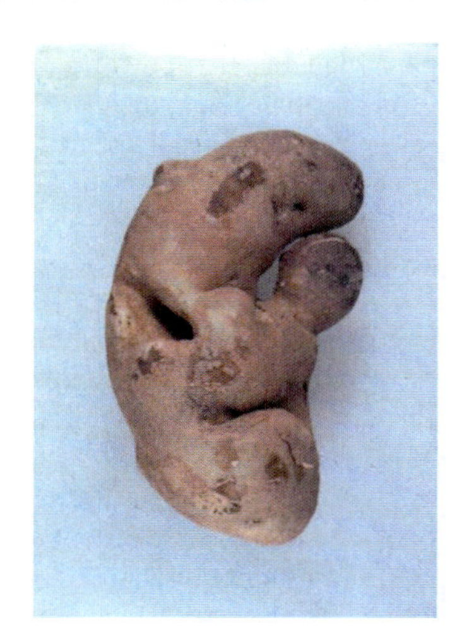

Fig. 7.2 An example of "Ugly Potato Syndrome" showing how wireworms damage products making them unsuitable for marketing (and peeling, etc.) (photo by B. Vernon)

established in Canada, they have largely replaced the indigenous species, and during the past several decades have become a primary pest of potatoes in the Fraser River Delta as well as in Atlantic Canada (Fig. 7.2).

During the 1950s and 1960s, the importance of wireworms as pests of many crops in Canada declined due to the widespread use of several chlorinated hydrocarbons on agricultural lands. The OC pesticides, aldrin and heptachlor, for example, not only controlled existing wireworm populations, but their persistent residues in soil continued to kill early instar wireworms for up to 13 years with a single application (Wilkinson 1963; Wilkinson et al. 1976). Other common OCs applied to the soil during those decades, including dieldrin and DDT, also likely had an immediate and residual impact on wireworms for years, even after they were banned in the early 1970s from use in Canada (Szeto and Price 1991). It is now widely accepted by many entomologists that the gradual increase in wireworm damage occurring throughout the northern hemisphere is due in part to the gradual decline of OC residues in agricultural soils.

By the late 1980s throughout the Fraser River Delta, wireworms once again began to be problematic for a number of crops, including potatoes. In infested potato fields, wireworms are attracted to potato tubers in the late summer and early autumn, and burrow deep into the potato flesh, leaving several darkened holes. Two or more of these conspicuous blemishes will render a tuber unmarketable, and entire crops can be downgraded or refused by retailers or processors. When that began to occur during the late 1980s in the Fraser River Delta, many growers resorted to use of soil-applied granular organophosphate insecticides for wireworm control, which once again put wildlife and agriculture unknowingly on a collision course.

This chapter relates how efforts to control an unwanted pest can have broad and unexpected consequences to the top of the food chain, how global market forces influence decision making processes, and finally how progress can be made both in controlling the pest and conserving associated wildlife by concerted efforts of concerned parties.

The Fraser Delta

The Fraser is the seventh largest of the North American rivers, draining an area of 234,000 km^2, mostly in British Columbia Canada, and flowing 1,375 km from headwaters in the continental divide. The river forms an alluvial floodplain extending back some 100 km from the mouth where it empties at the municipality of Delta with an average flow of 3,972 m^3/s. Beginning in 1894 following a catastrophic flood, most of the lower floodplain region, including the larger islands, was dyked and drained. The land base is now largely in agricultural or urban uses.

From at least 8,000 years before present, the Fraser Delta was inhabited by the peoples of the Coast Salish First Nations, where they harvested salmon and other marine and upland resources. Many descendants continue to live on lands of the Tsawwassen and other bands, and in other communities in and around Delta. Agriculture began in the area in the late 1860s. Initially, much of the land use was devoted to pasture and hay for feeding of horses and other livestock, and to support the expansion of timber harvesting throughout British Columbia. Until the late 1950s, agricultural production in the Delta continued to have a heavy emphasis on pasture and forage crops for dairy and livestock; at that time a tunnel was constructed

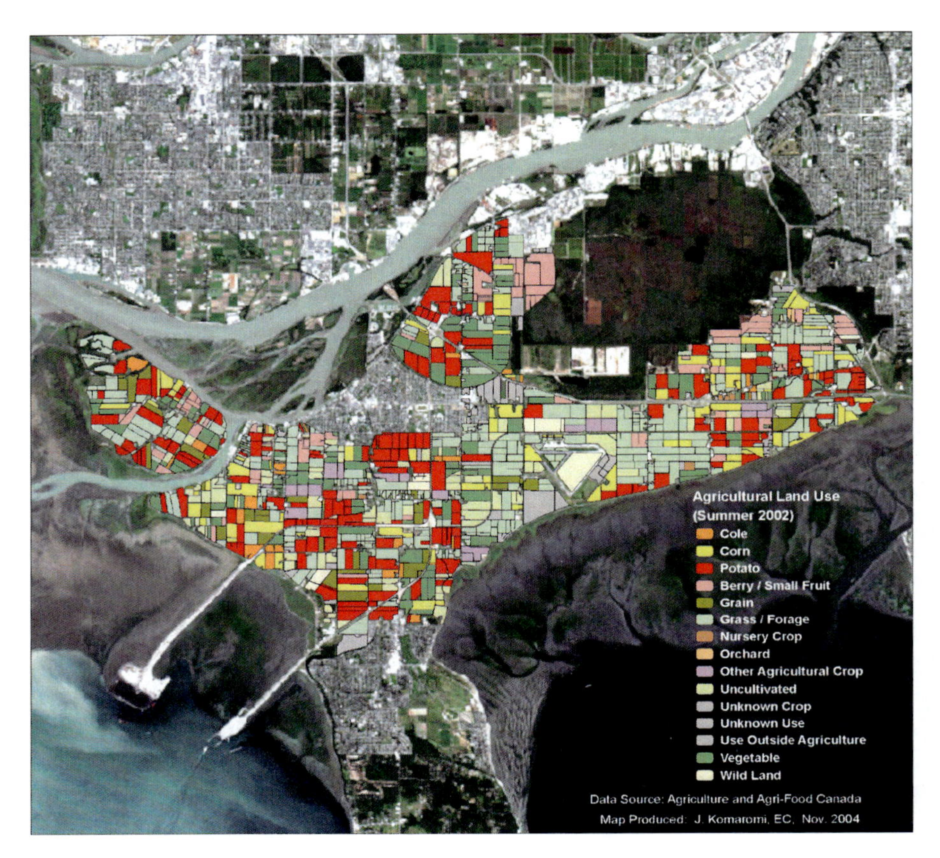

Fig. 7.3 The Fraser River Delta. The coloured overlay is of agricultural field use in 2002. Note the tidal mud flats surrounding the Delta farmlands, which provide rich wildlife habitat and a ready source of waterfowl which forage in agricultural fields

to link the Delta lands south of the Fraser River with Vancouver and other urban areas to the north. The rapid access to that market, along with changes in agricultural technology, particularly the development of new chemical pesticides and fertilizers, led to a diversification of crops. Local farmers supplied a thriving cannery industry in the area, with 15 canneries operating into the early 1960s. However, in the 1970s expropriation of farmland by government looking at future industrial use reduced the incentive for long-term care of the land, and led to an emphasis on production of intensive vegetable crops and dairy. Changing market conditions, such as the North American Free Trade Agreement (NAFTA), meant greater competition from producers in the USA and Mexico, and led in 1989 to the closing of the last local cannery. Through the 1990s and into the present century, agricultural land has been turned increasingly to greenhouse production and to cranberries and blueberries, as farmers have attempted to adapt to changing conditions (Fig. 7.3).

Changing Landscape: Changing Wildlife

The islands and marshes of the Fraser estuary once supported vast numbers of breeding and many more wintering waterfowl, waterbirds, shorebirds, raptors as well as land bird species. However, as the delta was dyked and the land increasingly cleared and turned to agriculture, the wildlife community changed. Nine species of birds[2] have been extirpated from the Fraser Delta or reduced to a tiny fraction of original numbers. The remaining protected areas of the Alaskan National Wildlife area and the associated Reifel Bird Sanctuary along with the South Arm Marshes Wildlife Management area provide a glimpse of the past habitat of the delta. Although 70% of the original wetlands of the lower Fraser Valley have been drained, the area still supports the highest numbers of wintering waterfowl and raptors in Canada (Butler and Campbell 1987; Breault and Butler 1992). Each year approximately 250,000 waterfowl and hundreds of raptors winter in the area, while millions of shorebirds pass through the estuary on migration. The current wildlife community in the delta comprises those species which have adapted to the changing landscape. Many waterfowl, particularly mallards, widgeon, snow geese and trumpeter swans now feed on the agricultural fields.

Some wildlife are viewed mainly as pests by farmers as they consume crops, while flocks, particularly of the larger geese and swans, can compact the soil. The Delta Farmland and Wildlife Trust (http://www.deltafarmland.ca/) works to solve such conflicts and to promote both sustainable agriculture and wildlife populations in the area.

History of Wildlife Mortality in the Delta

The earliest accounts of waterfowl mortality in the Fraser Delta which we can confidently attribute to labelled pesticide usage date from November, 1974. At least 50 dabbling ducks of various species were found dead in a flooded field, which had been treated the previous spring with a Furadan 10G, a granular formulation of carbofuran. Chemical analysis of gut contents and other tissues confirmed of the presence of substantial residues of carbofuran (Wilson et al. 1995). Other reports, similar in timing,

---

[2]Extirpated birds:
Sandhill Crane *Grus candensis*
Yellow-billed Cuckoo *Coccyzus americana*
Barn Owl *Tyto alba*
Western Screech-Owl *Megascops kennicottii*
Burrowing Owl *Athene cunicularia*
Purple Martin *Progne subis*
Horned Lark *Eremophila alpestris*
Western Bluebird *Sialia mexicana*
Yellow-headed Blackbird *Xanthocephalus xanthocephalus*

location and circumstances from the early 1970s described mortalities of ducks, but often due to budgetary constraints, no tissue residue chemistry was performed. There were some instances of unlabeled usage, where the distinct purple silicon granules were present in crop contents of the ducks and even evident in soil samples. Poisoning events, mainly attributed to labelled use, continued throughout the 1970s. Finally, in 1979 by agreement among the agricultural community, wildlife agencies and regulators, carbofuran was withdrawn from use in the lower Fraser Valley.

The Early Forensic Phase

The restrictions on carbofuran use in the Fraser Valley brought about a period of fewer reported waterfowl mortalities. Initial replacements for carbofuran included granular formulations of fensulfothion (Dasanit 10G) and diazinon (Basudin 10G). There were some incident reports, attributed mainly to diazinon (Wilson et al. 1995). The attention of local wildlife toxicologists had turned to other priority issues including lead shot and persistent chemical wastes, such as dioxin from the pulp and paper industry (see, e.g., Chaps. 2 and 12). Thus the first study of the ecotoxicology of birds of prey in the Fraser Delta initially did not have current use pesticides as a high priority, but rather focused on lead shot (Elliott et al. 1992) and chlorinated hydrocarbons (Elliott et al. 1996).

The focused studies of lead shot poisoning began in the autumn of 1989. Dr. Ken Langelier, a private veterinarian based in Nanaimo, British Columbia, proposed a province-wide effort to obtain and autopsy bald eagle carcasses. We focused on the toxicology component. A call for carcasses was circulated to provincial and federal biologists, enforcement officers, wildlife rehabilitators, veterinarians, taxidermists and the public at large.[3] Early in the first year of that study, February of 1990, a group of six bald eagles were found dead or debilitated on agriculture lands in the delta community of Richmond. All of the birds exhibited symptoms which eventually became familiar: poor coordination, clenched talons, constricted pupils, and a distended crop (Porter 1993) (Fig. 7.4). The symptoms were consistent with poisoning by anti-cholinesterase insecticides. Despite documenting the deaths of thousands of waterfowl and other birds from carbofuran and anti-cholinesterases during the 1970s and 1980s, my colleague, Phil Whitehead, had not previously encountered poisoned raptors, so we were surprised by the finding.

Based on recommended procedures, the crop contents of one eagle were surgically removed, and that bird recovered, as did two others, and were eventually released. The remaining raptors were dead on arrival or died shortly after, and in some cases crop contents were also removed. The ingesta of one dead bird, consisting mainly

[3]Lead poisoning of eagles: That work produced data which showed that a substantial portion of bald eagles found dead or debilitated in British Colombia had died of lead poisoning (x%) or had been significantly exposed to lead. The data were valuable for initiating the first provincial ban on lead shot use for waterfowl in Canada, and contributed to the eventual national ban (See Chapter, also Elliott et al 1992; Wayland et al 2003.

Simple figure of cholinesterase inhibition

Normal Reaction (happens very fast)

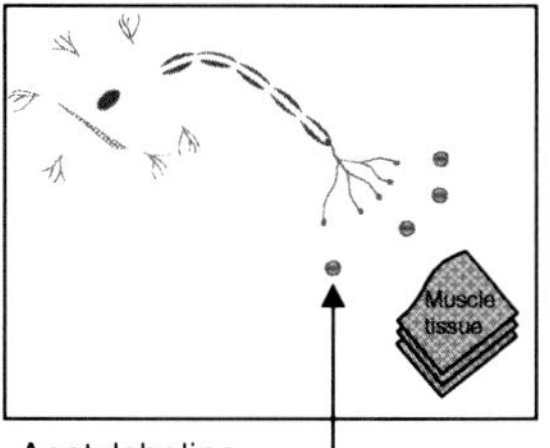

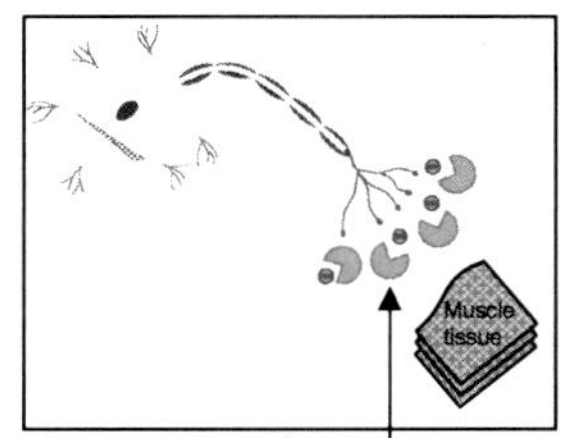

Reaction with

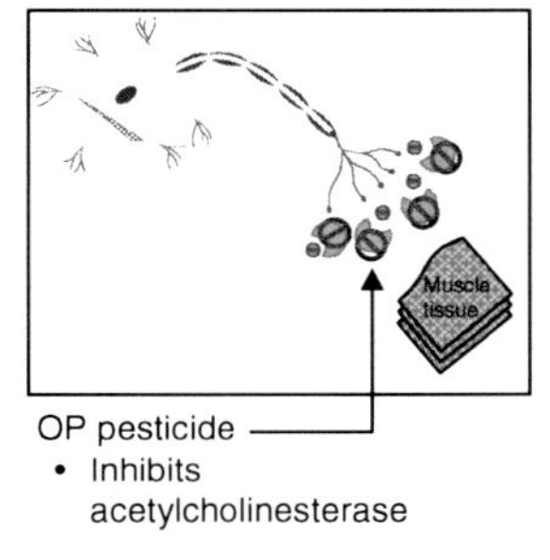

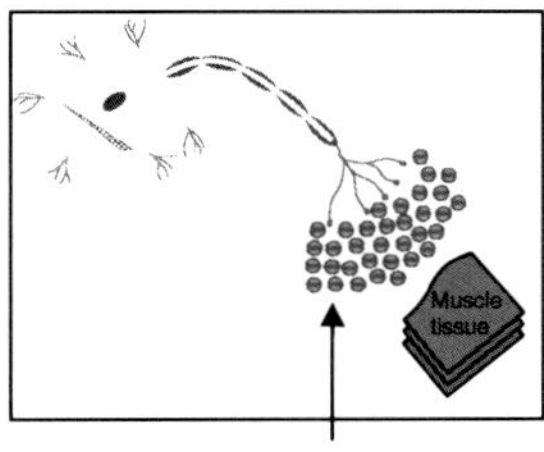

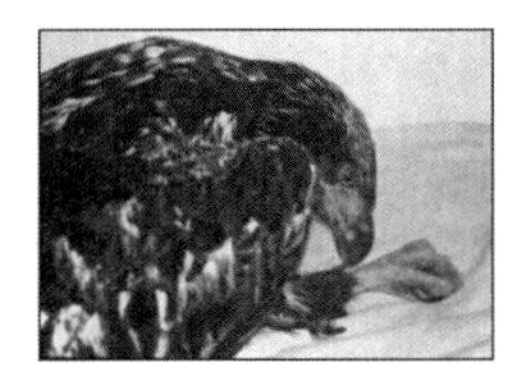

Fig. 7.4 How organophosphorus and carbamate insecticides cause toxicity by affecting the functioning of neurotransmitters, specifically by binding the enzyme, acetylcholinesterase. Birds exhibit classical symptoms of clenched talons, dilated pupils, inability to stand, swollen crop (photo by J.E. Elliott)

of duck remains, contained 200 μg/g carbofuran. Brain cholinesterase activity was inconclusive, likely due to a phenomenon referred to as "spontaneous reactivation." In the post-mortem brain of an animal poisoned by a carbamate insecticide, the inhibited acetylcholinesterase enzyme is decarbamoylated, and if assayed biochemically, will react quite normally, despite sufficient inhibition to cause death (Ecobichon 1991; Greig-Smith 1991). Later in mid-March of that year another group of red-tailed hawks and a bald eagle were also found in the same general area with similar symptoms, and with confirmed carbofuran residues of 2.2 μg/g in the crop contents of one bird (Elliott et al. 1996).

The information was communicated to the regulatory agencies; for the second time they withdrew carbofuran granular from the local market. It appeared that without any consultation, at least with the Canadian Wildlife Service, in 1986 there had been a decision to permit carbofuran granular to be used again in the Fraser Valley. Re-examination of an earlier kill report showed that in September of 1986, more than a thousand songbirds of different species had died in a nearby Richmond field. Carbofuran was found in ingesta and in intestines and gizzards of a number of birds. There were also reports of dead waterfowl in that field later in the winter (Mineau 1993; Wilson et al. 1995).

For the 1990 growing season, granular phorate, as Thimet-10G, was promoted by agricultural agencies as the preferred option for control of wireworm on a variety of crops, particularly potatoes. At that time, the federal government initiated a new funding envelope to support research into the environmental effects of pesticides, the "Pestfund." With funding from that source, we assembled a network of cooperators, particularly among the wildlife rehabilitator community, to obtain samples of birds of prey brought in sick and dead. New funding to the National Wildlife Research Centre provided for greater laboratory support for the field studies. We established a procedure to screen all blood and brains of raptors collected from the Lower Fraser Valley; the spatial scope eventually expanded to include areas of intensive agriculture on south-eastern Vancouver Island.

Over the three winters of 1990–1991 to 1992–1993, a number of raptors were diagnosed as poisoned by anti-cholinesterases; several had significant phorate residues in their crop contents (Elliott et al. 1997). In December of 1993, with repeated reports of sick eagles in fields and parks in the Delta area, and frustrated with 3 years of such problems and no apparent action, rehabilitators turned to the media. That resulted in widespread coverage including a front page article on December 23, 1993 in the Vancouver Sun daily with a colour picture of a poisoned eagle. Even wider media coverage ensued with articles in North American, European and Japanese television, radio and print.

In January of 1994 we received a phone call from a Canadian official of Cyanamid Corp of Princeton, New Jersey, the manufacturer of phorate. Cyanamid offered to cooperate in the kill investigations first by confirming the residue chemistry. In March of 1994, a meeting was held among wildlife and agricultural representatives, Environment Canada managers, regulatory officials and representatives of Cyanamid Corporation. While in Vancouver, the visiting representatives also toured the field sites and the rehabilitation facilities. We presented the evidence that labelled use of phorate was poisoning both waterfowl and raptors. There was some continued skepticism whether the poisonings were due to proper labelled use. Eventually there was an agreement that more work was needed on persistence of phorate in local soils, and possibly on application procedures or training of applicators.

The following week, however, and surprisingly, Cyanamid decided that their preferred course of action was to withdraw phorate from the British Columbia market. That decision was communicated to federal and provincial regulators and accompanied by a press release. Action extended to modifying the Phorate label to read, *Not for use in British Columbia*. That was the first label change in Canada based strictly on wildlife concerns and was an important precedent.

The labelling decision meant an end to phorate poisoning of raptors in the Fraser Delta. However, the local farmers were less than satisfied as they were now deprived of another pest control tool, still available to their competitors everywhere else in North America. Their farming situation was already complicated by increased access to local markets of produce from the USA and Mexico, including large corporate farming operations, made possible under NAFTA (North American Free Trade Agreement). Farming in the Fraser Valley was also subject to regulations under the British Columbia Agricultural Land Reserve which restricted land use options. The response by agricultural and regulatory officials was to recommend another organophosphorus compound for wireworm control, Fonofos, marketed in granular form as Dyfonate G.

Monitoring Fonofos

During the 1994 growing season granular dyfonate was applied to potatoes and other crops in the Fraser Valley. The 3-year Pestfund grant had run out, leaving diminished resources for the monitoring work. Given that avian LD_{50} values for fonofos were greater than for either carbofuran or phorate, there was some hope that it would kill less ducks and, therefore, raptors. However, that optimism was short lived, as rehabilitators soon encountered more pesticide-poisoned raptors. On December 31,1994 a moribund eagle was brought to Monica Tolksdorf's wildlife shelter in Surrey, British Columbia. Some rehabilitators were now sufficiently familiar with symptoms of anti-cholinesterase poisoning to make an accurate diagnosis. However, there were no crop contents for chemical analysis. Plasma cholinesterase activity level was only 27 μmol/min/L in a sample taken on arrival to rehab. In healthy eagles plasma cholinsesterase activity averages 824 μmol/min/L. When reactivated with 2-PAM (pyridine-2-aldoxime methochloride), an assay which indicates organophosphorus exposure, activity increased to 519 μmol/min/L. Fonofos was suspected but not confirmed. In early February, another debilitated eagle was found near agricultural fields. The duck remains from that eagle's gut had 26 μg/g fonofos.

Concerned by those findings, federal and provincial agencies provided new funding to increase monitoring and address priority research questions. Over the next 4 years, we documented 15 cases of raptors poisoned by fonofos, including 12 bald eagles (Elliott et al. 2008). The results were summarized and communicated to the agricultural agencies and regulators. In 1999 the Pesticide Management and Review agency, PMRA, decided that fonofos was no longer an acceptable option for wireworm control in the Fraser Valley. Thus, fonofos followed carbofuran, fensulfothion and phorate as being considered too hazardous to wildlife for use in the region.

Researching the Processes of Pesticide Poisoning

The first raptor poisonings in the winter of 1990 led to many discussions, presentations and meetings with farmers, wildlife biologists, chemists, agricultural and regulatory officials, environmental groups and the broader public. That dialogue led to a series of questions concerning the events. They are expressed below as the simple question, and presented as a research question:

1. How can these pesticides be poisoning birds in March when they were applied in June? Aren't they supposed to break down quickly? (What is the persistence in local Delta soils of granular formulations of the various wireworm control chemicals?)

2. Where are all the dead ducks? (What is the fate of duck carcasses during the Delta winter? Are they scavenged, and by which species and how quickly?)
3. What about the dump? Those birds are being poisoned in the landfill, not in my fields. (What are eagles doing at the Vancouver landfill? Do they feed there? Was there evidence of poisoning of raptors or other wildlife?)
4. Why is this happening now, why were we not finding dead eagles in the days of the carbofuran kills? (What are the long-term trends in eagle populations? What are the seasonal patterns of bald eagle use of Delta fields?)
5. Does it really matter to the eagles or hawks? (How important is pesticide poisoning as a cause of mortality of bald eagles? Is there any threat to population stability?)

How Are These Birds Being Poisoned in Winter? (Granular Persistence Study)

Environmental persistence was a key element that led to the eventual ban of most organochlorine insecticides. The half-life of DDE in soil can be greater than 20 years (Beyer and Krynitsky 1989), and appears to be even longer in some soil and climate conditions (Harris et al. 2000). In comparison, organophosphorus and carbamate insecticides were designed to degrade to non-toxic products within a few weeks of application (Matsumura 1975). Degradation kinetics appeared to be slower in the soils of the Fraser Delta, so we investigated the persistence of a select group of granular organophosphorus and carbamate insecticides. We developed a rather unique approach of placing fixed numbers of granule in small permeable bags, permitting their later retrieval, and planting the bags along with the seed potatoes. The experiment showed that phorate, fonofos and carbofuran persisted in soils well into the fall season, and that the granules retained sufficient activity through the winter to poison ducks ingesting relatively small amounts (Fig. 7.5, Wilson et al. 2002).

Where Are All the Dead Duck? (Scavenging Behaviour Study)

If raptors were poisoned from feeding on duck remains, where were the duck kills? Were carcasses disappearing that rapidly? Beginning in 1996, we devised a field study to investigate the process of scavenging. We mapped transects between the major bald eagle night-roosts in the delta, and used the intersection of those lines to define where the maximum number of eagles might encounter a duck carcass while foraging (Peterson et al. 2001). We then ran a transect line east from that point through the estuarine lands with study fields designated 5, 10 and 15 km from the central point. For a number or weeks, each morning before dawn, duck carcasses were placed randomly adjacent to the transect lines. We used motion-detector-activated still cameras to record the first species to arrive at the carcass. In 1998, we repeated the experiment

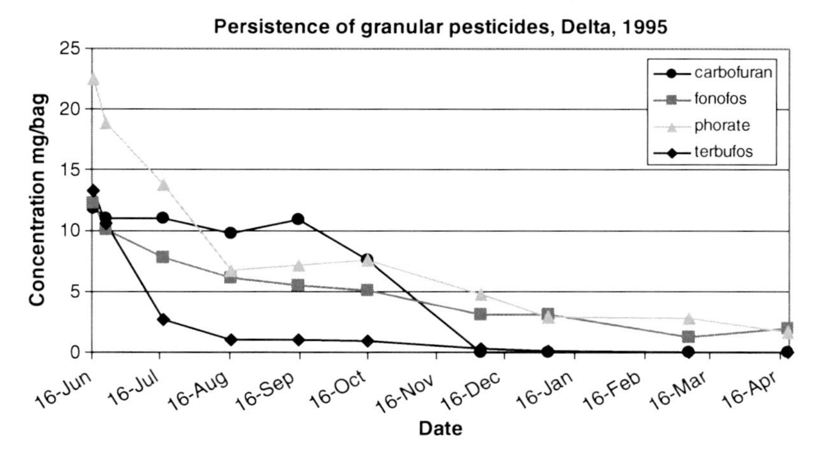

Fig. 7.5 Concentrations of various insecticides as determined in soils following seeding with potatoes in June in the Fraser River Delta of British Columbia, 1995. See Wilson et al. (2002) for details

Fig. 7.6 Adult bald eagle feeding on a duck carcass, while a juvenile waits its turn, in the Fraser River Delta. Photo taken using remote motion-detector cameras placed in fields. (Photo by S. Lee)

using time-lapse video to record the full history of each duck carcass (Figs. 7.6 and 7.7).

As shown in Table 7.1, 78% of the carcasses were found within 24 h, and usually soon after sunrise, with 91% discovered within 48 h (Peterson et al. 2001). Corvids, gulls and other raptors often located the carcass, but eagles soon appeared, and as the dominant predator, usually consumed it. Up to 30 eagles were recorded at a single duck. That study showed the intensity of competition and scavenging pressure.

Fig. 7.7 Scavenged duck remains from the Fraser Delta. (Photo by J.E. Elliott)

Table 7.1 Timing of arrival of first scavenger at duck carcasses placed on transects in the Fraser Delta (Peterson et al. 2001)

	Carcasses found (%)		
	1996	1998	Combined
0–24 h	80	75	78
24–48 h	13	13	13
48–72 h	7	4	6
>72 h	0	8	4
	$n=30$	$n=24$	$n=54$

A single dead duck could poison one or possibly more raptors, and very few duck remains would be left as evidence.

We also investigated primary poisoning of wildlife by insecticides. For 2 years during the fonofos period, 1996 and 1998, 420 ha of potato fields, half of which had been treated the previous spring with fonofos and the remainder untreated, were searched weekly from October to February for evidence of wildlife mortality (Elliott et al. 2008). Beginning at sunrise, fields were surveyed for wildlife activity. Searchers then walked transects, counting and attempting to identify remains and collecting any intact carcasses. Search efficiency was factored by placing duck carcasses the previous evening and unknown to the searchers.

Waterfowl outnumbered other species in field-use counts, and comprised the greatest proportion of birds found dead (Appendix). In total we located 211 wildlife remains; most were already scavenged. We retrieved 35 intact carcasses suitable for post-mortem examination and/or toxicology analyses. Cholinesterase activity was assayed in brains of 18 waterfowl, five of which had severely depressed activity (average inhibition 74%; range, 69–78%). The gastrointestinal tract of a mallard found in a field treated with granular product contained 49 μg/g fonofos residues, linking waterfowl mortality with labelled use of the product. Those findings demonstrated the risk of both primary and secondary poisoning of non-target wildlife by anti-cholinesterase insecticides.

Over the 5 months of field searching each year, we counted greater than 60,000 birds using the 420 ha study area. Those numbers clearly demonstrated the importance of farmland habitat in the Delta. Such intensity of field use by waterfowl showed that waterfowl would probably find even occasional and scattered "pesticide hotspots."

What About the Dump? (The Delta Landfill as a Poisoning Source)

The major urban landfill for Greater Vancouver is located in the municipality of Delta and adjacent to an area of intensive agriculture, particularly for potatoes. Farmers often complained that the landfill rather than their fields were a more likely source of wildlife poisonings. We knew that bald eagles used the landfill particularly in winter; therefore, in 1993 we began to count and observe the behaviour of bald eagles wintering around the refuse dump site. Contrary to expectations, we learned that the majority of the eagles present at the site did little active foraging (Elliott et al. 2006). During the winter upwards of 400 eagles were present at any time on the landfill site; however, only 10%, those we designated as "dump specialists," appeared to acquire their energetic needs by feeding there. The other 90% were roosting, sometimes interacting and possibly taking advantage of a local microclimate. We later confirmed those observations from daily radio-tracking a subset of eagles. Only 2 of 11 tagged birds, both juveniles, spent most of their time at the landfill. We concluded that the majority of eagles wintering in the Delta acquire their energetic needs principally from feeding on waterfowl, other birds and fish rather than the landfill.

We also examined our wildlife mortality records, and found no confirmed cases of eagles poisoned at the Delta landfill. There were surprisingly few records of any wildlife poisoning at the dump. One unusual case involved gulls poisoned by consuming chocolate dumped after Valentine's day (V. Bowes, person. comm.).

Why Is This Happening Now? (Bald Eagle Population Dynamics)

Why was there a sudden onset of bald eagle pesticide poisonings in the early 1990s? Was it due to increased numbers of eagles wintering in the area? Were they arriving earlier, staying longer? Had their behaviour changed, was there more feeding on fields rather than on fish? Or had the degree of surveillance increased with the population and the presence of wildlife rehabilitators?

We began by compiling any available data on trends in wintering bald eagles, such as the Audubon Christmas Bird Count data (Fig. 7.8). Since 1971, a count had taken place every year in the Ladner area of Delta. From the early 1970s until the early 2000s, largely the same group of experienced bird watchers did the Ladner count. Beginning about 1987, there was an exponential increase in the number of bald eagles wintering in the Fraser Delta. The likely reasons for that increase, mainly release from DDT and persecution, were discussed earlier in this chapter. The potential for an increase in poisoning is obvious. We showed from the scavenging and telemetry studies how eagles feed over the fields and locate and scavenge ducks. Thus it appeared that more eagles increased the speed and efficiency that carcasses, including poisoned ones, would be found and eaten. That seemed to account for the sudden appearance of poisoned eagles.

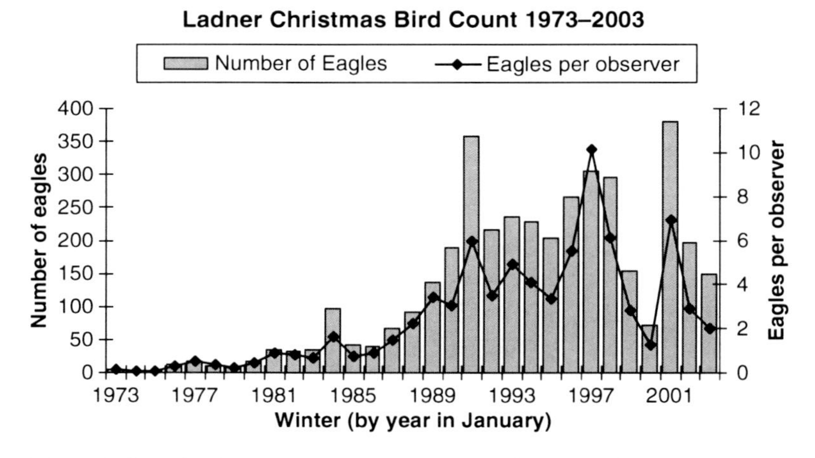

Fig. 7.8 Results from the Audubon Ladner (Delta, British Columbia) Christmas bird count over the period 1973 to 2003

Next we looked at the seasonality of eagle presence in the Delta. We made weekly counts throughout the winter at the major night-roosts in Delta and at the Harrison River, a site where eagles congregate in the fall to feed on spawning salmon, and a relatively short travel time of about 100 km from Delta. As expected, we found that eagles began to appear on the Harrison River in November with the maximum count in early December, coinciding with the peak in spawning of chum salmon *Oncorhynchus keta*. As the availability of salmon carcasses decreased on the Harrison and other regional salmon streams such as the Chekamus and Skagit Rivers, eagle numbers increased in the Delta in search of alternate food sources, particularly waterfowl and gulls (Elliott et al. 2011). When we plotted the timing of peak eagle counts at the night-roosts with the timing of eagle poisonings, we found a remarkable convergence; during the mid-1990s both numbers peaked in February.

Again we concluded that the apparent sudden appearance of an eagle poisoning problem in 1990 was at least part due to the rapid and ongoing increase in the number of eagles in the region, and, therefore, the likelihood of poisoned ducks being found and eaten by eagles. The post salmon-spawning influx of eagles into the Delta further supported this argument, as it explained the seasonal variation in incidence of poisonings, particularly why so many eagles were poisoned in February and on into early March. Increased use of waterfowl wintering areas by bald eagles may also be partly due to long-term decline in salmon stocks. As bald eagle populations recovered across the west, the options for feeding on salmon would have been greatly reduced from the pre-decline era, particularly as result of damming and habitat modifications on the Columbia River system (Elliott et al. 2011).

Bald eagles were not the only raptors poisoned during the course of the study. Not surprisingly, we found a number of red-tailed hawks (*Buteo jamaicansis*) which is the other larger raptor breeding and wintering in the Fraser Valley, and also one known to scavenge. One rough-legged hawk (*Buteo lagopus*) was poisoned by fonofos. Northern harriers (*Circus cyaneus*) were also observed feeding on placed duck carcasses (Peterson et al. 2001). It is likely that some harriers were poisoned over the years, but their tendency to roost in remote salt marshes would reduce the probability of finding carcasses. Based on limited data, the population trends of these other raptors in the Fraser Valley were relatively stable over the time period examined here (Sullivan 1992). Thus, for bald eagles the combination of a rapid increase in population, and their position as the dominant avian predator and scavenger likely accounts for the number of poisoned birds of that species found during the course of this investigation.

Does It Really Matter to Kill a Few Hawks or Eagles? (The Impact of Poisonings at the Population Level)

For birds of prey, like other longer-lived vertebrates, survival to breeding age is more critical to population stability compared, for example, to reductions in reproductive rates (Newton 1979). During the 1960s and 1970s in Britain, population decline of a number of raptor species was attributed to poisoning of adult birds by the insecticide dieldrin rather than reproductive failure from the effects of DDT (Newton et al. 1992; Sibly et al. 2000). Many birds of prey continue to be poisoned by human release of toxicants to the environment. Among the best documented examples is the continuing threat to the California condor posed by poisoning from ingestion of spent lead ammunition (Church et al. 2006). The most dramatic example of how poisoning of breeding age raptors can bring about population decline comes from Asia, and the impact of the veterinarian drug, diclofenac, recounted by Lindsay Oaks in another chapter in this book and in previous articles (e.g., Oaks et al. 2004).

In a previous assessment, we focused on the breeding population of bald eagles in the Fraser Delta and determined that the threat from pesticide poisoning was limited (Henny et al. 1999). That conclusion was reinforced by later evidence from radio-tracking studies which showed that resident breeding eagles spend most of the day in the vicinity of their nests, located in or near the estuarine environment (Elliott et al. 2006), and where they eat a primarily fish diet (Elliott et al. 1998). In contrast, non-resident wintering birds ranged widely over agricultural areas and, therefore, were potentially at greater risk of exposure to pesticides. Over the five winters when fonofos was the recommended chemical control (1994–1999), there were 19 bald eagles confirmed as pesticide poisoned; however, in the winter of 1995–1996, there were eight bald eagles poisoned by anti-cholinesterases, five of which were confirmed as poisoned by fonofos, and all from the Delta. Because of effective treatment by rehabilitators, all of those birds survived; however, we can likely assume they would have died otherwise. That leads to the next question: what is the probable ratio of confirmed to undiscovered mortalities?

As discussed above, we learned that greater than 90% of placed duck carcasses were scavenged within 48 h in the Fraser Delta (Peterson et al. 2001). We assumed that ducks were poisoned by direct consumption of insecticide granules while sieving for food; such direct exposure to the chemical caused rapid intoxication with the result that the birds normally died in the field at the site of exposure. However, dead or debilitated bald eagles most often were found at or near roost sites, demonstrating there was time for the eagle to travel some distance from the point of exposure, or duck carcasses were carried to roost sites for later feeding. We have found poisoned eagles concealed under hedgerow cover, as have workers trimming hedgerow set-asides in the spring. It is likely that sub-adult eagles with their plainer plumage and other more cryptic species would, therefore, largely go undetected. We believe, therefore, that many poisoned raptors are not found, despite the relatively open landscape and intensive human presence in the Fraser Delta. Our carcass collection has been almost entirely fortuitous, relying mainly on the public to report or retrieve birds. Our experience is consistent with other studies (e.g., Balcombe 1986; Wobeser and Wobeser 1992; Philibert et al. 1993; Madrigal et al. 1996; Vyas 1999; Kostecke et al. 2001; Prosser et al. 2008). Based on that literature and our own experience we estimated that 10% of eagles were found and tested. Therefore, in the worst years, 1995–1996, another 72 for a total of 80 wintering bald eagles may have died of anti-cholinesterase poisoning, mainly by fonofos.

We were questioned regularly whether the non-target impacts of pesticides or other commercial chemicals are significant "at the population level." Wildlife biologists are often hard pressed to demonstrate such evidence, due to lack of sufficient data on population trends and difficulty in defining meta-populations. The issue is further compounded for migratory, particularly wintering populations (Esler 2000). An alternative to population-based paradigm is to manage or conserve species of concern using a habitat-based approach. That supposes if there is a land base of adequate quality then the populations will take care of themselves, a kind of "build it and they will come" paradigm (see, e.g., Armstrong 2005). Clearly, however, the gains from creating a habitat stewardship program of set-asides, buffer zones and cover crops on agricultural land would be limited if pesticides then poison the wildlife attracted to the site. Thus, use of Integrated Pest Management (IPM) or other harm reduction programs are an essential component to any such stewardship programs.

We can, nonetheless, make a case for a potential impact on bald eagles at the local population level. Using a figure of 300 eagles on average wintering in the Fraser Delta (Elliott et al. 2011), at 80 mortalities per year, approximately 27% of the wintering population could have been poisoned primarily by that one compound, fonfofos. If the Fraser Delta was a closed population without immigration, and we assumed a continuous rate of duck poisoning and carcass finding and no recovery and rehabilitation of poisoned individuals, then bald eagles would have been extirpated from the area in less than 4 years. However, the bald eagles that winter on the Pacific coast constitute a large and fluid population of many thousands of birds, based on telemetry studies and counting of eagles (Hunt et al. 1992; Watson and Pierce 1997). Eagles would most likely continue to immigrate into this attractive wintering site, and replace poisoned birds. The impact of pesticides on the Delta bald eagle population could be modelled, although the confidence intervals around the output would be very wide and thus likely of limited utility.

The results of this work provide another example, like the American condor and the Asian *gyps* vultures, of the particular vulnerability of avian scavengers to poisoning by anthropogenic compounds. Because of the scavenging efficiency of some species, a significant number of individuals (on a population basis) can be exposed to a toxicant even from a low incidence of contaminated animal carcasses (e.g., Green et al. 2006).

Moving Forward

The 1999 decision to withdraw fonofos for wireworm control created a minor crisis. Efforts to develop non-chemical alternatives, such as trap crops, bio-barriers, pheromone traps and biological controls had shown some promising leads (Vernon 2005); however, the local farmers had lobbied for a more effective tool. In order to develop a consensus solution, The British Columbia Wireworm, Committee was formed. Chaired by Agriculture Canada with a secretariat provided by Environment Canada, the committee invited a broad range of members from governmental science and regulatory agencies, as well as non-government agencies, and farmers. The mandate was, as paraphrased in the title of this chapter, to find effective methods to control wireworm without causing significant harm to wildlife. In the spring of 1999, the members considered the use of yet another organophosphorus option, chlorpyrifos, provisionally available from the PMRA under a limited emergency registration. It may seem to the reader there were an endless number of anti-cholinesterase compounds available to replace those deemed to be a problem. Worldwide, at one time, at least 200 different organophosphorus ester and some 25 carbamic ester insecticides were in commerce; they were formulated into thousands of marketed products (Matsumura 1975). In a relatively regulated environment such as Canada, however, the number of products registered for a given use in a given area could be limited. As the Wireworm Committee learned, it can be a lengthy and frustrating process to register alternatives. That includes newer products with different chemical structures and modes of action (Van Herk et al. 2008), which may be safer for wildlife. The reasons may vary and include unwillingness of a registrant to pay the costs of extra testing necessary to register their compound for a small market use, or even concerns that the alternate product may compete with older and more profitable pesticides.

Thus, there were limited options for chemical control of wireworm in the Fraser Valley. Chlorpyrifos was available for local use in two forms, as a liquid Pyrinex 480EC and a granular Pyrifos 15G formulation. Site-specific experimental data showed the liquid formulation to be more effective. Environment Canada's position was that of the available effective options, chlorpyrifos with a lower avian LD_{50} relative to, for example, carbofuran, phorate or fonofos, would hopefully cause no, or at least fewer, non-target wildlife poisonings. There were concerns, however, as chlorpyrifos is very toxic to aquatic organisms. There was a perceived risk to the commercially and culturally important salmonid fishes, which used ditches and small streams draining agricultural land. Health regulators also were concerned about human exposure to chlorpyrifos in treated foods.

The issue became more complicated. A European study reported non-chemical suppression of wireworm populations could be achieved by not planting winter cover crops, and thus starving the over-wintering larvae (Parker and Howard 2001). For many years the Canadian Wildlife Service had actively supported winter cover cropping to provide alternate forage for waterfowl, and thus reduce damage to perennial grass fields. A reduction in winter cover was not a viable option to the wildlife managers, so based on a risk–benefits analysis, chlorpyrifos was the preferred solution at least for the short-term.

Fortunately, to date the decision has proven sound. Figure 7.9 summarizes the history of raptor poisonings from 1989 to 2003 in the Fraser Delta, and includes only those cases where a definitive diagnosis of anti-cholinesterase poisoning was made; the criteria were severe brain or plasma cholinesterase activity and/or detection of significant (>1 μg/g) residues of an anti-cholinesterase compound in crop contents. Since 2000, there have been no poisonings of raptors by registered wireworm control chemicals, particularly by the widely used chlorpyrifos. Each year there are a few cases of dead or debilitated raptors with symptoms of anti-cholinesterase poisoning, but a number of them were definitely caused by unlabeled use of chemicals, probably in attempts to control perceived pests such as crows or coyotes. Some cases have been investigated by enforcement officials.

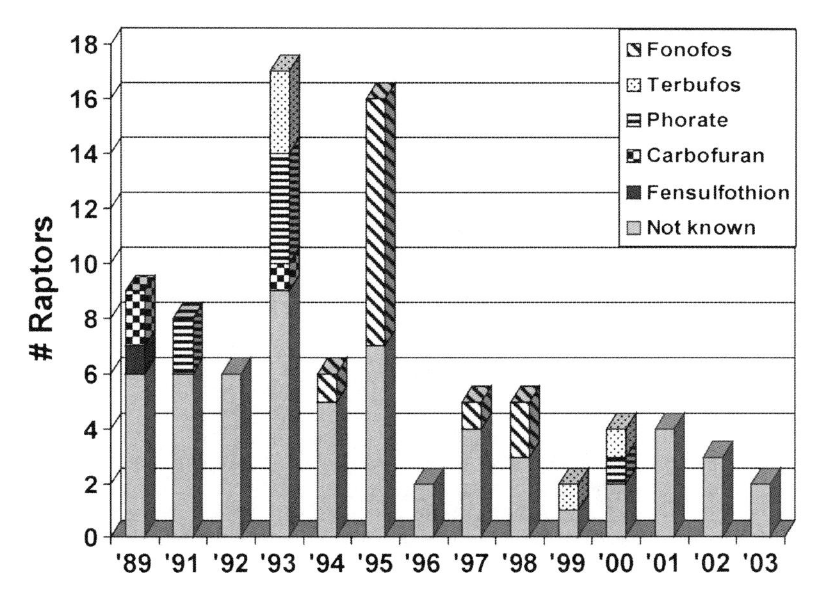

Fig. 7.9 Confirmed poisonings of raptors by different insecticides over time in the Lower Fraser Valley region of British Columbia. Poisonings were confirmed by significant inhibition of brain or plasma cholinesterase activity and/or presence of >1 μg/g pesticide residues in ingesta

Summary and Conclusions

Under the soil and climatic conditions of the Fraser River Delta, a number of widely used organophosphorus and carbamate insecticides, particularly as granular formulations, persist and remain toxic for as long as 8 months post-application. Wintering waterfowl move into agricultural fields from adjacent estuarine habitat and are exposed to pesticide residues, presumably while sieving soils in search of food or grit. Birds of prey, particularly eagles and hawks, are poisoned from feeding on dead or debilitated waterfowl. The circumstances in the Fraser River Delta may be somewhat unusual with the proximity of intensively farmed lands to prime estuarine wildlife habitat. On a global basis, however, the situation is hardly unique, and suggests the likelihood of similar problems elsewhere but where adequate surveillance is lacking. Such surveillance is clearly needed as long as agricultural methods rely on the use of biocides which are toxic to non-target vertebrates. A monitoring program also assumes availability of biologists and chemists with appropriate training and access to adequate laboratory resources.

Although the process was protracted and difficult at times, it is clear that mortality of non-target wildlife from agricultural pest control can be minimized or even eliminated while allowing profitable farming. A resolution to the problem was facilitated by establishment of a formal discussion forum through the wireworm committee.

The future of wildlife conservation and farming in the Fraser Delta is inextricably connected. The pressures on the habitat continue to increase. Yearly there is expansion of industrial-scale greenhouse and berry farm horticulture. New multi-lane transportation corridors are built through agricultural land. Such developments reduce the available habitat, while increasing road strikes to raptors, particularly owls (Hindmarch 2010). However, in future, trends of increasing costs for fossil fuels and fertilizers will likely increase costs of importing food (Mitchell 2008). There is also a growing consumer demand for food produced locally by sustainable methods, including minimal pesticide use (Mariola 2008). Those trends may prove to be positive for the farming community in the Fraser Delta. Provided there is any farmland left, local farmers would be well placed to serve local markets with a variety of crops, thus leading to both a healthier economy and ecosystem.

Acknowledgements Sandi Lee is thanked for her tireless efforts to visit sites, collect carcasses, bleed and otherwise sample debilitated birds, and handle samples. She is also thanked for her help with drafting the figures for this chapter. Many others have been involved over the years in data collection and investigations. Principle among them: Pam Sinclair, Phil Whitehead, Ken Langelier, Malcolm McAdie, Vicky Bowes, Craig Stevens, Harpreet Gill, Patti Dods, Christy Morrissey, Anna Birmingham. The following wildlife rehabilitators are also thanked: OWL, especially Bev Day, Lynn Short and Monika Tolksdorf. Karen Morrison and Sylvia Von Schuckman of the British Columbia Ministry of Environment were very supportive of this work. Jack Bates of the Delta Farmers Institute provided valuable information. Finally, we thank the many members of the public who willingly brought in or reported sick and dead raptors and waterfowl.

Appendix

Number of wildlife observed in agricultural fields that were treated or untreated with fonofos prior to surveying for carcass remains, by taxonomic group during Dec 1996–Jan 1997 and Oct–Dec 1997.

Taxonomic group	Treated	Perimeter treated	Untreated	Total
Hectares[a]	182.5	33.6	208	424.1
No.of surveys[b]	13	13	13	39
Waterfowl	29,268	1,901	21,245	52,414 (86%)
Shorebirds	531	860	3,403	4,794 (8%)
Seabirds	496	661	1,527	2,684 (4%)
Raptors	196	33	328	557 (1%)
Other birds[c]	74	6	131	211 (<1%)
Mammals	4	4	7	15 (<1%)
Total	30,569 (50%)	3,465 (6%)	26,641 (44%)	60,675

[a]1996–1997: 114.9 ha treated, 111.3 ha untreated; 1997–1998: 67.6 ha treated, 33.6 ha perimeter treated, 96.7 ha untreated
[b]1996–1997: 5–7 surveys depending on the field; 1997–1998: 6–7 surveys depending on the field
[c]Other birds include songbirds, game fowl and water birds

References

Armstrong DP (2005) Integrating the metapopulation and habitat paradigms for understanding broad-scale declines of species. Conserv Biol 19:1402–1410

Balcombe R (1986) Songbird carcasses disappear rapidly from agricultural fields. Auk 103:817–820

Beans BE (1996) Eagle's plume: the struggle to preserve the life and haunts of America's bald eagles. Simon and Schuster, New York

Beyer WN, Krynitsky AJ (1989) Long-term persistence of dieldrin, DDT, and heptachlor epoxide in earthworms. Ambio 18:271–273

Breault A, Butler RW (1992) Abundance, distribution and habitat requirements of American Wigeon, Northern Pintails and Mallards in farmlands. In: Butler RW (ed) Abundance, distribution and conservation of birds in the vicinity of Boundary Bay, British Columbia. Technical Report Series No. 155, Canadian Wildlife Service, Delta, BC, Canada, pp 19–41

Butler RW, Campbell RW (1987) The birds of the Fraser River Delta: populations, ecology and international significance. Canadian Wildlife Service, Occasional Paper No. 65, Ottawa, ON, Canada

Church ME, Gwiazda R, Risebrough RW, Sorenson K, Chamberlain CP, Farry S, Heirich W, Rideout BA, Smith DR (2006) Ammunition is the principal source of lead accumulated by California condors re-introduced to the wild. Environ Sci Technol 40:6143–6150

Ecobichon DJ (1991) Toxic effects of pesticides. In: Amdur MO, Doull J, Klaassen CD (eds) Casarett and Doul's toxicology. Pergamon, New York, NY, pp 565–622

Elliott JE, Harris ML (2001) An ecotoxicological assessment of chlorinated hydrocarbon effects on bald eagle populations. Rev Toxicol 4:1–60

Elliott JE, Langelier KM, Scheuhammer AM, Sinclair PH, Whitehead PE (1992) Incidence of lead poisoning in Bald Eagles and lead shot in waterfowl gizzards from British Columbia, 1988–1991. Canadian Wildlife Service, Progress Notes, Ottawa, ON, No. 200, p 7

Elliott JE, Langelier KM, Mineau P, Wilson LK (1996) Poisoning of Bald Eagles and red-tailed Hawks by Carbofuran and Fensulfothion in the Fraser Delta of British Columbia, Canada. J Wildl Dis 32:486–491

Elliott JE, Wilson LK, Langelier KM, Mineau P, Sinclair PH (1997) Secondary poisoning of birds of prey by the organophosphorus insecticide, phorate. Ecotoxicology 6:219–231

Elliott JE, Moul IM, Cheng KM (1998) Variable reproductive success of Bald Eagles on the British Columbia coast. J Wildl Manage 62:518–529

Elliott KH, Lee SL, Elliott JE (2006) Foraging ecology of Bald Eagles at an anthropogenic food source: does the Vancouver landfill affect local populations? Wilson Bull 118:380–390

Elliott JE, Birmingham A, Wilson LK, McAdie M, Mineau P (2008) Fonofos poisons raptors and waterfowl several months after labelled application. Environ Toxicol Chem 27:452–460

Elliott KH, Jones I, Stenersen K, Elliott JE (2011) Population trends for bald eagles in south-coastal British Columbia reflect reduced winter survival. J Wildl Manage (in press)

Esler D (2000) Applying metapopulation theory to conservation of migratory birds. Conserv Biol 14:366–372

Green RE, Taggert MA, Das D, Pain DJ, Sashikumar C, Cunningham MA, Cuthbert R (2006) Collapse of Asian vulture populations: risk of mortality from residues of the veterinary drug diclofenac in carcasses of treated cattle. J Appl Ecol 43:949–956

Greig-Smith PW (1991) Use of cholinesterase measurements in surveillance of wildlife poisoning in farmland. In: Mineau P (ed) Cholinesterase-inhibiting insecticides: their impact on wildlife and the environment. Elsevier, Amsterdam, Holland

Harris ML, Wilson LK, Elliott JE, Bishop CA, Tomlin AD, Henning KV (2000) Transfer of DDT and metabolites from fruit orchard soils to American robins (*Turdus migratorius*) twenty years after agricultural use of DDT in Canada. Arch Environ Contam Toxicol 39:205–220

Henny CJ, Mineau P, Elliott JE, Woodbridge B (1999) Raptor pesticide poisonings and current insecticide use: what do isolated kill reports mean to populations? Proceedings, 22nd International Ornithological Congress, Durban, South Africa, August 16–22, 1998, pp 1020–1033

Hindmarch S (2010) The effects of landscape composition and configuration on barn owl (Tyto alba) distribution, diet and productivity in the Fraser Valley, British Columbia. MSc. Thesis, Simon Fraser University, Burnaby, Canada

Hunt WG, Jackman RE, Jenkins JM, Thelander CG, Lehman RN (1992) Northward post-fledging migration of California Bald Eagles. J Raptor Res 26:1–21

Kostecke RM, Linz GM, Bleier WJ (2001) Survival of avian carcasses and photographic evidence of predators and scavengers. J Field Ornithol 72:439–447

Madrigal JL, Pixton GC, Collings BJ, Booth GM, Smith HD (1996) A comparison of two methods of estimating bird mortalities from field-applied pesticides. Environ Toxicol Chem 15:878–885

Mariola MM (2008) The local industrial complex? Questioning the link between local foods and energy use. Agric Hum Values 25:193–196

Matsumura F (1975) Toxicology of insecticides. Plenum, New York

Mineau P (1993) The hazard of carbofuran to birds and other vertebrate wildlife. Environment Canada, Canadian Wildlife Service, Tech Rep Ser. No.

Mitchell D (2008) A note on rising food prices. The World Bank, Development Prospects Group. Policy Research Paper 462. http://econ.worldbank.org. p 20

Newton I (1979) Population ecology of raptors. Buteo, Vermillion, ND, USA

Newton I (1986) The sparrowhawk. T&AD Poyser, Calton, UK, p 396

Newton I, Wyllie I, Asher A (1992) Mortality from the pesticides aldrin and dieldrin in British Sparrowhawks and Kestrels. Ecotoxicology 1:31–44

Oaks JL, Gilbert M, Virani MZ, Watson RT, Meteyer CU, Rideout BA, Shivaprasad HL, Ahmed S, Chaudhry MJI, Arshad M, Mahmood S, Ali A, Khan AA (2004) Diclofenac residues as the cause of vulture population decline in Pakistan. Nature 427:630–633

Parker WE, Howard JJ (2001) The biology and management of wireworms (Agriotes spp.) on potato with particular reference to the U.K. Agri For Entomol 3:85–98

Peterson CA, Lee SL, Elliott JE (2001) Scavenging of waterfowl carcasses by birds in agricultural fields of British Columbia. J Field Ornithol 72:150–159

Philbert H, Wobeser G, Clark RG (1993) Counting dead birds: examination of methods. J Wildl Dis 29:284–289

Poole AF (1989) Ospreys: a natural and unnatural history. Cambridge University Press, Cambridge UK

Porter SL (1993) Pesticide poisoning in birds of prey. In: Redig PT, Cooper JE, Hunter DB (eds) Raptor biomedicine. University of Minnesota Press, Minneapolis, MN, pp 239–245

Prosser P, Nattrass C, Prosser C (2008) Rate of removal of bird carcases in arable farmland by predators and scavengers. Ecotoxicol Environ Safety 71:601–608

Rattner BA (2009) History of wildlife toxicology. Ecotoxicology 18:773–783

Rattner BA, Scheuhammer A, Elliott JE (2011) History of wildlife toxicology and the interpretation of contaminant concentrations in tissues. In: Beyer WN, Meador J (eds) Environmental contaminants in wildlife, interpreting tissue concentrations. CRC, Boca Raton, FL, USA, pp 9–44

Sibley RM, Newton I, Walker CH (2000) Effects of dieldrin on population growth rates of sparrowhawks 1963–1986. J Appl Ecol 37:540–546

Stalmaster M (1987) The bald eagle. Universe, New York

Sullivan TM (1992) Populations, distribution and habitat requirements of birds of prey. In: Butler RW (ed) Abundance, distribution and conservation of birds in the vicinity of Boundary Bay, British Columbia. Technical Report Series No. 155, Canadian Wildlife Service, Delta, BC, Canada, pp 86–109

Szeto S, Price PM (1991) Persistence of pesticide residues and mineral and organic soils in the Fraser Valley of British Columbia. J Agri Food Chem 39:1679–1684

Van Herk WG, Vernon RS, Tolman JH, Ortiz Saavedra H (2008) Mortality of a Wireworm, *Agriotes obscurus* (Coleoptera: Elateridae), after Topical Application of Various Insecticides

Vernon RS (2005) Aggregation and mortality of *Agriotes obscura* (Coleoptera: Elateridae) at insecticide treated trap crops of wheat. J Econ Entomol 98:1999–2005

Vyas NB (1999) Factors influencing estimation of pesticide-related wildlife mortality. Toxicol Ind Health 15:186–191

Watson JW, Pierce DJ (1997) Skagit river bald eagles: movements, origins, and breeding population status. Washington Department of Fish and Wildlife Programs Report, Olympia, WA, USA

Wayland M, Wilson LK, Elliott JE, Miller MJR, Bollinger T, McAdie M, Langelier K, Keating J, Froese JMW (2003a) Lead poisoning and other causes of mortality and debilitation in bald and golden eagles in western Canada, 1986–98. J Rapt Res 37:8–18

Wayland M, Wilson LK, Elliott JE, Miller MJR, Bollinger T, McAdie M, Langelier K, Keating J, Froese JMW (2003b) Lead poisoning and other causes of mortality and debilitation in bald and golden eagles in western Canada, 1986–98. J Raptor Res 37:8–18

Wilkinson ATS (1963) Wireworm problems of cultivated land in British Columbia. Proc Entomol Soc Br Columbia 60:3–17

Wilkinson ATS, Finlayson DG, Campbell CJ (1976) Controlling the European wireworm, *Agriotes obscurus* L., in corn in Br. Columbia. J Entomol Soc Br Columbia 73:3–5

Wilson LK, Moul IE, Langelier KM, Elliott JE (1995) Summary of bird mortalities in British Columbia and Yukon 1963-1994. Technical Report Series No. 249, Canadian Wildlife Service, Delta, BC, Canada

Wilson LK, Elliott JE, Vernon RS, Smith BD, Szeto SY (2002) Persistence and retention of active ingredients in four granular cholinesterase-inhibiting insecticides in agricultural soils of the lower Fraser River Valley, British Columbia, Canada, with implications for wildlife poisoning. Environ Toxicol Chem 21:260–268

Wobeser G, Wobeser AG (1992) Carcass disappearance and estimation of mortality in a simulated die-off of small birds. J Wildl Dis 28:548–554

Chapter 8
Toxic Trees: Arsenic Pesticides, Woodpeckers, and the Mountain Pine Beetle

Christy A. Morrissey and John E. Elliott

Driving through the long stretches of quiet mountainous landscape, I could hardly believe what I was seeing. It looked more like autumn in northern Ontario not spring in the southern interior of British Columbia. The panorama of tall evergreen pines had turned varying shades of rusty brown - a revealing sign of another beetle attack. Once familiar terrain of lush green hills were now replaced by an odd checkerboard of multi-coloured dead stands next to blank empty

C.A. Morrissey (✉)
Department of Biology, University of Saskatchewan, Saskatoon,
Saskatchewan, SK, S7N 5E2, Canada
e-mail: christy.morrissey@usask.ca

J.E. Elliott et al. (eds.), *Wildlife Ecotoxicology: Forensic Approaches*,
Emerging Topics in Ecotoxicology 3, DOI 10.1007/978-0-387-89432-4_8,

squares. Could such devastation of an entire forest ecosystem really be brought about by a tiny insect? And what further damage may have been done in trying to control this mess?

C. Morrissey

Abstract During the mid-1990s, an unprecedented outbreak of the Mountain Pine Beetle (*Dendroctonus ponderosae*) in the economically valuable pine forests of British Columbia triggered a major campaign by forest managers to try and control the beetle damage. The result was a major change in the landscape from harvesting and prescribed burns in addition to the wide scale application of a systemic insecticide known as monosodium methanearsonate (MSMA). Since MSMA contains arsenic, is highly stable, and was being applied in increasingly large quantities, our study evaluated the potential impacts to forest birds, particularly woodpeckers. From 2002 to 2006, we investigated the exposure of breeding woodpeckers to MSMA ingested via contaminated bark beetles. We measured high levels of arsenic in beetles from treated trees, and found that significant amounts of debarking on many MSMA-treated trees indicating woodpeckers were feeding on them. Radio-telemetry confirmed that woodpeckers breeding near MSMA treatments regularly used those stands. Blood samples of woodpeckers and other forest birds revealed elevated arsenic concentrations. Through a concurrent laboratory dosing study of Zebra finches (*Taeniopygia guttata*), we estimated that woodpeckers were receiving enough MSMA through ingestion of contaminated beetles to cause poorer growth and mortality of young birds and mass loss in adults. We concluded that the combination of extensive harvesting in the region in combination with the large numbers of treated MSMA trees was potentially detrimental to forest bird populations. By the end of our study, MSMA approval for use in Canada was revoked and the Ministry of Forests did not pursue re-registration of MSMA. We clearly demonstrated that large-scale MSMA use had the potential to cause serious harm to forest birds, while appearing to have limited efficacy in beetle control. This was a landmark study in the field of wildlife ecotoxicology – a first to evaluate the exposure and effects of a toxic chemical to woodpeckers in forest ecosystems.

Background

Situated between the Pacific Ocean and the Rocky Mountains, the western Canadian province of British Columbia contains vast expanses of conifer forests. Across a range of soil and elevation conditions, the lodgepole pine tree (*Pinus contorta*) dominates the landscape, especially in the province's interior. In order to cultivate a new generation of pines, the "sealed" cones, which can survive for years even on the forest floor, need extreme heat to melt the cone's resin and release the seeds. Originating from catastrophic, stand-replacing fires, lodgepole pine forests typically generate a monoculture of even-aged trees (Shore et al. 2006).

Fig. 8.1 An adult mountain pine beetle and larvae occupying newly formed galleries in the phloem of a lodgepole pine tree. Photograph by Dion Manastyrski/British Columbia Ministry of Forests and Range

Together with fire, endemic insects such as the mountain pine beetle (*Dendroctonus ponderosae* Hopkins) (Fig. 8.1) play an important role in regulating pine forests by killing mature trees and allowing other species to thrive (Shore et al. 2006). Over time, early successional lodgepole would normally be interspersed with more shade and fire tolerant species such as spruce or fur, producing a varied forest of tree species and ages. In British Columbia, after many decades of forest fire suppression, stands have become highly uniform resulting in an expansive landscape of low diversity mature lodgepole pine forests. Prized for the tree's quick growing, high volume and straight timber, lodgepole pines have also been routinely planted and harvested by commercial loggers. By the end of the last century, lodgepole pine comprised 22% of the total forest in western Canada (Koch 1996) and the area covered by mature lodgepole pine was reported to be more than three times larger compared to the previous century (Taylor and Carroll 2004).

Beginning in the mid-1990s, the mountain pine beetle, an insect about the size of a grain of rice, began wreaking havoc in the lodgepole dominated forests of British Columbia's interior. The little beetle's success in killing pine trees and spreading at an astonishing rate were the result of events and forest management decisions that had been developing over many years. A combination of unusually warm winters and dry summers, a century of fire suppression and extensive monoculture plantations provided ideal conditions for the pine beetle to multiply to epidemic proportions.

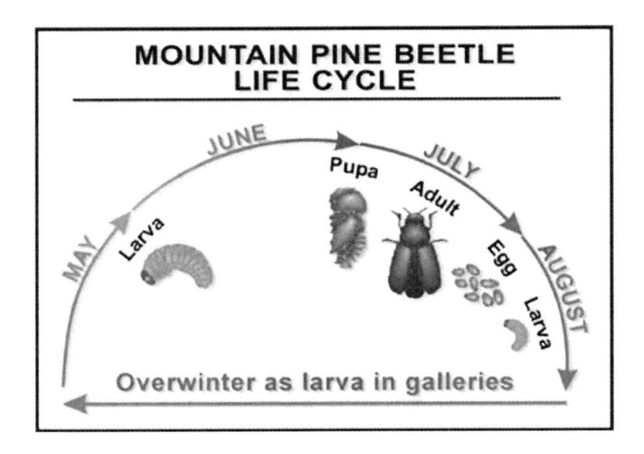

Fig. 8.2 The typical annual life cycle of a mountain pine beetle attacking pines in the temperate regions of the Pacific Northwest. Diagram from Alberta Sustainable Resource Development 2005

By 2009, the beetle wiped out more than 16.3 million hectares of healthy lodgepole pines, the most commercially harvested tree in western Canada. Future models have predicted that 80% of the remaining pine forests will be gone by 2013 (B.C. Ministry of Forests and Range, Canadian Forest Service 2009).

The mountain pine beetle is well adapted to exploit the conditions experienced during the early part of the twenty-first century across western North America. Over the course of its 1-year life cycle the beetle can rapidly multiply when conditions are favorable (Fig. 8.2). In mid to late summer, the adults take flight to attack live lodgepole pines, then drill holes and form tunnels or "galleries" under the tree bark. Following initial attack, the pioneering female beetle uses pheromones to attract males and encourage a mass attack, overwhelming even a healthy tree's natural defenses, which cause it to secrete resin to "pitch out" the beetles. The beetles also carry a blue-stain fungus (*Ophiostoma minus*) that can further weaken the tree's natural defenses as well as normal water and nutrient translocation. Beetles lay eggs in the newly formed galleries, which after 2 weeks hatch into small hungry larvae. Over the winter, the beetle larvae feed on the tree and mature, pupate, and emerge the following summer to attack again. The combined action of the beetles' mining activity and colonization of the fungus is ultimately fatal to the host tree as the girdled tree literally starves without nutrient or water translocation within a few weeks of attack (Safrynik and Carroll 2006).

Typically, the mountain pine beetle plays an important ecosystem role by attacking and killing old or weakened trees, which allows younger forests to grow. Infested trees also provide an important food source to predators such as woodpeckers and other cavity nesting birds. However, by the turn of the past century when climatic shifts caused warmer winters and drier summers, a massive growth in the mountain pine beetle population resulted in beetles turning their attack to healthy pines. Whole forests were destroyed as evidenced by the transformation from healthy green needles to shades of red (year 1), brown (year 2), and then gray (year 3) as the trees desiccate and die after attack.

Given the economic implications for the beetle's destructive behavior, the British Columbia forest industry deployed a range of tactics to combat the pest. In the early years of the outbreak in the mid-1990s, a combination of suppression methods were routinely used including large- and small-scale tree harvesting, controlled burning, and application of pesticides. In recent years, the pesticide of choice has been monosodium methanearsonate or MSMA. It is an organic arsenical, registered for widespread use as an herbicide in the USA for cotton and turf crops. In Canada, it was only registered for use in the forest industry – primarily as an insecticide to combat natural infestations of both mountain pine beetle and Spruce beetle (*Dendroctonus rufipennis* Kirby), and as a herbicide for tree thinning. MSMA was favored by foresters under certain conditions because it was relatively cheap and easy to use, especially in more remote locations where tree harvesting was not practical. Used by the industry for over 20 years, beginning in the 1980s, approximately 5,080 kg of MSMA was applied to B.C. forests amounting to some 500,000 treated trees in the most recent 10-year period (Morrissey et al. 2007).

Across the forests of North America there has been a long history of pesticide use in attempts to suppress and control insect pests. For example, during the 1950s and 1960s, millions of acres of eastern North American forests were aerially sprayed with DDT to try and control the Spruce budworm (*Choristoneura* species). The impact on wildlife populations was well documented, and even led to eggshell thinning and reproductive failure of a seabird, the northern gannet (*Sula bassana*) breeding in the Gulf of St. Lawrence (Elliott et al. 1988). Wide-scale spraying continued after DDT use was restricted and replaced by less persistent, but more acutely toxic organophosphorus compounds, such as fenitrothion, which poisoned birds over large areas of New Brunswick, Canada (Busby et al. 1990). Such broad scale spraying would be largely ineffective against a beetle, which spends most of its life under a layer of bark.

The use of pesticides to control bark beetle damage was not a recent development. Several chemical tactics for bark beetle control were developed during the late 1940s and early 1950s. Early methods involved application of insecticides (e.g., naphthalene, orthodichlorobenzene, or ethylene dibromide in diesel oil) to the bark of trees in order to penetrate affected trees or protect others from new infestation (Klein 1978). In the 1970s, there was movement toward water-based formulations of ethylene dibromide and also lindane or carbaryl in Canada and the USA (Klein 1978). While many of those early chemical insecticides were effective, their toxicity to the environment or to forestry workers led to discontinuation in their use (Carroll et al. 2006). Since the 1980s, MSMA was the only pesticide registered for bark beetle control in British Columbia.

Systemic insecticides, developed for use on bark beetles and other tree pests, involve injecting the tree to allow natural translocation of the chemical. They are in many ways a better design than those previous methods, which left exposed contamination. The MSMA method was designed to kill the beetle brood while still sealed under the tree bark. The overall approach required ground or aerial survey to identify isolated outbreaks of the beetle, and to select stands for MSMA treatment. The process typically involved placing pheromone baits to concentrate the beetle infestation into a targeted area. Sites were later revisited within 3–4 weeks of new beetle attack, and each infested pine (identified by characteristic resin "pitch tubes" and boring dust) was treated by cutting an axe frill around the base of the tree and squirting the frill with MSMA (commercial formulation Glowon®) (B.C. Ministry of Forests and Range 1995). The pesticide translocates up through the tree phloem, resulting in death of the tree and direct and indirect toxicity (through water loss) to the beetle brood (Maclauchlan et al. 1988a, b; Manville et al. 1988). For lodgepole pines, the treated trees are usually left standing, remaining intact for many years. Although the pesticide targets beetles under the bark, woodpeckers and other forest birds are also drawn to the same beetle infestations for food. During beetle outbreaks, diets of Hairy (*Picoides villosus*), Three-toed (*Picoides dorsalis*), and Black-backed Woodpeckers (*Picoides arcticus*) typically consist of 60–99% beetle larvae by volume (Steeger et al. 1998), with individuals capable of consuming many thousands of beetles in a single day (Koplin 1972).

Unlike its potently toxic cousin, inorganic arsenic, early indications suggested the organic form MSMA and related compounds DSMA or cacodylic acid had only mild toxicity. Some early studies on domestic mammals, while highly variable among species, appeared to have sublethal effects similar to inorganic arsenic especially gastritis, diarrhea, hematological changes, as well as growth and reproductive abnormalities (Judd 1979; Prukop and Savage 1986; Jaghabir et al. 1989). While the avian toxicity database was extremely limited, LD_{50} values for 17-week-old and 10-day-old bobwhite quail (*Colinus virginianus*) were reported to be 834 and 650 mg/kg MSMA respectively, while 9-day-old mallards were not affected even at the highest doses of 1,100 mg/kg/day (MMA Task Force Three 1993). Due to the high LD_{50} values in the animal species tested, high water solubility, and rapid excretion rate, most believed from the early laboratory studies that MSMA was relatively innocuous (see Dost 1995). Additionally, methylated arsenicals such as monomethylarsonic acid (MMAA), the unionized form of MSMA at gastrointestinal pH, were once believed to be detoxifying byproducts from the breakdown of inorganic arsenic. But it is the oxidation and valence state of arsenic that are important as well as the species and age of animal exposed (Cullen and Reimer 1989). Generally, inorganic arsenic and the trivalent forms (As^{3+}) are much more cytotoxic than organic and the pentavalent forms (i.e., $MMAA^{5+}$). What early researchers did not realize, however, was the metabolic process of breaking down the chemical after an animal ingests it is critical to understanding its toxicological action (Fig. 8.3). $MMAA^{5+}$ is often converted to dimethylarsinic acid ($DMAA^{5+}$). In this process, the original $MMAA^{5+}$ is reduced to the highly toxic, but unstable intermediate $MMAA^{3+}$ before another methyl group attaches to become $DMAA^{5+}$ – the relatively low toxic and most commonly excreted form. So in the process of metabolizing MSMA, the animal will be briefly exposed to the more toxic intermediate, $MMAA^{3+}$, with potential to cause harm, even though the final metabolite is relatively safe (Cullen and Reimer 1989).

The debate over the safety of MSMA began in the 1980s, largely because of initial concerns from high MSMA usage for brush control and selective thinning raising another issue of exposure to browsers such as Black-tailed deer. During the MPB epidemic in the 1990s and early 2000s, MSMA usage as a pesticide increased

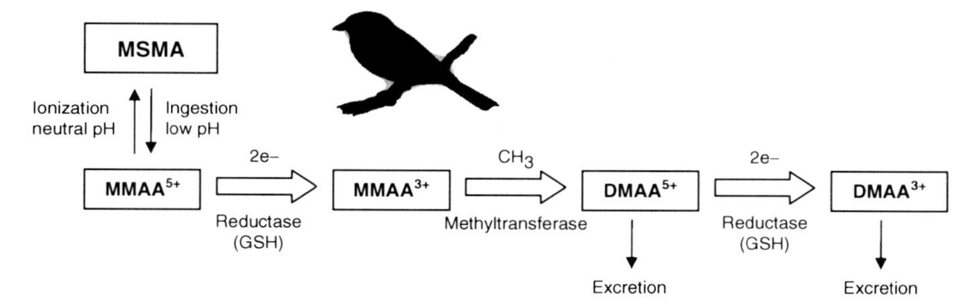

Fig. 8.3 Schematic diagram of the proposed metabolic pathway for monosodium methanearsonate (MSMA) following ingestion and uptake. Monomethylarsonic acid (MMAA) naturally ionizes and is associated with sodium to form the sodium salt, MSMA. If MSMA is ingested by an animal, the low pH (high acidity) of the stomach causes the hydrolization to MMAA. Following uptake, metabolism of the compound proceeds to dimethylarsinic acid (DMAA) as shown until excretion

in response to the beetle problem, resulting in continuous and widepread MSMA inputs to the forest environment. Given that MSMA contains arsenic, a stable element, and the pesticide does not degrade rapidly in the environment (aerobic soil half-life estimated at 240 days) (U.S. Environmental Protection Agency 2006), there was the potential for cumulative and substantial loadings to the forest ecosystem. The bark beetle outbreak presented an abundant food source to beetle specialists, like woodpeckers, which could depredate large numbers of beetles from treated and untreated trees. A 1990 preliminary analytical report of beetle larvae from a small sample of treated trees suggested concerns were unfounded with levels of only 4.5–5.0 parts per million (ppm) arsenic. But years of debate continued with mounting pressure from the public and other interest groups. We surmised that insectivorous birds, particularly woodpeckers, could be at risk of consuming MSMA if they regularly fed on treated trees. We questioned the extent of exposure: how often they were feeding in the pesticide treated stands, how much arsenical they were getting through their diet, and whether the chemical was particularly harmful to the adults or their growing young. Given the continued spread of the beetle epidemic, we also speculated whether the pesticide was as effective at killing the beetles as initially reported or was it just doing more harm than good. Over 4 years, our small team of researchers developed to address these important questions.

The Investigative Phase

Since at least the mid-1980s, the issue of MSMA and potential problems for avian insectivores had been on the radar of environmental toxicologists at the Canadian Wildlife Service. However, typical barriers such as lack of funding and other research priorities meant that efforts were focused elsewhere. Initial reports and toxicity information also seemed to indicate MSMA was not likely to be harmful. So the issue stayed on the back burner. By early 2002, the magnitude of the mountain pine beetle outbreak was beginning to hit the headlines, while some individuals and NGOs were advocating against use of MSMA based on human health and environmental concerns. It was a casual parking lot discussion between the manager of the Pacific Wildlife Research Centre, Rick McKelvey, and research scientist, John Elliott that catalyzed some action on MSMA and forest birds. McKelvey made the case that the mountain pine beetle was a major problem facing resource and environment agencies in western North America, that there was likely to be an increase in pesticide use and very little information existed about the potential impact to migratory birds. With a commitment for management support, Elliott decided to take the initial step and start making contacts and learn more about the beetle problem and insecticide use. With that, there was no turning back … Elliott hired a new Masters degree graduate, Patti Dods, to work under a federal internship program to gather information on the beetle issue and collect some preliminary data on MSMA. A year later, he offered Christy Morrissey a postdoctoral position administered through the Department of Forest Sciences at the University of British Columbia to head up a full scale study.

Quickly, the work became a journey of discovering the multitude of environmental problems associated with the beetle epidemic and MSMA use, particularly for forest birds. Like many other environmental issues – a long history of human attempts to manipulate nature was at the root.

The Evidence

Early Signs

In spring 2002, Patti Dods set about figuring out the who's who of the bark beetle community in British Columbia. She made contact with various experts in industry, academia, and various levels of government, and began to lay the groundwork to do what she really wanted most – get out into the field and examine the issue first hand. The bark beetle problem was enormous, but it was important to narrow down a study area to look for evidence of MSMA problems. The province of British Columbia is subdivided into regions or forest districts that are individually managed. After careful consideration, the Merritt (or Cascades) Forest District in southern interior of British Columbia was chosen. It had all the necessary components – lots of beetles and plenty of MSMA.

After the first field visits in the fall of 2002, it was clear from the preliminary data there was reason to be concerned. Dods had located numerous MSMA treatment areas and nearby reference patches that were also infested with the beetle but not treated. The following spring when she revisited the same marked trees, there were signs of woodpecker activity everywhere. In particular, she observed bark scaling – the typical feeding method of Three-toed (*Picoides tridactylus*), Black-backed (*P. arcticus*), and Hairy (*P. villosus*) woodpeckers who chip away the bark to reveal their beetle prey. Many trees were heavily debarked (scaled) including some of the MSMA treated trees. Dods also collected a number of adult and larval beetles from several of these trees to later test them for arsenic. The results were staggering.

In the fall of 2002, Elliott and Dods made initial contact with arsenic chemist, Bill Cullen at the University of British Columbia. Cullen's internationally recognized lab was already doing research in the area of organic arsenic toxicology so it was an immediately obvious collaboration. With their extensive experience, Cullen and his colleague, Vivian Lai, could offer analytical expertise on the speciation (organic and inorganic forms) of these complex arsenicals. They agreed to analyze the first batch of bark beetles that were collected from MSMA and reference trees. There was some degree of surprise and consternation at the degree of arsenic contamination of the adult and larval beetles collected from MSMA trees, with total dry weight concentrations up to 350 ppm. By comparison, the beetles from untreated reference trees had very low arsenic levels typically less than 1 ppm. The early data showed arsenic from MSMA was getting into the major prey of woodpeckers, in some cases at very high concentrations, and the data were of sufficient interest to convince Elliott to pursue the issue further. He set out a plan to get the necessary funding for a full scale study. Mounting pressure on government officials to make MSMA a priority combined with the new data helped to secure enough funding from the Canadian Wildlife Service and from the new federal Pesticide Science Fund to undertake field research on woodpeckers and a controlled laboratory toxicity study of MSMA to songbirds.

It was during the fall of 2003 that Christy Morrissey joined the team. She had heard plenty about the problems of the MPB taking over the forests and holding the forest industry hostage. However, like many, was unaware about the intensive use of organic arsenicals as pesticides to combat the beetle infestations. The project sounded intriguing. Fresh from a PhD working with contaminants in aquatic songbirds, she had lots of ideas, but also plenty of learning to do. The timing was right and it made for a seamless move from what Patti Dods had started to dig deeper into the issue.

MSMA on the Menu

Over the next 2 years, field crews set out to Merritt to look more closely at whether forest birds were using MSMA treated areas and to what extent. It soon became evident that a more direct wildlife toxicology field study was not going to work in this case. Traditionally, a study of this sort would involve locating nests, sampling exposure of the individuals in those nests via eggs (i.e., the "sample egg technique"; Blus 1984; Custer et al. 1999) or the adults or nestlings (e.g., Verreault et al. 2004). Exposure measurement would be integrated with measurements of effects such as reproductive success, behavior, chick growth, or physiological and biochemical markers (Elliott et al. 1996; Martinovic et al. 2003; Murvoll et al. 2006). However, woodpeckers posed a problem, particularly for the species of interest like the Three-toed woodpecker. Finding enough nests and accessing them high up in cavities of spindly dead snags was difficult and dangerous. Given those obstacles, we opted for a risk assessment approach: determine exposure indirectly by collecting and analyzing prey, more directly by trapping and blood sampling adults, and hopefully by accessing a handful of nests to sample nestlings. We then planned to determine use of treated stands by radio-tagging adult birds and comparing their use of MSMA and untreated areas. Sublethal toxicity would be determined in a model laboratory species, the zebra finch (*Taeniopygia guttata*), available to us at nearby Simon Fraser University. The results could then be incorporated into a risk assessment model.

During the first field season, we concentrated on catching woodpeckers and other cavity nesting species that were breeding within 1 km of recent MSMA treatment stands. Woodpeckers are vital to forest ecosystems for their ability to excavate nesting cavities in trees and snags, which are used by a range of secondary cavity nesters to breed and roost: songbirds, ducks, squirrels, and bats (Martin et al. 2004). Woodpeckers are also a reliable indicator of the bird species richness both at the landscape scale and the stand scale (Virkkala 2006; Drever et al. 2008). We therefore, focussed our efforts on Hairy and Three-toed woodpeckers – two important keystone species in forest ecosystems that reportedly consume the largest number of bark beetles (Steeger et al. 1998) and were relatively common in our study area.

Considered resident, or nonmigratory, Hairy and Three-toed woodpeckers are widely distributed across North America. They occupy primarily coniferous forests but Hairy woodpeckers are more flexible and can also be found in mixed forests and even suburban areas. Bark and wood-boring beetle larvae in dead and dying trees are the main food of both species while Hairy woodpeckers will also feed on sap, berries, nuts, seeds, and suet. The Three-toed woodpecker, a true bark beetle specialist, is often closely associated with mature boreal, montane forests, or spruce forests. Adult birds form monogamous pairs over the winter and both excavate a nest cavity in soft decaying trees – Hairy's prefer deciduous trees while Three-toed's select conifers. Previously treated "old" MSMA trees have been infrequently observed as suitable nest trees. Both males and females will excavate the cavity and incubate a clutch of four eggs. Typically, these species have one brood per season and the young remain in the nest for 3–4 weeks.

Catching woodpeckers is extremely difficult. We used mist nets, playback calls and even painted some dummy birds to place on trees to encourage high flying birds to swoop low enough to get caught in our nets. Although we caught some this way; it was a lot of effort for small returns. Another technique was to place a bag net (mist net on a loop at the end of a very long pole) over the entrance of the nest cavity and then lure the birds out by tapping on the tree or using playbacks. This technique worked well for some lower nests including those of another resident species, the Red-naped sapsucker (*Sphyrapicus nuchalis*). Each woodpecker we caught during the study was extremely valuable, so we gleaned as much information from it as possible.

We took data on each bird's mass and size, age using feather molt patterns, and took a blood sample to be later analyzed for arsenic residues. We fitted 12 of the

Hairy woodpeckers and 3 of the Three-toed woodpeckers with a 1.9-g radiotransmitter (<3% of the bird's lean mass) to the central tail feathers using SuperGlue®. The transmitters had a 14-week life span and the birds naturally molt their feathers at the end of the breeding season so would only need to carry the transmitter for the duration of a single study season. Over 2-study years, we successfully trapped 19 Hairy woodpeckers and four Three-toed's and made numerous observations of other unmarked birds within the study area. In the course of the work, we also captured other cavity nesting forest birds: 19 Mountain Chickadees (*Parus gambeli*) and 2 Red-breasted nuthatches (*Sitta canadensis*) from which we also took blood samples in order to look for evidence of arsenic exposure from feeding on insects in MSMA areas.

After chasing the radio-tagged woodpeckers through the forests three times a week across a range of landscapes, most scarred by the beetle, it was apparent how so much of the forest looked the same. A lack of fire meant endless stands of even-age lodgepole pines, many turning color from beetle attack. Those were interspersed by large clearcuts and isolated clumps of Trembling aspen (*Populus tremuloides*) and Engelmann spruce (*Picea engelmanni*). The birds led us around this matrix giving us some idea of how they managed to survive in their broken habitat.

Many of the Hairy woodpeckers nested in the aspen trees and made regular trips to the nearest beetle infestation site, while some of the Three-toeds nested within or near the affected areas. Both species routinely foraged in MSMA stands that were within their territories. Hairy woodpeckers spent 13% while Three-toed's spent 23% of their time there, despite the MSMA patches making up a tiny fraction (1–2%) of their core home range (Morrissey et al. 2008). The birds were impervious to the blood orange-stained axe frills marking the MSMA trees. They swooped from one tree to the next pecking, probing, and scaling, sometimes an untreated tree and sometimes an MSMA one. Typically, the birds fed low on the trees newly infested with mountain pine beetle (green attack) and higher up on the trees infested 2 or more years before (old attack) where secondary insects, such as *Ips* beetles usually colonize.

Over the course of the 4-year study, our team of researchers had evaluated woodpecker feeding patterns using debarking scores from 558 MSMA treated trees and another 597 untreated reference trees located within or in close proximity to MSMA trees across 50 different stands. Individual trees were scored based on the amount of bark removal from scaling where 0 was no feeding activity and 7 represented a tree with the majority of the bole stripped. Each tree was assessed about 10–12 months after they were attacked by the mountain pine beetle. Most of the reference trees were debarked to some extent and the average score was consistently higher than MSMA trees during all years of the study (Fig. 8.4). While a large number of the MSMA trees were left untouched, some 40% of the treated trees still had some evidence of debarking implying woodpeckers had been feeding on contaminated beetles. Some trees initially studied by Dods were followed over time from infestation in late summer until the beetles emerged the following year. She found that woodpeckers only started feeding on them in late winter and more heavily in spring/summer.

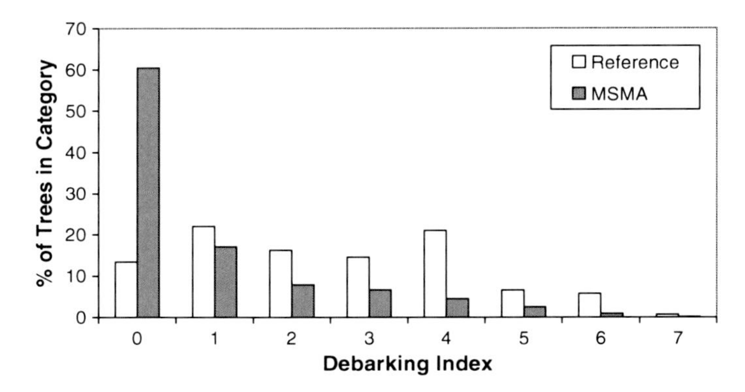

Fig. 8.4 Percent of trees surveyed (reference and MSMA) in each debarking category indicating level of foraging activity by woodpeckers 10–12 months after infestation and/or MSMA treatment. Data collected during 3 years (2002, 2003, 2005) using 597 untreated trees in proximity to MSMA (reference) and 558 MSMA treated trees in the Cascades Forest District. Debarking index refers to amount of woodpecker foraging (bark scaling) on a tree where 0 = no debarking and 7 = tree bark almost completely stripped

Table 8.1 Summary of total arsenic [As] (μg/g dry weight) measured in 90 composite samples (*n* = sample size) of bark beetles from MSMA treated and untreated reference trees in the Merritt Forest District, British Columbia from 2002 to 2005

Treatment	Age/species	*n*	Morbidity	Geometric mean total [As]	Range total [As]
MSMA	Adult MPB	15	Dead	155.50	57.3–354.1
MSMA	Adult MPB	6	Live	43.22	7.09–140.3
MSMA	Larval/pupal MPB	5	Dead	93.14	9.1–700.2
MSMA	Larval/pupal MPB	32	Live	20.00	1.3–327.4
MSMA	Larval Ips	4	Live	5.92	0.63–19.6
MSMA	Other insects	5	Live	10.15	0.22–62.9
Reference	Adult MPB	6	Dead	0.11	ND–1.06
Reference	Adult MPB	4	Live	0.75	0.32–1.06
Reference	Larval/pupal MPB	11	Live	0.15	0.04–0.79
Reference	Larval Cleridae	2	Live	0.36	0.08–1.62

MPB mountain pine beetle

Trees at different stages of MPB attack – first year (known as green attack), second year (red attack), and more than 2 years post attack (gray attack) had increasing debarking scores – revealing woodpeckers were continuing to feed on the affected pines even after the mountain pine beetles emerge in year one.

We continued to gather data on the arsenic levels in the beetles from MSMA trees where we also identified signs of woodpecker foraging (debarking). After collecting 90 beetle samples over 4 years, we were confident that our data showed arsenic from MSMA was getting into the prey of woodpeckers and in some instances, at very high concentrations (max 700 ppm) (Table 8.1). On average, the adult mountain pine beetles, which are the first line of attack for MSMA treatment accumulated the most arsenic (geometric mean = 108 ppm, range = 7–354 ppm). Larval and pupal stages tended to have lower amounts of arsenic (geometric mean = 25 ppm, range = 1–700 ppm) as did the beetles we extracted alive for the analysis. All the insects taken from reference trees had very low levels ranging from nondetectable up to a maximum of 1.6 ppm. Interestingly, regardless of the stage of attack (green attack or red attack), the life stage of the beetle (adult or larvae), or the species of insect we examined, 90–97% of the arsenic found was in the original, nonmetabolized form MMAA ($MMAA^{5+}$ and $MMAA^{3+}$ combined), the primary ingredient of MSMA.

We questioned how much the woodpeckers were feeding on those trees with the most contaminated beetles. This was important because it could give us an indication about the degree of risk to foraging birds. The amount of debarking on the trees was linked to the arsenic concentrations in the beetles. Trees that were heavily fed on (highest debarking) generally had the lowest beetle arsenic concentrations (Fig. 8.5) (Morrissey et al. 2007). This was very likely because these more heavily contaminated trees contained fewer live beetles to attract woodpeckers. Some ornithologists think woodpeckers can "sense" the beetles under the bark but no one knows if they can hear the beetles moving, or whether they can feel tiny vibrations (Backhouse 2005). As woodpeckers forage, they move around the tree, tapping it lightly until they locate their prey. They will then probe very purposefully through the bark and extract the larvae with exceptional precision. We cannot determine whether woodpeckers feed on dead beetles encountered while foraging – particularly

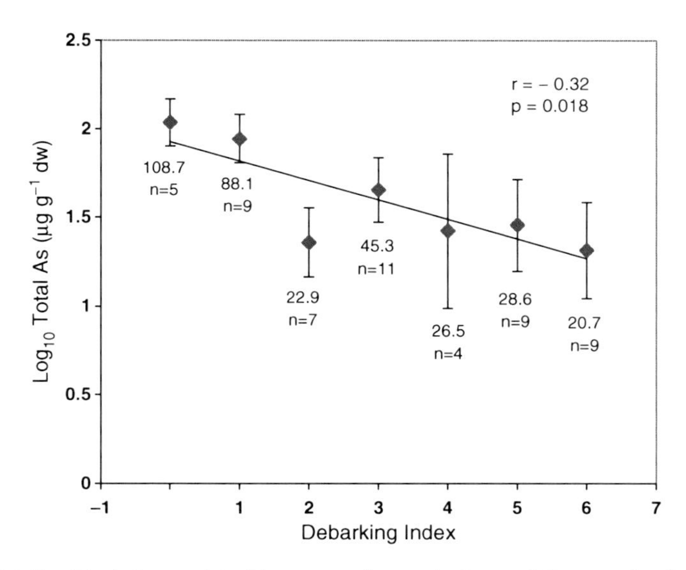

Fig. 8.5 Relationship between debarking scores (amount of cumulative woodpecker foraging 10–12 months post MSMA treatment) and mean dry weight (dw) total arsenic concentrations ($\log_{10} \pm$ SE) of all adult and larval MPB collected from individual MSMA treated pine trees in the Merritt Forest District, B.C. Values below points are the geometric mean arsenic concentrations (µg/g dw) and sample sizes

if there is a mixture of live and dead beetles under the bark as we observed. However, even the beetles collected dead must have accumulated the high arsenic load while still alive. So, while a woodpecker encountering an MSMA tree may sense there are fewer live prey and leave it more quickly in preference for another tree, the potential for exposure to the highest doses were still possible and were taken into account in assessing the risk.

Bird surveys carried out in MSMA and reference stands showed several species were responding to the effect of reduced beetle numbers from MSMA treatments. The species not highly dependent on bark beetles such as the Northern Flicker (*Colaptes auratus*), Red-naped sapsucker (*S. nuchalis*) and Downy woodpecker (*Picoides pubescens*) were encountered more frequently in MSMA stands. But the Three-toed woodpecker, a highly specialized bark beetle predator, was less abundant and detected less frequently in MSMA stands relative to untreated areas (Morrissey et al. 2008). Three-toed numbers were positively linked to the abundance of *Dendroctonus* prey (mostly mountain pine and Spruce beetle). Although this appeared to contradict our earlier observations of radio-tagged Three-toed woodpeckers using MSMA sites within their breeding territory, we concluded that while Three-toed woodpeckers were still present, their numbers were held at a lower density than would have been predicted by the survey information at untreated sites. We maintained that woodpeckers which breed within or in close proximity to MSMA-treated stands were still predisposed to MSMA since many birds are constrained to forage within their territory, the treated trees look the same as untreated ones, and beetle survival of MSMA treatment was highly variable allowing the stand to remain attractive to foraging birds.

Toxicity Concerns

Arsenic toxicity varies with the specific chemical form and among species. General effects on birds from exposure to inorganic arsenic have been reported: destruction of gut blood vessels, gastrointestinal irritation, blood-cell damage, muscular incoordination, debility, slowness, jerkiness, falling, hyperactivity, fluffed feathers, drooped eyelids, immobility, and seizures (Eisler 2000). For mallards (*Anas platyrhynchos*), Camardese et al. (1990) and Hoffman et al. (1992) also reported reduced growth of ducklings fed 30–300 ppm arsenic as sodium arsenate. Toxicity from organic arsenicals to domestic mammals, while highly variable among species, appear to have sublethal effects similar to inorganic arsenic especially gastritis, diarrhea, hematological changes, as well as growth and reproductive abnormalities (Judd 1979; Prukop and Savage 1986; Jaghabir et al. 1989). These studies provided some guidance for the sublethal effects, which might be expected in wild birds exposed to MSMA.

We needed to better understand what potential harm the woodpeckers might experience if consuming MSMA through their beetle diet. In response, two parallel laboratory studies were initiated at Simon Fraser University's Department of Biology. Master's student, Courtney Albert under supervision of Dr. Tony

Williams, agreed to take on the challenge of dosing adult and nestling songbirds to evaluate potential consequences of MSMA exposure for wild birds. We chose dosage levels based on data from the arsenic in the bark beetles as an estimate for woodpecker exposure in the wild population. Again, Bill Cullen's lab at University of British Columbia was a great asset by providing the pure active ingredient (MMAA) and testing all the dosing solutions and bird tissue samples throughout the experiments.

The initial dosing study was on adult Zebra finches, selected as a reasonable model for woodpeckers and other cavity nesting songbirds which might be exposed to MSMA through their beetle diet. While Zebra finches are granivores, not insectivores and sensitivity may differ between species – their body size and reproductive biology are more similar to our target wild species than most other captive birds used for toxicology studies (i.e., quail *Coturnix spp.* or mallard *A. platyrhynchos*). The finches are amenable to captive laboratory research and importantly an experimental breeding colony was already available through Tony William's lab at nearby Simon Fraser University.

Adult zebra finches were allocated into groups and orally dosed for 2 weeks with the pesticide (in the form MMAA) at 0, 8, 24, or 72 μg/g body weight/day. Doses were calculated based on what an average adult Three-toed or Hairy woodpecker would be ingesting if feeding regularly on contaminated beetles. In light of the information on beetle arsenic levels, the 0, 8, 24, and 72 μg/g/day dose groups were equivalent to a woodpecker ingesting a beetle diet containing 0, 50, 150, and 450 ppm MMAA, respectively. Given that we recorded beetle arsenic residues up to 700 ppm, the range of dosages tested should have encompassed the average and worst case scenarios. While adults did not suffer mortality, initial pilot studies for young nestling birds revealed dosage levels above 24 μg/g/day proved fatal, and so had to be adjusted in repeated studies to look for sublethal effects (Albert et al. 2008a).

Courtney Albert's thesis played a big role in convincing us that MSMA use in the forest industry was potentially harmful to birds. Her two studies with both adult and nestling Zebra finches showed detrimental effects from oral exposure at doses relevant to what woodpeckers could be ingesting if regularly consuming the more contaminated beetle larvae. While most of the chemical was rapidly excreted in the feces, adult and nestling birds were accumulating the compound in the blood, liver, kidney, brain, and carcass in a dose-dependent manner (Table 8.2). The adults in the two higher dose groups lost weight, up to 15% of their original mass (Albert et al. 2008b). We found high mortality (69–100%) of nestling birds in the dose groups at or above μg/g, and the developing young at lower doses also experienced poorer growth (shorter tarsi and wing length) relative to controls (Fig. 8.6) (Albert et al. 2008a). Albert's study was in sharp contrast to the early avian lethal toxicity studies on mallards and Bobwhite quail suggesting passerines were far more sensitive to organic arsenicals.

Based on the results of the laboratory study, we determined that of the five cavity-nesting bird species we tested who were breeding near MSMA treatment stands, 79% (42/53) had blood arsenic levels greater than 0.07 μg/g – the value we assumed from the lab study as a background arsenic concentration (Fig. 8.7). Nineteen

Table 8.2 Concentrations of total arsenic detected in tissue samples of adult zebra finches dosed with monomethylarsonic acid (MMA) and relative change in mass of the course of the study

Dose group	Tissue residue (μg/g)					Mean mass
(μg/g/day)	Blood	Liver	Kidney	Brain	Carcass	change (g)
0 (control)	0.07 ± 0.2 (3)	0.05	0.1	0.05	0.24 ± 1.91 (3)	−0.47 ± 0.59
8 (low)	0.40 ± 0.1 (6)	0.1	0.3	0.2	1.33 ± 1.36 (6)	−0.53 ± 0.42
24 (medium)	1.25 ± 0.35 (6)	0.1	0.4	0.7	3.13 ± 1.49 (5)	−1.17 ± 0.42
72 (high)	3.39 ± 0.74 (6)	1.1	1.6	3.7	12.31 ± 1.36 (6)	−2.24 ± 0.42

Liver, kidney, and brain values represent a pooled sample of two birds. Blood and carcass values are expressed as the mean ± SE (sample sizes)

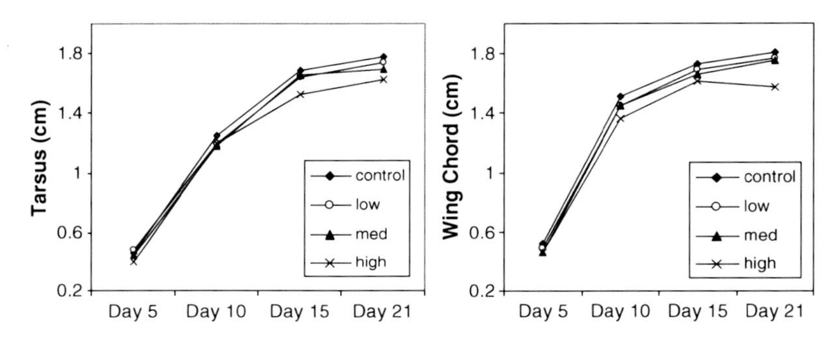

Fig. 8.6 Measurements of average nestling Zebra finch tarsus length and wing chord length over time from 5 to 21 days old while being administered different MMAA doses from hatching to fledging: controls = no MMAA, low = 8 μg/g body wt/day, medium = 12 μg/g body wt/day, and high = 24 μg/g body wt/day. Significant effects at day 21 were seen in the 8 and 24 μg/g doses (tarsus length) and 12 and 24 μg/g dose groups (wing chord length) compared to controls

percent (10/53) of the bird blood samples were above the no observable effect concentration (NOEC) of 0.4 μg/g and typically within the range of blood arsenic values of Zebra finches dosed with 8 μg/g. The highest concentration recorded (3.73 μg/g) was from a female Red-naped sapsucker breeding in an MSMA stand with 74 treated trees. This was not a species we expected to be susceptible to MSMA; implying tree sap was also a potentially important exposure route that we had not previously considered. Although there was large variability among individuals, the data suggested low but widespread exposure to MSMA.

The Risk Assessment Phase

By this point, we had confirmed that there were toxicologically relevant concentrations of arsenic in the bark beetles with low levels in the blood of the wild adult birds sampled. At the end of the third field season in 2005, we had sufficient

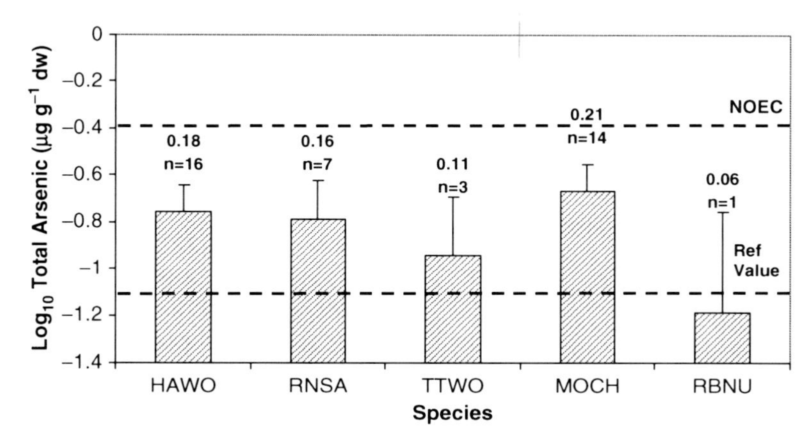

Fig. 8.7 Log_{10} mean total (±SE) arsenic concentrations in blood of five species of forest birds breeding within 1 linear km of MSMA treatment area in the Merritt Forest District, B.C. Numbers above bars represent geometric means (µg/g dw) and effective sample sizes for each species. The *dashed lines* represent a mean reference value for unexposed birds and the no observable effect concentration (NOEC) based on dosing study by Albert et al. 2008a. (American Ornithologist Union (AOU) bird species codes: HAWO = Hairy Woodpecker, RNSA = Red-naped sapsucker, TTWO = Three-toed woodpecker, MOCH = Mountain chickadee, RBNU = Red-breasted nuthatch)

information from our lab and field studies to conduct an ecological risk assessment. Using known information on diet, feeding rates, and body masses for our two key species, Hairy and Three-toed woodpeckers, we estimated how much food adult and nestling woodpeckers of both sexes and species should be ingesting. Since we also had quantitative information on the concentrations of arsenic in bark beetles and the time spent foraging in MSMA stands from the radio-telemetry study, we were able to estimate the daily dose of MSMA to adult and nestling woodpeckers under moderate and relatively high exposure conditions.

Equation 1:

$$\text{Dose} = \frac{[(C_{\text{food}} \times IR_{\text{food}} \times AUF)]}{BW}$$

where Dose = daily dose resulting from ingestion (µg/g/day); C_{food} = concentration of total arsenic in prey (moderate = 50 or high = 350 µg/g dry weight); IR_{food} = estimate of daily ingestion rate of food for each species by age (range 7–12.4 g/day dry weight); AUF = area use factor (moderate = 13 or 23%, high = 60 or 66%); BW = body weight (g).

Daily doses of adult and nestling woodpeckers of the two main study species (Hairy and Three-toed woodpeckers) were estimated independently and ranged from 0.95 to 1.94 µg/g/day under moderate exposure conditions or 27.9–38.9 µg/g/day under high exposure conditions.

Toxicity reference values (TRV), measures used by risk assessors to quantify the concentration of a chemical which causes harm to a given animal species (Sample et al. 1996), were established from the Zebra finch laboratory dosing study. The dose estimates were then used to derive a series of hazard quotients (HQ=dose/TRV) to compare with a "No observable effect level" (NOAEL) and "Lowest observable effect level" (LOAEL). Albert's studies showed significant mass loss in adults at 24 and 72 mg/kg/day and a reduction in growth of nestlings at 8 and 12 mg/kg/day. Considering all of this information, we predicted that under the high exposure scenario, both species would be subject to toxic effects with adults and particularly nestlings of Three-toed woodpeckers exhibiting the greatest hazard (Morrissey et al. 2006). We therefore, concluded that there was potential risk for adverse ecological effects to woodpeckers breeding in MSMA treatment areas which warranted further action.

Efficacy of MSMA Use

Early in our study we had developed some concerns about the effectiveness of MSMA in controlling MPB and other bark beetle pests due to the degree of debarking on MSMA trees. Many adult and larval beetles from treated trees were still alive when we collected them and the beetles were spreading from treated areas. Several early studies, which tested the toxicity of organoarsenicals (MSMA and cacodylic acid) showed high mortality and brood reduction of bark beetles under controlled conditions (Newton and Holt 1971; Maclauchlan et al. 1988a; Manville et al. 1988). In 2004, we were given a clue that under operational conditions, perhaps MSMA was not as effective as it had demonstrated under those experimental conditions, or possibly the beetle was developing resistance to the pesticide. Data from felled MSMA trees in the Merritt Forest District showed MSMA killed on average only 53% of beetle broods by tree (MoFR unpublished data 2003). The Ministry of Forests speculated that this was due to poor translocation of the pesticide due to drought conditions or from treating the trees too late when the larval mining was well-developed. Lab studies have also shown that in order to be effective, MSMA needs to be applied during the egg or early in-star stages of larval development – typically 3–4 weeks post attack (Maclauchlan et al. 1988a, b)

We were concerned that the level of efficacy reported in the field was not high enough to warrant continued use of MSMA, given its potential toxicity to birds. A reduction in the beetle brood emergence by more than 75% is necessary to keep a small population of mountain pine beetle (<500 infested trees) from increasing (Carroll et al. 2006). Therefore, a reduction in beetle emergence by one-half would not be sufficient to reduce the population size or control the spread of the beetle. It also creates habitat conditions with toxic trees containing surviving beetles that remain attractive to predators. In 2005, we conducted our own study using insect flight traps set at three MSMA and three reference sites in Merritt. In following the course of beetle emergence over 14 weeks, we found a peak in emergence during a warm spell in late May – a full 6 weeks before any anticipated beetle flights should occur. Those results confirmed that in some years the beetle was acting too unpredictably to time the MSMA treatments accurately. At the stand level, we found MSMA treatments reduced mountain pine beetle emergence on average by 88% and other beetles from the same genus (*Dendroctonus*) by 80% (Morrissey et al. 2008). Though efficacy appeared much higher than the previous year's data suggested, it still meant that perhaps 6,000 adult beetles might escape the trap trees to spread again (assuming a conservative estimate of 1,000 beetle larvae normally emerge and there are 50 treated trees in a stand). It further confirmed why we were finding live beetles under the bark of MSMA trees containing high levels of arsenic. Even if woodpeckers do not feed on dead beetle larvae, it appeared that sufficient beetles were surviving treatment to attract the predators to the contaminated trees.

Science for Policy Change

Our research began during 2002 at the height of the beetle outbreak, and when there were expressions of public concerns about pesticide use for suppression of the rapidly expanding pine beetle problem. The publicity originally was not centred on birds or more generally, the environment, but rather human health. The campaign was led by a retired pediatrician from Smithers, British Columbia, Dr. Josette Weir. She spent the better part of 4 years writing letters, filing complaints and making phone calls. In the course of launching complaints to B.C.'s Environmental Appeal Board, Supreme Court, and the Forest Practices Board, to name a few, she also made her appeal to Environment Canada, the Canadian Wildlife Service (CWS), and the Pesticide Management Regulatory Agency (PMRA) to stand up and listen. She claimed that the MSMA pesticide program was poorly managed - treated trees could not be accounted for and routinely were being logged, milled and burned.[1] Gradually, her message got out that MSMA use in B.C. forests was putting humans and possibly the environment at risk.

Each year there is an interactive forum, The Pesticide Information Exchange, where scientists and regulators from various levels of government, academia and industry meet to discuss environmental and health aspects of pesticide use in British Columbia. In both 2003 and 2004, we shared our preliminary data on MSMA,

[1]The Health Canada Pest Management Regulatory Agency (PMRA) is responsible for pesticide regulation in Canada. Pesticides are stringently regulated in Canada to ensure they pose minimal risk to human health and the environment. Under authority of the *Pest Control Products Act*, Health Canada: (1) registers pesticides after a stringent, science-based evaluation that ensures any risks are acceptable; (2) re-evaluates the pesticides currently on the market on a 15-year cycle to ensure the products meet current scientific standards; and (3) promotes sustainable pest management. *Information supplied by PMRA website* available at http://www.hc-sc.gc.ca/ahc-asc/branch-dirgen/pmra-arla/index-eng.php.

thereby informing the agencies in attendance responsible for issuing permits for pesticide use (B.C. Ministry of Environment), and for registering pest control products in Canada (Health Canada Pesticide Management and Review Agency).[2] In November of 2004, we also presented our results at the Society for Environmental Toxicology and Chemistry (SETAC) World Congress in Portland, Oregon (Morrissey et al. 2004).

By the end of 2004, there was considerable momentum from our field and laboratory research. Several agencies were also now involved in concerns raised over the product's impact on forestry workers (Workers Compensation Board), the public (Health Canada), and forest birds (Environment Canada). An Environmental Science and Technology online news article published on February 9, 2005 reinforced the mounting concerns about MSMA use in the USA and in Canada, and highlighted our research findings of arsenic in the blood of forest birds as a serious environmental issue. The deadline for the expiration of the pesticide use permit in B.C. was looming the following year while the manufacturer had already made the decision not to pursue re-registration.[3] The Ministry of Forests was also currently assessing the role of MSMA in bark beetle control given the nature of the expanding beetle epidemic.[4]

In January 2005, we assembled a workshop to exchange information on the state of knowledge of MSMA use in B.C. Initially, we were uncertain of the response to the proposed workshop, which eventually had 28 participants from around the province, including representatives from Ministries of Forests and Environment, Canadian Forest Service, Health Canada (PMRA), Workers Compensation Board, Simon Fraser University, University of British Columbia, Canadian Wildlife Service, Environmental Protection (Environment Canada), independent consultants, and public activists. The workshop revealed there were still gaps in knowledge about arsenic residues in the media (soil, water, and air) in proximity to MSMA treatments or near wood burners and the potential for birds to use MSMA trees years after treatment. The participants agreed there was a need for better integration of the research priorities and to exchange information available among the agencies.

By the end of the 2005 field season, we were further convinced there was now sufficient information to show that MSMA use posed a significant risk to forest insectivores. Time in the forests also added to concerns about MSMA's limited efficacy in the face of an epidemic that had already spiraled out of control. We hiked into in numerous watersheds where MSMA treated stands stood amongst a landscape of beetle killed trees, stark evidence of a failed approach.

---

[2]"Poisoned bark worse than beetles' bite?" The Globe and Mail News, December 21, 2004.

[3]Various personal communications from PMRA officials, and BC Ministry of Environment, Pesticide Administrator, D. Cronin to J. Elliott. See also ES&T Science News, February 5, 2005 concerning manufacturer, United Agri Products Canada Inc.

[4]Comment by Doug Konklin, Deputy Minister, Ministry of Forests in Times Colonist (Victoria), January 11, 2005.

Meanwhile, the B.C. government's permit to use MSMA was due to expire at the end of 2005, while the compound was not being used elsewhere in Canada. Organic arsenicals were also under regulatory review by the US Environmental Protection Agency (EPA). It soon became apparent that some winds of change were already blowing. On June 27, 2005, there was a conference-call meeting of the British Columbia Integrated Pest Management (IPM) Committee. The committee decided to use the mechanism afforded by the transition from the old B.C. *Pesticide Control Act*, under which MSMA use had been regulated, to a new *Integrated Pest Management (IPM) Act*[5] to recommend that all uses of MSMA in the province be revoked, and to bring all future use of MSMA into consistency with the new act. Later that summer, the administrator of the IPM Act, Christine Houghton, wrote to the Field Operations Manager in the Prince George Forest District and the Forest Entomologist in the Southern Interior Forest Region to inform them that approvals for use of Glowon® MSMA had been revoked.[6] Subsequent to that decision, the Ministry of Forests decided not to pursue further registration or permits for MSMA.

In order to emphasize Environment Canada's position, on February 23, 2006, Paul Kluckner, regional director of the Canadian Wildlife Service, wrote to the Chief Forestor, Jim Snetsinger, stating the results of the woodpecker and laboratory dosing study and recommending initiation of a program for locating and mapping treated trees and monitoring future woodpecker foraging of existing MSMA trees. We believed it was necessary to maintain pressure given there were still proponents for ongoing use of MSMA likely from the Ministry of Forests and from the forest timber operators.

In 2006, the U.S. EPA released its decision to regulate arsenical pesticides, including MSMA, because they posed significant risks to human and environmental health U.S. EPA 2006. The report determined that organic arsenicals had the potential to transform into the more toxic inorganic form in soil, and enter drinking water through run-off. The ecological risk assessment also determined that MSMA (and other arsenical pesticides) failed to meet safety standards for insectivorous and herbivorous birds and mammals. The EPA also concluded they posed a risk to consumers of meat and milk products from animals fed cotton byproducts treated with MSMA. That decision provided strong support for our arguments that forestry use of MSMA in British Columbia was hazardous to insectivorous birds. Faced

---

[5]The Ministry of Environment now administers the British Columbia *Integrated Pest Management Act* and Regulations, which regulates the sale, use and handling of pesticides in the province and supports an Integrated Pest Management (IPM) approach to managing pests. The IPM approach involves a science-based process including prevention through good ecosystem and crop practices, identification and monitoring of the pest problem, combining a range of control followed by regular evaluation of the effectiveness of the treatments.

[6]Letters from Christine Houghton to L. MacLauchlan, Forest Entomologist, Southern Interior Forest Region, and to B. Doerksen, Compliance and Enforcement Field Operations Supervisor, Prince George Forest District, sent August 31, 2005

with our evidence and official concerns, a degree of public pressure through the media, the EPA decision, and given that local suppression of beetle expansion was futile in face of a run-away epidemic; the British Columbia Ministry of Forests did not oppose the forces aligned against continuing use of MSMA. The ministry subsequently agreed to cancellation of the registration of MSMA and to destroy remaining stocks of the chemical.

What remained outstanding was the issue of the "MSMA trees," the 500,000 or so treated trees in the province left standing and still containing arsenic residues. Those trees pose a lower risk to avian predators as the arsenic is bound in the tree and the mountain pine beetles are either dead or dispersed, thereby diminishing the major cause of food chain transfer. However, little is known about the risk posed by secondary woodboring insects attacking decaying MSMA trees or the fate of the arsenic in surrounding soil or nearby water from dropped and decaying needles. We still had concerns that secondary insects could potentially accumulate MSMA and transfer it to predators, particularly when the beetle epidemic subsides and birds will switch to other prey. The Ministry of Forests took the initiative to develop a policy to deal with "legacy tree" problem. In January 2007, there was a public forum in Smithers, B.C. attended by the major MSMA stakeholders from government, academia, industry and nongovernment agencies. The intent was to develop a policy on the management of MSMA-treated trees in British Columbia. In July 2007, the policy was successfully released and represented an important achievement in concluding the MSMA story. The policy provided clear procedures for potential introductions of any new similar pesticides for bark beetle control and also set out guidelines for identifying and researching existing legacy trees – some half a million treated trees that the committee decided to mark and leave where they were. Today, the B.C. Ministry of Forests and Range have a website[7] to inform the public about health and environmental hazards to MSMA, and have created a database and maps of the locations of the legacy MSMA trees as a tool to better protect environment and human health. There is also a monitoring program in place to determine the extent to which arsenic from past MSMA use may continue to cause contamination of forest ecosystems.

While organic arsenicals have been banned in Canada, advocates for continued use in the USA have lobbied effectively to keep MSMA on the market specifically for use on cotton. Data submitted by stakeholders showed the chemical was not getting into the meat and milk products of animals fed cotton byproducts grown in MSMA treated fields. They also documented the increasing spread of Palmer amaranth or pigweed, a glyphosate-resistant and economically significant pest, which only MSMA could control at present. As a result of these appeals and new information following the 2006 decision, MSMA was retained for use only on cotton, but with tighter restrictions to better protect drinking water. In early 2009,

[7]Ministry of Forests and Range set up a website giving details of the MSMA policy, information on toxicity, maps of treatment areas and other related information for the public available at http://www.for.gov.bc.ca/hfp/health/MSMA.htm.

the U.S. EPA reached a voluntary agreement in principle with the major manufacturers of the organic arsenicals in an effort to steadily remove them from the market for all other uses especially on turf and lawns. A phasing out of other uses is expected to accelerate a transition to new, lower risk herbicides to replace all uses of organic arsenicals in the near future (U.S. EPA 2009).

Lessons Learned and Conservation Gains

We revealed multiple lines of evidence that MSMA was getting into the food chain of insectivorous birds, and had the potential to cause harm. Bark beetles could accumulate and survive high concentrations of the original MMAA compound and birds were identified feeding on those trees. Laboratory studies showed that the concentrations the birds were predicted to be ingesting under certain conditions could cause weight loss in adults, reduced growth in nestling birds, and even mortality. There was also potential for other long term, sublethal effects that we did not measure, given it accumulated in tissues such as the liver, kidney, and brain.

MSMA, like many pesticides, has the potential to affect wildlife populations both directly through toxic action and indirectly by limiting food availability. Both mechanisms can act independently or in combination to influence the survival and reproduction of birds. In the course of our study, we attempted to address both contaminants and reduced bark beetle prey abundance as stressors to forest bird populations (Morrissey et al. 2008).

The forest management practices we witnessed in the study area included large scale harvesting in combination with MSMA applications. Alternative treatments such as controlled burning and single tree removal had long been abandoned for their high cost and limited scope once the beetle epidemic had taken hold. There is strong evidence that the forest practices used in central British Columbia are capable of reducing available habitat for birds and placing beetle predators at even greater risk. By limiting the available habitat, it can make birds more likely to feed on MSMA trees, which also contain fewer prey that are also contaminated. Food limitation in combination with pollutants is known to cause physiological stress, depressed metabolic rates and poorer growth in birds (Hutton 1980; Di Guilio and Scanlon 1985; Eeva et al. 2003). We concluded a dangerous combination of beetle management practices that use pesticides and large-scale salvage logging in the same area had the potential to limit woodpecker populations through a combination of effects – increased susceptibility to exposure, direct toxicity and indirect nutritional stress (Morrissey et al. 2008).

It had been known for decades that woodpeckers play an important ecological role in regulating bark beetle populations (Hutchison 1951; Baldwin 1960; McCambridge and Knight 1972; Goggans et al. 1989; Steeger et al. 1998; Fayt et al. 2005). They are considered one of the most important predators of insect pests throughout North America (Buckner 1966) and have also been observed increasing the amount of bark beetles in their diet in response to past beetle epidemics (Koplin and Baldwin 1970; Koplin 1972; Crocket and Hansley 1978). Recommendations for managing beetle infested stands have cautioned against the use of large scale salvage logging to remove all damaged trees because of the potential for reducing woodpecker populations and ultimately causing increases in beetle numbers (Kroll et al. 1980).

Following our study, we still had reason to believe that extensive salvage logging to remove beetle infested trees in combination with legacy MSMA trees could be potentially harmful to woodpeckers and other cavity nesting birds. The MSMA policy written in 2007, continues to allow MSMA areas to be included in "wildlife patches" or "set asides" for forest management purposes. The danger is that these areas can count towards the timber supply companies' mandatory requirements for retaining suitable habitat. At present, the decision to leave MSMA trees standing means wildlife are still at risk, albeit a reduced risk. There is a possibility that the future situation may reveal birds' struggling to find suitable, uncontaminated foraging and breeding areas in a fragmented landscape – especially after the beetle population has crashed and even more of the remaining pine forest has been harvested.

Our research clearly showed that MSMA was an important environmental problem and we have hopefully contributed to the prevention of future poisoning of insectivorous birds from MSMA. In the end, the chemical was removed from the market, but in part because the government had accepted their loss to the beetle. The pesticide was not going to help their cause and at the end of the day, few were fighting to keep it. Would the outcome have been the same if we had done the study 5 years earlier when they were still fighting the beetle feverishly? Would they have surrendered one of their tools so easily? On the broader landscape scale, involvement with this study drew our attention to the problems associated with forest management in British Columbia. One factor was continually evident – fire suppression. However, the problems associated with fire suppression extend beyond forestry policy as there are broader societal issues associated with development and insurance in fire-prone zones. These are common not only to regions of British Columbia, but to jurisdictions world-wide.

We only hope that our study contributes to future decisions on how commercial forests are managed and the guidelines for use of insecticides against forest pests. Perhaps future governments might consider managing forests for woodpeckers: a strategy that maintains the integrity of the forest ecosystem and naturally regulates beetle populations to increase yields.

Acknowledgments We are indebted to a previous generation of Canadian Wildlife Service biologists, who studied the problems of forest insecticide use and effects on birds, principally, Dan Busby, Neville Garrity, Peter Pearce, and David Peakall. Pierre Mineau, who researched the effects of pesticides on wildlife for many years, first raised the potential issue of MSMA at meetings of the Wildlife Toxicology Division. The following people made important contributions to the project namely Patti Dods, Courtney Albert, Tony Williams, Bill Cullen, and Vivian Lai. We thank field and laboratory research assistants Alicia Newbury, Mark Wong, Sandi Lee, Sheila Carroll, Tracy Sutherland, and Jason Berge. We also thank the many personnel at the Ministry of Forests and Range (Merritt office) and several timber operators (BCTS, Tolko and Weyerhauser), who gave advice and assistance on the logistical aspects of the project. John Borden, Kathy Martin, Josette Wier, and Judy Strachan also offered helpful insight into the research.

References

Albert CA, Williams TD, Morrissey CA, Lai VW, Cullen WR, Elliott JE (2008a) Tissue uptake, mortality, and sublethal effects of monomethylarsonic acid (MMA(V)) in nestling Zebra finches (*Taeniopygia guttata*). J Toxicol Environ Health A 71:353–360

Albert CA, Williams TD, Morrissey CA, Lai VW-M, Cullen WR, Elliott JE (2008b) Dose-dependent uptake, elimination, and toxicity of monosodium methanearsonate in adult Zebra finches (*Taeniopygia guttata*). Environ Toxicol Chem 27:605–611

B.C. Ministry of Forests and Range (1995) Bark beetle management guidebook. http://www.for.gov.bc.ca/tasb/legsregs/fpc/fpcguide/beetle/betletoc.htm. Accessed 12 Jan 2006

B.C. Ministry of Forests and Range (2007) Treatment of trees with monosodium methanearsenate (MSMA) for bark beetle control. http://www.for.gov.bc.ca/hfp/health/MSMA.htm. Accessed 1 Mar 2009

B.C. Ministry of Forests and Range. Mountain pine beetle. http://www.for.gov.bc.ca/hfp/mountain_pine_beetle/facts. Accessed 1 Mar 2009

Backhouse F (2005) Woodpeckers of North America. Firefly Books Ltd., Buffalo

Baldwin OH (1960) Overwintering of woodpeckers in bark-beetle infested spruce-fir forests in Colorado. In: Proceedings of the 12th international ornithological congress, Helsinki, Finland, pp 71–84

Blus LJ (1984) DDE in birds eggs: comparison of 2 methods for estimating critical levels. Wilson Bull 96:268–276

Buckner CH (1966) The role of vertebrate predators in the biological control of forest insects. Annu Rev Entomol 11:449–470

Busby DG, White LM, Pearce PA (1990) Effects of aerial spraying of fenitrothion on breeding abundance of White-throated sparrows. J Appl Ecol 27:743–755

Camardese MB, Hoffman DJ, LeCaptain LJ, Pendleton GW (1990) Effects of arsenate on growth and physiology in mallard ducklings. Environ Toxicol Chem 9:785–795

Canadian Forest Service, Pacific Forestry Centre. The mountain pine beetle. http://mpb.cfs.nrcan.gc.ca/index. Accessed 1 Mar 2009

Carroll AL, Shore TL, Safrynik L (2006) Direct control: theory and practice. In: Safranyik L, Wilson WR (eds) The mountain pine beetle: a synthesis of biology, management, and impacts on lodgepole pine. Natural Resources Canada, Canadian Forest Service, Pacific Forestry Centre, Victoria, BC, 304 pp

Crocket AB, Hansley PL (1978) Apparent response of Picoides woodpeckers to outbreaks of the pine bark beetle. Western Birds 9:67–70

Cullen WR, Reimer KJ (1989) Arsenic speciation in the environment. Chem Rev 89:713–764

Custer TW, Custer CM, Hines RK, Gutreuter S, Stromborg KL, Allen PD, Melancon MJ (1999) Organochlorine contaminants and reproductive success of double-crested cormorants from Green Bay, Wisconsin, USA. Environ Toxicol Chem 18:1209–1217

Di Guilio RT, Scanlon PF (1985) Effects of cadmium ingestion and food restriction on energy metabolism and tissue metal concentrations in mallard duck. (*Anas platyrhynchos*). Environ Res 37:433–444

Dost FN (1995) Public health and environmental impacts of MSMA as used in bark beetle control in British Columbia 1995; FS 48 HSI 95/2. Silviculture Practices Branch, B.C. Ministry of Forests, Victoria, BC

Drever MC, Aitken KEH, Norris AR, Martin K (2008) Woodpeckers as reliable indicators of bird richness, forest health and harvest. Biol Conservat 141:624–634

Eeva T, Lehikoinen E, Nikinmaa M (2003) Pollution-induced nutritional stress in birds: an experimental study of direct and indirect effects. Ecol Appl 13:1242–1249

Eisler R (2000) Arsenic. In: Handbook of chemical risk assessment: health hazards to humans, plants, and animals, vol 3. Lewis, Boca Raton, pp 1501–1566

Elliott JE, Norstrom RJ, Keith JA (1988) Organochlorines and eggshell thinning in Northern gannets (*Sula bassanus*) from Eastern Canada, 1968–1984. Environ Pollut 52:81–102

Elliott JE, Norstrom RJ, Lorenzen A, Hart LE, Philibert H, Kennedy SW, Stegeman JJ, Bellward GD, Cheng KM (1996) Biological effects of polychlorinated dibenzo-*p*-dioxins, dibenzofurans, and biphenyls in bald eagle (*Haliaeetus leucocephalus*) chicks. Environ Toxicol Chem 15:782–793

Fayt P, Machmer M, Steeger C (2005) Regulation of spruce bark beetles by woodpeckers – a literature review. Forest Ecol Manage 206:1–14

Goggans R, Dixon RD, Seminara LC (1989) Habitat use by three-toed and black-backed woodpeckers, Deschutes National Forest, Oregon. Oregon Department of Fish and Wildlife, Technical Report 87-3-02, 43 pp

Hoffman DJ, Sanderson CJ, LeCaptain LJ, Cromartie E, Pendleton GW (1992) Interactive effects of arsenate, selenium, and dietary-protein on survival, growth, and physiology in mallard ducklings. Arch Environ Contam Toxicol 22:55–62

Hutchinson FT (1951) The effects of woodpeckers on the Engelmann spruce beetle, *Dendroctonus engelmanni* Hopk. MSc thesis, Colorado State University, Fort Collins

Hutton M (1980) Metal contamination of feral pigeons *Columba livia* from the London area. Part 2: biological effects of lead exposure. Environ Pollut Series A 22:281–293

Jaghabir MTW, Abdelghani AA, Anderson AC (1989) Histopathological effects of monosodium methanearsonate (MSMA) on New Zealand White rabbits (*Oryctalagus cuniculus*). Bull Environ Contam Toxicol 42:289–293

Judd FW (1979) Acute toxicity and effects of sublethal dietary exposure of monosodium methanearsonate herbicide to *Peromyscus leucopus* (*Rodentia – Cricetidae*). Bull Environ Contam Toxicol 22:143–150

Klein WH (1978) Strategies and tactics for reducing losses in lodgepole pine to the mountain pine beetle by chemical and mechanical means, pp 54–63. In: Kibbee DL, Berryman AA, Amman GD, Stark RW (eds) Theory and practice of mountain pine beetle management in lodgepole pine forests. Symposium proceedings. University of Idaho, Moscow, ID, 224 pp

Koch P (1996) Lodgepole pine in North America, vol 2, Chapter 11. Forest Products Society, Madison, 763 pp

Koplin JR (1972) Measuring predator impact of woodpeckers on spruce beetles. J Wildl Manage 36:308–320

Koplin JR, Baldwin PH (1970) Woodpecker predation on an endemic population of Engelmann spruce beetles. Am Midl Nat 83:510–515

Kroll JC, Conner RN, Fleet RR (1980) Woodpeckers and the southern pine beetle. USDA Agriculture Handbook No. 564

Maclauchlan LE, Borden JH, D'Auria JM, Wheeler LA (1988a) Distribution of arsenic in MSMA-treated lodgepole pines infested by the mountain pine beetle *Dendroctonus ponderosae* (Coleoptera: Scolytidae), and its relationship to beetle mortality. J Econ Entomol 81:274–280

Maclauchlan LE, Borden JH, D'Auria JM (1988b) Distribution of arsenic in lodgepole pines treated with MSMA. J Appl Forest 3:37–40

Manville JF, McMullen LH, Reimer KJ (1988) Impact and role of monosodium methanearsonate on attack and progeny production by the Douglas-Fir Beetle (Coleoptera: Scolytidae) in lethal trap trees. J Econ Entomol 81:1691–1697

Martin K, Aitken KEH, Wiebe KL (2004) Nest sites and nest webs for cavity-nesting communities in interior British Columbia, Canada: nest characteristics and niche partitioning. Condor 106:5–19

Martinovic B, Lean DRS, Bishop CA, Birmingham E, Secord A, Jock K (2003) Health of tree swallow (*Tachycineta bicolor*) nestlings exposed to chlorinated hydrocarbons in the St Lawrence River Basin. Part I. Renal and hepatic vitamin A concentrations. J Toxicol Environ Health 66A:1053–1072

McCambridge WF, Knight FB (1972) Factors affecting spruce beetles during a small outbreak. Ecology 53:830–839

MMA Task Force Three (1993) SARA Title III Section 313 Delisting petition for monosodium methanearsonate (MSMA) and disodium methanearsonate (DSMA). ISK Biotech, Mentor, OH

Morrissey C, Dods P, Albert C, Wilson L, Cullen W, Williams T, Elliott J (2004) Assessing avian exposure to monosodium methanearsonate (MSMA) as used for bark beetle control in British Columbia forests. Poster presentation, Society for Environmental Toxicology and Chemistry (SETAC). http://abstracts.co.allenpress.com/pweb/setac2004/document/43214. Accessed 1 Mar 2009

Morrissey C, Dods P, Albert C, Cullen W, Lai V, Williams T, Elliott J (2006). Assessing forest bird exposure and effects from monosodium methanearsonate (MSMA) during the mountain pine beetle epidemic in British Columbia. Technical Report Series No. 460. Environment Canada, Canadian Wildlife Service, Pacific and Yukon Region, British Columbia

Morrissey CA, Albert CA, Dods PL, Cullen WR, Lai VW-M, Elliott JE (2007) Arsenic accumulation in bark beetles and forest birds occupying mountain pine beetle infested stands treated with monosodium methanearsonate (MSMA). Environ Sci Technol 41:1494–1500

Morrissey CA, Dods PL, Elliott JE (2008) Pesticide treatments affect mountain pine beetle abundance and woodpecker foraging behaviour. Ecol Appl 18:172–184

Murvoll KM, Skaare JU, Anderssen E, Jenssen BM (2006) Exposure and effects of persistent organic pollutants in European shag (*Phalacrocorax aristotelis*) hatchlings from the coast of Norway. Environ Toxicol Chem 25:190–198

Newton M, Holt HA (1971) Scolytid and buprestid mortality in ponderosa pines injected with organic arsenicals. J Econ Entomol 64:952–958

Prukop JA, Savage NL (1986) Some effects of multiple, sublethal doses of monosodium methanearsonate (MSMA) herbicide on haematology, growth, and reproduction of laboratory mice. Bull Environ Contam Toxicol 36:337–341

Safrynik L, Carroll AL (2006) The biology and epidemiology of the mountain pine beetle in lodgepole pine forests. In: Safranyik L, Wilson WR (eds) The mountain pine beetle: a synthesis of biology, management, and impacts on lodgepole pine. Natural Resources Canada, Canadian Forest Service, Pacific Forestry Centre, Victoria, BC, 304 pp

Sample BE, Opresko DM, Suter GW (1996) Toxicological benchmarks for wildlife: 1996 revision. Oak Ridge National Laboratory, Oak Ridge, TN, ES/ER/TM-86/R3

Shore TL, Safranyik L, Hawkes BC, Taylor SW (2006) Effects of the mountain pine beetle on lodgepole pine stand structure and dynamics. In: Safranyik L, Wilson WR (eds) The mountain pine beetle: a synthesis of biology, management, and impacts on lodgepole pine. Natural Resources Canada, Canadian Forest Service, Pacific Forestry Centre, Victoria, BC, 304 pp

Steeger C, Machmer MM, Gowans B (1998) Impact of insectivorous birds on bark beetles: a literature review. Pandion Ecological Research Report, Nelson, BC

Taylor SW, Carroll AL (2004) Disturbance, forest age, and mountain pine beetle outbreak dynamics in BC: a historical perspective, pp 41–51. In: Shore TL, Brooks JE, Stone JE (eds) Mountain pine beetle symposium: challenges and solutions, Kelowna, British Columbia, 30–31 Oct 2003, Natural Resources Canada, Canadian Forest Service, Pacific Forestry Centre, Information Report BC-X-399, 298 pp

U.S. Environmental Protection Agency (2006) Reregistration eligibility decision for MSMA, DSMA, CAMA, and cacodylic acid. EPA-HQ-OPP-2006-0201

U.S. Environmental Protection Agency (2009) Agreement in principle to implement the organic arsenicals reregistration eligibility decision (RED) between EPA and the MAA research task force. EPA-HQ-OPP-2009-0191

Verreault J, Skaare JU, Jenssen BM, Gabrielsen GW (2004) Effects of organochlorine contaminants on thyroid hormone levels in Arctic breeding glaucous gulls, *Larus hyperboreus*. Environ Health Perspect 112:532–537

Virkkala R (2006) Why study woodpeckers? The significance of woodpeckers in forest ecosystems. Ann Zool Fenn 43:82–85

Chapter 9
Amphibians Are Not Ready for Roundup®

Rick A. Relyea

As I peered into the water, I recall being shocked at the sight. This tank was one of hundreds we were using as experimental ponds, each simulating a real pond by containing leaf litter, algae, zooplankton, and tadpoles. However, this particular tank did not contain very many tadpoles - at least not many live ones. Instead, the bodies of dead tadpoles were littered across the bottom. This tank and a few others of similar macabre appearance had been exposed to the most popular herbicide in the world. It was at that moment that we learned that the herbicide could be highly lethal to amphibians. As I think back upon that day, I am struck by the fact that one simple experiment led us to a discovery that would take me into years of debate.

Abstract The herbicide glyphosate, sold under a variety of commercial names including Roundup® and Vision®, has long been viewed as an environmentally friendly herbicide. In the 1990s, however, after nearly 20 years of use, the first tests were conducted on the herbicide's effects on amphibians in Australia. The researchers found that the herbicide was moderately toxic to Australian amphibians. The leading manufacturer of glyphosate-based herbicides, Monsanto, declared that the researchers were wrong. Nearly 10 years later, my research group began examining the effects of the herbicides on North American amphibians. Based on an extensive series of experiments, we demonstrated that glyphosate-based herbicides can be highly toxic to larval amphibians. Monsanto declared that we were also wrong. These experiments have formed the basis of a spirited debate between independent, academic researchers, and scientists that either work as consultants for Monsanto or have a vested interest in promoting the application of the herbicide to control undesirable plants in forests and agriculture. The debate also moved into unexpected

---

R.A. Relyea (✉)
Department of Biological Sciences, University of Pittsburgh, Pittsburgh, PA, 15260, USA
e-mail: relyea@pitt.edu

J.E. Elliott et al. (eds.), *Wildlife Ecotoxicology: Forensic Approaches*,
Emerging Topics in Ecotoxicology 3, DOI 10.1007/978-0-387-89432-4_9,

arenas, including the use of glyphosate-based herbicides in the Colombian drug war in South America where a version of Roundup is being used to kill illegal coca plantations. In 2008, the US EPA completed a risk assessment for the effects of glyphosate-based herbicides on the endangered California red-legged frog (*Rana aurora draytonii*) and concluded that it could adversely affect the long-term persistence of the species. More recent data from Colombia have confirmed that the herbicides not only pose a risk to tadpoles in shallow wetlands, but that typical applications rates also can kill up to 30% of adult frogs. As one reflects over the past decade, it becomes clear that our understanding of the possible effects of glyphosate-based herbicides on amphibians has moved from a position of knowing very little and assuming no harm to a position of more precise understanding of which concentrations and conditions pose a serious risk.

Introduction

As a community ecologist studying aquatic organisms, my research for many years focused on understanding how animals respond to natural stressors including predation and competition. I spent my doctoral years focusing on how tadpoles responded to these natural stressors by changing their behavior, morphology, and life history. I certainly had no interest in toxicology and no goal of working in this field of research. I never had and still do not have a personal antipesticide agenda. I grew up in a rural community and spent a good deal of my younger life working on a farm and spraying a lot of different pesticides. As a result, I recognized the benefits of pesticides for feeding the world and protecting human health. So years later, why did I find myself debating toxicology with a multinational corporation like Monsanto and their consultants?

After completing my Ph.D. and before moving to a faculty position at the University of Pittsburgh, I spent the summer of 1999 collaborating with Ray Semlitsch at the University of Missouri-Columbia. Ray's research group not only shared my interest in studying natural stressors, but also was developing an exciting research program in amphibian toxicology, primarily with the insecticide carbaryl (commercial name: Sevin®). At the time, few amphibian biologists were interested in venturing into this realm, but the University of Missouri was located just a few miles away from the U.S.G.S. Toxicology Lab in Columbia, making it an ideal collaborative effort. An impromptu conversation with one of Ray's graduate students, Nathan Mills (now a professor at Harding University), raised an interesting question: If tadpoles could respond to predators by altering their behavior and morphology, how might they respond to the smell of predators in the presence of an insecticide like carbaryl that is designed to interfere with an animal's nervous system? Could sublethal insecticide concentrations interfere with a tadpole's ability to smell or respond to a predator in the water? We decided to conduct an experiment to find the answer. We placed predators in small cages so that the smells of predators could scare the tadpoles but the predators could not actually eat the tadpoles. In regard to that particular question, the experiment was a complete failure – but not for lack of an adequate experimental design. It turned out that sublethal concentrations of carbaryl, when combined with the smell of predators in the water, became quite lethal to tadpoles (Relyea and Mills 2001). Instead of the tadpoles changing their behavior or morphology, they simply died.

For reasons that we still do not understand, adding just one element of the tadpole's natural world – the smell of a predator – made the pesticide deadlier than anyone had ever suspected. In the years that followed, we found that several commercial pesticide formulations including carbaryl, malathion, and Roundup had a similar effect on a diverse group of amphibians (Relyea 2003, 2004a, 2005c). No one had previously described such a synergistic interaction. As I began to learn about the field of toxicology, I realized that most toxicology research is conducted on species in the laboratory that are isolated from all other species with which they coexist in nature. I realized that as a community ecologist, I could approach toxicology from a different perspective than traditionally trained toxicologists. A community ecologist asking toxicology questions would focus more on how pesticides affect complex communities rather than a single species. Not necessarily a better perspective, just different. It was this opportunity to offer a unique perspective that led me to pursue research into community ecotoxicology and brought me face to face with a tank full of dead tadpoles that had been treated with the popular herbicide Roundup®.

The Rise of Roundup

In 1974, Monsanto Corp. (St. Louis, MO, USA) first commercially produced the chemical glyphosate after discovering that the chemical could prevent plants from making essential amino acids, thereby causing the plants to die. However, most herbicides are not good at penetrating the waxy outer coating of plant leaves. This coating, known as the cuticle, acts as a natural barrier to prevent foreign organisms and compounds, including herbicides, from entering the plant's tissues. To penetrate the cuticle, manufacturers frequently include additional chemicals to help the herbicide to penetrate into the leaf. These chemicals, called surfactants, are good at dissolving in water and at cutting through wax and grease (e.g., dish soap is a surfactant). In the case of Roundup, the most common surfactant is polyethoxylated tallowamine (POEA), a derivative of animal fat. Together, glyphosate and POEA make a very effective herbicide. Some glyphosate formulations use different surfactants or blends of surfactants. Others contain no surfactants, but applicators are encouraged to add an after-market surfactant if glyphosate alone proves to be ineffective at killing weeds.

In the early years, glyphosate had a relatively small share of the herbicide markets. In the late 1980s and early 1990s, however, glyphosate sales began to grow rapidly for a variety of uses including homes, gardens, no-till agriculture, and forestry

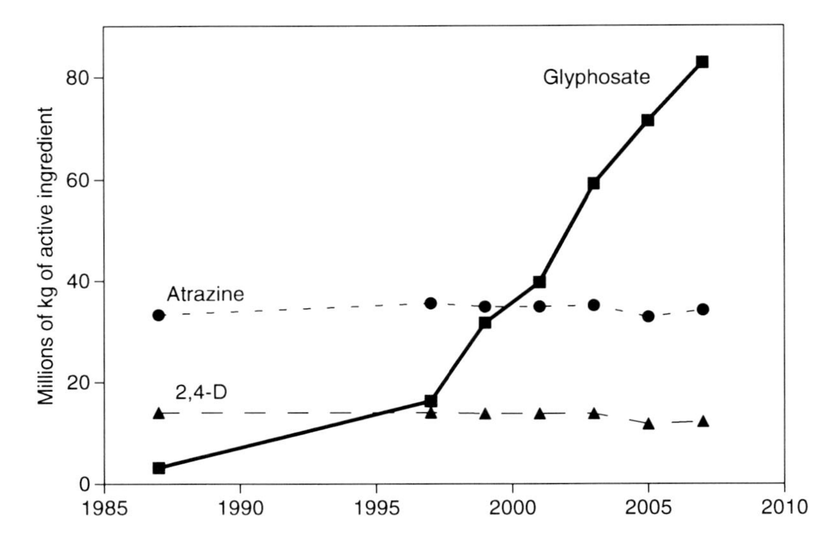

Fig. 9.1 The growth of glyphosate-based herbicides relative to other globally common herbicides (Kiely et al. 2004; Grube et al. 2011)

(where herbicides are used to kill broadleaf trees and favor the growth of conifer trees). In 1996, however, Monsanto took advantage of new genetic techno-logies which allowed the insertion of glyphosate-resistant genes into crop plants, thereby making the plant much more resistant to the herbicide than surrounding weeds. This discovery allowed Roundup to kill weeds that compete with crops without harming the crops. This marked the birth of Roundup-Ready® crops that currently span a variety of common crops including corn, soybeans, cotton, canola, and sugar beets. The genetically modified seeds were patented and farmers, eager to plant a seed variety that would tolerate a powerful herbicide, signed contracts with Monsanto. The pairing of Roundup herbicide and Roundup-Ready seeds produced an exponential increase in the use of glyphosate-based herbicides around the world (Fig. 9.1) and it is now the top selling herbicide in the world (Baylis 2000). Although Monsanto is one of the largest manufacturers of glyphosate-based herbicides, many other companies now produce them as well.

How Do We Assess the Risk of Pesticides?

Whenever a new pesticide is developed, it must be tested to determine its impact on nontarget organisms before it can be registered. Amphibians, however, have never been part of the testing process for the vast majority of pesticides used today, including glyphosate. Instead, the standard protocol of the US Environmental Protection Agency (EPA) requires testing of each pesticide on mammals, birds, fish, and tiny crustaceans such as water fleas (e.g., *Daphnia* spp.).

In traditional toxicology studies, individuals are exposed to a range of concentrations for 1–4 days under laboratory conditions (i.e., acute toxicity tests). Providing that mortality in the control (i.e., no pesticide) is less than 10%, one can use these experiments to determine the lethal concentration that will kill 50% of the population (i.e., the LC50 value). The LC50 values for a new pesticide can be compared to a large amount of past data from these four animal groups to assess its relative toxicity. Using US EPA classifications,[1] the toxicity of a pesticide is then categorized as either *highly toxic* ($0.1\ mg/L < LC50 < 1\ mg/L$), *moderately toxic* ($1\ mg/L < LC50 < 10\ mg/L$), or *slightly toxic* ($10\ mg/L < LC50 < 100\ mg/L$).[2] In addition to these acute toxicity tests, some organisms are also tested over longer time intervals to examine potential reproductive impacts (i.e., chronic toxicity tests).

When the pesticide registration rules were designed long ago, there was little concern about amphibians and no sense that amphibians were experiencing the massive global population declines that are well documented today (Stuart et al. 2004). For aquatic stages of amphibians (e.g., tadpoles and larval salamanders), the EPA currently assumes that fish can serve as a surrogate group. This is based on the assumption that the sensitivity of tadpoles is similar to that of fish. This is often a reasonable assumption. When the assumption is correct, regulating pesticides at levels that protect the most sensitive species of fish simultaneously protects the aquatic stages of all amphibians. If amphibian data on aquatic stages of amphibians are available, these data can also be considered by the EPA.

For terrestrial stages of amphibians, the EPA assumes that birds can serve as a surrogate group. These are not studies of birds being over sprayed, but birds ingesting food that has been contaminated by a given chemical (Jones et al. 2004). Thus, exposure via the skin (i.e., *dermal exposure*) or respiratory system is currently not part of the risk assessment process. Unfortunately, it is unknown whether regulating pesticides at levels that protect birds (based on ingestion studies) is also protective of terrestrial amphibians that can be exposed to pesticides via ingestion, dermal exposure, and respiration.

In assessing risk, we need to know the concentrations of the chemical that cause harm (based on LC50 studies) and then compare these values to the concentrations of the chemical that occur in nature. To set a safe upper limit for the concentrations of a chemical in nature, the EPA uses an index known as the "Risk Quotient." The Risk Quotient is calculated by dividing the LC50 value for the most sensitive species in a group (e.g., the most sensitive fish) by the expected concentration of the chemical in the nature. To not harm wildlife, this ratio should not exceed 0.05 (Jones et al. 2004). More simply put, the concentration that we expect to see in nature should not exceed 5% of the most sensitive species' LC50 value. The logic is that if the LC50 concentration kills 50% of the animals, 5% of this number should provide a concentration that kills few or none of the animals.

[1] http://www.epa.gov/espp/litstatus/effects/redleg-frog/.

[2] In the case of glyphosate-based herbicides, the most common reported units are milligrams of acid equivalents per liter (mg a.e./L).

When a pesticide is being considered for approval by the EPA, the agency also can require data from toxicity data on inert ingredients if there is evidence that the inert ingredients pose a risk (Jones et al. 2004). Inert ingredients are not designed to kill the target pest, but are added to the commercial formulation to make the active ingredient more effective. Inert ingredients are considered trade secrets and therefore do not have to be listed on the container's label. Many people assume that the "inert ingredient" category implies that these chemicals are not toxic to any organism. For surfactants such as POEA, this is not the case.

The First Studies of Roundup's Impact on Amphibians

Because the EPA and the regulatory agencies in other countries have not required amphibian testing as part of their process for registering most pesticides, little was known about the effect of Roundup formulations on amphibians. Indeed, the herbicide had been on the market for nearly 20 years before the first amphibian study was ever conducted. In the 1990s, this began to change.

In the early 1990s, the East Kimberly shire council submitted a proposal to the Western Australia Department of Environmental Protection to aerially spray an emergent weed in Lake Kunnunurra (Mann et al. 2003). At the time, Roundup with POEA could be applied to control aquatic plants (this was not the case in the USA). However, there had been numerous reports of dead amphibians following the spraying of herbicides, so the government funded Joe Bidwell and John Gorrie (then at the Curtin University of Technology) to conduct acute toxicity tests using both the active ingredient and the commercial formulation (which contained the POEA surfactant). Their first experiments found that the commercial formulation (Roundup Herbicide®) was much more toxic to amphibians than the active ingredient alone, likely because of the POEA surfactant (Bidwell and Gorrie 1995; Fig. 9.2). Similar results had previously been reported in fish (Folmar et al. 1979). Bidwell and Gorrie's research prompted a special review of how glyphosate-based products were being used over water. Eighty-four products were deemed no longer safe for application to plants growing in and around water. These products now were required to carry a new label in Australia, "Do NOT apply to weeds growing in or over water. Do NOT spray across open water bodies, and do NOT allow spray to enter the water."

A series of follow-up studies (Mann and Bidwell 1999) demonstrated for the first time that Roundup containing the POEA surfactant had $LC50_{2\text{–}d}$ values (measured in units of acid equivalents; a.e.) that ranged from 2.9 to 11.6 mg a.e./L for four species of tadpoles: sign-bearing froglet (*Crinia insignifera*), moaning frog (*Heleioporus eyrei*), western bullfrogs (*Limnodynastes dorsalis*), and golden bell frogs (*Litoria moorei*; Fig. 9.2). This meant that Roundup with POEA could be classified as slightly to moderately toxic to amphibians. Because the commercial formulation was moderately toxic but the active ingredient was not, there was the suggestion that aquatic plants could be sprayed with glyphosate that was combined with a separately purchased surfactant. Subsequent studies found that, similar to the POEA

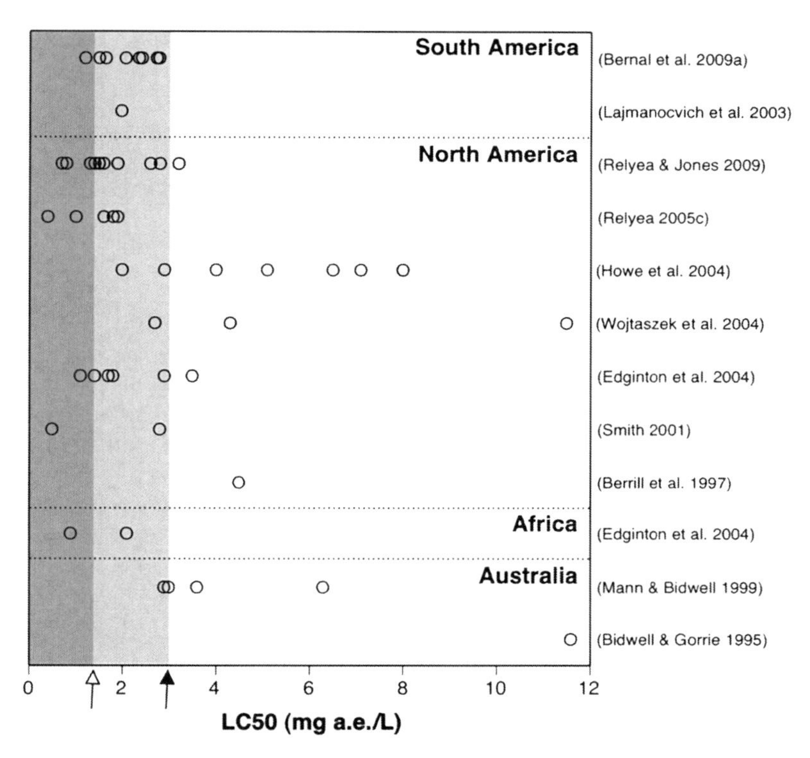

Fig. 9.2 All larval amphibian LC50 studies of glyphosate-based herbicides including the POEA surfactant or an unknown surfactant with similar toxicity as POEA. LC50 values were estimated from data in two cases (Berrill et al. 1997; Smith 2001) and all data were converted to the common units of mg a.e./L. Each *circle* represents an individual test of an amphibian species under a given set of conditions (pH, predator cues, etc.). The *closed arrow* indicates a worst-case scenario concentration of 3 mg a.e./L and the *open arrow* indicates the Canadian government's worst-case scenario in forestry applications of 1.43 mg a.e./L

surfactant, other leading surfactants that were effective at helping glyphosate kill plants were also effective at killing amphibians (Mann and Bidwell 2000, 2001). Unfortunately, the same properties that allow the surfactant to penetrate leaf cuticles also make it particularly good at rupturing the gill cells of fish and larval amphibians, leading to suffocation.

This was bad news for companies selling glyphosate-based herbicides. Media coverage at the time in Australia highlighted the threat that applications of Roundup around water posed to amphibians. In the Sydney Morning Herald, Leigh Dayton (1995) reported, "Mr. Nicholas Tydens, regulatory and environmental affairs manager for Monsanto Australia Ltd, said yesterday that the scientists were wrong. 'We have investigated every claim about Monsanto and Roundup affecting aquatic organisms. All of the evidence to date shows there is no adverse effect,' he said." Despite the conclusions of the scientists and the agreement from the Australian government that Roundup posed a risk to amphibians, the Monsanto representative continued to proclaim that Roundup posed no risk. After they had reported their findings, Joe Bidwell had lunch with a Monsanto representative who was quite upset with the amphibian results and warned that Joe would never land a job in the chemical industry. As an academic scientist with no desire to work in the chemical industry, this warning amused Joe greatly (J. Bidwell, personal communication).

Interest in Testing Amphibians Slowly Builds

Following the Australian studies, there were three additional studies on glyphosate and amphibians in three countries that received considerably less media attention. One reason may be that two of the studies did not calculate LC50 values, which makes it difficult to compare these studies to past results and limits their utility for assessing general patterns of toxicity and risk.

As part of a large screening of several amphibians against a wide variety of pesticides in Canada, Berrill et al. (1997) conducted tests on green frog tadpoles (*Rana clamitans*) exposed to Roundup. After 4 days of exposure, they found very little death with 3 mg a.e./L but nearly 100% death with 6 mg a.e./L. While the LC50 value was not estimated, the data suggest that the $LC50_{4\text{-d}}$ was ~4.5 mg a.e./L (Fig. 9.2).

Smith (2001) tested a formulation sold as Kleeraway® on larval western chorus frogs (*Pseudacris triseriata*) and plains leopard frogs (*Rana blairi*) in the USA. Based on his reported ratios of formulation to water and the nominal concentration of glyphosate in the bottle, one can estimate Smith's concentrations in units of mg a.e./L. For the chorus frogs, he found no deaths in the controls, 45% death in his lowest concentration (~0.5 mg a.e./L), and 100% death in all higher concentrations (>5 mg a.e./L). This would suggest that chorus frogs had an $LC50_{1\text{-d}}$ of ~0.5 mg a.e./L. For the leopard frogs, he found no deaths in the control or in the lowest concentration (~0.5 mg a.e./L), but 100% death in all higher concentrations (>5 mg a.e./L). This would suggest that plains leopard frogs had an $LC50_{1\text{-d}}$ of ~2.8 mg a.e./L (Fig. 9.2).

Finally, Lajmanovich et al. (2003) conducted a study in Argentina on tadpoles of *Scinax nasicus* and found an $LC50_{4\text{-d}}$ value of 1.98 mg a.e./L (Fig. 9.2). Together, these three studies produced LC50 values that were well within the range of values that would later be published on a wide variety of amphibian species.

The First Roundup Studies by the Relyea Lab

At about the time that the latter two glyphosate studies were being published, my lab was expanding its own research in toxicology. Our initial discovery that adding a bit of ecological reality – the smell of predators in the water – could make pesticides more lethal to tadpoles, turned out to be a common outcome in a variety of tadpole species (Relyea and Mills 2001; Relyea 2003). Clearly, there was a need for more pesticide studies that incorporated ecological reality.

Laboratory Experiments

With funding from the US National Science Foundation in 2001, we began exploring how amphibians reacted to simultaneous exposure to several pesticides. Whereas most studies were examining one pesticide at a time, the reality in nature was that animals were being exposed to suites of chemicals. Using commercial formulations of four different pesticides under laboratory conditions (carbaryl, malathion, diazinon, and Roundup), we measured survival and growth in five species of tadpoles when exposed to the four pesticides separately and in pairwise combinations (Relyea 2004b). The impacts of the paired pesticides were largely additive in these experiments, but the individual effects of Roundup were quite interesting. Based on the earlier Australian research, we did not expect any substantial mortality at the concentrations that we used (0.75 and 1.5 mg a.e./L). However, at the higher concentration we found 48% mortality in American toads (*Bufo americanus*), 60% mortality in green frogs, and 30% mortality in bullfrogs (*Rana catesbeiana*).

The following year we decided to examine the synergy between predator cues and pesticides using a wider range of pesticides, including Roundup. Because these experiments were conducted using a range of pesticide concentrations, they also allowed us to estimate the LC50 values for the pesticides during the 16-day experiments. For Roundup, we found that $LC50_{16\text{-}d}$ values ranged from 0.4 to 1.9 mg a.e./L (Relyea 2005c). This was consistent with the mortality we observed in the previous paired-pesticide experiment. This was important because it meant that Roundup could now be classified by the EPA as moderately to highly toxic to amphibians. Because Roundup had been on the market for nearly three decades, I had assumed that this high toxicity must be well documented. I was wrong. Only a handful of geographically scattered studies existed in the world (as described above).

Outdoor Mesocosm Experiments

As a community ecologist, finding high rates of mortality with Roundup under laboratory conditions was interesting, but testing the effect under more natural conditions was the more relevant question. Given that testing pesticides in natural wetlands has a number of logistical and ethical issues (i.e., few people are interested in intentionally contaminating their wetlands), often the best compromise is to use simulated wetlands. In aquatic ecology experiments, a standard technique is to fill large plastic water tanks with hundreds of liters of water and add many components of a natural wetland including leaf litter, algae, and zooplankton. In essence, these wetland "mesocosms" are intended to simulate (although not exactly mimic) real wetlands.

At the same time we were conducting the Roundup lab experiments, we initiated our first mesocosm experiment to test the effects of different pesticides under more seminatural conditions. In 2002, we decided that it might be insightful to assemble diverse communities designed to mimic simple natural wetland communities. To that end, we established mesocosms consisting of algae, nine species of zooplankton, three species of snails, five species of tadpoles, and eight species of predators. Once these communities were set up, we could apply different insecticides and herbicides to each tank. The idea behind the experiment was to ask two basic questions. First, would insecticides have "top-down" effects on the community (Polis and Strong 1996) by knocking out many of the insect predators and allowing their prey to flourish? Second, would herbicides would have "bottom-up" effects on the community by removing some fraction of the algae which forms the base of the food web thus decreasing food availability for grazers and ultimately decrease food availability for predators as well? To ask these questions, we used five treatments: a control, one of two popular insecticides (carbaryl and malathion) and one of two popular herbicides (Roundup Original and 2,4-D). We applied each pesticide at the rate recommended on the back of the bottles assuming that each was directly over sprayed on a wetland. For Roundup, this translated to a nominal (i.e., expected) concentration of 3 mg a.e./L (Relyea 2005a).

This was how I came to stare into that tank one summer day and see dead tadpoles littered across the bottom. The day after applying the pesticides we found very high tadpole mortality in the tanks treated with Roundup. Based on the Australian work, we expected some death, but nothing so widespread. In the end, the study produced a number of very interesting insights about the direct and indirect effects of pesticides on aquatic communities; however, the effect of Roundup on tadpoles was the most striking. Compared to the controls, mesocosms receiving Roundup experienced a 70% decline in amphibian species richness and an 86% decline in tadpole biomass (Fig. 9.3). For example, there was 100% mortality in both leopard frogs (*R. pipiens*) and gray tree frogs (*Hyla versicolor*) and 98% mortality in wood frogs (*R. sylvatica;* Relyea 2005a). These effects were similar to the effects we were observing in the lab that same summer (Relyea 2005c). It was immediately clear that we needed to do more research on the toxicity of Roundup. Thus, we made plans to examine the effects of Roundup under a variety of ecological conditions.

Our Roundup Results Hit the Fan

On April 1, 2005, the mesocosm study was published in the journal *Ecological Applications*. It was on this day that my career took an unanticipated turn.

In preparing for the paper's publication, it occurred to me that the impact of Roundup on tadpoles was so devastating that I needed to inform more than my fellow scientists. I needed to inform the public about the lethal consequences of applying glyphosate formulations containing POEA around wetlands that contain tadpoles. After giving this a good deal of consideration, I contacted the University's press office and inquired about how a scientist might conduct a press release.

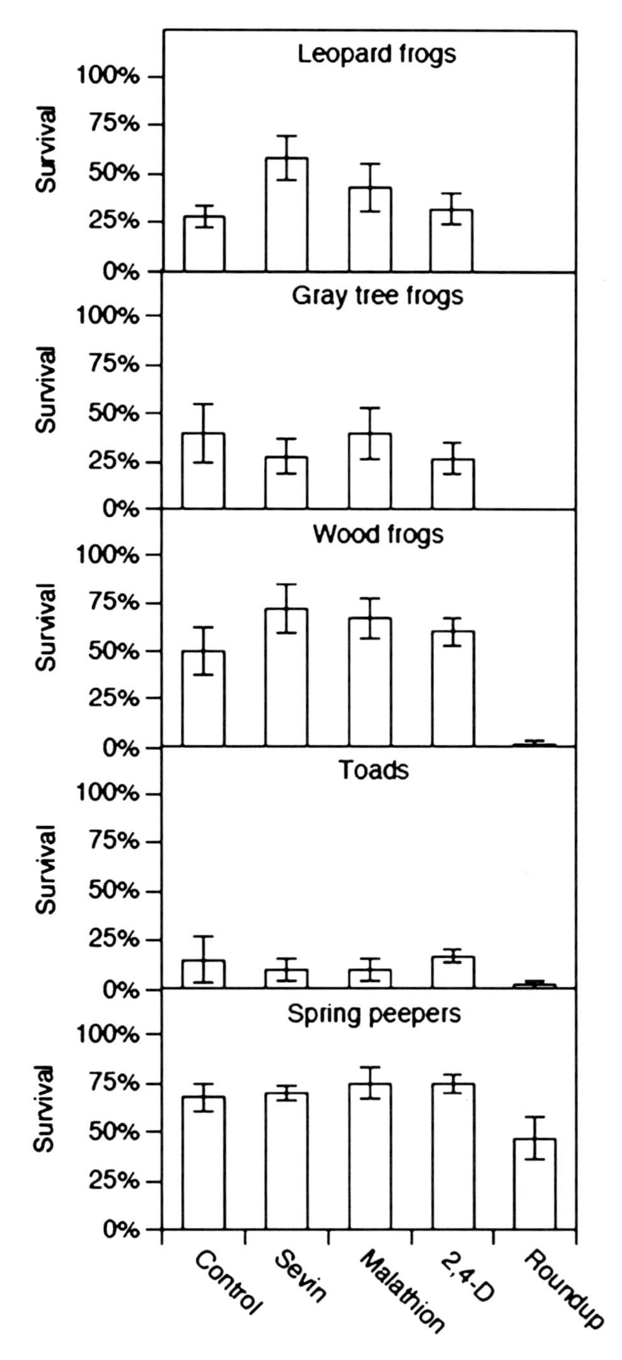

Fig. 9.3 The impact of a worst-case application of Roundup on pond mesocosms containing five species of North American tadpoles (Relyea 2005a)

As it is for most scientists, this was a shaky leap into a completely foreign world and I had no idea where such attention to my work might take me and my research group. The press release was issued on the same day the paper was published and it highlighted the lethal effects of Roundup on tadpoles that we observed.[3]

It would be an understatement to say that I was unprepared for what happened next. In April 2005, I still remember sitting in our small cabin at the University's field station when the journal article and press release were issued. I was bombarded by hundreds of e-mails per hour, many of which were very positive and supportive. Others cautioned me about what I was getting myself into. A few individuals were quite unhappy with our results.

Most of the e-mails that questioned our work were very professional, stating valid points from a variety of perspectives including from individuals who applied pesticides for a living. Some voiced genuine concerns about the intricacies of the potential for Roundup to impact tadpoles. For example, one individual said, "Roundup breaks down very rapidly once applied and is a very safe chemical. It literally does its job and degrades." This latter statement is certainly true. Roundup does break down relatively rapidly compared to other longer lasting environmental contaminants. The half-life for both the active ingredient (glyphosate) and the surfactant (POEA) ranges from 7 to 70 days in water, depending on environmental conditions (Giesy et al. 2000). However, the problem for tadpoles is that they die on the first day or two of exposure.

A few e-mails were personal attacks. In reading such e-mails, I noticed several patterns. First, some of the most outspoken critics read the press release and voiced strong criticisms without even reading the original journal article. Second, many people knew that pesticides went through extensive testing for registration, but had no idea that amphibians had not been tested in North America for the first 25 years of Roundup's existence. Third, most people knew that Roundup formulations were designed for terrestrial applications, so applying these formulations to water is an illegal act. Therefore, in their view, asking what would happen if Roundup got into water would be pointless. During a period of e-mail exchanges, a Fish and Wildlife Service biologist defended our work by noting "People are missing the point of this research. Adding the chemical of concern directly to water is a standard first step in research on chemicals and their impacts to aquatic life. Now research is needed on what concentrations may be problematic, does the chemical get transported to water bodies, and what are the methods of transport." He was correctly arguing that we need to first know the effects of Roundup *if* it gets into water, then we need to assess the probability of Roundup actually getting in the water. Together, these two facts help assess risk. Finally, some people saw our research as a serious threat to their jobs that involved promoting the use of glyphosate-based herbicides.

---

[3]http://mac10.umc.pitt.edu/m/FMPro?-db=ma.fp5&-format=d.html&-lay=a&-sortfield=date&-sortorder=descend&keywords=relyea&-max=50&-recid=36262&-find= (last accessed June 2010).

Monsanto's Response

Shortly after the article's publication and the press release, Monsanto put out its own press release.[4] They had three main criticisms. First, the application rate that we used was higher than used in agricultural settings. This is true, when compared to only *agricultural* uses. The application rate we used was based on home and garden use, which came straight from the bottle of Roundup Original. Regardless of the application rate used, what really matters, of course, is the concentration that ended up in the water. Our nominal concentration was 3 mg a.e./L and this represented a worst-case scenario for a direct over spray of a wetland. Past researchers had estimated the worst-case scenario for an over spray to be between 2.7 and 7.6 mg a.e./L, depending on assumptions about the depth of the water and whether vegetation intercepts any of the pesticide prior to landing on the water (Mann and Bidwell 1999; Giesy et al. 2000; Solomon and Thompson 2003). Observed worst-case scenarios range from 1.7 to 5.2 mg a.e./L (Edwards et al. 1980; Giesy et al. 2000; Thompson et al. 2004). The value in asking whether the worst-case scenario harms an organism is that if it does not, future testing at lower concentrations is not necessary. If it does cause harm, we need to consider lower concentrations of the pesticide to determine the lowest concentration which poses no harm.

Monsanto also argued for the irrelevancy of any aquatic experiment. They said, "This study does not represent realistic use conditions for Roundup brand herbicides for applications to aquatic environments. In fact, there are no Roundup brand formulations approved in the US or Canada for application over water." Thus, the argument being made is that Roundup formulations containing the POEA surfactant are not approved for application over water; therefore, Roundup should not be in wetlands and any tests of Roundup's effects on aquatic organisms are irrelevant. This argument stands up only if the terrestrial use of Roundup never causes the herbicide to end up in water bodies. While there are few data on concentrations of glyphosate in natural water bodies, there are enough data to confirm that the herbicide, while not intentionally sprayed on water, is either unintentionally sprayed on water, leaches out of the soil, or washes off vegetation and into bodies of water (Thompson et al. 2004). Monsanto's press release goes on to cite a single study of Roundup that found no effect of Roundup on tadpoles as the one definitive study (Thompson et al. 2004; discussed below), even though they claimed that any test of Roundup on aquatic life is irrelevant and despite contradictory studies showing that the Roundup posed a risk to amphibians (Mann and Bidwell 1999).

In hindsight, it was probably Monsanto's press release that brought more attention to the issue of Roundup's toxicity to amphibians than anything I could have done. The strong negative reaction from Monsanto brought a great deal of media attention to the issue and I found myself in a strange position thrust before the press. Thankfully, the vast majority of the reporters were quite professional and objective,

[4] http://www.monsanto.com/monsanto/content/products/productivity/roundup/bkg_amphib_05a.pdf (last accessed June 2010).

although there was one interesting case in which a newspaper reporter from Nebraska wrote a criticism of the research for her local farm community. In the article, she interviewed an extension agent who had not read the article, and, when I contacted her, she admitted that she had neither read the article nor tried to contact me to get a balanced story. Fortunately, this type of interaction was a rare exception.

One of the most significant events happening during this time happened behind the scenes. In June 2005, 2 months after the release of the paper, I attended the World Congress of Herpetology in Stellenbosch, South Africa. Herpetologist Ron Heyer (from the Smithsonian Museum) informed me that when my article came out, he had been contacted by Monsanto to see if he would criticize the article. He read the article, found no reason to criticize it, and declined Monsanto's request. Seven months later, while attending the Illinois Crop Protection Conference, toxicologist Allan Felsot (from Washington State University) told the same story to the conference audience. He said that when our article first came out, Monsanto had called him and asked him to publicly criticize our work. As he told the audience, he read the article and did not agree with Monsanto's criticisms, and refused to speak out against the study. I do not know how many other scientists were contacted, but clearly Monsanto was trying to find scientists that would criticize our research and these scientists were not going along.

The Roundup Studies of 2004

What is particularly interesting about the strong opposition from Monsanto in 2005 is that this opposition was not voiced a year earlier when a series of papers were published by Canadian researchers saying very similar things. In 2004, after our article was already in press, these researchers published a series of four papers examining tadpole exposure to Vision® (Monsanto's Canadian version of Roundup which also contains POEA). In the first paper, Edginton et al. (2004) examined the toxicity of Vision to four species of tadpoles under multiple pH conditions (pond pH in nature typically ranges from 4 to 9). They found that all four species had similar sensitivities and that sensitivity increased with pH. Indeed, the LC50 values they produced (1.8–3.5 mg a.e./L at pH = 6; 0.9–1.7 mg a.e./L at pH = 7.5; Fig. 9.2) were very similar to those that we published the following year ($LC50_{16\text{-}d}$ = 0.4–1.9 mg a.e./L at pH = 8; Relyea 2005c). In the end, the authors state, "we concluded that, at EEC [environmentally expected concentration] levels, *there was an appreciable concern of adverse effects to larval amphibians in neutral to alkaline wetlands*. The finding that the mean pH of Northern Ontario wetlands is 7.0 further compounds this concern" (Edginton et al. 2004, p. 821).

In the second paper, Chen et al. (2004) examined how the herbicide interacted with different levels of pH (5.5 vs. 7.5) and different levels of food stress. After conducting lab experiments on a single zooplankton species and a single amphibian species, they found that higher pH caused significantly more mortality. Indeed, even their lowest herbicide concentration (0.75 mg a.e./L) cause 100% mortality under conditions of high pH and low food. As a result, the authors concluded that "For both species, significant effects of the herbicide were measured *at concentrations lower than the calculated worst-case value* for the expected environmental concentration" (Chen et al. 2004, p. 823).

In the third paper, Wojtaszek et al. (2004) investigated the impact of Vision when green frog and leopard frog tadpoles were living inside enclosures that had been set up in natural wetlands. The enclosures had polyethylene sidewalls that were anchored to the bottom of the wetland. Two wetlands were chosen for the experiment, one with a lower pH (6.4) and one with a higher pH (7.0). They then added different amounts of Vision to each enclosure to determine the $LC50_{4-d}$ values for the two species under natural conditions. Overall, they found that the two tadpole species appeared to be less sensitive in these two wetlands compared to past lab studies, but there was substantially greater tadpole mortality in the pond with higher pH. The $LC50_{4-d}$ for green frogs was 4.3 mg a.e./L in the lower pH wetland, but 2.7 mg a.e./L in the higher pH wetland. Similarly, the $LC50_{4-d}$ for leopard frogs was 11.5 mg a.e./L in the lower pH wetland, but 4.3 mg a.e./L in the higher pH wetland (Fig. 9.2). Although these LC50 values were a bit higher than the companion lab studies of Chen et al. (2004) and Edginton et al. (2004), it was still clear that the herbicide would cause tadpoles to die at environmentally relevant concentrations, whether one considers worst-case concentrations of 3 mg a.e./L (as used in our studies) or even if one uses the more conservative estimates of 1.43 mg a.e./L used by the Canadian government for forest applications (Wojtaszek et al. 2004). Based on the LC50 estimates, 50% of the tadpoles would not die, but 10 or 20% of the tadpoles would die. Despite this expectation, the authors concluded, "The results of this in situ enclosure study provide no evidence to conclude that environmentally relevant concentrations of Vision cause significant mortality, abnormal avoidance, or reduced growth in native larval amphibians used in this study."(Wojtaszek et al. 2004, p. 841).

The final paper in this series set out to determine the concentrations of Vision that could actually occur in wetlands when a forest is sprayed from a helicopter to kill broadleaf trees and favor the more marketable conifer trees. A second goal was to determine the impact on tadpole mortality during these sprayings (Thompson et al. 2004). To achieve these goals, helicopters applied Vision either directly over wetlands, adjacent to wetlands, or with a buffer of vegetation separating the sprayed area and the wetlands. In terms of the average concentration found in each scenario, there were no surprises; buffered wetlands had the least herbicide (0.03 mg a.e./L), adjacent wetlands had moderate amounts of herbicide (0.18 mg a.e./L), and over sprayed wetlands had the most herbicide (0.33 mg a.e./L). It is also not surprising that the data were quite variable among wetlands due to differences in pond depth. For example, while the average concentration of an over sprayed wetland was 0.3 mg a.e./L, the values range from undetectable (<0.01 mg a.e./L) to quite high (2.0 mg a.e./L), exceeding the Canadian government's own worst-case scenario estimates (1.43 mg a.e./L). In assessing risk, however, one typically uses the *average* natural concentration and ignores the fact that some wetlands have much higher concentrations than the average. If we were to consider the wide range of observed concentrations in Canadian wetlands, we would predict that some wetlands would experience no mortality while other wetlands would experience high rates of mortality. Curiously, in a subsequent paper published 4 years later, Thompson and colleagues (Struger et al. 2008) report glyphosate concentration data from a survey of streams and wetlands in Ontario. These samples, which were not taken immediately after the herbicide was sprayed, never exceeded 0.041 mg a.e./L. As a result, they concluded that glyphosate concentrations in wetlands are far below any concentrations that would cause harm to amphibians. They further concluded, "Aqueous environmental exposure concentrations for glyphosate residues as observed in this study were very similar to other surface water concentration values published in monitoring studies in the USA and in various European countries." (Struger et al. 2008, p. 380). This conclusion was made despite the fact that Thompson and colleagues reported detections of up to 2.0 mg a.e./L in wetlands, a concentration that is nearly 50 times higher (Thompson et al. 2004).

To assess the impact of these three spray treatments on tadpoles, the researchers caged groups of five tadpoles (green frogs and leopard frogs) in each wetland for 2 days and then counted how many tadpoles survived. As noted earlier, the standard in toxicology studies is that not more than 10% of the control animals should die in short-term trials; more than 10% death in the controls suggests either that the animals are in poor condition or the environment is not hospitable enough for a reliable experiment. Their investigation found that the buffered wetlands, which contained almost no herbicide and thus could serve as control, experienced 15% death of leopard frogs and 26% death of green frogs. While it is unclear what was killing such a high fraction of tadpoles after only 2 days in a, tadpole mortality in the over sprayed wetlands, which experienced 14 and 36% death of leopard frogs and green frogs, respectively, was not significantly different from the buffered wetlands. Because the differences in mortality were not significantly different among the three application treatments ($p=0.19$ for leopard frogs, $p=0.13$ for green frogs), the authors concluded that Vision poses a low risk to amphibians. The EPA had a different interpretation of these data (Carey et al. 2008, p. 104):

> The results suggest that there was a large amount of variability that could have obscured detecting treatment effects especially given that these were naturally occurring wetlands that represented a range of environmental conditions.

In the end, Thompson et al. (2004) concluded, "Overall, results of this tiered research program confirm that amphibian larvae are particularly sensitive to Vision herbicide and that these effects may be exacerbated by high pH or concomitant exposure with other environmental stressors. Although results from laboratory studies were very useful in the comparative sense and in understanding mechanisms of interaction, they tended to overestimate effects as observed under natural exposure scenarios" (Thompson et al. 2004, p. 848). This final conclusion may be why Monsanto voiced no public opposition to these studies.

Counter to this result, one additional paper – produced by an independent research group (Howe et al. 2004) – examined the lethal and sublethal effects of Roundup Original (containing POEA) and several other glyphosate formulations to four species of tadpoles under several exposure scenarios. Using a tadpole developmental stage similar to past studies (Gosner stage 25; Gosner 1960), the researchers found $LC50_{4\text{-}d}$ values that were similar to what others had found (2.0–5.1 mg a.e./L; Fig. 9.2). Even more interesting were the effects observed when tadpoles were reared under longer (i.e., chronic) exposures, from hatchling tadpole through metamorphosis. They found that even low concentrations (0.6 mg a.e./L) could cause 30% mortality. The authors concluded, "The present results indicate that formulations of the pesticide glyphosate that include the surfactant POEA at environmentally relevant concentrations found in ponds after field applications can be toxic to the tadpole stages of common North American amphibians." (Howe et al. 2004, p. 1933).

What is striking about these five papers is that three of them expressed clear concerns that glyphosate products containing the POEA surfactant posed a clear risk to amphibian survival, and the other two demonstrated substantial death rates, although the rates happened to be less than 50%. Yet, to my knowledge, not one of these studies was denounced by Monsanto.

Our Follow-Up Studies of 2005

A few months after our article and press release were issued in April 2005, we published three more articles reporting the results of four additional experiments. The first article was our work that documented interactions between the herbicide and predator cues (Relyea 2005c). Not only did we confirm that the herbicide could become more toxic when combined with the stress of predation, but we also estimated the LC50 values of Roundup for six species of North American tadpoles. We exposed six amphibian species to a wide range of glyphosate concentrations (0–15 mg a.e./L). To look for the synergy with predator cues, we also ran the experiments for a longer duration (16 days) than typical LC50 lab experiments (1–4 days). We estimated the $LC50_{16\text{-}d}$ values as 0.4–1.8 mg a.e./L (Fig. 9.2). These experiments predicted that we should observe 80–90% tadpole death at 3 mg a.e./L. This was consistent with the conclusions of the concurrent mesocosm study that a concentration of 3 mg a.e./L caused very high rates of tadpole mortality.

The second study tested the long-touted claim that, because glyphosate and POEA strongly bind to soil in agricultural fields, the presence of soil in aquatic environments should ameliorate the effects of Roundup on amphibians by removing the herbicide from the water. The extrapolation to wetlands sounded reasonable, but had not been tested. So we conducted an experiment to determine if the addition of soils made Roundup less toxic. Using 3 mg a.e./L again in outdoor mesocosms, we examined whether the survival of three tadpole species improved if we manipulated the soil type in experimental mesocosms to include either no soil, sand, or loam (the soil type in our region). Our results were clear: adding sand or loam did not improve tadpole survival; with a single application of Roundup, 98% of the animals died (Relyea 2005b; Fig. 9.4).

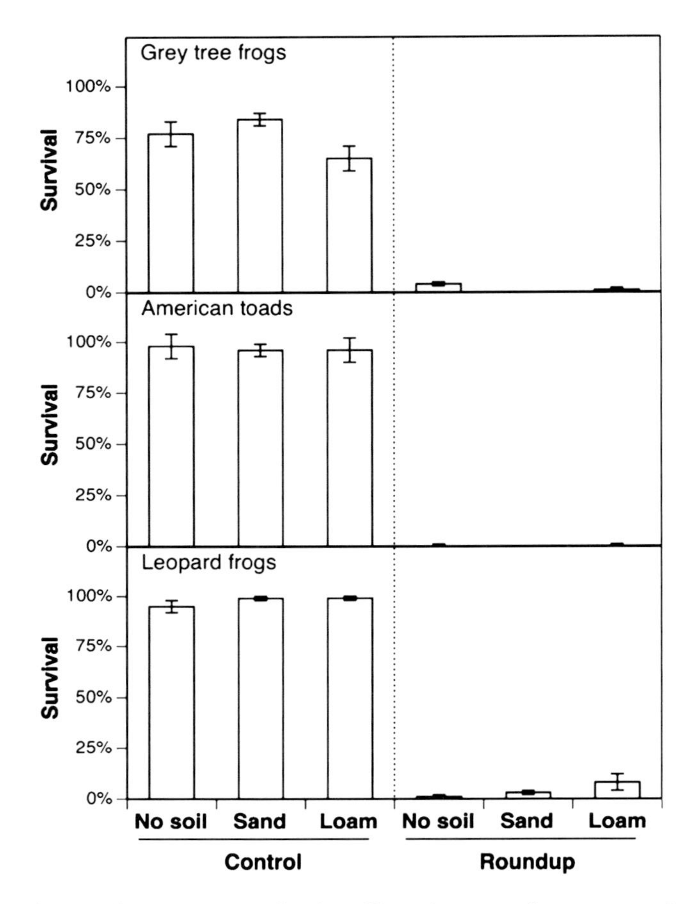

Fig. 9.4 The impact of a worst-case application of Roundup on pond mesocosms when combined with either no soil, a sand substrate, or a loam substrate (Relyea 2005b)

Later, we learned why soil plays less of a role than previously hypothesized. Although Roundup does strongly bind to soil, it is also highly soluble in water. The critical factor concerns what medium Roundup contacts first. If Roundup hits water first, as would happen during an over spray, the herbicide stays dissolved in the water and consequently remains toxic in the environment. While we currently do not know how much Roundup is bound by wetland soil or the rate at which this binding occurs in equilibrium with the water, it is clear that one cannot assume Roundup will be immediately removed by the sediments of a wetland and make the water safe for amphibians. We certainly need much more research on this issue.

The third study examined the direct and indirect effects of Roundup and malathion in communities with different species of lethal predators (Relyea et al. 2005). With regard to Roundup, the important aspect of this study is that it examined a much lower concentration of Roundup (1 mg a.e./L) rather than the worst-case scenario concentration used previously (3 mg a.e./L). The resulting effect was not as lethal, but dramatic nonetheless. In the absence of predators, one-third as much Roundup caused no significant mortality in gray tree frogs, but caused 29% mortality in leopard frogs, and 71% mortality in American toads. In other words, even a much lower concentration caused substantial amphibian mortality.

In the fourth study, we examined the effects of Roundup on the terrestrial stage of amphibians (Relyea 2005c).[5] Almost nothing was known about the susceptibility of the terrestrial stage, except for a few Australian species that had been tested while living in water (Mann and Bidwell 1999). For our study, we collected newly metamorphosed gray tree frogs, wood frogs, and American toads. We then placed groups of ten frogs or toads into laboratory tubs containing wet paper towels and conducted an over spray at the manufacturer's recommended application rate. Across the three species, 79% of the animals died within 24 h.

As we approached the publication date of these four studies, it became clear that we had a number of new and important messages to convey to the public. We produced a second press release[6] to highlight the discovery that soils had no ameliorating effect on Roundup's lethality and that terrestrial applications could cause very high rates of death in the terrestrial stage of frogs and toads. This prompted another round of press coverage including an article in the St. Louis Dispatch (Hand 2005), a newspaper that resides in the hometown of Monsanto's international headquarters. As part of that interview, I asked reporter Eric Hand to inquire why Roundup Biactive®, Monsanto's new formulation that was less toxic to fish and amphibians, was being used in Australia and Europe but was not available in North America. This safer formulation was brought to market in direct response to the Australian research that determined Roundup was moderately lethal to Australian amphibians. Their spokesperson, Donna Farmer, gave two answers. First, Roundup Biactive has a weaker surfactant, so although it was safer for amphibians, it was less effective at penetrating plant cuticles and therefore less effective at killing North American weeds. Second, she said that registering the formulation in North America would be "subject to a cumbersome EPA approval process." This is the same approval process that the company had gone through for countless other Roundup formulations, and the active ingredient (glyphosate) was already approved in North America.

It was at this time that the EPA Environmental Fate and Effects Division started making inquiries. The initial inquiries asked for reprints of my papers. I obliged. When reporters asked the EPA for comment, their representatives generally gave the same answer: they were aware of our studies and were examining them.

---

[5]See errata Ecol Appl 19:276.

[6]http://mac10.umc.pitt.edu/m/FMPro?-db=ma.fp5&-format=d.html&-lay=a&-sortfield=date&-sortorder=descend&keywords=relyea&-max=50&-recid=36355&-find= (last accessed June 2010).

A Letter to the Editor

Shortly after our first mesocosm study was published in April 2005, a group of scientists from Canada wrote a letter to the editor of *Ecological Applications*. The journal's editor contacted me and asked if I thought the letter criticizing our study should be published as long as I had an opportunity to publish a response. After reading the letter, I felt strongly that it should be published.

The letter was written by a group of scientists from the University of Guelph (Barbara Wojtaszek, Andrea Edginton, Gerald Stephenson, Keith Solomon, and Dean Thompson [now with the Canadian Forest Service]). This group included many of the authors who conducted the studies on Vision that were published in 2004. One individual, Keith Solomon, was not an author of the studies published in 2004, but had conducted multiple risk assessments of glyphosate and had been the graduate advisor for many of the other authors. The lead author of the letter, Dean Thompson, was recommending the widespread use of Vision across Canadian forests and was some facing local opposition to this plan. Our research results, if they stood the test of time, would make this recommendation more difficult to promote.

Their letter to the editor was sent to an anonymous reviewer and me as an 11-page manuscript. It appeared to both of us that the criticism was quite hurriedly written and we asked that the authors correct a number of fundamental errors. Given time to rethink their arguments, the revised letter expanded from 11 to 21 manuscript pages. The final version of the letter was published in October 2006 and made four major claims (Thompson et al. 2006): (1) our application rates were too high, (2) the concentrations we used were not relevant to those found in nature, (3) there were potential methodological errors, and (4) past risk assessments had already concluded that the herbicide to pose no risk to nontarget organisms. In my response (Relyea 2006) I systematically identified the flaws in each argument.

First, the authors provided a list of application rates that were all lower than the application rate used in my study. However, the authors failed to acknowledge that higher application rates existed. Regardless of differences in application rates for different uses, the real issue, of course, is the concentration that was in the water. As noted earlier, the aquatic concentration in our first mesocosm experiment was 3 mg a.e./L, which fell within the range of both estimated and observed worst-case scenarios in nature.

Thompson et al.'s (2006) second claim was that our concentration of glyphosate in the water was unusually high. To support this claim, they provided a list of aquatic concentrations that had been observed in nature. In my response, I pointed out that the authors tried to make the case that only low concentrations occur in nature by including a lot of data from habitats such as water wells and streams. These two water sources are known for having low concentrations of glyphosate. More importantly, these two water sources and are largely irrelevant for North American species of tadpoles – tadpoles do not live in underground wells and most species of tadpoles do not live in streams. They live in wetlands. I also pointed out that several well-known studies that documented high concentrations of glyphosate in nature were omitted from their list. I made it clear that the authors were surely aware of these studies. They had co-authored several of them.

Their third claim was that our study contained a number of potential flaws that made the study unreliable. For example, they claimed that the absence of soil in our study detracted from its applicability to nature because soil would have taken the herbicide out of the water column. Of course, by this time we had already published our follow-up study demonstrating that adding soil did nothing to improve tadpole survival with Roundup (Relyea 2005b). They were also concerned that our mesocosms did not contain any aquatic plants (i.e., macrophytes) For example, they claimed that aquatic plants might add oxygen to the water, speculating that oxygen concentrations in the water were low and may have unnaturally stressed our animals. Repeated experiments have found very high oxygen levels in these tanks, with or without macrophytes. All of these arguments were weak attempts at finding flaws that were easy to refute. In response to these speculations, I came to the following conclusion, "The authors propose a number of methodological flaws that are not only without support, but, in many cases, demonstrate a lack of knowledge of aquatic ecology" (Relyea 2006, p. 2033).

The final claim took an approach that is a classic response to criticisms of a manufacturer's product. The authors argued that Roundup has been used to kill weeds for a very long time and multiple risk assessments have been conducted that have found "no unacceptable risk," so the product poses no risk to amphibians. This is a strong argument, until one investigates the details of the past risk assessments. Thompson et al. (2006) begin by citing reports from the EPA (1993) and WHO (1994). There was one small problem – neither of these reports included any amphibian data because no amphibian data existed in the world until Bidwell and Gorrie's (1995) work in Australia. They also cited the risk assessment by Giesy et al. (2000); this assessment included the Australian data and actually concluded that Roundup posed a potential risk to amphibians that needed further evaluation. Finally, they cited the risk assessment of Solomon and Thompson (2003); an assessment that included fish and invertebrates, but not amphibians. Collectively, these risk assessments actually tell us little about the risk of Roundup to larval and adult amphibian populations.

I concluded my response by quoting excerpts from the 2004 papers in which several of the authors agreed that Roundup/Vision posed a risk to amphibians in nature (as quoted above) and that the weight of the evidence from all studies continued to support this assessment. I further noted that a very likely reason some of the Canadian studies had produced lower (and more variable) death rates from the herbicide was due to the lower (and more variable) levels of pH in those studies. I suggested we should look more carefully at the role of pH to get a better idea of Roundup's impacts on amphibian mortality.

While writing my response, there was one issue that still bothered me. At the same time that my lab was conducting research on Roundup and amphibians, Tyrone Hayes (University of California, Berkeley) was in the midst of a debate over the effects of the herbicide atrazine on amphibians. Tyrone was finding that atrazine could cause male frogs to become feminized via endocrine disruption (Hayes 2004, see Chap. 10).

This debate involved the Syngenta pesticide company and EcoRisk, a consortium of pesticide companies that hires academic scientists to work as consultants. One of the people hired by EcoRisk was Keith Solomon, one of the authors of the letter to the editor. Several months after their letter to the editor was published, Keith Solomon admitted to a reporter that he had received funding from Monsanto (Lubick 2007). According to the University of Guelph,[7] Dean Thompson was also receiving funding from Monsanto. Accepting research money from a pesticide manufacturer is not a problem. Debating the safety of the company's product without full disclosure that the company is funding your research is a problem. It can affect the world's assessment of your independence and objectivity.

Roundup and the Colombian Drug War

While the Roundup-amphibians debate was ongoing, there was a war on drugs occurring in South America and these two issues came together quite unexpectedly. Beginning in the 1990s, a rebel group in Colombia known as FARC (Fuerzas Armadas Revolucionarias de Colombia) was obtaining a large fraction of its financial support from coca production. The coca plants were processed into cocaine and smuggled to other countries, including the USA. Since the United States considered FARC a terrorist group, US officials recognized that if they could curtail coca production in Colombia it would have the twin benefit of taking away the financial foundation of a terrorist group and reducing the amount of cocaine being smuggled into the USA.

Plan Colombia

In 1999, "Plan Colombia" was developed between the USA and Colombian governments that included funds for aerial fumigation of illegal coca fields. The herbicide of choice was a glyphosate formulation called Glyphos-Cosmo-Flux, which used the POEA surfactant. However, after a large number of complaints related to legal crops being sprayed as well as worries over unintentional impacts of glyphosate application to humans and wildlife, the Organization of American States in 2004 agreed to assemble an independent panel of outside experts to assess the potential risk that these aerial spraying might pose to amphibians and humans. The person selected to lead this independent panel was Keith Solomon, the same person who was funded by Monsanto.

The US government was particularly concerned about potential impacts to wildlife in Colombia, especially after our research published in April 2005 showed that inadvertent over sprays of wetlands could be highly lethal to amphibians and many

[7] https://www.uoguelph.ca/research/summaries/2006/table1-oac.shtml. (last accessed June 2010).

of the coca fields had adjacent wetlands. This concern was exacerbated by the fact that Colombia has an incredibly high diversity of amphibians (746 species). Nearly a third of them (255 species) are currently classified as either vulnerable, near threatened, endangered, or critically endangered.[8] There also was a concern that the eradication program was having no effect on coca production. The Senate appropriations bill (US Senate 2005) stated the following:

> The Committee reaffirms its commitment to assist the efforts of Colombian President Uribe in destroying the threats of terrorism and narcotics in that country .... The Committee is increasingly concerned, however, that the aerial eradication program is falling far short of predictions and that coca cultivation is shifting to new locations. Since the start of Plan Colombia, over 525,000 hectares of coca crops have been sprayed, yet coca cultivation has decreased by only 7%. Last year alone, 136,555 ha were sprayed, but the total area under cultivation, estimated by the State Department at 114,000 ha, remained essentially unchanged from the previous year. There is no indication that the quantity of cocaine entering the United States has decreased .... The Committee directs the Secretary of State, in consultation with the EPA and appropriate Colombian authorities, to submit a report not later than 180 days after enactment of the Act, with the following information: the results of a GIS analysis of the proximity of small, shallow water bodies to coca and poppy fields and of tests to determine the toxicity of the spray mixture to Colombian amphibians; and, an assessment of potential impacts of the spray program on threatened species, including in Colombia's national parks.

In 2007, reporter Naomi Lubick contacted me about the risk assessment being conducted in Colombia. At the time, no formal assessment had yet been produced, although I had seen preliminary reports and I knew that Keith Solomon's research team in Colombia was finding similar levels of tadpole death using Glyphos-Cosmo-Flux as we had reported for Roundup. In her story that was published in *Environmental Science and Technology*, Naomi reported, "The team concludes that the glyphosate mixtures used in the program *are potentially harmful to tadpoles, particularly those living in shallow pools* [italics added]." (Lubick 2007, p. 3404). Moreover, she reported, "Solomon and his colleagues, however, predict moderate toxicity of glyphosate mixtures at levels of exposure *similar to some of Relyea's lowest concentrations* [italics added]" (Lubick 2007, p. 3405). At this point, I realized even our staunchest critic had finally arrived at the inescapable conclusion that glyphosate formulations containing POEA posed a significant risk to amphibians.

The Initial Risk Assessment for Colombia

In 2007, Solomon and colleagues produced a preliminary risk assessment for Plan Colombia (Solomon et al. 2007). This risk assessment did not contain any data on Colombian amphibian species, but took what was known about the sensitivity of amphibians in North America and Australia and applied it to the herbicide application

[8] Database search of AmphibiaWeb conducted on 11/02/09; http://www.amphibiaweb.org/.

Table. 9.1 The summary of potential environmental impacts associated with coca production as determined by Solomon et al. (2007; see their Table 14)

Impacts	Intensity score	Recovery time (years)	Impact score	Impact %
Clear cutting and burning	5	60	300	96.9
Planting the coca or poppy	1	4	4	1.3
Fertilizer inputs	1	0.5	0.5	0.2
Pesticide inputs (by farmers)	5	0.5	2.5	0.8
Cosmo-Flux spray (for eradication)	1	0.5	0.5	0.2
Processing and refining	2	1	2	0.6

Cosmo-Flux spray is the herbicide that contains the same active ingredient (glyphosate) and surfactant (POEA) as many commercial formulations of Roundup

rates that were being conducted in Colombia. Based on the application rates used to eradicate coca, the estimated concentration of glyphosate in a shallow wetland (15 cm deep) with no initial absorption to sediments was 2.47 mg a.e./L. Actual concentrations in sprayed fields and wetlands could not be obtained because coca fields are often protected by armed guards. Because published LC50 values for many tadpole species were considerably lower than 2.47 mg a.e./L, an inadvertent over spray of wetlands would kill more than 50% of all tadpoles living in shallow wetlands. In agreement with what Naomi Lubick had reported earlier, the research team concluded, "Moderate risks to some aquatic wildlife may exist in some locations where shallow and static water bodies are located in close proximity to coca fields and are accidentally oversprayed." (Solomon et al. 2007, p. 104). Our research completely agreed with this conclusion.

What Solomon et al. (2007) did next was an attempt to relativize the risk of Roundup on amphibians. They proposed subjective "intensity scores" (from 1 to 5) for how various steps in the coca production process would harm nontarget plants and animals (Table 9.1). They decided that the cutting and burning of the forest and the use of other pesticides by coca farmers should both receive an intensity score of 5 whereas the spraying of the herbicide should only receive an intensity score of 1. Next, they assigned values for how many years it would take the plants and animals to recover from each impact. They decided that amphibian habitats could recover from deforestation after 60 years and amphibian populations experiencing death from the herbicide could recover from the spraying of the herbicide in only 0.5 years. Given the high rates of mortality that can occur at environmentally expected concentrations of the spray program, however, it is not reasonable that an amphibian population could recover in 0.5 years (especially for amphibians that only breed once per year). Finally, they computed "impact scores" (calculated as the product of an intensity score and recovery time). Using these numbers, they came to the conclusion that clear cutting the forest for coca production was responsible for 96.9% of the effect on the environment whereas the herbicide spraying program was responsible for only 0.2% of the effect on the environment. As a result, they concluded, "When taken in the context of the environmental risks from other activities associated with the production of coca and poppy, in particular, the uncontrolled and unplanned clearing of pristine lands in ecologically important areas for the purposes of planting the crop, the added risks associated with the spray program are small." (Solomon et al. 2007, p. 104). This conclusion, of course, is only correct if the subjective estimates are correct. Moreover, this conclusion does not address the question that the US Congress asked. The team was charged with the task of assessing the risk of the spray program to amphibians in coca fields *where deforestation has already occurred*. In other words, given that there amphibians living in and around coca fields, what is the impact of the spray program on them?

The Final Risk Assessment for Colombia

In August 2009, Solomon's research team published a series of articles in the *Journal of Toxicology and Environmental Health* in which they presented their research findings from the Plan Colombia experiments. This included studies of herbicide drift, impacts on human health, and impacts on amphibians. Because the eradication program flies airplanes over coca fields, often while being fired upon by the coca farmers (Lubick 2007), the researchers were unable to determine how much glyphosate lands on the fields, forests and wetlands. As in the initial assessment, they could only estimate the amounts based on application rates. They also never produced the valuable GIS analysis of wetland proximity to coca and poppy fields that the US Congress had requested.

The research team did conduct a number of experiments. In laboratory tests on eight species of Colombian tadpoles, they found that the $LC50_{4\text{-d}}$ for Glyphos-Cosmo-Flux ranged from 1.2 to 2.8 mg a.e./L (Bernal et al. 2009a; Fig. 9.2). These values are very much in line with values that we published on six species in North America ($LC50_{16\text{-d}}$ = 0.4–1.8 mg a.e./L; Relyea 2005c). Indeed, the team concluded, "Data suggest that sensitivity to Roundup-type formulations of glyphosate in these species is similar to that observed in other tropical and temperate species." (Bernal et al. 2009a, p. 961) and, in a companion paper, state, "There are some potential risks to amphibians from direct overspraying of shallow waters." (Marshall et al. 2009, p. 930). This seemed like a logical and reasonable conclusion from the data. The team, however, was not finished.

The next step they took was to examine the effect of the herbicide under "field conditions" to determine whether laboratory LC50 values were predictive of impacts under field conditions (Bernal et al. 2009b). They set up six outdoor mesocosms, each containing soil and 15 cm of water. To each mesocosm they added between 165 and 200 tadpoles (representing two species in each of two experiments), exposed each mesocosm to a different concentration of herbicide, and then determined survival 4 days later. In a fashion unlike most peer-reviewed scientific studies, their study included only one replicate of each treatment, thereby preventing any assessment of repeatability. Importantly, their mesocosm experiments had a lower pH (average pH = ~7) than their lab studies (pH = 8.2). Not surprising, the researchers found less tadpole death in the mesocosm experiments ($LC50_{4\text{-d}}$ = 9–11 mg a.e./L) than in the lab experiments ($LC50_{4-d}$ = 1.2–2.8 mg a.e./L). As noted earlier, the herbicide is well known to be less lethal at lower pH (Chen et al. 2004, Wojtaszek et al. 2004). What was surprising is that the researchers attributed the lower death rates in the mesocosm experiments not to the lower pH, but to the presence of soil, despite the fact that studies had already shown that adding soil has no ameliorating effect on the lethality of Roundup (Relyea 2005b). Using their interpretation that soil was the underlying cause of the reduced mortality, the researchers concluded that the herbicide poses a low risk under "field conditions."

As part of this same paper, the team investigated the lethal effects of the herbicide on postmetamorphic amphibians (i.e., metamorphosed juvenile and adult frogs). Using tubs containing moist soil and leaf litter, they sprayed the amphibians at a variety of application rates. At the typical application rate for spraying coca, they found that up to 30% of the animals died. Because fewer than 50% of the adults died at the typical application rate, they concluded that, "Data indicate that, under realistic, worst-case exposure conditions, the mixture of Glyphos and Cosmo-Flux … exerts a low toxicity to aquatic and terrestrial stages of anurans." (Bernal et al. 2009b, p. 966).

It is important to ask, however, whether a 30% loss of an adult population is unimportant to the persistence of the population. In amphibians, larvae typically experience high rates of death from natural causes (e.g., predation, competition, disease) whereas adults have much higher rates of natural survival. As a recently published life-history model of amphibian populations demonstrates, this means that the persistence of the population is much more dependent upon adult survival than larval survival (Taylor et al. 2006). This model found that a 20% decrease in annual adult survival would cause the population size of breeding females to decline by 45% while a 40% decrease in annual adult survival would cause the population size of breeding females to decline by 87%. Hence, 30% annual mortality of adults following terrestrial spraying of Roundup is expected to cause a substantial decline in the population. Moreover, given that the frequency of spraying the coca fields is every 6–12 months, this means that up to 30% of the adult population could be killed as frequently as every 6 months (Solomon et al. 2007).

Given the unavoidable conclusion now that glyphosate with POEA is moderately to highly toxic to amphibians in both the aquatic and terrestrial stages, the team again relativized their results. First, they examined the species ranges of numerous amphibians and found that the mortality caused by the herbicide would only impact a portion of many species' ranges. As a result, "… populations as a whole are at low risk." (Lynch and Arroyo 2009, p. 974). This means that the authors accepted that areas being sprayed will cause amphibian populations to decline locally, but argued that the remaining unsprayed areas would ensure the persistence of the species. For those species with only part of their range in Colombia, "the consequences of coca production may be more serious and may have placed several species of frogs at risk." (Lynch and Arroyo 2009, p. 974). In other words, if a species of frog had a small piece of its natural range in Colombia (and the rest of its range in neighboring countries), the eradication program might have a negative impact on the persistence of the species in Colombia.

As they did in their initial assessment in 2007 (Solomon et al. 2007), the team in 2009 again compared the impact of the herbicide against other factors that are known a priori to have much more devastating effects. For example, the team compared the toxicity of the herbicide against the toxicity of other pesticides and found that some insecticides are orders of magnitude more toxic (Brain and Solomon 2009). Relative to a very highly toxic insecticide such as endosulfan, Glyphos-Cosmo-Flux is not as toxic (Jones et al. 2009). This is analogous to saying that rat poison can be deadly to humans, but not as deadly as arsenic. While true, the comparison does not make rat poison any less toxic to humans and Brain and Solomon's (2009) comparison between the herbicide and endosulfan does not make the Glyphos-Cosmo-Flux any less toxic to amphibians. They went on to suggest that the deforestation caused by coca farmers causes habitat loss to amphibians and that this impact is also much larger than the impact of spraying the herbicide. Thus, while Roundup poses a risk, "the uncontrolled deforestation for the production of illicit crops such as coca will have a major effect on amphibians in Colombia through habitat alteration." (Brain and Solomon 2009, p. 944). While no real data are brought to bear on this question, it is likely true. Nevertheless, this conclusion had little to do with the assessment requested by the US Congress, which funded the researchers to determine the impact of the herbicide on amphibians living in and near cleared fields – not to assess the impact of the herbicide relative to other factors. Comparing the effect of the herbicide to complete deforestation of amphibian habitat allowed the team to conclude, "In summary, there are a number of human activities associated with the production of coca that present greater risks to amphibians than the glyphosate and Cosmo-Flux mixture used in the aerial eradication spraying." (Brain and Solomon 2009, p. 945). Although this is certainly correct, it only served to distract attention from the real issue at hand. All evidence from the Plan Colombia assessment pointed to the fact that the typical applications rate of the herbicide can kill larval, juvenile, and adult amphibians in large numbers.

An Independent Assessment of Plan Colombia

Nearly a decade has passed since the implementation of Plan Colombia using aerial fumigation with Glyphos-Cosmo-Flux. The original objective in 1999 was to reduce coca production by 50%. In 2008, the independent Congressional General Accounting Office (GAO) examined the program and found that from 2000 to 2006, the US government had spent over $6 billion to improve security and fight the drug war (US GAO 2008). By 2007, they were spraying 160,000 ha of coca fields annually. What was the result? The GAO found that between 2000 and 2007, coca production had *increased* by 15% (Fig. 9.5). Farmers quickly learned that they could replant new coca shrubs, prune their perennial shrubs after being sprayed, and move to new areas that can be cleared for new coca farms. Indeed, the aerial eradication program had pushed farmers out of their existing fields and motivated them to clear

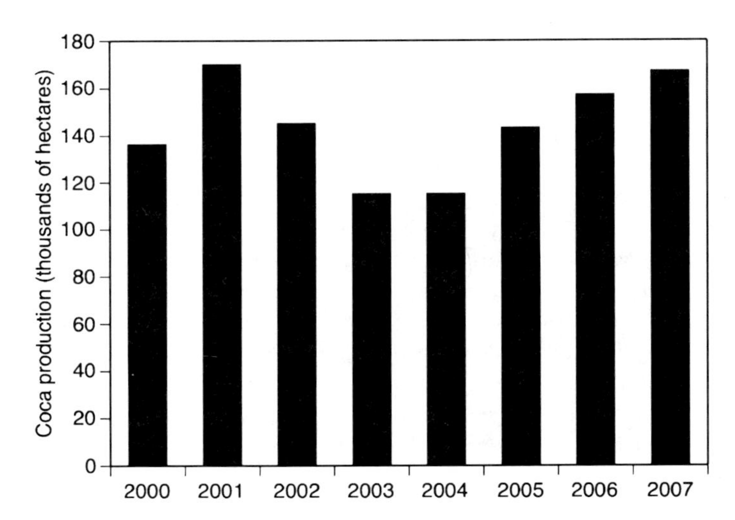

Fig. 9.5 Coca cultivation from the beginning of Plan Colombia (2000) until 2007 US GAO (2008)

new areas of rainforest habitat in promised no-spray zones in Colombian national parks and along the border with Ecuador. Thus, deforestation, one of the major threats to amphibian populations highlighted by Brain and Solomon (2009), is increasing *because* of the coca fumigation program. In summary, while providing no benefit of reduced coca production, the program ultimately contaminated large areas of Colombia and caused additional deforestation of amphibian habitats. Even the relativistic arguments of Solomon's group held no ground this time; directly or indirectly, the spraying of glyphosate was causing the loss of amphibian habitat in Columbia. While spraying Glyphos with Cosmo-Flux has been a complete failure in eradicating coca, it may be making excellent progress in eradicating Colombia's amphibians.

Ironically, this outcome was actually predicted by the US government. In 2000, the US Central Intelligence Agency was asked to assess the probability of being successful in the antidrug war in Colombia (US CIA 2000). In a document that was declassified in 2003, the CIA assessment stated, "A 50% decline in coca cultivation in the south over the next 5 years likely would encourage substantial new cultivation in other parts of Colombia. Farmers would probably be able to compensate for their losses by growing elsewhere in Colombia." Thus, at the start of Plan Colombia, it was estimated that the plan would fail to reduce coca production and would push farmers out to clear new areas for coca farms. Despite this prediction, the Plan still proceeded.

In 2009, the US State Department was asked about the lack of success in reducing coca production by aerially spraying. Naomi Lubick (2009) reported, "Despite this, a spokesman from the US State Department says that the USA will support Colombia's government as long as it chooses to continue the herbicide-spraying strategy to tackle its cocaine problem."

The Case of the California Red-Legged Frog

At the same time that Roundup's impact on amphibians was being evaluated in Colombia, there was an effort to also have its risk to amphibians assessed in California. In 2002, the Center for Biological Diversity sued the EPA, alleging that the EPA violated the Endangered Species Act by approving the registration of multiple pesticides to be used in California without considering the potential impacts of pesticides on the endangered California red-legged frog (*Rana aurora draytonii*). In 2005, the US District Court agreed, and an agreement was drawn up requiring the EPA to assess the potential impact of 66 pesticides on the red-legged frog and its habitat.[9] One of these pesticides was glyphosate.

In 2008, the EPA completed its assessment for glyphosate risks to the California red-legged frog (Carey et al. 2008). Because the California red-legged frog occupies both aquatic and terrestrial habitats, separate assessments were conducted for each habitat. Within each habitat, they assessed the potentially direct effects of the herbicide on survival, reproduction, and growth as well as the indirect negative effects that might occur if the herbicide application altered the frog's habitat.

The first risk assessment was conducted on the aquatic stage. The assessment includes lists of published studies on glyphosate toxicity on both fish and amphibians. When calculating the Risk Quotient, however, they used a study of bluegills (*Lepomis macrochirus*) that had an LC50 value of 3.17 mg a.e./L. They also calculated an expected concentration in the water based only on drift; they did not include inadvertent over spray that occurs during forestry applications. This produced an expected environmental concentration of 0.095 mg a.e./L. Using the bluegill LC50 value and the forestry expected environmental concentration value, they calculated a Risk Quotient of 0.095/3.17=0.03. Because the Risk Quotient value did not exceed the EPA's Level of Concern (i.e., 0.05), the EPA concluded there was no direct risk of Roundup applications to aquatic stages of amphibians. As I discuss below, more recent studies have discovered that at least two species of tadpoles have LC50 values that are much lower than bluegills (Relyea and Jones 2009), making the Risk Quotient now exceed 0.05.

Next, the risk assessment evaluated the impact on the terrestrial stage of the frog. As noted earlier, the sensitivity of terrestrial amphibians is assumed to be similar to the sensitivity of birds eating herbicide-contaminated food. It does not allow an assessment of potential impacts when the frog's skin is sprayed with the herbicide (we now know from the Colombian research that terrestrial applications of Roundup also can directly kill up to 30% of adult frogs; Bernal et al. 2009b). Using this approach, the EPA concluded that there were no concerns over direct effects of the herbicide on the terrestrial stage of the California red-legged frog.

Although the EPA did not identify sufficient evidence for direct negative effects, they did identify concerns regarding indirect negative effects on the frog's habitat

[9] http://www.biologicaldiversity.org/news/press_releases/rlf-10-19-2006.html (accessed June 2010).

and prey. In the end, they concluded using standard EPA categories, "Based on the best available information, the Agency makes a May Affect, and Likely to Adversely Affect determination for the CRLF [California red-legged frog] from the use of glyphosate." (Carey et al. 2008, p. 10). In plain language, the EPA concluded that the use of glyphosate in the habitats of the California red-legged frog would have a negative effect on the long-term persistence of the endangered frog's population. To my knowledge, this was the first time that the EPA had concluded that a glyphosate-based herbicide could negatively affect amphibians.

Reflecting on the Future of Amphibians and Roundup

As I reflect on the research that has been conducted on Roundup and amphibians, I am struck by the path that the research has taken and the attention it has received. In the mid 1990s, the Australian researchers determined that glyphosate-based herbicides containing the POEA surfactant were moderately toxic to amphibians. Monsanto declared that the researchers were wrong. My group has followed in their path confirming that the herbicide can be classified as highly toxic to amphibians in North America. Monsanto has declared that we also were wrong. Our latest work continues to confirm our initial studies. We recently conducted $LC50_{4-d}$ tests on 13 species of amphibians including – for the first time – tests on larval salamanders (Relyea and Jones 2009). We have also conducted additional mesocosm experiments that have demonstrated that the amount of mortality differs with tadpole age and that the herbicide becomes more lethal when combined with the stress of competition from other tadpoles (Jones et al. 2010, 2011). These studies have used a newer formulation that has been heavily marketed to farmers (Roundup Original Max®). A spokesperson for Monsanto, Scott Mortenson, called me in the summer of 2009 to insist that this formulation does not contain the POEA surfactant. He would not divulge what surfactant was added, claiming that it was a trade secret. In the end, the identity of the surfactant did not matter. The toxicity of Roundup Original Max was very much the same as the older formulations that contain POEA. Newer formulations, including Roundup PowerMAX® and WeatherMAX®, have received little testing on amphibians. However, a recent study found that chronic exposure to a relatively low concentration (0.572 mg a.e/L) of WeatherMAX caused 80% mortality in western chorus frog tadpoles (Williams and Semlitsch 2010).

In June 2009, the EPA announced it is seeking data on glyphosate toxicity as part of Monsanto's application to renew the herbicide's registration. The challenges will be familiar. The EPA has to assume that the herbicide is only used according to label directions. Hence, the inadvertent and unavoidable over sprays of herbicide on wetlands that are known to occur when spraying Colombian coca fields or Canadian forests will not likely factor into the risk assessment, although the impact of herbicide drift into water bodies will likely be considered. A large number of assumptions can be made about exposures in nature to obtain a variety of risk assessment outcomes (the depth of the water, how much herbicide is intercepted by overhanging vegetation, how much herbicide is held by the soils, etc.). Unfortunately, there is a paucity of data on natural concentrations to evaluate the accuracy of these estimates. As Renier Mann (University of Technology in Sydney) recently said in an interview with Lubick (2009), "Everyone agrees on concentrations that cause toxicity. The argument is over whether the frogs are exposed to those concentrations." This is a critical area for future research.

For the aquatic stage of amphibians, we now have substantially more toxicity studies covering a broader range of species. The studies definitively show that glyphosate formulations containing POEA and other formulations that may or may not contain POEA (i.e. Roundup OriginalMAX, Roundup WeatherMAX) are highly toxic to tadpoles. Based on these most recent studies, the most sensitive tadpole species are American bullfrogs and the spring peepers (*P. crucifer*). Both have $LC50_{4-d}$ values of 0.8 mg a.e./L, which is considerably lower than the bluegill data that have been used in past amphibian risk assessments. If we were to follow the EPA risk assessment protocol (Jones et al. 2004) and divide this value by 20, the upper limit that would protect larval amphibians would be 0.04 mg a.e./L. In the case of the EPA's risk assessment of the California red-legged frog (Carey et al. 2008), the Risk Quotient would now be 0.095/0.8 = 0.12. This means that the Risk Quotient would exceed the Level of Concern value of 0.05 that has been set to protect amphibians. We also now have new data on the concentrations found in nature. For example, Battaglin et al. (2009) surveyed vernal pools in the Rock Creek National Park and found concentrations up to 0.328 mg a.e./L, which substantially exceeds the 0.065 mg/L freshwater standard for aquatic life exposed to glyphosate.

We still have few data on the terrestrial stages of amphibians. The only two published experiments (Relyea 2005b; Bernal et al. 2009b) have found that aerial oversprays at expected application rates can cause substantial death in adult frogs (30–86% death in 1–4 days). With these data now available, it is unclear whether the EPA will continue to use estimates of dietary ingestion in birds as surrogates to assess the effects of Roundup on terrestrial stages of amphibians. If this turns out to be the case, then any concerns about amphibian exposures to Roundup will likely follow the outcome of the California red-legged frog and emphasize the indirect effects rather than direct lethal effects.

It will be quite interesting to follow the re-registration process for glyphosate-based herbicides in the years to come. With Monsanto's share of the glyphosate sales market reaching $3.5 billon in 2009 (Monsanto 2009), there is been a lot of corporate profit hanging in the balance. Interestingly, however, increased foreign competition following the end of the glyphosate patent has forced large reductions in Monsanto's price point for its line of Roundup products (by approximately two-thirds) and Monsanto has announced that its long-term strategy is to reduce its manufacturing of chemicals and focus its efforts on its manufacturing of genetically modified seeds (Leonard 2010). Regardless of who makes glyphosate-based herbicides and seeds, there will be a strong push by the corporations to continue the use of glyphosate. Based on the history of such debates across many industries, the corporate strategy will likely be to proclaim that decades of past studies have determined there is no risk, to discredit any scientists who question past risk assessments with new data, and to ultimately delay any decision that could impact profits (Michaels 2008). If I ran a corporation that made a great deal of money from a popular herbicide and associated genetically modified seeds, I suppose I might be tempted to do the same thing. As a scientist, however, my goal has always been to better understand how pesticides might be impacting our environment, to inform the public about the science, and to let the public and policy makers decide what, if anything, should be done next. What is clear based on published data is that unlike many species of Roundup-Ready crop plants, amphibians are not ready for Roundup.

Acknowledgments This chapter was improved by the comments of Christine Bishop, Will Brogan, Rickey Cothran, Maya Groner, John Hammond, Jason Hoverman, Jessica Hua, Heather Shaffery, Aaron Stoler, and several anonymous reviewers. I also thank the National Science Foundation for their continued support of my research. For the record, I have never accepted any funding from antipesticide groups.

References

Battaglin WA, Rice KC, Focazio MJ, Salmons S, Barry RX (2009) The occurrence of glyphosate, atrazine, and other pesticides in vernal pools and adjacent streams in Washington, DC, Maryland, Iowa, and Wyoming, 2005–2006

Baylis AD (2000) Why glyphosate is a global herbicide: strengths, weaknesses and prospects. Pest Manag Sci 56:299–308

Bernal MH, Solomon KR, Carrasquilla G (2009a) Toxicity of formulated glyphosate (glyphos) and cosmo-flux to larval colombian frogs 1. Laboratory acute toxicity. J Toxicol Environ Health A 72:961–965

Bernal MH, Solomon KR, Carrasquilla G (2009b) Toxicity of formulated glyphosate (glyphos) and cosmo-flux to larval and juvenile colombian frogs 2. Field and laboratory microcosm acute toxicity. J Toxicol Environ Health A 72:966–973

Berrill M, Bertram S, Pauli B (1997) Effects of pesticides on amphibian embryos and larvae. In: Green D (ed) Amphibians in decline: Canadian studies of a global problem. Society for the Study of Amphibians and Reptiles, St. Louis, MO, pp 233–245

Bidwell JR, Gorrie JR (1995) Acute toxicity of a herbicide to selected frog species. Department of Environmental Protection. Technical Series 79, Perth

Brain RA, Solomon KR (2009) Comparisons of the hazards posed to amphibians by the glyphosate spray control program versus the chemical and physical activities of coca production in Colombia. J Toxicol Environ Health A 72:937–948

Carey S, Crk T, Flaherty C, Hurley P, Hetrick J, Moore K, Termes SC (2008) Risks of glyphosate use to federally threatened California red-legged frog (*Rana aurora draytonii*). US EPA, Washington, DC

Chen CY, Hathaway KM, Folt CL (2004) Multiple stress effects of Vision® herbicide, pH, and food on zooplankton and larval amphibian species from forest wetlands. Environ Toxicol Chem 23:823–831

Dayton L (1995) Our frogs: are they heading for the last round-up? Sydney Morning Herald, 13 September

Edginton AN, Sheridan PM, Stephenson GR, Thompson DG, Boermans HJ (2004) Comparative effects of pH and Vision® herbicide on two life stages of four anuran amphibian species. Environ Toxicol Chem 23:815–822

Edwards WM, Triplett GB Jr, Kramer RM (1980) A watershed study of glyphosate transport in runoff. J Environ Qual 9:661–665

Folmar LC, Sanders HO, Julin AM (1979) Toxicity of the herbicide glyphosate and several of its formulations to fish and aquatic invertebrates. Arch Environ Contam Toxicol 8:269–278
Giesy JP, Dobson S, Solomon KR (2000) Ecotoxicological risk assessment for Roundup® herbicide. Rev Environ Contam Toxicol 167:35–120
Gosner KL (1960) A simplified table for staging anuran embryos and larvae with notes on identification. Herpetologica 16:183–190
Grube A, Donaldson D, Kiely T, Wu L (2011) Pesticide industry sales and usage: 2006 and 2007 market estimates. US EPA, Washington, DC
Hand E (2005) Study finds herbicide is deadly for amphibians. St. Louis Dispatch. August 10
Hayes TB (2004) There is no denying this: defusing the confusion about atrazine. Bioscience 54:1138–1149
Howe CM, Berrill M, Pauli BD, Helbring CC, Werry K, Veldhoen N (2004) Toxicity of glyphosate-cased pesticides to four North American frog species. Environ Toxicol Chem 23:1928–1938
Jones R, Leahy J, Mahoney M, Murray L, Odenkirchen E, Petrie R, Stangel C, Sunzenauer I, Vaituzis Z, Williams AJ (2004) Overview of the ecological risk assessment process in the Office of Pesticide Programs, U.S. Environmental Protection Agency. US EPA, Washington, DC
Jones DK, Hammond JR, Relyea RA (2009) Highly lethal effects of endosulfan across nine species of tadpoles: lag effects and family-level sensitivity. Environ Toxicol Chem 28:1939–1945
Jones DK, Hammond JR, Relyea RA (2010) Roundup® and amphibians: the importance of concentration, application time, and stratification. Environ Toxicol Chem 29:2016–2025
Jones DK, Hammond JR, Relyea RA (2011) Competitive stress can make the herbicide Roundup® more deadly to larval amphibians. Environ Toxicol Chem 30:446–454
Kiely T, Donaldson D, Grube A (2004) Pesticides industry sales and usage: 2000 and 2001 market estimates. US EPA, Washington, DC
Lajmanovich RC, Sandoval MT, Peltzer PM (2003) Induction of mortality and malformation in Scinax nasicus tadpoles exposed to glyphosate formulations. Bull Environ Contam Toxicol 70:612–618
Leonard C (2010) Monsanto 3Q net income sags on weak Roundup sales. Associated Press. June 30, 2010
Lubick N (2007) Drugs, pesticides, and politics – a potent mix in Colombia. Environ Sci Technol 41:3403–3406
Lubick N (2009) Environmental impact of cocaine strategy assessed. Nature News. November 12
Lynch JD, Arroyo SB (2009) Risks to Colombian amphibian fauna from cultivation of coca (*Erythroxylum coca*): a geographical analysis. J Toxicol Environ Health A 72:974–985
Mann RM, Bidwell JR (1999) The toxicity of glyphosate and several glyphosate formulations to four species of southwestern Australian frogs. Arch Environ Contam Toxicol 36:193–199
Mann RM, Bidwell JR (2000) Application of the FETAX protocol to assess the developmental toxicity of nonylphenol ethoxylate to *Xenopus laevis* and two Australian frogs. Aquat Toxicol 51:19–29
Mann RM, Bidwell JR (2001) The acute toxicity of agricultural surfactants to the tadpoles of four Australian and two exotic frogs. Environ Poll 114:195–205
Mann RM, Bidwell JR, Tyler MJ (2003) Toxicity of herbicide formulations to frogs and the implications for product registration: a case study from Western Australia. Appl Herpetol 1:13–22
Marshall EJP, Solomon KR, Carrasquilla G (2009) Coca (*Erythroxylum coca*) control is affected by glyphosate formulations and adjuvants. J Toxicol Environ Health A 72:930–936
Michaels D (2008) Doubt is their product. Oxford University Press, New York, NY
Monsanto (2009) Annual Report to Shareholders
Polis GA, Strong DR (1996) Food web complexity and community dynamics. Am Nat 147:813–846
Relyea RA (2003) Predator cues and pesticides: a double dose of danger for amphibians. Ecol Appl 13:1515–1521

Relyea RA (2004a) Synergistic impacts of malathion and predatory stress on six species of North American tadpoles. Environ Toxicol Chem 23:1080–1084
Relyea RA (2004b) The growth and survival of five amphibian species exposed to combinations of pesticides. Environ Toxicol Chem 23:1737–1742
Relyea RA (2005a) The impact of insecticides and herbicides on the biodiversity and productivity of aquatic communities. Ecol Appl 15:618–627
Relyea RA (2005b) The lethal impact of Roundup® on aquatic and terrestrial amphibians. Ecol Appl 15:1118–1124
Relyea RA (2005c) The lethal impacts of Roundup and predatory stress on six species of North American tadpoles. Arch Environ Contam Toxicol 48:351–357
Relyea RA (2006) Response to Thompson et al. Letter to the editor, "The impact of insecticides and herbicides on the biodiversity and productivity of aquatic communities". Ecol Appl 16:2027–2034
Relyea RA, Jones DK (2009) The toxicity of Roundup Original MAX® to 13 species of larval amphibians. Environ Toxicol Chem 28:2004–2008
Relyea RA, Mills N (2001) Predator-induced stress makes the pesticide carbaryl more deadly to grey treefrog tadpoles (*Hyla versicolor*). Proc Natl Acad Sci USA 98:2491–2496
Relyea RA, Schoeppner NM, Hoverman JT (2005) Pesticides and amphibians: the importance of community context. Ecol Appl 15:1125–1134
Smith GR (2001) Effects of acute exposure to a commercial formulation of glyphosate on the tadpoles of two species of anurans. Bull Environ Contam Toxicol 67:483–488
Solomon KR, Thompson DG (2003) Ecological risk assessment for aquatic organisms from over-water uses of glyphosate. J Toxicol Environ Health B Crit Rev 6:289–324
Solomon KR, Anadón A, Carrasquilla G, Cerdeira AL, Marshall J, Sanin L-H (2007) Coca and poppy eradication in Colombia: environmental and human health assessment of aerially applied glyphosate. Rev Environ Contam Toxicol 190:43–125
Struger J, Thompson D, Staznik B, Martin P, McDaniel T, Marvin C (2008) Occurrence of glyphosate in surface waters of southern Ontario. Bull Environ Contam Toxicol 80:378–384
Stuart SN, Chanson JS, Cox NA, Young BE, Rodrigues ASL, Fischman DL, Waller RW (2004) Status and trends of amphibian declines and extinctions worldwide. Science 306:1783–1786
Taylor BE, Scott DE, Gibbons JW (2006) Catastrophic reproductive failure, terrestrial survival, and persistence of the marbled salamander. Conserv Biol 20:792–801
Thompson DG, Wojtaszek BF, Staznik B, Chartrand DT, Stephenson GR (2004) Chemical and biomonitoring to assess potential acute effects of Vision® herbicide on native amphibian larvae in forest wetlands. Environ Toxicol Chem 23:843–849
Thompson DG, Solomon KR, Wojtaszek BF, Edginton AN, Stephenson GR (2006) Letter to the editor, "The impact of insecticides and herbicides on the biodiversity and productivity of aquatic communities". Ecol Appl 16:2022–2027
US EPA (1993) R.E.D. Facts – Glyphosate. United States Environmental Protection Agency. EPA-739-F-93-011
US CIA (2000) Plan Colombia's potential impact on the Andean cocaine trade: an examination of two scenarios. Intelligence Report 19 September 2000. DCI Crime and Narcotics Center
US GAO (2008) Plan Colombia: drug reduction goals were not fully met, but security has improved; U.S. agencies need more detailed plans for reducing assistance. Report 09–71
US Senate (2005) Department of State, Foreign Operations, and Related Programs Appropriations Bill, 2006. Report 109–96
WHO (1994) International programme on chemical safety. Environmental Health Criteria 159 – Glyphosate. World Health Organization, Geneva, Switzerland
Williams BK, Semlitsch RD (2009) Larval responses of three Midwestern anurans to chronic, low-dose exposures of four herbicides. Archiv Environ Contam Toxicol 58:819–827
Wojtaszek BF, Staznik B, Chartrand DT, Stephenson GR, Thompson DG (2004) Effects of Vision® herbicide on mortality, avoidance response, and growth of amphibian larvae in two forest wetlands. Environ Toxicol Chem 23:832–842

Chapter 10
Atrazine Has Been Used Safely for 50 Years?

Tyrone B. Hayes

At the time, I had never heard of atrazine. Now, you google "tyrone" and "atrazine" and you get tens of thousands of hits. I guess you can say we are joined at the carbon bond.

Tyrone B. Hayes

Abstract The herbicide atrazine is a potent endocrine disruptor, active in fish and amphibians in the low ppb range. Among other effects, atrazine impairs reproductive development and function including decreased testosterone levels, impaired testicular development, and low fertility/sperm production in male fish, amphibians, and in some reptiles. Atrazine also feminizes fish, amphibians and reptiles resulting in the development of oocytes in the testes and complete feminization. In addition to laboratory experiments, similar effects have been associated with animals in the wild. Although there is some question about how to compare the doses, adverse effects are also observed in laboratory rodents: including prostate disease, low sperm production, and decreased testosterone levels in males and mammary cancer, abortion, and impaired mammary development in females. These effects are all consistent with the induction of aromatase, the enzyme that converts testosterone into estrogen, a mechanism that has been demonstrated across vertebrate classes. Despite well over 150 publications from at least 50 independent laboratories showing adverse reproductive effects in all vertebrate classes examined, and recent epidemiological studies associating atrazine exposure with low sperm counts in men, breast and prostate cancer, and birth defects, the major manufacturer still maintains that "atrazine has been used safely for 50 years" and the US EPA still struggles with how to evaluate pesticides for endocrine disrupting effects.

T.B. Hayes (✉)
Department of Integrative Biology, University of California, Berkeley, CA, 94720, USA
e-mail: tyrone@berkeley.edu

J.E. Elliott et al. (eds.), *Wildlife Ecotoxicology: Forensic Approaches*,
Emerging Topics in Ecotoxicology 3, DOI 10.1007/978-0-387-89432-4_10,

Introduction

Atrazine is a triazine herbicide. Up until 1999, it was the number one selling pesticide in the world. Now it has been displaced by another broad spectrum herbicide, glyphosate (Roundup) (Relyea 2010). Atrazine was initially introduced by CIBA in 1958. Eventually, through various mergers and acquisitions, NOVARTIS and then SYNGENTA took over as the major manufacturer. Over the last couple of decades, approximately 80 million pounds of atrazine per year have been applied (primarily to corn) in the USA. Atrazine is a water soluble, persistent, and highly mobile herbicide. As much as 500,000 pounds of atrazine are transported in rainfall and 1.2 million pounds collect in rivers and flow into the Gulf of Mexico each year. Atrazine kills weeds by disrupting electron transport in photosynthesis, but monocot crops, such as corn, have the capacity to remove atrazine via the glutathione reaction and excretion (Solomon et al. 1996). Thus, atrazine can be applied to corn fields and kills weeds but not corn, which is naturally resistant. This natural resistance and the resulting ability to apply atrazine without harming the crop is what made atrazine the big seller (number one in fact), until Round up-Ready corn came along. It was about that time (1997) that I was introduced to atrazine, when I was a new assistant professor at the University of California Berkeley.

History and Background

I'm still just a little boy who likes frogs.

Tyrone B. Hayes

As long as I can remember, I have been fascinated by amphibians. As an undergraduate at Harvard, I explored the effects of temperature on growth and development rates and sexual differentiation in woodfrogs (*Rana sylvatica*). As I developed my undergraduate honor's thesis, I became increasingly interested in endocrine mechanisms that regulate growth and development and how environmental changes regulate endocrine signals. How are environmental factors (changes in the external world) translated, through hormones, into changes in physiology, development and growth? These were the types of questions that I pursued in my doctoral work, examining interactions between temperature and hormone production and function during growth, metamorphosis, and sex differentiation in amphibian larvae.

Fulfilling another childhood dream, I began travelling to Africa during my doctoral studies. While working in Kenya, in the Arabuko Sokoke Forest, I was introduced to the species *Hyperolius argus*. The species caught my eye and drew my interest as a result of its unusual sexual dimorphism. The males and females were so differently colored, that initially I thought they were different species. As adults, the males are a bright lime-green color, whereas the females are reddish brown with large white spots. I brought the species back to Berkeley and developed a breeding colony. In our first studies, we followed the development and examined hormonal regulation of pigmentation (Hayes and Menendez 1999). Our studies revealed that the green coloration was the default pattern. Both males and females metamorphosed with the green (male-type) coloration and females changed color at sexual maturity. In addition, we examined hormonal regulation of color change and found that only estrogens induce the female color pattern. Exogenous estrogen not only prematurely induced color change in juvenile females, but induced the female color pattern in males as well. So males *H. argus* have the receptor to respond to estrogen and the ability to synthesize the pigments to produce the female color pattern, but in the absence of ovaries do not synthesis the estrogens necessary to initiate color change.

My laboratory screened dozens of steroids and other estrogenic chemicals, and as it turns out, the list of estrogens that induced color change in *H. argus*, correlated directly with estrogens that induce breast cancer cell growth (Hayes 2000b; Noriega and Hayes 2000). While giving a lecture on our findings, my wife, Katherine Kim, happened to be in the audience once. Having recently completed her MBA and MPH, she saw an opportunity. "You should patent that," she said. That simple comment led to my relationship with NOVARTIS and my introduction to atrazine (Hayes 2000b).

From the Wild to the Regulatory World

SYNGENTA means bringing people together.

Syngenta

At first, I served as a consultant only. Along with Dr. Tim Gross (at the time a researcher at the University of Florida at Gainesville and a USGS employee) and Dr. James Carr (a professor at Texas Tech University) and a number of other scientists hired as consultants and NOVARTIS employees, we were tasked with evaluating whether atrazine was an endocrine disruptor in fish, amphibians, and reptiles. The panel met several times per year and was a joint effort between NOVARTIS and ECORISK Inc. ECORISK was a consulting firm run by Dr. Ron Kendall (then at Clemson University) and the CEO Robert Bruce, who occasionally joined the panel meetings. As a developmental endocrinologist (not an ecotoxicologist), I found myself learning a whole new vocabulary. Whereas I was used to talking about hormone concentrations measured in ng/ml and pg/ml, I now had to translate chemical concentrations into ppb and ppt.

After completing a report that amounted to a large multi-authored review paper, I assumed that I had served my duty to ECORISK and NOVARTIS and moved on. A few weeks later, however, I was asked to evaluate a proposal written by Jim Carr to examine atrazine effects in frogs using the African clawed frog (*Xenopus laevis*). I reviewed the proposal, offered my criticisms, my praises, and made suggestions for improvement. Eventually, I was asked by Alan Hosmer, representing NOVARTIS, to describe how I would approach this question. I responded and was offered a contract.

I submitted several proposals that utilized a diversity of species (including my patent pending *Hyperolius argus* Endocrine Screen (Hayes 2000a)), but the company decided on *Xenopus laevis* because it was a well-known model.

Given the lack of data at the time, there was no real hypothesis, only a question, "Does atrazine affect thyroid hormone, androgen, or estrogen-dependent development in larval amphibians?" I designed a study that exposed larvae throughout development (from hatching to metamorphosis) with several atrazine concentrations ranging from 0.1 to 200 ppb in a static renewal system. Using positive controls that included thyroid hormone, estradiol, and dihydrotestosterone (DHT), we proposed to monitor growth, development, and metamorphosis as measures of thyroid hormone interference or activity by atrazine; gonadal development as a measure of estrogen-like activity; and laryngeal growth and size as a measure of androgen (DHT) activity or interference. In the absence of any preliminary or background data on which to base predictions, the best hypothesis was, "atrazine does not affect endocrine-regulated end points in amphibians."

The first set of data suggested some effects on the larynx. We analyzed a few samples only, but even this small sample size revealed that atrazine-exposed animals had smaller larynges than controls. We increased the sample size slightly, and the problem got worse approaching statistical significance.

The ECORISK panel requested a few samples at a time, but after a couple of meetings and conference calls, I completed the work. I suggested to the panel that we repeat the experiment with a different population and that we consult Dr. Darcy Kelley (Columbia University), the world's expert on development and hormonal regulation of the larynx in *Xenopus laevis*. I requested to present the data at the Western Regional Conference on Comparative Endocrinology. Although, I was told I could present the data, corporate review took so long (more than a month) that I missed the abstract deadline. I was told that I could repeat the experiment. The proposed experiment was exactly the same as the first, but compared frogs from two separate sources: our lab colony of mixed origin and animals from NASCO Inc. (Fort Atkinson, Wisconsin). A second part of the study proposed to examine the ultimate effects on reproductive potential after metamorphosis. By the time the committee reviewed the proposal, however, November was approaching. ECORISK and SYNGENTA did not provide a review, and eventually suggested that we should postpone starting the project to avoid it extending beyond the Christmas holidays. No funds came forward, but on Oct 28, 2000, I initiated the experiment without funding from SYNGENTA or ECORISK.

It was during this experiment, that I observed that several of the exposed animals had gonads with unusual morphologies. Histological analysis revealed that some animals from the atrazine treatments had a mix of testicular and ovarian tissues. We then re-examined all of the animals in more detail (including animals from the previous experiment) and discovered that three types of gonadal morphologies were observed only in the atrazine-exposed animals: multiple (or lobed) testes, nonpigmented ovaries, and hermaphrodites (animals with both testes and ovaries). We would later go on to show that these abnormalities could be induced by an androgen blocker (cyproterone acetate) and/or by brief periods of estrogen exposure, suggesting that some morphologies represented demasculinization and that others represented partial (incomplete) feminization (Hayes et al. 2006b).

On January 26, 2001, representatives of Ecorisk and Syngenta came to Berkley and I arranged everything as if it were a local conference. I organized the hotels, local ground transportation, seminar rooms, and arranged all of the meals and refreshments. I developed an agenda. I would present a summary of the consultation and research conducted under the direction of the ECORISK panel, SYNGENTA would present on their additional needs, and then I would present my laboratory's new data. In addition to the students from my laboratory and the staff who worked on the project, I invited Beth Burnside (then, Vice Chancellor for Research), Paul Licht (then, Dean), Marvalee Wake (then, Chair of my department), David Wake (then, Director of the Museum of Vertebrate Zoology, where I had an appointment), and the University's legal counsel (Michael Smith). Beth Burnside opted not to attend and Smith determined that he did not feel he needed to be there. Lastly, my wife, Katherine Kim, was also in attendance.

They arrived: The ECORISK panel (Ron Kendall, Keith Solomon, John Giesy, Earnest Smith, and Jim Carr) along with Alan Hosmer (SYNGENTA), with Charles Breckenridge arriving later that afternoon. They immediately requested to meet with me in private. I refused. I presented our introduction and they again asked for a private meeting. I refused. Then they introduced Robert Sielken (Sielken and Associates Consulting Inc.) who attended with them.

In his presentation, Sielken presented a critique of our previous experiments. He pointed to corrections made in the hand-written data, some as trivial as recording the time in standard instead of military time, but, in fact, corrected. In his opinion, our data on the larynx was of no significance. He had gone back to our initial data sets. After we realized that the data showed a decrease in the size of the male larynx, we focused on the males because there was no effect (in this regard) on females. Thus, we did not analyze any more than the first three estrogen-treated frogs because all estrogen-treated animals were female. Sielken felt that because the smallest sample size was three, then all of the data had to be simulated, using a Monte Carlo simulation, to approximate a sample size of three. Even after this alternate approach to analysis of the data, Sielken obtained a P value of 0.045. Sielken and his firm only accepted P values of <0.04 as significant (Parshley 2000).

Shortly after the meeting at Berkeley, ECORISK would drop out of the picture. Alan Hosmer joined by Peter Hertl and Janis McFarland at SYNGENTA sought to draw up a contract directly between myself and SYNGENTA. Hosmer worked with SYNGENTA Vice President, Gary Dickson. I would stay in touch with John Giesy for a while, who assured me that Gary Dickson would be helpful.

We never worked out a new contract, however. Just 1 day before I was to speak with SYNGENTA, National Geographic News, to my surprise, aired footage that they shot in my lab showing the hermaphrodites produced by atrazine. That was the end of my relationship with SYNGENTA. My laboratory subsequently prepared the manuscript. Although I had no idea it would cause the stir it did, I knew the data deserved to be in a widely read journal. I submitted it to the Proceedings of the National Academy of Sciences (PNAS).

I suggest you get yourself a working phone.

David Wake, after our first paper was accepted into PNAS

Even before our first paper came out in PNAS, I received phone calls from both the Natural Resources Defense Council and Tom Steeger from the US Environmental Protection Agency. Tom Steeger told me that ECORISK, assuming that I had reported data to the EPA, were submitting a critique of my work (Parshley 2000).

Tom Steeger forwarded SYNGENTA's data and their unpublished manuscripts. He was similarly forwarding my unpublished data, submitted manuscripts, etc., to others even though he told me he was holding them in confidence and using them only in his internal review. The Natural Resources Defense Council (NRDC) would later obtain, through the freedom of information act, email exchanges between Steeger and Hosmer during this time (2000–2003), discussing my data, information that I gave Steeger during a visit to my laboratory and other confidential issues. The noise that SYNGENTA and ECORISK made would stimulate a debate, however, that would culminate in the 2003 EPA/FIFRA Scientific Advisory Panel meeting. With the exception of Carr's paper, none of the SYNGENTA data were published at the time, but through Steeger, I had seen most of it.

When SYNGENTA's first paper finally appeared (Carr et al. 2003), it was indeed everything I had seen previously from communications with Steeger. Although they would later claim that they "could not" reproduce my work, the fact is they "did not" reproduce my work. With the exception that we both used static renewal, no other efforts to replicate my study were made: the rearing medium, the food type and amount, the container size, the volume of rearing medium, the number of animals per tank, the amount of atrazine added, the temperature, the animal source, etc., virtually everything was different. Further, the health of their animals under these conditions was questionable. Greater than 50% of the animals in their study had suffered mortality (compared to 10% on average in my laboratory). The animals that survived showed an inverse relationship between time to metamorphosis and size at metamorphosis, an indication that they were in dire health. Atrazine not only affected gonadal differentiation ($P=0.0003$ for intersex and $P=0.0042$ for discontinuous or lobed testes), but the effects they reported were of greater statistical significance than in our studies. Furthermore, the metamorphs in Carr's study were smaller than any I had ever seen, indicating their poor health. Carr also reported several other significant adverse effects on growth and development: Atrazine reduced the number of animals reaching metamorphosis ($P=0.03$ for foreleg emergence and $P=0.04$ for tail reabsorption). Atrazine also induced edema ($P=0.02$) and abnormal erratic swimming ($P=0.004$). Despite these very significant effects, however, they repeatedly referred to their results as "weak trends" even though the effects were well below the significance value of ($P<0.04$).

Carr's work was followed by a study conducted in Giesy's laboratory in Michigan State (Coady et al. 2005). Here, the authors suggested that hermaphroditism was normal in *X. laevis*. That conclusion was based on their finding that hermaphrodites were also observed in their "controls." They conducted their experiments at Michigan State using uncovered tanks. In their report to the EPA (Hecker et al. 2003) and in their publication, they reported that frogs jumped from tank to tank (frogs would appear in previously empty tanks, etc.) and in some cases, even disappeared from the experiment (escaped from the tanks). In addition, due either to the aspiration of atrazine-contaminated water, or atrazine contamination in their water source, their "controls" had more than two times the biologically effective concentration of atrazine based on our studies (Hayes et al. 2002a) thus explaining why they found abnormalities in their "controls."

Carr's laboratory would go on to show that atrazine and especially atrazine + nitrates resulted in testicular oocytes in leopard frogs, *Rana pipiens* (Orton et al. 2006). Their finding mirrored our studies that showed testicular oocytes and even vitellogenic eggs produced in male gonads in *Rana pipiens* larvae exposed to atrazine (Hayes et al. 2002b, c).

SYNGENTA and ECORISK also followed with a series of field studies. In one of their studies in the USA, they showed that newly metamorphosed frogs collected from areas with atrazine contamination were more likely to have testicular oocytes ($P<0.04$) (Murphy et al. 2006), yet they concluded "given the lack of a consistent relationship between atrazine concentrations and testicular oocytes (TO) incidence, it is more likely the TOs observed in this study result from natural processes in development rather than atrazine exposure" (Murphy et al. 2006). In South Africa, they conducted another study, where they reported that (again) gonadal abnormalities were found, but were equally likely to occur in frogs from "non-corn-growing" areas as they were to occur in frogs collected from "corn-growing areas." (Du Preez et al. 2005b). In the study in South Africa, they examined adult frogs and claimed that frogs were confined to the areas where they were captured and were not likely to have moved from elsewhere (e.g, from a pond in an agricultural area to a pond in a nonagricultural area). However, they reported losing all of their animals in one study, because catfish moved into the pond and ate all of the frogs. As pointed out in my review in *Biosciences*, "Thus, the authors' claims rely on the unlikely assumption that fish (which are truly aquatic and have no lungs or legs) are capable of moving between ponds, but frogs (which have lungs and legs) are not"(Hayes 2004). What's more, the atrazine levels reported from the ponds in their "non-corn-growing" area were as high as those reported from their corn-growing regions, and both exceeded biologically active levels.

The ECORISK group also conducted a microcosm study. They reported gonadal abnormalities in the controls (Jooste et al. 2005b). Specifically, they reported testicular oocytes in newly metamorphosed *Xenopus laevis*. This is an interesting finding, because my laboratory has never observed testicular oocytes in *Xenopus laevis* even in atrazine-exposed or estrogen-exposed frogs at metamorphosis. Females do not normally have oocytes at metamorphosis in *Xenopus laevis* in my laboratory's experience. The figure, eventually published by the ECORISK group (Jooste et al. 2005a), in response to a letter to the editor from me (Hayes 2005), did not contain a scale bar, but the size of the gonad in the photomicrograph can be estimated based on the average oocyte diameter for the species. It is not likely that a gonad of this size with oocytes as advanced as those in the photomicrograph could have come from a newly metamorphosed animal. Considering that the population that they examined simply might develop differently from others, my laboratory travelled to Potchefstrum in 2006 and brought our own animals into the laboratory and established a breeding colony. Like other populations we worked with, even females did not have oocytes at metamorphosis from that population. The only possibilities were a contaminant in their rearing water, or that the unusual gonadal development was an artifact of the near lethal temperatures (Jooste et al. 2005b) used in their studies.

To date, 100% of the studies conducted by SYNGENTA claimed no affect of atrazine (Coady et al. 2005; Du Preez et al. 2002, 2005a, 2005b, 2008; Hecker et al. 2004, 2005; Jooste et al. 2005b; Kloas et al. 2009a, b, c; Preez et al. 2009), even SYNGENTA studies that showed effects (Carr et al. 2003; Murphy et al. 2006). On the other hand, studies not funded by Syngenta report significant effects of atrazine on amphibian reproductive development (Hayes et al. 2002a, b; 2006b; Langlois et al. 2009; McDaniel et al. 2008; Oka et al. 2008; Orton et al. 2006; Reeder et al. 1998; Tavera-Mendoza et al. 2002a; Tavera-Mendoza et al. 2002b). But, once again, SYNGENTA took the opportunity to say that they were unable to repeat our studies, and again, focused on critiquing my studies and ignoring the fact that several other independent laboratories have shown adverse effects of atrazine on gonadal development in frogs, in addition to my laboratory: Solomon (Solomon et al. 2008) concluded in his review that "with rare exceptions, the only studies that report adverse effects on amphibian development and reproduction are those from the Hayes laboratory." That view contradicts his long-term coauthor, Carr's statement, "The important issue is for everyone involved to come to grips with (and stop minimizing) the fact that independent laboratories have demonstrated an effect of atrazine on gonadal differentiation in frogs. There is no denying this" (Hayes 2004; 2009a). Furthermore, as Rohr and McKoy pointed out (Rohr and Mckoy 2009), the review by Solomon et al. "regularly dismissed significant effects of atrazine."

Eventually, SYNGENTA published what they reported as "the definitive" atrazine studies in *Xenopus laevis* (Kloas et al. 2009a). Along with the EPA, SYNGENTA oversaw two atrazine studies in *Xenopus laevis*. One of the studies contracted by SYNGENTA was conducted in the laboratory of Werner Kloas, a member of the EPA's 2003 FIFRA Scientific Advisory Panel (SAP). An SAP member (Kloas), who affected EPA decisions, benefitted financially from the EPA's decision (at the recommendation of the SAP) that "there is not sufficient scientific evidence to indicate that atrazine consistently produces effects across the range of amphibian species examined" (Kloas et al. 2009a). This scenario is analogous to a jury coming back with a decision that there is not enough evidence to convict, with one of the jurors then receiving a lucrative contract to investigate further and bring the case back to court.

Kloas et al. concluded that "it seems likely that no further endocrine mechanism of atrazine affecting sexual differentiation in *X. laevis* exists" (Kloas et al. 2009b). As in the past, it was not that SYNGENTA was unable to repeat our studies, but rather that they *did not* repeat our studies. Not only were many variables different in their studies (Kloas et al. 2009a, b, c), including constant flow-through vs. static exposure, different food and different feeding patterns, different sized containers, and different animal densities, but in addition both of the new SYNGENTA studies used a new and unexamined population of frogs from a company called Xenopus 1 (Dexter, Michigan, 48130) that reports that their adults come from Chile.

It was, in fact, this single Syngenta-funded study that formed the basis for the EPA's second Scientific Advisory Panel review of atrazine. Among others, the New York Attorney General's Office raised concern prior to the EPA's review: "EPA's approach of evaluating the effects of atrazine only on frog gonads, and relying almost entirely on only one unpublished industry-sponsored study (Hosmer et al. 2007) is clearly not based on the available science and may lead to a biased outcome." (http://www.regulations.gov/search/Regs/home.html#documentDetail?R=09000064802ff8e0), The EPA not only ignored this advice, but even ignored the concerns of their advisory panel. Similar to the concern of the New York State Attorney General's office, the panel expressed "concern that the Agency did not utilize any of these studies" (from the open literature) and "noted that these studies may have provided some added value in evaluating the conclusions drawn from the data provided in response to the DCI." The panel noted several concerns with the DCI (Data call in study, provided by Syngenta) including the use of a flow-through system, the appropriateness of the animal model, and concern that the population used in the study was a resistant strain: "The strain used in the DCI studies was apparently an insensitive strain. Panel members were concerned that this apparent insensitivity may have resulted in insensitivity of the apical endpoints to atrazine in general." The panel recommended that additional statistical analyses be applied, that histopathalogical results be evaluated by an independent laboratory, that metabolites of atrazine be examined, and that "studies with *X. laevis* be followed up with comparable studies using a North American species as soon as possible." Further, "Some panel members expressed concern that EPA completely rejected its own hypothesis based solely on the negative results of the DCI study" and concluded that "From the scientific perspective, the Panel agreed that the relevance of the uncertainties justifies the generation of additional data." (http://www.epa.gov/oscpmont/sap/meetings/2007/100907_mtg.htm#transcripts). Despite these concerns, the EPA concluded, "At this time, EPA believes that no additional testing is warranted to address this issue." (http://www.epa.gov/pesticides/reregistration/atrazine/atrazine_update.htm)

Moving the Science Forward

> For the EPA, Hayes' work is interesting but irrelevant to any decision to regulate the pesticide.
>
> Stephen Bradbury, US EPA; Oakland Tribune, Dec 3, 2006

Rather than focus my immediate attention on EPA recommendations and answering SYNGENTA's challenges at the time, I focused my research efforts on the remaining important scientific questions. In my mind, the issue of atrazine's effects on gonadal development was no longer an open question, even if there were population variation in the response or with variability in environmental parameters or study conditions. Several important questions superseded this one, however. Are the hermaphrodites produced by atrazine exposure males with ovaries or females with testes? And what was the ultimate outcome? Do animals that display hermaphroditism at metamorphosis remain hermaphrodites throughout life or do they transform into one sex or the other at sexual maturity? Further, what are the long-term effects on reproduction in exposed animals? Though SYNGENTA previously funded a study where they claimed no long-term effects of atrazine on adult males, their studies did not examine morphology at all, did not examine competition between atrazine-exposed males and control males, or allow female choice. Finally, their animals were treated with hormones prior to examinations of fertility (Du Preez et al. 2008) effectively providing hormone replacement therapy for the exposed animals.

We hypothesized it was more likely that atrazine affected genetic (ZZ) males only. *Xenopus laevis* is a female heterogametic species. Females are ZW and males are ZZ, therefore females determine the sex of the offspring . Applying estrogen to the rearing water of larvae will produce 100% females (Chang and Witschi 1955; Gallien 1953; Hayes 1997a, b; 1998; Villapando and Merchant-Larios 1990), but adding androgens, or any other steroid for that matter, will not alter the sex ratio (Hayes 1998). In other words, the W chromosome is dominant and environmental factors or exogenous hormones do not override the genetics of being female. Genetic males (ZZ), however, can be manipulated. In addition, it has now been revealed that the W chromosome in females carries at least one unique gene, DMW (Okada et al. 2009; Yoshimoto et al. 2008). As it turns out DMW is a transcription factor that induces aromatase expression, which leads to estrogen production and subsequent ovarian differentiation (Yoshimoto et al. 2008). Atrazine induces aromatase in fish (Suzawa and Ingraham 2008), reptiles (Crain et al. 1997; Keller and McClellan-Green 2004), and mammals (Fan et al. 2007a, b; Heneweer et al. 2004; Sanderson et al. 2000, 2001, 2002; Suzawa and Ingraham 2008) and results in increases in estrogen in fish (Spano et al. 2004), amphibians (Hayes et al. 2010b), and mammals (Stoker et al. 2000; Wetzel et al. 1994). Thus, a plausible mechanism for sex reversal of males by atrazine was available. Further, other studies suggested that atrazine induces complete feminization in not only *Xenopus laevis* (Oka et al. 2008) but also in *Rana pipiens* (McDaniel et al. 2008) and in zebra fish (*Danio rerio*) (Suzawa and Ingraham 2008). These studies showed a dose-dependent decrease in males, shifting the sex ratio in favor of females, but it was not known if the shift in the sex ratio was truly the result of sex reversal of genetic males.

To answer these questions, I took advantage of a population of *X. laevis* generated in my laboratory 18 years ago. A feral population of *X. laevis* originally collected in San Diego was treated with estradiol, resulting in all females. Sex-reversed genetic males (ZZ-females) were identified by crossing each animal back to unexposed ZZ males from the same population. Females that produced only male offspring were isolated and had been maintained in my laboratory since 1992. By crossing ZZ-females with ZZ males, we produced a population that contained only genetic male (ZZ) larvae. After exposing these larvae to atrazine, any hermaphrodites that were produced would have to be genetic males with ovaries and any females produced would have to be truly sex-reversed males (ZZ females).

This study revealed that atrazine indeed completely sex-reversed genetic males (Hayes et al. 2010b). At sexual maturity, 10% of the animals had protruding cloacae typical of females. Dissection or laparatomy revealed that these animals also had fully developed ovaries and vitellogenic eggs. These animals expressed aromatase and produced estrogen. These neo-females were capable of copulating with males and producing viable eggs. The majority of the remaining males treated with atrazine, were also demasculinized. These males suffered from suppressed testosterone levels and their androgen-dependent breeding glands were reduced, and sperm production suppressed. Further, atrazine-exposed males were unable to compete with control males for females and had severely reduced fertility when paired with females in the absence of control males (Hayes et al. 2010c).

As to what these laboratory studies tell us about effects in the real world it is difficult to say. Again our work (Hayes et al. 2002b, c) showed a strong correlation between atrazine contamination and testicular oocytes in the field, as did Reeder et al.'s study (1998). Similarly, McKoy et al. found demasculinized and feminized frogs in areas where atrazine was used in Florida (Mckoy et al. 2002, 2008) and a mesocosm study found that atrazine exposure resulted in sex ratios skewed toward females (McDaniel et al. 2008). Though they state differently, SYNGENTA's data showed an association between atrazine contamination and frogs with feminized gonads also (Murphy et al. 2006). Alone, these findings would still only represent correlations and could not establish a cause–effect relationship; however, together with controlled laboratory studies consistent with these field studies, the case for atrazine as a causative agent in the wild is strengthened.

> ...use of atrazine according to the label instructions will not likely result in harm to human health or the environment.
>
> Office of Pesticide Programs, US EPA (2008)

In addition to the effects on reproductive development and function in frogs, atrazine has a number of other important effects that could impact wild amphibians and contribute to population declines. The numerous nonreproductive effects of atrazine are best summarized in a qualitative meta-analysis published recently by Rohr and McKoy (2009). Rohr and McKoy reported that atrazine reduced size at metamorphosis in 15/17 studies and in 14/14 amphibians examined. They reported that atrazine increased activity in fish and amphibians in 12/13 studies and decreased predator avoidance behaviors or defense in 6/7. Such behavioral effects can have dramatic effects on amphibian survival. For example, Rohr et al. (2004) reported that atrazine increased activity (and thus energy expenditure), reduced shelter use, decreased the larval period, and reduced size at metamorphosis in exposed salamanders. Interactions with food limitations and drying conditions decreased the chances of survival for exposed individuals in this study, demonstrating the importance of examining atrazine exposure in combination with other stressors (Rohr et al. 2004). The work of Rohr et al. (2004), also demonstrated that the multiple effects of atrazine are quite complex and there are significant interactions when atrazine is applied to species communities (Rohr and Crumrine 2005). One must also consider the immediate and long-term ("carry over") effects of atrazine. Rohr et al. (2006), showed that not only did atrazine exposure induce mortality in the streamside salamander *Ambystoma barbouri*, but atrazine had a carry-over effect 14 months later, reducing the ability of the surviving salamanders to recover (Rohr et al. 2006). In a later study, Rohr and Palmer showed that atrazine resulted in a greater risk of mortality from water loss in salamander up to 8 months after exposure (Rohr and Palmer 2005). All of these well-documented effects have the potential to contribute to amphibian declines, but in addition likely impact other exposed wildlife as well.

In particular, atrazine's negative effects on immune function raise concern, given the focus on disease-driven mortality as a driving factor in amphibian decline. Rohr and McKoy report that atrazine decreased 33/43 immune function parameters and 13/16 infection endpoints. Atrazine exposure increases *Rana* virus infection in exposed salamanders (*Ambystoma tigrinum*) (Forson and Storfer 2006a, 2006b). Atrazine also increases trematode infections that result in limb deformities in exposed amphibians (Kiesecker 2002). In fact, Rohr et al. (2008), showed that atrazine was the best predictor (out of 240 factors examined) for trematode infections and that atrazine and phosphate accounted for most (74%) of the variation in trematode abundance. Further, atrazine, in combination with other pesticides is associated with decreased immune function in a number of studies in several species and including a wide range of disease pathogens (Bishop et al. 2010, Hayes et al. 2010a).

With regards to mechanisms, atrazine exposure alters gene expression of a number of genes, including genes involved in growth, development and immune function (Langerveld et al. 2009). It is also possible that atrazine (and other pesticides) decrease immune function by increasing stress hormone (corticosterone) levels. At least one study from our laboratory showed that a mixture of pesticides containing atrazine increases corticosterone levels (Hayes et al. 2006a). Studies in fish, show that atrazine increases the related glucocortioid, cortisol (Cericato et al. 2009; Nieves-Puigdoller et al. 2007).

> There is no direct scientific information to assess this hypothesis.
>
> Anne Lindsay, US EPA, before the Agriculture and Rural Development Committee of the Minnesota House of Representatives February 16, 2005

In addition to the multitude of effects of atrazine on amphibians, the number of taxa affected by atrazine also provides strong evidence for the impact of atrazine in the environment. Glen Fox wrote, "In ecoepidemiology, the occurrence of an association in more than one species and species population is *very strong* evidence for causation" (Fox 1991). Perhaps, then, the most compelling case against atrazine is the fact that similar effects are produced across not just species, or species populations, but across vertebrate classes.

The evidence for demasculinization is available from several studies. Atrazine decreases androgen production in fish and mammals, in addition to amphibians (references below). In addition, atrazine exposure causes a decrease in androgen-dependent development, morphology, and behavior in fish, reptiles, and mammals, similar to its demasculinizing effects in amphibians (references below). In salmon, atrazine caused a dose-dependent decrease in androgens, resulting in the absence of male reproductive behaviors (response to the female pheromone) and a decline in milt (semen) (Moore and Waring 1998). These effects occurred at exposure levels in the low ppb range and were similar to our recent reports in *Xenopus laevis*. Atrazine caused a decline in androgens and sperm production in caiman (*Caiman latirostris*) (Stoker et al. 2008). Atrazine-exposed caiman developed testes with reduced sperm content and histological appearance nearly identical to our observations in adult *X. laevis* (Hayes et al. 2010b). Similarly, atrazine demasculinizes laboratory rodents, consistent with findings in wildlife: atrazine causes a decline in testosterone in exposed rats, *Rattus norvegicus*, (Friedmann 2002; Stoker et al. 2000; Trentacoste et al. 2001). Atrazine also suppresses sperm production and fertility in rats (Friedmann 2002; Stoker et al. 2000; Trentacoste et al. 2001). EPA laboratories also showed that atrazine delayed puberty in male rats (Stoker et al. 2000), and in fact the EPA laboratories were among the first to conclude that atrazine is an endocrine disruptor, "ATR tested positive in the pubertal male screen that the Endocrine-Disrupter Screening and Testing Advisory Committee (EDSTAC) is considering as an optional screen for endocrine disrupters" (Stoker et al. 2000). Cross-generational effects have also been demonstrated in rodents: exposure of pregnant dams to atrazine reduced androgen levels in male pups. Male pups also had a decreased anal-genital distance and delayed preputial separation, both androgen-dependent aspects of development (Rosenberg et al. 2008)

In addition to atrazine's feminizing effects reported in amphibians, similar effects have been reported in every vertebrate examined, including fish, reptiles, birds, and mammals. Atrazine caused an increase in estrogens in goldfish (Spano et al. 2004). Suzawa and Ingraham showed that atrazine upregulated gonadal aromatase in zebrafish and that atrazine exposure resulted in a dose-dependent increase in females in exposed fish (Suzawa and Ingraham 2008). The latter effect was similarly reported in *X. laevis* (Oka et al. 2008) and *Rana pipiens* (McDaniel et al. 2008), but my laboratory was the first to show directly that atrazine completely feminizes truly genetic males (Hayes et al. 2010b).

Aromatase induction and estrogen production have also been shown in amniotes. The induction of aromatase by atrazine was first proposed in a study on alligators (*Alligator mississippiensis*) (Crain et al. 1997), but the effect reported here was not statistically significant. Atrazine induced aromatase in gonadal cells from turtles as well (Keller and McClellan-Green 2004). Fewer studies have been conducted in birds, but at least one study reported sex reversal effects on the gonads of chickens, possibly through aromatase induction (Matsushita et al. 2006).

There are supporting findings for the induction of estrogen synthesis and subsequent estrogenic effects in laboratory rodents. In addition to declines in circulating androgens, atrazine exposure through food results in an increase in circulating estrogens in exposed rats (Stoker et al. 2000). In female Sprague–Dawley rats, atrazine exposure results in increased mammary tumors (Eldridge et al. 1994; Stevens et al. 1994; Ueda et al. 2005), studies funded by SYNGENTA argue that the higher tumor incidence is rather an earlier onset (Eldridge et al. 1994). Nevertheless, the tumors are estrogen responsive (Ueda et al. 2005) and thus, likely caused by excess estrogen. This effect of atrazine does not occur in Fischer rats. Eldridge et al. (1994), argues that Sprague–Dawley (SD) rats have higher rates of mammary tumor anyway, and thus are abnormally sensitive to atrazine. However, this two taxon statement, does not negate the effects in the SD rat.

In addition, laboratory rodents show a number of other reproductive effects when exposed to atrazine that do not necessarily occur through alterations in aromatase expression and/or activity. Perhaps most concerning are a number of studies that have examined effects of atrazine on pregnant dams. Atrazine induced abortion in four strains of rats in a series of studies conducted in EPA laboratories (Cummings et al. 2000; Narotsky et al. 2001). The authors suggest that the effect is due to disturbance in the brain, hypothalamus and pituitary rather than direct effects on the gonads (Cooper et al. 1999, 2000). This same laboratory showed that male pups exposed in utero and perhaps through the dam's milk, developed prostatitis (Stoker et al. 1999). Yet another EPA laboratory showed cross-generational effects as well. Female pups exposed in utero suffered from severe inhibition of mammary growth (Rayner et al. 2004, 2005). Follow-up studies showed that even a second generation (not exposed directly) were affected (Rayner et al. 2004). At sexual maturity, the females exposed in utero lacked sufficient mammary growth to provide nutrients for their offspring. The second generation, though never exposed directly, displayed retarded growth and development as result of the effect on their mother's who were exposed in utero.

> Syngenta assumes no obligation to update forward-looking statements to reflect actual results.
>
> Sherry Ford, SYNGENTA Crop Protection, August, 2009,
> "Syngenta Cautionary Statement Regarding Forward-Looking Statements"

Although SYNGENTA's scientists disagree that atrazine is an endocrine disruptor, it is difficult to imagine how so many studies, conducted in laboratories around the world, in every vertebrate class examined, can coincidentally and wrongfully come to similar conclusions. Although the EPA has held two SAPs to evaluate the effects of atrazine on amphibians, their questions asked have been fairly narrow, focused on variability in the response of different amphibian population under varying environmental conditions and have never asked the fundamental question: "Is atrazine an endocrine disruptor in vertebrates?" This question would draw all of the data available from all vertebrates examined.

I have not stated the case for human health effects of atrazine related to its demasculinizing and feminizing effects here; however, data do exist. Atrazine detection in the urine of men in Missouri (≥0.1 ppb) was associated with decreased sperm count, semen quality, and low fertility (Swan et al. 2003). Men who work in agricultural fields applying atrazine can have levels up to 2,400 ppb atrazine in their urine (Lucas et al. 1993). In addition, men who worked in SYNGENTA's (NOVARTIS at the time) atrazine production facility in San Gabriel, Louisiana (so called "Cancer Alley"), experienced an eightfold increase in prostate cancer (Maclennan et al. 2002; Sass 2003). Though SYNGENTA has tried to downplay these findings excluding cases from the analysis that were identified after the study and claiming that the incidence of prostate cancer only appear higher because of their careful screening of their employees (Sass 2003), several features of their report are important and worth quoting here. The authors reported: "The increase in all cancers combined seen in the overall study group was concentrated in the company employee group." (page 1052); "The increase in prostate cancer in male subjects was concentrated in company employees" (page 1052); "The prostate cancer increase was further concentrated in actively working company employees" (page 1053); "all but one of the cases occurred in men with 10 or more years since hire" (page 1052); and "analyses restricted to company employees also found that the prostate cancer increase was limited to men under 60 years of age" (page 1053). The induction of prostatitis (Stoker et al. 1999) and prostate cancer (Pintér and al 1980) in laboratory rodents supports the findings in humans.

Atrazine also upregulates aromatase expression and estrogen production in several human cell lines (Fan et al. 2007a, b; Heneweer et al. 2004; Sanderson et al. 2000, 2001, 2002; Suzawa and Ingraham 2008). In fact, the strongest case for the induction of aromatase comes from these studies in human cells. The mechanism was initially proposed by Sanderson et al. (2000), who showed that atrazine inhibits phosphodiesterase (PDE). PDE regulates cAMP levels in the cytoplasm. By blocking PDE, the exposed cell experiences increased cAMP levels. cAMP regulates a number of phosphorylation events in the cell cytoplasm as well as binds to a receptor protein (cyclic-AMP response element binding protein: CREBP) that is translocated to the nucleus. After binding the promoter for the gene cyp19-aromatase, aromatase expression is enhanced. Interestingly, this regulatory pattern for aromatase is observed for aromatase expression in all vertebrate gonads and in fibroblasts associated with breast cancer. An increase in local estrogen production, could explain the increase in breast cancer associated with atrazine contamination of drinking water in at least one study (Kettles et al. 1997). Interestingly, NOVARTIS and ASTRAZENECA both market aromatase blockers as first-line treatments for breast cancer (Hayes 2009b).

Immune-suppressive effects of atrazine (discussed earlier) are not restricted to amphibians. In rodents, atrazine has a number of detrimental effects that have been more fully characterized than effects in amphibians. In mice, atrazine causes a decrease in white blood cells (Pruett et al. 2003). Atrazine increased neutrophils and T lymphocytes, and decreased lymphocytes, Natural Killer cells and B cells. Interestingly, atrazine also elevated corticosterone in this study (Pruett et al. 2003). In other studies, atrazine increased spleenic T cells, decreased cytotoxic T cell function and decreased mixed leukocyte responses. The result was a decrease in resistance to melanoma (Karrow et al. 2005). The thymus and spleen weight, splenic cell number, and macrophage function were reduced in mice exposed to atrazine (Karrow et al. 2005). Atrazine also inhibited lytic granule release, thus blocking Natural Killer cell function in mice (Rowe et al. 2006, 2007, 2008). Many of these effects are worsened when rodents are exposed during early development. For example, prenatal and lactational exposure reduced humoral and cell-mediated immune function in mice (Rowe et al. 2008). Depression of immune function was pronounced in females (Rowe et al. 2008), which is interesting because similar sex-specific effects on immune-suppression have been reported in frogs (Langerveld et al. 2009). Some of the effects may also involve the inhibition of dendritic cell maturation, via inhibition of several signaling molecules (Filipov et al. 2005).

> He's taking his information to people who don't have enough independent information to make a truly independent decision.
>
> Tim Pastoor, SYNGENTA, Minnesota Star Tribune 2005

The manufacturer's position, even if we accept the proposed mechanism of endocrine disruption, is that the effective doses in mammalian studies are far beyond what humans would be exposed to. The data, or a standard way to translate data from wildlife studies to human exposures, do not exist.

Environmental Relevance?

The effective doses used in amphibian and fish studies are in the low ppb range in most cases (0.1 ppb in our studies). Those levels are the concentrations applied to and measured in the rearing medium. For reptile studies (Crain et al. 1997; Stoker et al. 2008) and for birds (Matsushita et al. 2006), the atrazine levels represent the concentration of solutions painted onto or injected into the eggs, although Keller and McClellan-Green used cell lines exposed to atrazine in the incubation medium (Keller and McClellan-Green 2004). For the rodent studies, the effective levels reported are in the ppm range, but those are the levels added to the food, dissolved in the drinking water, or in some cases delivered by gavage to the test animals. There is no indication in the rodent studies of how much food or how much water was consumed by the test animals, so we have no idea how much atrazine was consumed by the test animals, or (even if we did) how to compare consumption rates with absorption from the rearing medium in the case of cell lines or aquatic organisms, or across an egg shell in the case of reptiles and birds. Perhaps constantly absorbing low ppb amounts across the skin, gills and digestive tract in fish and amphibians, is equivalent to periodic exposure to higher concentrations in consumed food and water or by gavage (Although it should be noted that atrazine is absorbed across the skin of humans and by inhalation but transfer is much more significant across amphibian skin). In human cell line studies, atrazine is also effective in the low ppb range (Fan et al. 2007b) similar to effects in fish and amphibians. The levels in those cell line studies reflect concentrations in the rearing medium, and thus might be most appropriately compared to amphibian and fish studies.

On the other hand, levels in humans are measured in the urine (not in the blood and not at the tissue level) (Barr et al. 2007; Lucas et al. 1993; Swan et al. 2003). Atrazine levels in human urine, can be in ppm and are probably underestimates due to the metabolite measured in those studies (Barr et al. 2007), but it is not clear how those exposures can be compared to levels in rearing medium or how urine levels can be used to estimate tissue level exposure. Further, though we know what concentrations were supplied in food and water in rodent studies, there are no blood or urine measurements from exposed rodents available to compare to urine levels in humans.

Two other commonly misunderstood and misrepresented issues regarding dose are at play, in addition to the fact that "doses" (or concentrations) used in aquatic organisms cannot be translated into doses used in rodent studies and concentrations to which wildlife and humans are exposed. First, the effective dose of atrazine in amphibians, fish and cell lines (low ppb), often referred to as "extremely low" is not low at all. Estrogens are also active at that dose range in amphibians. Depending on the population, estradiol is very effective below 1 ppb, feminizing male larvae in amphibians. Thus, although atrazine (and other pesticides) may not be toxic otherwise at such levels, they are potent endocrine disruptors. Secondly, it is a mistake to assume that rodents (smaller than humans) would respond to lower amounts of atrazine relative to humans: i.e., it is not true that an effect in the ppm range in a rodent would have to be scaled up to an enormous exposure to get the same effect in a larger animal, such as humans. The contrary is true. I am reminded of the tragic story of Tusko the elephant, given an overdose of LSD, because the researchers simply scaled up the amount of LSD based on human responses (West et al. 1962). In fact, within endotherms, smaller species, with higher metabolisms and variations in enzymes, would likely tolerate higher exposures not lower, relative to humans.

Summary and Conclusions

> The ultimate decision of whether or not to ban atrazine is much bigger than science...It weighs in public opinion.
>
> Stephen Bradbury (US EPA), Oakland Tribune, Dec 3, 2006

Regardless of the mechanism(s), the number of adverse effects of atrazine that have been shown across wildlife, in laboratory rodents, and associations in human epidemiological studies, demonstrate its significant impact on environmental health and public health. In addition to the many effects on wildlife and the effects of in utero exposure documented in rodents, increasing studies are examining the impacts of atrazine on the unborn human fetus. Atrazine and other agrichemicals are associated with birth defects (Winchester et al. 2009) and low birth weight and small for gestational age in humans (Ochoa-Acuna et al. 2009; Villanueva et al. 2005). Ultimately, for chemicals like atrazine, the question becomes a cost–benefit analysis. Are the health risks to the environment and humans worth the benefits? Some estimates suggest that atrazine increases corn yield by less than 1% (Ackerman 2007), others suggest no effect at all. Although it is commonly argued that chemicals in agriculture help produce food economically, less than 1.5% of the corn grown in the USA is directly consumed as food (http://usda.mannlib.cornell.edu/usda/current/FDS/FDS-11-12-2010.pdf), while 15% of the world's population is faced with starvation (http://www.fao.org/publications/sofi/en/), in a world where amphibian are declining globally (likely along with other vertebrates) with atrazine and other environmental contaminants likely key players (Bishop et al. 1999; Blaustein and Kiesecker 2002; Boone et al. 2007; Carey and Bryant 1995; Hayes et al. 2010a). It appears that the concerns of Sanderson et al (2000): "A logical concern would be that exposure of wildlife and humans to triazine herbicides, which are produced and used in large quantities, and are ubiquitous environmental contaminants, may similarly contribute to estrogen-mediated toxicities and inappropriate sexual differentiation." have been borne out, as predicted. Furthermore, the impact on human health remains a concern: "The observed induction of aromatase, the rate-limiting enzyme in the conversion of androgens to estrogens, may be an underlying explanation for some of the reported hormonal disrupting and tumor promoting properties of these herbicides in vivo." (Sanderson et al. 2000). Considering these concerns along with the many other mechanisms of action and effects produced by atrazine, has atrazine truly been used safely for over 50 years? Or have financial considerations masked a "no-brainer"?

References

Ackerman F (2007) The economics of atrazine. Int J Occup Environ Health 13:437–445

Barr DB, Panuwet P, Nguyen JV, Udunka S, Needham LL (2007) Assessing exposure to atrazine and its metabolites using biomonitoring. Environ Health Perspect 115:1474–1478

Bishop C, Mahony N, Struger J, Ng P, Petit K (1999) Anuran development, density, and diversity in relation to agricultural activity in the Holland River watershed, Ontario, Canda (1990–1992). Environ Monit Assess 57:21–43

Bishop C, McDaniel T, DeSolla S (2010) Atrazine: effects on amphibians and reptiles. In: Sparling DW, Linder GL, Krest S, Bishop CA (eds) The ecotoxicology of amphibians and reptiles. Society of Environmental Toxicology and Chemistry Press, Pensacola, FL, p 944

Blaustein A, Kiesecker J (2002) Complexity in conservation: lessons from the global decline of amphibian populations. Ecol Lett 5:597–608

Boone MD, Semlitsch RD, Little EE, Doyle MC (2007) Multiple stressors in amphibian communities: effects of chemical contamination, bullfrogs, and fish. Ecol Appl 17:291–301

Carey C, Bryant CJ (1995) Possible interactions among environmental toxicants, amphibian development and decline of amphibian populations. Environ Health Perspect 103:13–17

Carr J, Gentles A, Smith E, Goleman W, Urquidi L, Thuett K, Kendall R, Giesy J, Gross T, Solomon K et al (2003) Response of larval *Xenopus laevis* to atrazine: assessment of growth, metamorphosis, and gonadal and laryngeal morphology. Environ Toxicol Chem 22:396–405

Cericato L, Neto JGM, Kreutz LC, Quevedo RM, da Rosa JGS, Koakoski G, Centenaro L, Pottker E, Marqueze A, Barcellos LJG (2009) Responsiveness of the interrenal tissue of Jundia (Rhamdia quelen) to an in vivo ACTH test following acute exposure to sublethal concentrations of agrichemicals. Comp Biochem Physiol C Toxicol Pharmacol 149:363–367

Chang C, Witschi E (1955) Genic control and hormonal reversal of sex differentiation in *Xenopus*. Proc Soc Exp Biol Med 93:140–144

Coady KK, Murphy J, Villeneuve DL, Hecker MJ, Carr J, Solomon K, Van Der Kraak G, Smith E, Kendall RJ, Giesy JP (2005) Effects of atrazine on metamorphosis, growth, laryngeal and gonadal development, aromatase activity, and plasma sex steroid concentrations in *Xenopus laevis*. Ecotoxicol Environ Saf 62:160–173

Cooper RL, Stoker TE, McElroy WK (1999) Atrazine (ATR) disrupts hypothalamic catecholamines and pituitary function. Toxicologist 42:60–66

Cooper RL, Stoker TE, Tyrey L, Goldman JM, McElroy WK (2000) Atrazine disrupts the hypothalamic control of pituitary-ovarian function. Toxicol Sci 53:297–307

Crain D, Guillette LJ, Rooney AA, Pickford D (1997) Alterations in steroidogenesis in alligators (*Alligator mississippiensis*) exposed naturally and experimentally to environmental contaminants. Environ Health Perspect 105:528–533

Cummings A, Rhodes B, Cooper R (2000) Effect of atrazine on implantation and early pregnancy in 4 strains of rats. Toxicol Sci 58:135–143

Du Preez L, Solomon K, Jooste A, Jansen G, van Rensburg P, Smith E, Carr RJ, Kendall GJP, Gross T, et al. (2002) Exposure characterization and responses to field exposures of *Xenopus laevis* to atrazine and related triazines in South African corn growing regions. In 23 rd Annual Meeting in North America, Soc. Environ. Toxicol. Chem. Salt Lake City, UT

Du Preez LH, Rensburg PJJV, Jooste AM, Carr JA, Giesy JP, Gross TS, Kendall RJ, Smith EE, Kraak GVD, Solomon KR (2005a) Seasonal exposures to triazine and other pesticides in surface waters in the western Highveld corn-production region in South Africa. Environ Pollut 135:131–141

Du Preez LH, Solomon K, Carr J, Giesy J, Gross C, Kendall RJ, Smith E, Van Der Kraak G, Weldon C (2005b) Population structure characterization of the clawed frog (*Xenopus laevis*) in corn-growing versus non-corn-growing areas in South Africa. Afr J Herp 54:61–68

Du Preez LH, Kunene N, Everson GJ, Carr JA, Giesy JP, Gross TS, Hosmer AJ, Kendall RJ, Smith EE, Solomon KR et al (2008) Reproduction, larval growth, and reproductive development in African clawed frogs (*Xenopus laevis*) exposed to atrazine. Chemosphere 71:546–552

Eldridge J, Tennant M, Wetzel L, Breckenridge C, Stevens J (1994) Factors affecting mammary tumor incidence in chlorotriazine-treated female rats: Hormonal properties, dosage, and animal strain. Environ Health Perspect 102:29–36

Fan W, Yanase T, Morinaga H, Gondo S, Okabe T, Nomura M, Hayes TB, Takayanagi R, Nawata H (2007a) Herbicide atrazine activates SF-1 by direct affinity and concomitant co-activators recruitments to induce aromatase expression via promoter II. Biochem Biophys Res Commun 355:1012–1018

Fan W, Yanase T, Morinaga H, Gondo S, Okabe T, Nomura M, Komatsu T, Morohashi K-I, Hayes T, Takayanagi R et al (2007b) Atrazine-induced aromatase expression is SF-1- dependent: implications for endocrine disruption in wildlife and reproductive cancers in humans. Environ Health Perspect 115:720–727

Filipov N, Pinchuk L, Boyd B, Crittenden P (2005) Immunotoxic effects of short-term atrazine exposure in young male C57BL/6 mice. Toxicol Sci 86:324–332

Forson D, Storfer A (2006a) Atrazine increases *Ranavirus* susceptibility in the tiger salamander, *Ambystoma tigrinum*. Ecol Appl 16:2325–2332

Forson D, Storfer A (2006b) Effects of atrazine and iridovirus infection on survival and lifehistory traits of the long-toed salamander (*Ambystoma macrodatylum*). Environ Toxicol Chem 25:168–173

Fox G (1991) Practical causal inference for epidemiologists. J Toxicol Environ Health 33:359–373

Friedmann A (2002) Atrazine inhibition of testosterone production in rat males following peripubertal exposure. Reprod Toxicol 16:275–279

Gallien L (1953) Inversion totale du sexe chez *Xenopus laevis* Daud. À la suite d'un traitment gynogène par le benzoate of oestradiol, administré pendant la vie larvaire. C R Hebd Seances Acad Sci 237:1565–1566

Hayes TB (1997a) Steroid-mimicking environmental contaminants: their potential role in amphibian declines. Herpetologia Bonnensis. SEH Bonn., 145–150

Hayes TB (1997b) Steroids as modulators of thyroid hormone activity in amphibian development. Am Zool 37:185–195

Hayes TB (1998) Sex determination and primary sex differentiation in amphibians. J Exp Zool 281:373–399

Hayes T (2000a) Hyperolius argus endocrine screen test. In: Official Gazette of the United States Patent and Trademark Office Patents, USA

Hayes TB (2000b) Hyperolius argus endocrine screen test. In: Official Gazette of the United States Patent and Trademark Office Patents, USA

Hayes TB (2004) There is no denying this: defusing the confusion about atrazine. Bioscience 54:1138–1149

Hayes TB (2005) Comment on "Gonadal development of larval male *Xenopus laevis* exposed to atrazine in outdoor microcosms". Environ Sci Technol 39:7757–7758

Hayes TB (2009a) More feedback on whether atrazine is a potent endocrine disruptor chemical. Environ Sci Technol 43:6115

Hayes TB (2009b) The one stop shop: chemical causes and cures for cancer, President's Cancer Panel – October 21, 2008. Rev Environ Health 24:297–307

Hayes TB, Menendez K (1999) The effect of sex steroids on primary and secondary sex differentiation in the sexually dichromatic reedfrog (*Hyperolius argus*: Hyperolidae) from the Arabuko Sokoke Forest of Kenya. Gen Comp Endocrinol 115:188–199

Hayes TB, Collins A, Lee M, Mendoza M, Noriega N, Stuart AA, Vonk A (2002a) Hermaphroditic, demasculinized frogs after exposure to the herbicide atrazine at low ecologically relevant doses. Proc Natl Acad Sci USA 99:5476–5480

Hayes TB, Haston K, Tsui M, Hoang A, Haeffele C, Vonk A (2002b) Atrazine-induced hermaphroditism at 0.1 ppb in American leopard frogs (*Rana pipiens*): Laboratory and field evidence. Environ Health Perspect 111:568–575

Hayes TB, Haston K, Tsui M, Hoang A, Haeffele C, Vonk A (2002c) Feminization of male frogs in the wild. Nature 419:895–896

Hayes TB, Case P, Chui S, Chung D, Haefele C, Haston K, Lee M, Mai V-P, Marjuoa Y, Parker J et al (2006a) Pesticide mixtures, endocrine disruption, and amphibian declines: are we underestimating the impact? Environ Health Perspect 114:40–50

Hayes TB, Stuart A, Mendoza G, Collins A, Noriega N, Vonk A, Johnston G, Liu R, Kpodzo D (2006b) Characterization of atrazine-induced gonadal malformations and effects of an androgen antagonist (cyproterone acetate) and exogenous estrogen (estradiol 17b): support for the demasculinization/feminization hypothesis. Environ Health Perspect 114:134–141

Hayes TB, Falso P, Gallipeau S, Stice MJ (2010a) The cause of global amphibian declines: a developmental endocrinologist's perspective. J Exp Biol 213:921–933

Hayes TB, Khoury V, Narayan A, Nazir M, Park A, Brown T, Adame L, Chan E, Buchholz D, Stueve T et al (2010b) Atrazine induces complete feminization and chemical castration in male African clawed frogs (*Xenopus laevis*). Proc Natl Acad Sci USA 107:4612–4617

Hecker MJ, Coady KK, Villeneuve DL, Murphy MB, Jones PD, Giesy JP (2003) Response of Xenopus laevis to atrazine exposure: assessment of the mechanism of action of atrazine

Hecker MJ, Giesy JP, Jones P, Jooste AM, Carr J, Solomon KR, Smith EE, Van Der Kraak G, Kendall RJ, Du Preez LH (2004) Plasma sex steroid concentrations and gonadal aromatase activities in African clawed frogs (*Xenopus laevis*) from South Africa. Environ Toxicol Chem 23:1996–2007

Hecker M, Kim W, Park J-W, Murphy M, Villeneuve D, Coady K, Jones P, Solomon K, Van Der Kraak G, Carr J et al (2005) Plasma concentrations of estradiol and testosterone, gonadal aromatase activity and ultrastructure of the testis in *Xenopus laevis* exposed to estradiol or atrazine. Aquat Toxicol 72:383–396

Heneweer M, van den Berg M, Sanderson J (2004) A comparison of human H295R and rat R2C cell lines as *in vitro* screening tools for effects on aromatase. Toxicol Lett 146:183–194

Hosmer A, Kloas W, Lutz I, Springer T, Wolf J, Holden L (2007). Atrazine. Response of larval Xenopus laevis to atrazine exposure: assessment of metamorphosis and gonadal morphology. Final Report. Conducted by the Leibniz Institute of Freshwater Biology and Inland Fisheries (IGB), Wildlife International, Ltd., and Experimental Pathology laboratories, Inc. Sponsor: Syngenta Crop Protection. Unpublished. (MRID 471535–01)

Jooste A, Du Preez L, Carr J, Giesy JP, Gross T, Kendall R, Smith E, Van Der Kraak G, Solomon K (2005a) Response to comment on "Gonadal development of larval male *Xenopus laevis* exposed to atrazine in outdoor microcosms". Environ Sci Technol 39:5255–5261

Jooste AM, Du Preez LH, Carr J, Giesy JP, Gross C, Kendall RJ, Smith E, Van Der Kraak G, Solomon KR (2005b) Gonadal development of larval male *Xenopus laevis* exposed to atrazine in outdoor microcosms. Environ Sci Technol 39:5255–5261

Karrow N, McCay J, Brown R, Musgrove D, Guo T, Germolec D, White KJ (2005) Oral exposure to atrazine modulates cell-mediated immune function and decreases host resistance to the B16F10 tumor model in female B6C3F1 mice. Toxicology 209:15–28

Keller J, McClellan-Green P (2004) Effects of organochlorine compounds on cytochrome P450 aromatase activity in an immortal sea turtle cell line. Mar Environ Res 58:347–351

Kettles MA, Browning SR, Prince TS, Hostman SW (1997) Triazine exposure and breast cancer incidence: an ecologic study of Kentucky counties. Environ Health Perspect 105:1222–1227

Kiesecker J (2002) Synergism between trematode infection and pesticide exposure: a link to amphibian limb deformities in nature? Proc Natl Acad Sci USA 99:9900–9904

Kloas W, Lutz I, Springer T, Krueger H, Wolf J, Holden L, Hosmer A (2009a) Does atrazine influence larval development and sexual differentiation in *Xenopus laevis*? Toxicol Sci 107:376–384

Kloas W, Lutz I, Urbatzka R, Springer T, Krueger H, Wolf J, Holden L, Hosmer A (2009b) Does atrazine affect larval development and sexual differentiation of South African clawed frogs? Trends Comp Endocrinol Neurobiol 1163:437–440

Kloas W, Lutz I, Urbatzka R, Springer T, Krueger H, Wolf J, Holden L, Hosmer A (2009c) Does atrazine affect larval development and sexual differentiation of South African clawed frogs? (vol 1163, 437, 2009). Natural Compounds and Their Role in Apoptotic Cell Signaling Pathways vol. 1171, 660

Langerveld AJ, Celestine R, Zaya R, Mihalko D, Ide CF (2009) Chronic exposure to high levels of atrazine alters expression of genes that regulate immune and growth-related functions in developing Xenopus laevis tadpoles. Environ Res 109:379–389

Langlois V, Carew A, Pauli B, Wade M, Cooke G, Trudeau V (2010) Low levels of the herbicide atrazine alters sex ratios and reduces metamorphic success in *Rana pipiens* tadpoles raised in outdoor mesocosms. Environ Health Perspect 118:552–557

Lucas A, Jones A, Goodrow M, Saiz S, Blewett C, Seiber J, Hammock B (1993) Determination of atrazine metabolites in human urine: development of a biomarker of exposure. Chem Res Toxicol 6:107–116

Maclennan P, Delzell E, Sathiakumar N, Myers S, Cheng H, Grizzle W, Chen V, Wu X (2002) Cancer incidence among triazine herbicide manufacturing workers. J Occup Environ Med 44:1048–1058

Matsushita S, Yamashita J, Iwasawa T, Tomita T, Ikeda M (2006) Effects of *in ovo* exposure to imazalil and atrazine on sexual differentiation in chick gonads. Poult Sci 85:1641–1647

McCoy KA, Bortnick LJ, Campbell CM, Hamlin HJ, Guillette LJ, St Mary CM (2008) Agriculture alters gonadal form and function in the toad *Bufo marinus*. Environ Health Perspect 116:1526–1532

McDaniel TV, Martin PA, Struger J, Sherry J, Marvin CH, McMaster ME, Clarence S, Tetreault G (2008) Potential endocrine disruption of sexual development in free ranging male northern leopard frogs (*Rana pipiens*) and green frogs (*Rana clamitans*) from areas of intensive row crop agriculture. Aquat Toxicol 90:82

Mckoy KA, Sepulveda MS, Gross TS (2002) Atrazine exposure and reproductive system abnormalities in field collected *Bufo marinus*. In Soc. Environ. Toxicol. Chem., 23 rd Annual Meeting in North America, Salt Lake City, UT

Moore A, Waring C (1998) Mechanistic effects of a triazine pesticide on reproductive endocrine function in mature male Atlantic salmon (*Salmo salar* L.) parr. Pestic Biochem Physiol 62:41–50

Murphy MB, Hecker M, Coady KK, Tompsett AR, Jones PD, Du Preez LH, Everson GJ, Solomon KR, Carr JA, Smith EE et al (2006) Atrazine concentrations, gonadal gross morphology and histology in ranid frogs collected in Michigan agricultural areas. Aquat Toxicol 76:230–245

Narotsky M, Best DS, Guidici DL, Cooper RL (2001) Strain comparisons of atrazine-induced pregnancy loss in the rat. Reprod Toxicol 15:61–69

Nieves-Puigdoller K, Bjornsson BT, McCormick SD (2007) Effects of hexazinone and atrazine on the physiology and endocrinology of smolt development in Atlantic salmon. Aquat Toxicol 84:27–37

Noriega N, Hayes TB (2000) DDT congener effects on secondary sex coloration in the reed frog *Hyperolius argus*: a partial evaluation of the *Hyperolius argus* estrogen screen. Comp Biochem Physiol B Biochem Mol Biol 126B:231–237

Ochoa-Acuna H, Frankenberger J, Hahn L, Carbajo C (2009) Drinking-water herbicide exposure in Indiana and prevalence of small-for-gestational-age and preterm delivery. Environ Health Perspect 117:1619–1624

Oka T, Tooi O, Mitsui N, Miyahara M, Ohnishi Y, Takase M, Kashiwagi A, Shinkai T, Santo N, Iguchi T (2008) Effect of atrazine on metamorphosis and sexual differentiation in *Xenopus laevis*. Aquat Toxicol 87:215–226

Okada E, Yoshimoto S, Ikeda N, Kanda H, Tamura K, Shiba T, Takamatsu N, Ito M (2009) Xenopus W-linked DM-W induces Foxl2 and Cyp19 expression during ovary formation. Sex Dev 3:38–42

Orton F, Carr J, Handy R (2006) Effects of nitrate and atrazine on larval development and sexual differentiation in the northern leopard frog *Rana pipiens*. Environ Toxicol Chem 25:65–71

Parshley T (2000) Report of an alleged adverse effect from atrazine: Atrazine technical, EPA, Reg. No. 100–529: Environmental Protection Agency

Pintér A et al (1980) Long-term carcinogenecity bioassay of the herbicide atrazine in F344 rats. Neoplasma 37:533–544

Preez LHD, Kunene N, Hanner R, Giesy JP, Solomon KR, Hosmer A, Kraak GJVD (2009) Population-specific incidence of testicular ovarian follicles in Xenopus laevis from South Africa: a potential issue in endocrine testing. Aquat Toxicol 95:10–16

Pruett S, Fan R, Zheng Q, Myers L, Hebert P (2003) Modeling and predicting immunological effects of chemical stressors: Characterization of a quantitative biomarker for immunological changes caused by atrazine and ethanol. Toxicol Sci 75:343–354

Rayner J, Wood C, Fenton S (2004) Exposure parameters necessary for delayed puberty and mammary gland development in Long–Evans rats exposed in utero to atrazine. Toxicol Appl Pharmacol 195:23–34

Rayner J, Enoch R, Fenton S (2005) Adverse effects of prenatal exposure to atrazine during a critical period of mammary gland growth. Toxicol Sci 87:255–266

Reeder A, Foley G, Nichols D, Hansen L, Wikoff B, Faeh S, Eisold J, Wheeler M, Warner R, Murphy J et al (1998) Forms and prevalence of intersexuality and effects of environmental contaminants on sexuality in cricket frogs (*Acris crepitans*). Environ Health Perspect 106:261–266

Relyea RA (2010) Amphibians are not ready for Roundup

Rohr J, Crumrine P (2005) Effects of an herbicide and an insecticide on pond community structure and processes. Ecol Appl 15:1135–1147

Rohr JR, Mckoy KA (2010) A qualitative meta-analysis reveals consistent effects of atrazine on freshwater fish and amphibians. Environ Health Perspect 118:20–32

Rohr J, Palmer B (2005) Aquatic herbicide exposure increases salamander desiccation risk eight months later in a terrestrial environment. Environ Toxicol Chem 24:1253–1258

Rohr J, Elskus A, Shepherd B, Crowley P, McCarthy T, Niedzwiecki J, Sager T, Sih A, Palmer B (2004) Multiple stressors and salamanders: Effects of an herbicide, food limitation, and hydroperiod. Ecol Appl 14:1028–1040

Rohr JR, Sager T, Sesterhenn TM, Palmer BD (2006) Exposure, postexposure, and density-mediated effects of atrazine on amphibians: breaking down net effects into their parts. Environ Health Perspect 114:46–50

Rohr JR, Schotthoefer AM, Raffel TR, Carrick HJ, Halstead N, Hoverman JT, Johnson CM, Johnson LB, Lieske C, Piwoni MD et al (2008) Agrochemicals increase trematode infections in a declining amphibian species. Nature 455:1235–1239

Rosenberg BG, Chen HL, Folmer J, Liu J, Papadopoulos V, Zirkin BR (2008) Gestational exposure to atrazine: effects on the postnatal development of male offspring. J Androl 29:304–311

Rowe A, Brundage K, Schafer R, Barnett J (2006) Immunomodulatory effects of maternal atrazine exposure on male Balb/c mice. Toxicol Appl Pharmacol 214:69–77

Rowe AM, Brundage KM, Barnett JB (2007) In vitro atrazine-expo sure inhibits human natural killer cell lytic granule release. Toxicol Appl Pharmacol 221:179–188

Rowe AM, Brundage KM, Barnett JB (2008) Developmental immunotoxicity of atrazine in rodents. Basic Clin Pharmacol Toxicol 102:139–145

Sanderson JT, Seinen W, Giesy JP, van den Berg M (2000) 2-chloro-triazine herbicides induce aromatase (CYP19) activity in H295R human adrenocortical carcinoma cells: A novel mechanism for estrogenicity? Toxicol Sci 54:121–127

Sanderson JT, Letcher RJ, Heneweer M, Giesy JP, van den Berg M (2001) Effects of chloro-s-triazine herbicides and metabolites on aromatase activity in various human cell lines and on vitellogenin production in male carp hepatocytes. Environ Health Perspect 109:1027–1031

Sanderson J, Boerma J, Lansbergen G, van den Berg M (2002) Induction and inhibition of aromatase (CYP19) activity by various classes of pesticides in H295R human adrenocortical carcinoma cells. Toxicol Appl Pharmacol 182:44–54

Sass J (2003) Letter to the editor. J Occup Environ Med 45:1–2

Solomon K, Baker D, Richards R, Dixon K, Klaine S, LaPoint T, Kendall R, Weisskopf C, Giddings J, Giesy J et al (1996) Ecological risk assessment of atrazine in North American surface waters. Environ Toxicol Chem 15:31–76

Solomon KR, Carr JA, Du Preez LH, Giesy JP, Kendall RJ, Smith EE, Van Der Kraak GJ (2008) Effects of atrazine on fish, amphibians, and aquatic reptiles: a critical review. Crit Rev Toxicol 38:721–772

Spano L, Tyler C, van Aerle R, Devos P, Mandiki S, Silvestre F, Thome J-P, Kestemont P (2004) Effects of atrazine on sex steroid dynamics, plasma vitellogenin concentration and gonad development in adult goldfish (*Carassius auratus*). Aquat Toxicol 66:369–379

Stevens J, Breckenridge C, Wetzel L, Gillis JH, Luempert L III, Eldridge JC (1994) Hypothesis for mammary tumorigenesis in Sprague-Dawley rats exposed to certain triazine herbicides. J Toxicol Environ Health 43:139–154

Stoker TE, Robinette CL, Cooper RL (1999) Maternal exposure to atrazine during lactation suppresses suckling-induced prolactin release and results in prostatitis in the adult offspring. Toxicol Sci 52:68–79

Stoker T, Laws S, Guidici D, Cooper R (2000) The effect of atrazine on puberty in male Wistar rats: an evaluation in the protocol for the assessment of pubertal development and thyroid function. Toxicol Sci 58:50–59

Stoker C, Beldomenico PM, Bosquiazzo VL, Zayas MA, Rey F, Rodriguez H, Munoz-de-Toro M, Luque EH (2008) Developmental exposure to endocrine disruptor chemicals alters follicular dynamics and steroid levels in Caiman latirostris. Gen Comp Endocrinol 156:603–612

Suzawa M, Ingraham H (2008) The herbicide atrazine activates endocrine gene networks via non-steroidal NR5A nuclear receptors in fish and mammalian cells. PLoS One 3:2117

Swan S, Kruse R, Liu F, Barr D, Drobnis E, Redmon J, Wang C, Brazil C, Overstreet J (2003) Semen quality in relation to biomarkers of pesticide exposure. Environ Health Perspect 111:1478–1484

Tavera-Mendoza L, Ruby S, Brousseau P, Fournier M, Cyr D, Marcogliese D (2002a) Response of the amphibian tadpole (*Xenopus laevis*) to atrazine during sexual differentiation of the testis. Environ Toxicol Chem 21:527–531

Tavera-Mendoza L, Ruby S, Brousseau P, Fournier M, Cyr D, Marcogliese D (2002b) Response of the amphibian tadpole *Xenopus laevis* to atrazine during sexual differentiation of the ovary. Environ Toxicol Chem 21:1264–1267

Trentacoste S, Friedmann A, Youker R, Breckenridge C, Zirkin B (2001) Atrazine effects on testosterone levels and androgen-dependent reproductive organs in peripubertal male rats. J Androl 22:142–148

Ueda M, Imai T, Takizawa T, Onodera H, Mitsumori K, Matsui T, Hirose M (2005) Possible enhancing effects of atrazine on growth of 7,12-dimethylbenz(a) anthracene induced mammary tumors in ovariectomized Sprague–Dawley rats. Cancer Sci 96:19–25

Villanueva CM, Durand G, Coutte MB, Chevrier C, Cordier S (2005) Atrazine in municipal drinking water and risk of low birth weight, preterm delivery, and small-for-gestational-age status. Occup Environ Med 62:400–405

Villapando I, Merchant-Larios H (1990) Determination of the sensitive stages for gonadal sex-reversal in *Xenopus laevis* tadpoles. Int J Dev Biol 34:281–285

West LJ, Pierce CM, Thomas WD (1962) Lysergic acid diethylamide – its effects on a male Asiatic elephant. Science 138:1100–1103

Wetzel LT, Luempert LG III, Breckenridge CB, Tisdel MO, Stevens JT, Thakur AK, Extrom PJ, Eldridge JC (1994) Chronic effects of atrazine on estrus and mammary gland formation in female Sprague-Dawley and Fischer-344 rats. J Toxicol Environ Health 43:169–182

Winchester PD, Huskins J, Ying J (2009) Agrichemicals in surface water and birth defects in the United States. Acta Paediatr 98:664–669

Yoshimoto S, Okada E, Umemoto H, Tamura K, Uno Y, Nishida-Umehara C, Matsuda Y, Takamatsu N, Shiba T, Ito M (2008) A W-linked DM-domain gene, DM-W, participates in primary ovary development in *Xenopus laevis*. Proc Natl Acad Sci USA 105:2469–2474

Chapter 11
Selenium, Salty Water, and Deformed Birds

Harry M. Ohlendorf

Abstract Selenium was identified in the 1930s as the cause of embryo mortality and severe embryo deformities when chickens were fed grains grown on seleniferous soils in South Dakota. There had been no documented occurrences of such effects in wild birds before 1983, when we studied the effects of agricultural irrigation drainage water contaminants on birds feeding and nesting at Kesterson Reservoir, located within the Kesterson National Wildlife Refuge in the San Joaquin Valley of California. The Reservoir was used for disposal of subsurface saline drainage waters from agricultural fields and was intended to provide beneficial habitat for wildlife, particularly waterfowl and other aquatic birds. Analyses of food-chain biota (plants, aquatic invertebrates, and fish) and bird tissues or eggs showed that selenium was the only chemical found at concentrations high enough to cause the observed adverse effects on bird health or reproduction. Results of the studies at Kesterson Reservoir stimulated interest and concern about the effects of selenium in agricultural drainage throughout the western USA where similar scenarios might exist (as well as in industrial settings such as mining and power generation). Those studies showed that selenium-related problems with agricultural drainage were widespread and locally significant. Problems of managing selenium in agricultural drainwater are difficult to solve or mitigate, despite intensive efforts to do so. This chapter briefly describes the field and laboratory studies that documented the effects of selenium in birds using wetlands receiving seleniferous agricultural drainage, the linkages between those studies and subsequent efforts to address the issue of selenium contamination in agricultural drainage water, and the consequent conservation gains.

H.M. Ohlendorf (✉)
CH2M Hill, Sacramento, CA, 95833-2937, USA
e-mail: Harry.Ohlendorf@CH2M.com

J.E. Elliott et al. (eds.), *Wildlife Ecotoxicology: Forensic Approaches*,
Emerging Topics in Ecotoxicology 3, DOI 10.1007/978-0-387-89432-4_11,

Introduction

In the 1930s, selenium was identified as the cause of embryo mortality and severe developmental abnormalities ("deformities" or "terata") in chickens (*Gallus domesticus*) that were fed grains grown on seleniferous (i.e., high-selenium) soils in South Dakota (Franke and Tully 1935). Documented occurrences of such effects in wild birds were not known until 1983, when my US Fish and Wildlife Service (USFWS) colleagues and I studied the effects of agricultural irrigation drainage water contaminants on birds feeding and nesting at Kesterson Reservoir, located in the San Joaquin Valley of California. At that time, the Reservoir consisted of a series of 12 ponds (totaling about 500 ha) within the Kesterson National Wildlife Refuge (NWR) that were used for disposal of subsurface saline drainage waters from agricultural fields. The studies initially considered various inorganic and organic chemicals that were known or suspected to occur in the wastewater. However, analyses of food-chain biota (plants, aquatic invertebrates, and fish) and bird tissues or eggs showed that selenium was the only chemical found at concentrations high enough to cause the observed adverse effects on bird health or reproduction.

The goal of this chapter is to briefly describe the field and laboratory studies that documented the effects of selenium in birds using wetlands receiving seleniferous agricultural drainage, the linkages between those studies and subsequent efforts to address the issue of selenium contamination in agricultural drainage water, and the consequent conservation gains. I also briefly describe some of my personal experiences (and the consequences) of having done studies related to selenium and agricultural drainage waters at Kesterson Reservoir where the saga began.

The references cited in this chapter and other summaries (e.g., Ohlendorf 1989, 1996, 2002, 2003; Ohlendorf and Hothem 1995; Heinz 1996; Ohlendorf and Heinz 2011) provide more details about the studies of birds mentioned in this chapter. Other studies at Kesterson Reservoir focused on fish exposed to selenium in agricultural drainage waters (see, e.g., Saiki 1986; Saiki and Lowe 1987; Saiki and Ogle 1995; Saiki and Schmitt 1985; Hamilton et al. 1990) and are not described in this chapter, because my main focus was on birds and their food-web exposures to selenium.

Background

After serving as the Assistant Director of the USFWS Patuxent Wildlife Research Center in Maryland from 1973 to 1980, I had an opportunity to open a new field research station under the Center on the campus of the University of California, Davis. During 1971 to 1973, I had conducted research on the effects of environmental contaminants on wildlife as a Wildlife Research Biologist at the Center and continued to do so as the Assistant Director. Because I was more interested in returning to life as a full-time researcher than in continuing up the administrative line within the USFWS, I voluntarily took a grade reduction and moved west. It was expected that one focus of the new field research station would be a study of the effects of contaminants in saline agricultural drainage at the Kern NWR (located near historic Tulare Lake in the southern San Joaquin Valley of California), where plans had been made to construct experimental ponds to evaluate salinity management practices for wetlands. My plans were to add to the overall evaluation of those saline wetlands by conducting studies of the other constituents in drainwater that could affect health or reproduction of birds using the wetlands. However, the Kesterson Reservoir events (described below) arose before the plans at Kern NWR could be implemented, and those studies were never conducted.

During the 1970s and 1980s, there was much interest in the potential for using agricultural drainage water for wetland management in California's Central Valley. Most of the Valley's wetlands had been altered by flood control, drainage, or water diversion projects, and by agricultural development (Gilmer et al. 1982; USFWS 1982). Those wetlands that persisted had inadequate water supplies, and the problem was expected to worsen as more water was needed for agriculture.

In their review of published and unpublished literature, Jones and Stokes (1977) suggested it might be feasible to use irrigation drainage water to create new or restore former wetland habitats. However, they recognized the potential negative consequences for fish and wildlife resources from using low-quality water (because of high salinity and potential for pesticides and trace elements in the water) for marsh management, and the need for a comprehensive assessment. The harmful direct and indirect effects of environmental contaminants (such as pesticides and trace elements that may be leached from the soil) on fish and wildlife populations were among the significant unknowns.

Natural drainage (specifically, the downward percolation of applied irrigation water through the plant root zone) is inadequate for long-term crop production in some western portions of the San Joaquin Valley because saline water accumulates in the plant root zone (IDP 1979; Hanson 1982). To maintain productivity on those irrigated lands it is necessary to install subsurface drains to collect shallow saline groundwater (which contains salts imported with the irrigation water, but also contains selenium leached from soil) and carry it away from the fields. However, disposal of the drainage water then becomes a problem, so installation of the required subsurface drains was limited by the amount of subsurface drainage that could be discharged. To solve that problem, the US Bureau of Reclamation (Reclamation or USBR) received authorization to construct the San Luis Drain, a concrete-lined canal that was intended to carry the drainwater to the San Francisco Bay Estuary (also referred to as the Sacramento/San Joaquin Delta) for discharge (Fig. 11.1). Construction of the Drain was discontinued in 1975, however, after about 40% of it had been completed. The terminus of the Drain was at Kesterson Reservoir, which was intended to be one of the "regulating reservoirs" along the Drain where water would be stored until it could be discharged into the Estuary during high winter flows from the Sacramento and San Joaquin rivers. At Kesterson Reservoir, the water was impounded in 12 shallow ponds for evaporation but it also was intended to provide beneficial habitat for wildlife. Based on that anticipated beneficial use of the water, USFWS and Reclamation had signed a cooperative agreement in July 1970 for the management of Kesterson Reservoir and associated lands for

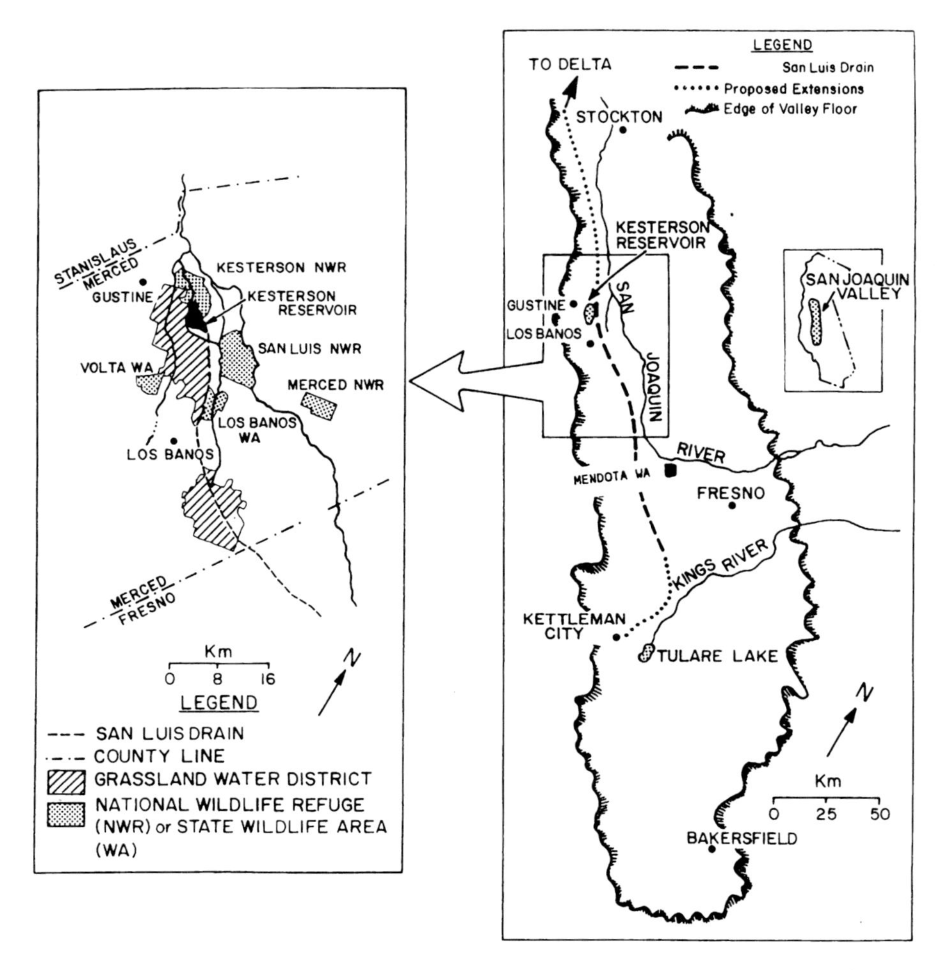

Fig. 11.1 Location of Kesterson Reservoir, Volta Wildlife Area, the Grasslands, and other features in the San Joaquin Valley, California

the conservation and management of wildlife under authority of the Fish and Wildlife Coordination Act.

The California State Water Resources Control Board (State Board) required that Reclamation complete several studies of hydrology, water quality, and potential effects on fisheries resources at the proposed discharge site) before it would approve agricultural drainwater discharge to the Estuary, but there was no requirement from the State Board for studies of wildlife using Kesterson Reservoir or the planned discharge site in the Estuary. However, the USFWS was concerned about the potential effects of chemicals in the drainwater on fish and wildlife, so the agency recommended to Reclamation that studies should be conducted at Kesterson Reservoir, the proposed discharge site, and a reference site near Kesterson that did not receive subsurface agricultural drainwater (Volta Wildlife Area; see Fig. 11.1). Because there was no requirement from the State Board for studies of wildlife at Kesterson Reservoir or the planned discharge site, Reclamation declined to fund the proposed studies, and the USFWS undertook the studies with its own funding.

From 1971 to 1977, inflow to Kesterson Reservoir was fresh water, including a mixture of irrigation supply water and surface runoff from agricultural fields that provided high-quality wetland habitat for birds and other wildlife. Increasing proportions of the flow during 1978 to 1980 were from subsurface drains. During 1971 until about 1980, the San Luis Drain supported fish such as striped bass (*Morone saxatilis*), largemouth bass (*Micropterus salmoides*), green sunfish (*Lepomis cyanellus*), and channel catfish (*Ictalurus punctatus*). By 1982, only mosquitofish (*Gambusia affinis*) were common in the Drain and in Kesterson Reservoir, and their high abundance suggested that few predatory fish survived.

As part of our research planning effort, Mike Saiki (USFWS fishery biologist with whom I worked on the initial Kesterson study) collected a few samples of mosquitofish from the San Luis Drain, Kesterson Reservoir, and Volta Wildlife Area (the nearby reference area that did not receive subsurface agricultural drainage water). Concentrations of several metals/metalloids (arsenic, cadmium, chromium, copper, mercury, nickel, lead, and zinc) in these preliminary samples were generally similar among locations, and organochlorine concentrations were very low (Saiki 1986). Of the 23 organochlorine compounds in the analytical suite, only *p,p'*-DDE (22–44 μg/kg, wet weight [ww]) and Aroclor 1248 (60 μg/kg, ww) were detected in the mosquitofish from the Drain or Kesterson. However, there were marked differences in selenium concentrations between mosquitofish from the Drain (27–30 μg/g, ww) and Kesterson Reservoir (26–31 μg/g, ww) in comparison to those from the Volta Wildlife Area (0.39 μg/g, ww). This suggested that there were elevated levels of selenium in the Kesterson food chain for birds, and that mosquitofish are very tolerant of selenium and can accumulate high levels.

My review of the literature available at that time showed that bird embryos are very sensitive to selenium toxicity (Franke and Tully 1935; Moxon and Olson 1974; NAS 1976; Ort and Latshaw 1977, 1978). Hatchability of chicken eggs was significantly reduced when dietary selenium concentrations were 5 μg/g;[1] when dietary concentrations were between 5 and 10 μg/g, there was a high incidence of grossly deformed embryos. Typical deformities included missing eyes and beaks, edema of the head and neck, and distorted wings and feet. Because selenium in the diet of chickens at much less than 10% of the concentration found in mosquitofish from Kesterson Reservoir (converted to about 130–150 μg/g dry-weight basis to be consistent with the poultry diet) produced significant reproductive effects in chickens (described above), it seemed important to evaluate embryo mortality and deformities in a field study of aquatic birds inhabiting the Kesterson ponds.

---

[1] All selenium concentrations for biota in this paper are given on dry-weight basis, unless otherwise noted.

This background information provided the rationale and approach for a study to determine whether Kesterson Reservoir provided favorable habitat for aquatic wildlife, especially because of the potential for drainwater to be used elsewhere for marsh management or to be discharged into the San Francisco Bay Estuary.

The Research Phase

Research on the effects of selenium (and other contaminants) in drainwater at Kesterson began with our field study in 1983 (summarized below) that strongly implicated selenium as the chemical responsible for high rates of embryo mortality and embryo/chick deformities in wild birds that were similar to those observed previously in poultry. Following the first year of field research at Kesterson, laboratory studies were conducted at the Patuxent Wildlife Research Center and elsewhere with several species of birds that corroborated findings from the field study (see Ohlendorf 2002, 2003; Ohlendorf and Heinz 2011). Additional studies conducted at Kesterson and nearby areas in 1984 and 1985 showed the problem of selenium in agricultural drainage water was not isolated to Kesterson Reservoir.

The findings of the Kesterson studies have been reported in more detail in other publications, some of which provide illustrations of the specific embryo deformities observed in the field (e.g., Ohlendorf et al. 1986a, 1988, 1989, 1990; Hoffman et al. 1988; Hothem and Ohlendorf 1989; Ohlendorf and Skorupa 1989; Williams et al. 1989) and in several reviews that also incorporated the results of studies in other areas and those of related experimental studies (e.g., Ohlendorf 1989, 1996, 2002; Skorupa and Ohlendorf 1991; Ohlendorf et al. 1993; Ohlendorf and Hothem 1995; Heinz 1996; O'Toole and Raisbeck 1998; Ohlendorf and Heinz 2011).

The studies of selenium effects in birds have been recognized as one of the "gold standards" of retrospective ecological risk assessment, as discussed by Suter (1993). The integration/combination of related laboratory studies with the Kesterson field studies fulfilled the equivalent of Koch's postulates by recognizing a disease syndrome, identifying the causative agent, and reproducing the disease syndrome in healthy individuals by administration of the putative agent (selenium) in their diet.

Field Studies

The Beginning: Kesterson in 1983

Based on information available in 1983, my objectives in the first study at Kesterson Reservoir were to "determine whether selenium or heavy metals occur at harmful levels in aquatic birds or their food chains in the evaporation ponds at Kesterson NWR and interpret the significance of these contaminants to wildlife" (Ohlendorf 1983). This study focused on reproductive effects in aquatic birds and on contaminants in their food chains at Kesterson Reservoir and the Volta Wildlife Area reference site (Fig. 11.1).

Methods and Study Areas

We studied several species of birds that were common at one or both of the study sites, including eared grebe (*Podiceps nigricollis*), pied-billed grebe (*Podilymbus podiceps*), American coot (*Fulica americana*), mallard (*Anas platyrhynchos*), gadwall (*A. strepera*), cinnamon teal (*A. cyanoptera*), American avocet (*Recurvirostra americana*), black-necked stilt (*Himantopus mexicanus*), and killdeer (*Charadrius vociferus*). We collected adult birds of several species at each site in spring and early summer to determine which foods they were eating and what concentrations of selenium were in their tissues at the beginning and end of the nesting season. In addition, we sampled juveniles of some of these species before fledging to determine concentrations of chemicals in their tissues, because their exposure (as nonflying young) could be more clearly associated with the area in which they were collected.

We evaluated nesting success of these species by searching for nests in favorable nesting habitats at Kesterson and Volta, marking the nests, and monitoring them (usually weekly) to determine their fate. We collected randomly selected eggs from some of the monitored nests and we collected nonrandom eggs from other nests (such as those that had failed to hatch, or sibling eggs from nests where chicks with deformities were found) for analysis. Searching for, marking, and monitoring the nests of the study species was a major effort; however, I believed the effort was warranted because impaired reproduction was likely to be a significant effect of the birds' exposure to selenium and the effects could be determined best by monitoring nests. It is unlikely that we would have understood the significance of the effects of selenium on bird reproduction without using this study approach, though some had suggested during the planning phase that it would be sufficient to do brood counts and compare "productivity" between the two study sites.

In a parallel study, Mike Saiki sampled mosquitofish, aquatic invertebrates, and plants of the types eaten by the aquatic birds (Saiki and Lowe 1987). The food-chain sampling included spatial and seasonal comparisons to evaluate changes related to inflows of agricultural drainwater to Kesterson and its evaporation during the summer.

Results and Discussion

Eared grebes, coots, ducks, stilts, and avocets nested at Kesterson in sufficient numbers for us to monitor 347 nests to late stages of incubation or hatching and thereby evaluate reproductive success (Ohlendorf et al. 1986a). Forty percent of those nests had one or more dead embryos, and 20% had at least one embryo or chick with

Fig. 11.2 Newly hatched American coot chicks in a nest at Kesterson Reservoir in 1983. Three of the chicks are normal but the one in the foreground has no eyes (note absence of blue coloration where eyes are missing). Other eggs in the nest failed to hatch, and the one that was analyzed had 44 μg Se/g (about 20 times the normal concentration). Photo by author

severe deformities (Figs. 11.2–11.4). Some chicks that hatched were blind, had no lower beak, or had no feet. In contrast, observations at Volta and findings in studies with these species elsewhere showed that the incidences of embryo mortality or deformities at Kesterson were much higher than expected. The deformities in embryos and chicks were similar to those previously found in chickens, and almost all of them were severe enough to be fatal. The deformities were often multiple, including (roughly in decreasing order of frequency) missing or abnormal eyes (anophthalmia and microphthalmia), beaks (missing, reduced, or crossed), legs and wings (micromelia and amelia), feet (ectrodactyly and clubfoot), and brain (hydrocephaly and exencephaly), and were typically bilateral.

Selenium concentrations in livers of adult and juvenile birds from Kesterson were about ten times those found at the Volta reference site (Ohlendorf et al. 1986a, 1990), and selenium concentrations in bird eggs at Kesterson usually averaged at least 20 times higher than at Volta (Ohlendorf and Hothem 1995). Mean selenium concentrations in eggs of all species at Volta were less than 3 μg/g, which is typical of normal background (U.S. Department of the Interior [USDI] 1998), whereas means at Kesterson ranged up to nearly 70 μg/g (the geometric mean for eared grebes; see Table 11.1).

Food-chain organisms such as those eaten by the aquatic birds we studied also contained highly elevated selenium concentrations (Ohlendorf et al. 1986a; Saiki 1986; Saiki and Lowe 1987). Leaves and seeds of some wetland plants had selenium

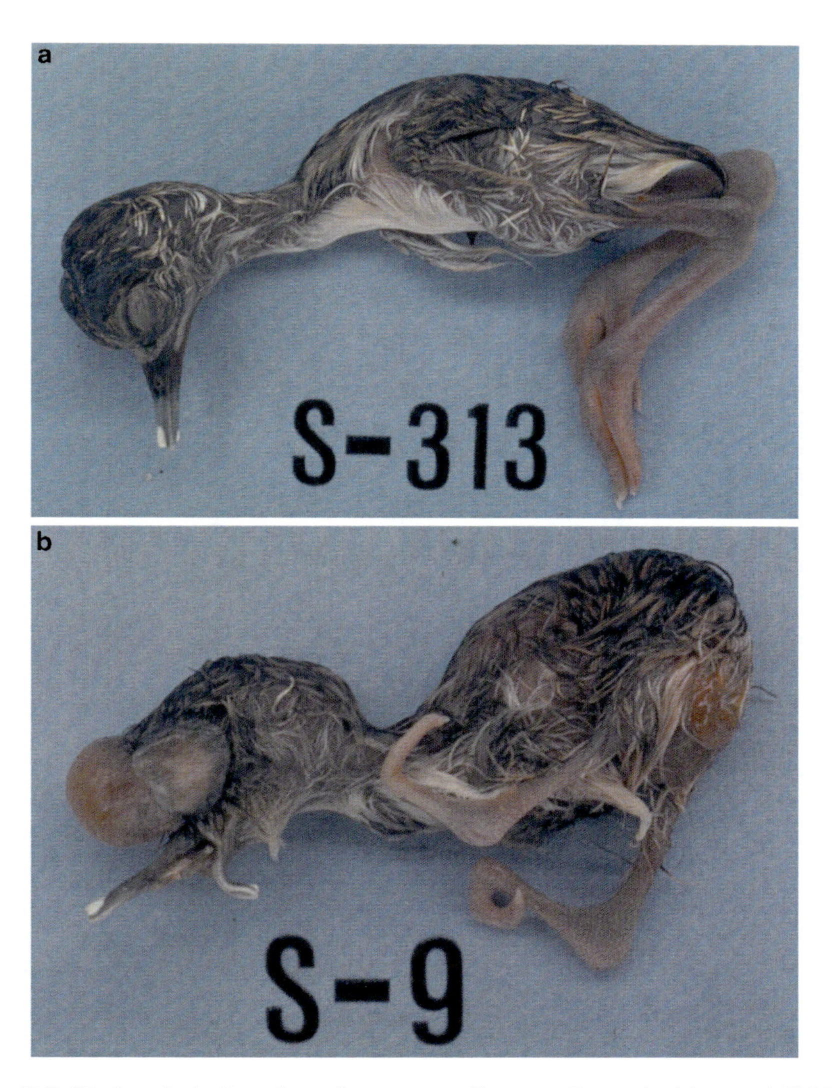

Fig. 11.3 Black-necked stilt embryos from nests at Kesterson Reservoir: (**a**) normal (S-313); (**b**) eyes missing, severe exencephaly (protrusion of brain) through orbits, lower beak curled, upper parts of legs shortened and twisted, and only one toe on each foot (S-9); (**c**) eyes missing, upper beak elongated and eroded at nostrils, lower beak missing, legs missing, and only one (small) wing (S-35). Photos courtesy of US Fish and Wildlife Service

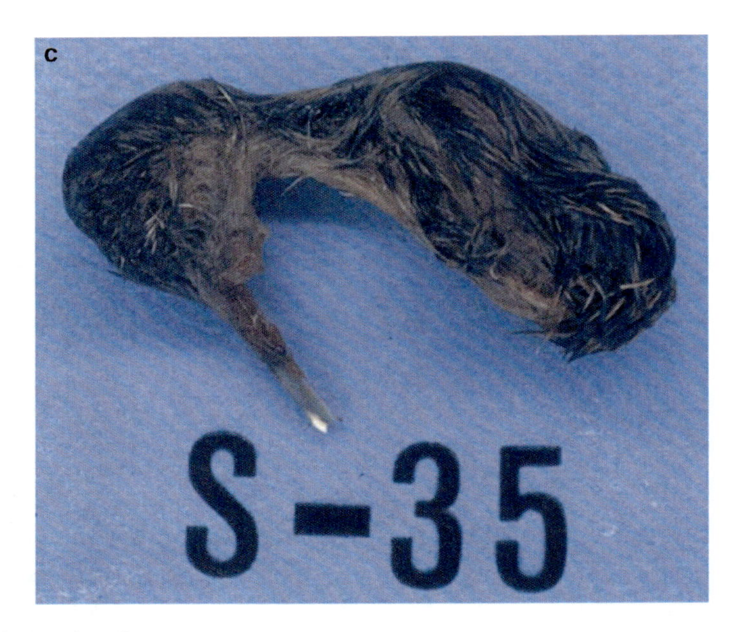

Fig. 11.3 (continued)

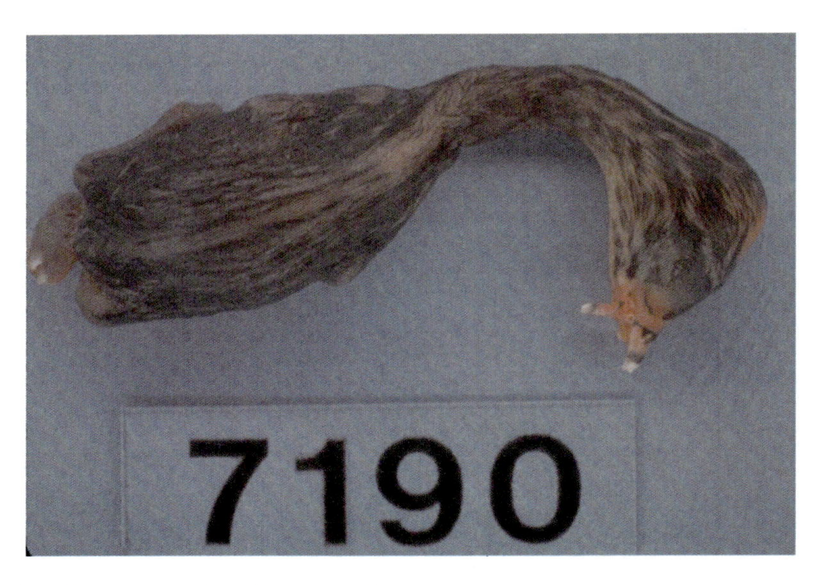

Fig. 11.4 Eared grebe embryo from nest at Kesterson Reservoir showing severely deformed lower beak. Photo courtesy of US Fish and Wildlife Service

Table 11.1 Selenium Concentrations (μg/g, dry wt.) in Randomly Collected Eggs of Aquatic Birds from Kesterson Reservoir and Reference Sites, 1983–1985

			Selenium concentrations					
		Moisture[a]	Kesterson Reservoir			Reference sites[b]		
Species	Year	content (%)	*N*	Mean[c]	Min.–Max.	*N*	Mean[c]	Min.–Max.
Pied-billed grebe (*Podilymbus podiceps*)	1983	75.7	–	–	–	1	–	1.9
	1984	77.1	–	–	–	2	1.70	1.6–1.8
	1985	76.8	–	–	–	3	2.76	2.6–3.0
Eared grebe (*Podiceps nigricollis*)	1983	75.6	18	69.7	44–130	–	–	–
Mallard (*Anas platyrhynchos*)	1983	64.8	5	15.2	9.3–31	1	–	1.2
	1984	70.6	5	10.4	3.6–19	2	1.55	1.5–1.6
	1985	69.4	11	11.5	4.3–23	7	2.52	1.3–5.0
Cinnamon teal (*A. cyanoptera*)	1983	65.5	2	6.85	6.6–7.1	–	–	–
	1984	69.0	5	13.5	7.7–37	–	–	–
	1985	67.8	5	9.91	3.9–22	2	1.54	1.4–1.7
Gadwall (*A. strepera*)	1983	66.2	6	18.8	9.6–32	2	0.84	0.64–1.1
	1984	69.6	6	21.4	18–26	1	–	1.6
	1985	68.5	10	19.6	7.3–33	5	1.60	1.5–1.7
American coot (*Fulica americana*)	1983	73.6	17	32.4	17–74	–	–	–
	1984	76.7	–	–	–	5[d]	0.756	0.59–1.1
Killdeer (*Charadrius vociferus*)	1984	74.4	9	33.1	16–50	–	–	–
	1985	72.4	21	46.4	14–180	4	2.07	1.8–2.2
Black-necked stilt (*Himantopus mexicanus*)	1983	69.9	11	28.2	14–58	2[e]	1.61	1.3–2.0
	1984	71.5	37	24.8	5.2–64	10	2.42	1.6–3.4
	1985	73.4	70	35.5	4.3–100	19[f]	2.65	1.4–3.6
American avocet (*Recurvirostra americana*)	1983	68.4	7	6.65	3.2–22	1	–	1.4
	1984	72.9	26	16.4	3.4–61	5	1.86	1.2–3.0
	1985	73.6	25	32.2	3.4–88	14[g]	1.92	1.4–2.4

[a] Arithmetic mean
[b] All eggs were collected at Volta Wildlife Area; additional samples from other areas are noted below
[c] Geometric mean (GM) calculated when two or more eggs were analyzed
[d] One additional egg from Mendota Wildlife Area (0.82 μg Se/g)
[e] Ten additional eggs from Carson Lake, Nevada (GM = 1.0 μg Se/g; range = 0.44–1.7 μg Se/g)
[f] Eight additional eggs from San Francisco Bay (GM = 1.8 μg Se/g; range = 1.4–3.6 μg Se/g)
[g] One additional egg from San Francisco Bay (1.4 μg Se/g)

concentrations below 5 μg/g, but algae, rooted plants, net plankton, aquatic insects, and mosquitofish from the Kesterson ponds and the San Luis Drain typically had average selenium concentrations of 20 to more than 300 μg/g, and concentrations often were 100 times those found at Volta (Fig. 11.5). These dietary selenium

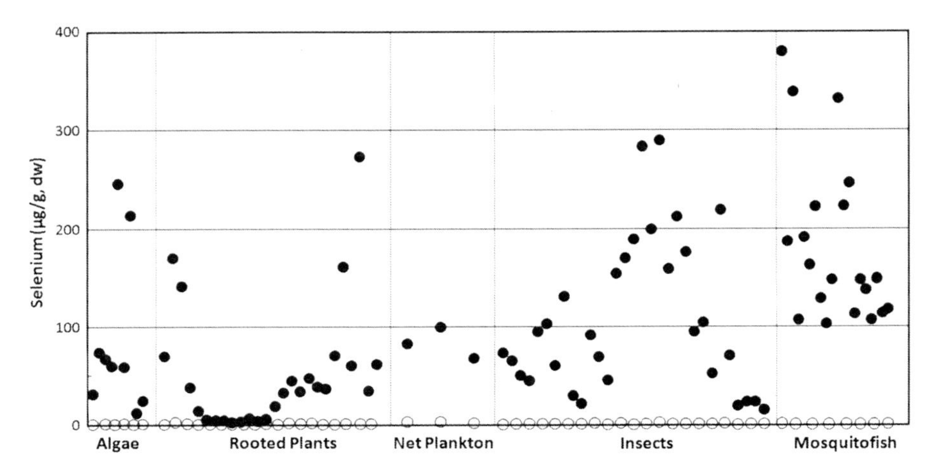

Fig. 11.5 Geometric mean selenium concentrations (µg/g, dry weight) in algae, rooted plants, net plankton, insects, and mosquitofish from Kesterson Reservoir (*filled circles*) and Volta Wildlife Area (*open circles*), 1983–1985 (data from Saiki and Lowe 1987; Hothem and Ohlendorf 1989; Schuler et al. 1990). The individual points in the figure represent values for biota at different times and locations within the two study areas.

concentrations were many times those that had caused severe reproductive impairment in earlier studies with chickens (5–10 µg/g, as summarized by Moxon and Olson 1974; NAS 1976).

The average selenium concentration in the Kesterson inflow water from the Drain was about 300 µg/L, compared to less than 2 µg/L for the Volta ponds (Presser and Ohlendorf 1987; Saiki and Lowe 1987). One of the important and unexpected observations about selenium in water and food-chain biota was that concentrations were highest in samples from the San Luis Drain and the "upstream" (i.e., near the inflow) ponds where the drainwater entered Kesterson, rather than in the "downstream" ponds (farther along the flow path within the series of ponds). (Inflow from the San Luis Drain to the upstream ponds was the only source of water to Kesterson Reservoir, so there was no dilution due to inputs of fresh water). Unlike conservative constituents present in the drainwater (such as boron and salinity) that increased in concentration through evaporative water loss along the flow path, selenium concentrations in water and biota decreased through the series of ponds (because of biological uptake and deposition to sediment), which is a pattern that has subsequently been observed in other wetland systems.

My colleagues at the Patuxent Wildlife Research Center used the results of the 1983 field study to plan a series of laboratory studies to test the effects of selenium on several species of wild birds under controlled conditions (summarized by Heinz 1996; Ohlendorf and Heinz 2011), and I used them to plan subsequent field studies at Kesterson and in the surrounding Grassland Water District (Ohlendorf et al. 1987) to increase our understanding of the nature and extent of selenium contamination issues related to agricultural drainage water in the San Joaquin Valley.

Continuing Study: Kesterson in 1984–1985

We continued to measure selenium bioaccumulation in food-chain organisms at Kesterson and Volta and to evaluate the effects of selenium on aquatic birds at the two sites in 1984 and 1985. Food-chain organisms at the two sites had selenium concentrations similar to those in 1983 (Hothem and Ohlendorf 1989; Schuler et al. 1990). Selenium concentrations in aquatic plants, invertebrates (mostly insects), and mosquitofish at Kesterson typically averaged 50 to 150 μg/g, and were far higher than the threshold dietary levels associated with impaired reproduction for poultry (Moxon and Olson 1974; NAS 1976) or mallards (as determined by studies conducted at the Patuxent Wildlife Research Center; see Ohlendorf and Heinz 2011). By comparison, food-chain organisms from Volta generally averaged 2 μg/g or less, similar to other uncontaminated areas. Bioaccumulation factors (i.e., tissue concentration [μg/g] divided by waterborne selenium concentration [μg/L]) for many food-chain organisms often were about 1,000. However, the bioaccumulation factors decreased significantly as the waterborne selenium concentrations increased, so the generalized bioaccumulation factor of 1,000 did not apply to all waterborne selenium concentrations.

Many coots were present at Kesterson during the 1984 and 1985 nesting seasons, but we found no coot nests in areas where we had found 92 nests in 1983. Unlike 1983, we found many dead coots (as well as some ducks, grebes, and other birds) at Kesterson in 1984 and 1985. We attributed this mortality and failure of coots to nest to the debilitating effects of excess selenium in the birds' diets and tissues (Ohlendorf et al. 1988, 1990; Ohlendorf 1989, 1996). The coots we collected at Kesterson, whether live or found dead, were about 25% below their normal body weights; their livers contained more selenium than in 1983, and they had lesions associated with selenium toxicosis.

In addition to high incidences of embryo mortality and abnormalities, we found poor survival of chicks that hatched for several of the aquatic bird species in 1984 and 1985 (Ohlendorf 1989; Ohlendorf et al. 1989; Williams et al. 1989). Overall, nearly 40% of the 578 nests that we could evaluate during 1983–1985 contained one or more dead or deformed embryos or chicks. Some embryo mortality was expected to occur normally, but only about 1% of the eggs from Volta contained dead embryos, and we found no abnormal embryos there. Among the young that we presumed to have hatched, there was poor survival in grebes and coots (about 2%) in 1983, and in 1984–1985 our observations indicated there was no survival among the nearly 440 avocets and stilts that apparently hatched.

Beyond Kesterson Reservoir

Grasslands of the San Joaquin Valley

In addition to continuing the study at Kesterson in 1984, we sampled birds and their eggs from the nearby Grassland Water District, which includes two areas; one is located adjacent to Kesterson (the "North Grasslands") and the other is farther south (the "South Grasslands," nearer the source areas of the agricultural drainage water; see Fig. 11.1 and Ohlendorf et al. 1987). The water supply to the Grasslands was a mixture of subsurface agricultural drainage and other water, with an average selenium concentration in the inflow of about 50 μg/L. We found elevated selenium concentrations in several species of aquatic birds and their eggs during this study. The spatial pattern of selenium concentrations was similar to that at Kesterson; selenium concentrations were higher in the South Grasslands, near the inflow areas, than downstream in the North Grasslands. As a result of our findings, the water supply to the Grasslands was changed to exclude use of irrigation drainage water during the autumn of 1985.

Following this initial study Roger Hothem led a 2-year study of reproductive success of ducks and shorebirds nesting in the Grasslands and contaminant concentrations in their eggs during 1986 and 1987 (Hothem and Welsh 1994). Selenium concentrations were higher in eggs from the Grasslands than at an uncontaminated reference site, and concentrations in some species and locations exceeded reproductive effect levels. Nevertheless, no effects on reproduction were found in any of the ducks or shorebirds nesting in the Grasslands in 1986 or 1987, probably because of the small sample size from the more contaminated areas.

In parallel studies, birds also were sampled throughout the Grasslands for tissue analysis in 1985–1988 (Paveglio et al. 1992) and in 1989 and 1994 (Paveglio et al. 1997) to further define the spatial and species-specific patterns of contamination and to determine improvements resulting from the change to a freshwater supply for the Grasslands in 1985. Not surprisingly, these studies showed that the South Grasslands continued to be more contaminated than the North Grasslands. In addition, the livers of birds collected during the breeding and wintering periods contained selenium at concentrations associated with reproductive impairment. Selenium contamination (as measured in bird livers) had been reduced as a result of the change in water source for the Grasslands, but concentrations still were higher than background, and sometimes exceeded those associated with adverse effects.

Elsewhere in the Western USA

Following our discoveries of irrigation drainage-related contamination impacts at Kesterson Reservoir, the USDI initiated the National Irrigation Water Quality Program (NIWQP) in late 1985 to study the effects of irrigation drainage on water resources and on fish and wildlife elsewhere (Deason 1986; Engberg et al. 1998; Seiler et al. 1999; NIWQP 2009). During Phase 1, the NIWQP screened existing data from about 600 irrigation project areas and national and state wildlife refuges, primarily in the western USA, to identify other areas where agricultural drainage might be causing contamination problems. Based on the initial screening step (which was completed in 1989), the NIWQP identified 26 areas for investigation. During Phase 2, reconnaissance-level sampling at those 26 sites confirmed that harmful effects either had occurred or were likely to occur at eight of those areas (in

California, California/Oregon [along the border], Colorado, Nevada, New Mexico, Utah, and Wyoming), so more detailed studies were conducted in those areas in Phase 3. Bird embryos with multiple severe deformities similar to those in birds from Kesterson Reservoir were found in 4 of the 26 NIWQP areas (in California, Colorado, Utah, and Wyoming; see Seiler et al. 1999 for summary). Although waterborne selenium concentrations in those areas were greatly elevated (sometimes greater than 100 μg/L), they were lower than the average inflow concentrations to Kesterson Reservoir (about 300 μg/L, as described above).

Results of these investigations were published in a series of US Geological Survey (USGS) reports for the individual study areas, and they were analyzed and synthesized as the overall results for the NIWQP. The synthesis reports, such as the one by Seiler et al. (1999), also identified areas that are susceptible to irrigation-induced selenium contamination of water and biota, based on a combination of several environmental factors (such as rainfall, geology, etc.).

Laboratory Studies

As a result of our findings at Kesterson Reservoir, biologists at the Patuxent Wildlife Research Center conducted a series of laboratory studies to determine whether the effects observed in the field were related to the levels of selenium in the birds' diets (as measured by analysis of typical food items eaten by the birds at the Reservoir), and to evaluate possible interactions of selenium with other contaminants (arsenic, boron, and mercury; Stanley et al. 1994, 1996; Heinz and Hoffman 1998) that were found in drainwater at Kesterson or other sites. Most of those studies were conducted with mallards (e.g., Heinz 1993a, b; Heinz and Fitzgerald 1993a, b; Heinz et al. 1987, 1988, 1989, 1990; Hoffman and Heinz 1988, 1998; Hoffman et al. 1989, 1991; Albers et al. 1996) but studies also were conducted with birds of other species (such as black-crowned night-herons [*Nycticorax nycticorax*], Smith et al. 1988; and eastern screech-owls [*Megascops asio*], Wiemeyer and Hoffman 1996). Experimental studies also have been conducted elsewhere with mallards (e.g., O'Toole and Raisbeck 1997) and other species (such as American kestrels [*Falco sparverius*]; Yamamoto et al. 1998; Santolo et al. 1999; Yamamoto and Santolo 2000). Although these various experimental studies are mentioned only briefly here, they were closely integrated with the field studies, and they were very important in validating the field-based observations. The results from the experimental studies support the conclusion that chronic dietary selenium exposure such as that observed at Kesterson Reservoir causes severe effects on health, survival, and reproduction of birds under realistic field conditions.

Taken together, these studies support the conclusion that the health and reproductive effects observed in the field – including mortality and abnormal development of embryos and chicks and the weight loss and mortality of adults – could be attributed to the toxicity of selenium. The various studies clearly showed that a combination of observed effects (such as characteristic multiple deformities of embryos/chicks, pathological tissue changes of the integument [including feather loss] and liver, or death) and tissue analyses (especially of food, eggs, and livers) can be used to diagnose selenium toxicosis in birds. Results of many of the laboratory studies published through 2008 were described in other reviews (e.g., Heinz 1996; Ohlendorf 1996, 2003; O'Toole and Raisbeck 1998; Ohlendorf and Heinz 2011), and they are not reiterated here.

Management/Regulatory and Public Response

Kesterson Reservoir

From the beginning, regional USFWS resource and national wildlife refuge managers were very interested in our field study at Kesterson. Soon after we began to find apparent effects in our nest monitoring, Felix Smith, a senior biologist at the USFWS office in Sacramento, California, accompanied me on one of our weekly monitoring trips on June 7, 1983. He had recently been assigned the responsibility of looking into the emerging issues involving agricultural drainage and wastewater. He has said that the experience of holding a coot egg as the chick was in the process of hatching, although it had no eyes, lower beak, or feet, greatly affected his life (Smith 2007). Thereafter, events unfolded quickly to progressively notify the agencies with either responsibility for conditions at Kesterson or for regulation of water quality in the San Joaquin Valley.

On July 8, 1983, the USFWS Field Supervisor in Sacramento sent the Regional Director of Reclamation (also in Sacramento) and other offices within USFWS a "Concern Alert" to advise them of USFWS's concern about the high incidence of embryo deformities observed in our ongoing field study. The "Concern Alert" stressed the need for additional information and outlined a course of action to begin the collection of that information. It also suggested that Reclamation and USFWS meet to discuss the work and answer any questions Reclamation may have about the study.

I had been informing my research supervisor at the Patuxent Wildlife Research Center of the unfolding events from the field study, and on July 13, 1983, I forwarded the "Concern Alert" along with a transmittal memo in which I noted "There is likely to be a lot of activity related to drain-water problems in the future." What an accurate prediction that turned out to be!

Throughout the summer I participated in meetings and briefings to provide USFWS managers with updates on the results of the field study and to accompany the USFWS managers in meetings with Reclamation as related to the study findings as they emerged. Reclamation was generally not convinced that selenium contamination of the drainage water being discharged to Kesterson was a significant issue or that it should affect plans for discharge of subsurface agricultural drainwater to the San Francisco Bay Estuary. Within the USDI, this series of meetings culminated in a briefing for the USDI Regional Environmental Coordinator on September 12, 1983.

Then, on September 15, 1983, USFWS convened a meeting in Sacramento to advise agencies outside USDI of the preliminary study results. More than 40 individuals attended the meeting, representing Reclamation, USFWS, USGS, National Marine Fisheries Service, the State Board, Central Valley and San Francisco Bay Regional Water Quality Control Boards, California Department of Fish and Game, and California Department of Water Resources. Mike Saiki and I presented the results that were available at that time and various courses of action were discussed. Efforts of the USFWS to get a news release issued about the study findings by the time of this meeting were stalled by individuals high up in the USDI, so there was no public release of information until someone (later identified as Felix Smith) leaked some information to the media.

A barrage of news media coverage of the findings from Kesterson began when the preliminary results of this study were reported in an article in the Fresno Bee on September 21, 1983 (Blum 1983). Many other newspaper articles (though not always accurate), radio and television reports, and documentaries appeared during the next several years. I believe the research findings received so much public attention because (1) photographs of the embryo deformities provided dramatic evidence of a problem, (2) there was a plan for the agricultural drainage to be discharged to the San Francisco Bay Estuary (for which there was much public interest), and (3) the Reservoir was part of the Kesterson NWR, so many people were shocked to learn about the nature and magnitude of effects that we had observed. The NWR was supposed to provide high-quality wetland habitat for wildlife, but the contaminated water caused severe impacts on the wildlife it was supposed to benefit. The photographs of embryos with deformities clearly showed that something was wrong at Kesterson, and the images quickly circulated through the regional and network news media. Some who had opposed completion of the San Luis Drain to discharge agricultural drainage to the Estuary for other reasons used the findings of the Kesterson studies to fight that controversial project. In addition, there were substantial disagreements among involved federal agencies as to the significance of the findings (see review by Garone 1999).

Following the initial release of information about the Kesterson bird studies, the rest of 1983 was hectic, but resulted in significant movement in late 1983 and 1984 toward better understanding and taking action toward solving the irrigation drainage issues (though many still remain to this day).

On October 28, 1983, US Congressman George Miller sent USFWS Director Robert Jantzen a letter asking for information about the Kesterson studies, expressing concern about the issue of contaminants in irrigation drainwater, and asking about the level of cooperation between Reclamation and USFWS on the management of Kesterson. Congressman Miller was very interested in the ongoing studies that were important to his constituents and he remained engaged in the issue over the years. Some of the results for the bird study as well as results of Mike Saiki's fish study were incorporated into the Director's reply letter dated November 29, 1983.

On December 3, 1983, University of California, Berkeley hosted a conference on "Contaminants in Agricultural Waste Water: A Problem for the S. F. Bay Region as Well as the Central Valley?" for which I was an invited speaker to talk about "What has been observed at Kesterson National Wildlife Refuge." I was able to provide information that was available at that time, and I noted a high degree of interest among the conference participants. Congressman Miller also was a speaker at this conference and spoke about "political perspectives" of the issue.

Then on December 5–7, 1983, Reclamation held a conference on "Research Issues Associated with Toxicity Problems at Kesterson Reservoir, California" to which they invited individuals familiar with selenium issues in locations such as South Dakota and those with experience in studying the effects of selenium in poultry (e.g., Dr. J. D. Latshaw). The outside experts concluded that selenium likely was causing the effects we had observed in birds at Kesterson, as was subsequently confirmed by the studies described in previous sections. In addition to supporting our preliminary assessment of the relationship between selenium and the observed effects, this conference resulted in forward movement among the federal agencies. At the conclusion of the conference, Reclamation Regional Director David Houston outlined his proposed "action plan" that included the following elements:

- Immediate action – reduce exposure of birds at Kesterson to selenium.
- Short-term action – continue data collection and reanalyze options to minimize drainage water production.
- Long-term action – establish an interagency study program, within Interior, consisting of Reclamation, USFWS, and the USGS to further define and resolve the selenium issue.

San Joaquin Valley

In August 1984, USDI Secretary William Clark and California Governor George Deukmejian established the San Joaquin Valley Drainage Program (SJVDP), with Reclamation, USFWS, USGS, California Department of Fish and Game, and California Department of Water Resources as the principal participating agencies. Over the next 6 years, the SJVDP studied the problems of subsurface agricultural drainage in the western San Joaquin Valley intensively and produced a management plan (SJVDP 1990) that recommended a combination of components to address the problem, including the following:

- Source control, mainly on-farm improvements in application of irrigation water to reduce deep percolation
- Drainage reuse, which consisted of a planned system of drainage water reuse on progressively more salt-tolerant plants
- Evaporation systems, including evaporation ponds planned for storage and evaporation of drainage water remaining after reuse on salt-tolerant plants

- Land retirement – cessation of irrigation of areas in which underlying shallow groundwater contains elevated levels of selenium and soils are difficult to drain
- Discharge to the San Joaquin River, as controlled and limited discharge of drainage water from the northern portion of the valley, while meeting water quality objectives
- Protection, restoration, and provision of substitute water supplies for fish and wildlife habitat, intended to replace drainage-contaminated water previously used for wetlands and to allow protection and restoration of contaminated fisheries and wetland habitat
- Institutional changes that included tiered pricing for irrigation water, improved scheduling of water deliveries, water transfers and marketing, and formation of regional drainage management organizations to aid in implementing other plan components

The USFWS initiated a waterfowl "hazing" program at Kesterson Reservoir in September 1984 in an attempt to minimize exposure of birds to the high-selenium environment of the ponds. Because it was important for the birds to have some other habitat to move to, USFWS requested that Reclamation help provide alternate habitat to lure the birds away from Kesterson. At about this same time, the California Department of Fish and Game and the California Department of Health Services warned that coots from Kesterson should not be eaten because of selenium contamination.

On February 5, 1985, the State Board adopted Cleanup and Abatement Order WQ 85-1 that required USDI to close, upgrade, clean up, and/or otherwise abate the nuisance from Kesterson Reservoir within 3 years. The Order further required Grassland Water District to develop a plan to control selenium concentrations in agricultural drainage water entering the District.

USDI Secretary Donald Hodel ordered the immediate closure of Kesterson Reservoir and the San Luis Drain on March 15, 1985, and termination of irrigation water deliveries to 42,000 acres of land within the Westlands Water District (WWD) where the subsurface drainage water originated. This decision by the Secretary was based on a legal analysis that determined that continued operation of the Reservoir might constitute a violation of the Migratory Bird Treaty Act.

When the decision by Secretary Hodel was announced during a public meeting in Los Banos, California, on March 15 (just 5 days after 60 minutes [the TV news program] had aired a story about Kesterson and the Drain), there was an immediate response of anger and concern from the growers attending the meeting. Because of that response and reported threats for our safety, the USDI scientists attending the meeting were abruptly evacuated to return to Sacramento. Growers were shocked about the impending loss of irrigation water and their inability to discharge subsurface drainage from their fields, which threatened their livelihood.

On April 3, 1985, USDI and WWD signed an agreement ensuring continued delivery of irrigation water by USDI to WWD, but requiring phased reductions in drainage flows by WWD through the Drain to Kesterson, with zero inflow by June 30, 1986.

Reclamation undertook several studies and control actions to reduce the hazard of selenium exposure to aquatic birds (USBR 1986; USDI 1989). In particular, Reclamation halted inflows of agricultural drainage to Kesterson in 1986 and subsequently, in 1988, de-watered the Reservoir and filled all areas to at least 15 cm above the expected seasonal (winter) elevation of high-selenium groundwater. These actions effectively transformed the Reservoir into terrestrial habitats, as described by Ohlendorf and Santolo (1994). To mitigate for the loss of wetlands at the Reservoir, Reclamation provided additional fresh water to the Grassland Water District as alternative habitat and for nuisance abatement in 1988.

Monitoring of Kesterson has been conducted since 1987 to measure selenium concentrations in representative plants and animals using the site and to determine whether adverse effects occurred (Ohlendorf and Santolo 1994; Byron et al. 2003). The data from the monitoring program during 1989 through 1991 were used to estimate likely future levels of selenium in various biota, assess risks of adverse effects to animals using the site, identify needs for contingency plans for site management, and recommend further research and monitoring to provide the information needed to improve management of the site.

The risk assessment based on the 1989–1991 data concluded that selenium concentrations in terrestrial plants and animals were not expected to change markedly in the next 20 years (Ohlendorf and Santolo 1994). No adverse effects had been documented in terrestrial birds or mammals, and no adverse effects were expected in the future.

Surface ponding of rainwater occurs in some portions of Kesterson Reservoir during very wet winters (Ohlendorf and Santolo 1994; Byron et al. 2003). Ephemeral pools that form under current conditions, and those expected to occur in the future, result from the accumulation of rainwater, rather than from rising groundwater. Although these pools are formed by rainwater, selenium concentrations in the pools averaged less than 10 µg/L, but concentrations up to 162 µg/L were measured in some pools. These concentrations and those in aquatic invertebrates collected from them were high enough to be of concern for aquatic birds (such as waterfowl and shorebirds). The greatest concern was that these pools may form during late winter or spring and persist during the nesting season of aquatic birds, which might feed on aquatic plants and invertebrates in the pools if they persisted long enough to support these food-chain components.

As a result of the ecological risk assessment described by Ohlendorf and Santolo (1994), the monitoring program was revised in 1995 to be less costly (e.g., sampling fewer kinds of biota, and sampling every 3 years) but still sufficient to determine whether the model predictions (described above) were correct. In addition, further studies with terrestrial birds such as the American kestrel (e.g., Yamamoto et al. 1998; Santolo et al. 1999; Yamamoto and Santolo 2000) were conducted during subsequent years to resolve some of the unknowns identified by the risk assessment.

This new information, along with monitoring results since 1989, was evaluated in a subsequent ecological risk assessment that focused on the terrestrial and aquatic wildlife using the site (CH2M HILL and LBNL 2000; Byron et al. 2003). The overall conclusions of the risk assessment were that future soluble (i.e., bioavailable) soil selenium levels will be similar to those observed over the past decade, waterborne selenium concentrations in rainwater pools can be predicted on the basis of soil selenium where the pools form, and selenium concentrations in terrestrial and rainwater-pool biota will continue to be elevated above normal background-site levels, but there is low risk to wildlife for several reasons. A model developed using the historic (127-year) Kesterson-area rainfall data and observed ponding on the site since 1989 (including winter of 1997–1998, the year of maximum rainfall) indicated that there is low probability that substantial portions of the site would be flooded during the spring breeding season for birds, so their exposure during that critical period would be minimal. Selenium concentrations in terrestrial birds and mammals also will be greater than normal background, but they are expected to be below levels known to cause health or reproductive effects. As a result of the findings from this risk assessment, the frequency of monitoring was further reduced to a 5-year frequency for 2001, 2006, and 2011. Thereafter, if conditions were stable, long-term management but no further monitoring would be recommended.

Following the closure of Kesterson Reservoir as a disposal site for subsurface agricultural drainage, growers were required to find on-farm ways of managing salinity-affected lands and any associated subsurface drainage water that was collected from the lands. Examples include increasing the efficiency of irrigation (thereby generating less drainage water for disposal), re-cycling/blending the drainage water with fresh water and using it for irrigation, and growing more salt-tolerant crops.

In other portions of the San Joaquin Valley (outside of Kesterson) and the western USA, management actions have been taken as a result of the SJVDP and NIWQP, as described above. However, management of selenium in agricultural drainwater remains a difficult problem. In the northern San Joaquin Valley, dischargers formed the San Luis & Delta-Mendota Water Authority in 1992. This is an administrative body consisting of several water agencies representing about 2,100,000 acres supplied with water by 29 federal and exchange water service contractors. The Authority oversees use of the San Luis Drain and administers conveyance of agricultural drainwater from several irrigation/drainage districts through adjacent wildlife management areas to Mud Slough, a tributary of the San Joaquin River. This activity is identified as the Grassland Bypass Project, and it has demonstrated success in lowering the concentrations of selenium being discharged to the river. Their current Use Agreement for the San Luis Drain for this project expired at the end of 2009, so they prepared an Environmental Impact Statement/Environmental Impact Report (Entrix 2008) to support receiving a new Use Agreement that would allow the Grassland Bypass Project to continue through 2019. Discharges to the San Joaquin River are regulated under a Total Maximum Daily Load (TMDL) allocation (CVRWQCB 2001).

In the southern San Joaquin Valley, options for disposal of subsurface drainwater are limited, because the water must be managed within the Valley (i.e., no discharge to the river). In the 1980s there were more than 20 groups of evaporation pond basins situated in the southern portion of the Valley, but several of those systems were closed by 1995 so that the total area of evaporation ponds was reduced from more than 7,000 acres to an area of less than 5,000 acres (EPTC 1999). The remaining ponds are regulated by the State of California through Waste Discharge Requirements (WDRs) that include mitigation measures for pond design and operation such as minimum water depth of 2 ft, steepening of side-slopes to 3:1, vegetation control, removal of windbreaks/islands, disease surveillance and control program, invertebrate sampling, and hazing to reduce bird use and the impacts of selenium in the ponds on birds.

In addition to the mitigation measures described above, the USFWS developed protocols for two mitigation approaches for unavoidable impacts by providing foraging habitats for targeted waterbirds so that selenium exposure of birds could be reduced (Skorupa and Schwarzbach 1995a, b). They included development of compensation habitat (distant from the evaporation ponds being mitigated) and creation of alternative wetland habitats (near the evaporation ponds, to "dilute" the exposure of birds using the ponds). Based upon preliminary estimates of unavoidable losses and required compensation and/or alternative habitats, WDRs were adopted for several evaporation basin systems (EPTC 1999). Monitoring of bird use and selenium concentrations in eggs demonstrated that one of the large compensation habitats was very attractive to shorebirds (Davis et al. 2008) and that alternative freshwater habitat adjacent to an evaporation pond resulted in lower selenium exposure of shorebirds (Gordus 1999).

Reclamation has completed an Environmental Impact Statement (USBR 2006) and a Record of Decision (USBR 2007) for agricultural drainage from the San Luis Unit (which was the source of the drainage that was discharged to Kesterson Reservoir) and from lands immediately adjacent to the Unit (total drainage study area of about 379,000 acres). They describe the drainage problem and the limited potential solutions for the area, and they recommend an approach for management of seleniferous drainwater. Briefly, Reclamation developed and evaluated alternatives to meet the following four objectives:

- Drainage service will consist of measures and facilities to provide a complete drainage solution, from production through disposal, and avoid a partial solution or a solution with undefined components.
- Drainage service must be technically proven and cost effective.
- Drainage service must be provided in a timely manner.
- Drainage service should minimize adverse environmental effects and risks.

After considering several alternatives, Reclamation selected an alternative for implementation (called "In-Valley/Water Needs Land Retirement Alternative"). Due to the expected costs for implementation of that alternative, it was considered likely that new authorizing legislation would be needed to increase the appropriations ceiling beyond what was authorized by the San Luis Act of 1960 (USBR 2007).

The selected alternative (like all the Action alternatives) includes on-farm/in-district actions to reduce drainwater through irrigation system improvements, seepage reduction, shallow groundwater management, and drainwater recycling; drainage collection systems; and regional reuse facilities (USBR 2007). In addition, the selected alternative would treat drainwater from the regional reuse facilities with reverse osmosis and biological selenium treatment before containment in evaporation basins (which would total about 1,900 acres). Reverse osmosis would reduce the drainwater volume by half and produce an equivalent amount of clean product water. The alternative also would target retirement of lands with selenium concentrations in the shallow groundwater above 20 μg/L (about 194,000 acres). This alternative was formulated to retire sufficient land to balance water needs with the lands remaining in production in the Unit.[2]

Presser and Schwarzbach (2008) evaluated constraints on drainage management and the implications of various approaches to management that included the San Luis Drainage Feature Re-evaluation (USBR 2006, 2007), approaches recommended by the SJVDP (1990), other recent conceptual plans, and other USGS models and analyses relevant to the western San Joaquin Valley. They concluded that rather than a lack of information, there is a lack of decision analysis tools to enable meeting the combined needs of sustaining agriculture, providing drainage service, and minimizing impacts to the environment. A more formal decision-making process may better address uncertainties inherent in the analyses, help optimize combinations of specific drainage management strategies, and document underlying data analysis for future use. Among the benefits of such a process of decision analysis are that it provides the flexibility to move forward in the face of uncertainty. They noted, however, that it requires long-term collaboration among stakeholders and a commitment to formalized adaptive management.

All of the proposed management strategies seek to reduce the amount of drainage water produced (Presser and Schwarzbach 2008). Land retirement also is a key strategy to reduce drainage, because it can effectively reduce drainage to zero if all drainage-impaired land is retired. Within the proposed area of drainage there are, for all practical purposes, unlimited reservoirs of salt (imported with irrigation water) and selenium stored within the aquifers and in Valley soils that were formed from the Coast Ranges on the west side of the Valley. A selenium mass balance analysis by USGS quantified the approximate amount of selenium exposed on the landscape and contained in each waste-stream component for the land retirement alternatives in the San Luis Drainage Feature Re-evaluation Environmental Impact Statement. A third of the selenium in drainage water is expected to be lost in the first step of the agricultural water reuse areas and selenium bio-treatment, if successful, is expected to reduce the selenium concentration to 10 μg/L in water to be evaporated. Elsewhere in the western USA, remediation planning was conducted for five sites identified as posing risks for

[2]Costs of the selected alternative were evaluated by Reclamation in a Feasibility Report (USBR 2008) that considered technical feasibility and environmental feasibility along with economic and financial feasibility. The selected alternative was found to be technically and environmentally feasible. However, the economic feasibility evaluation found that the selected alternative could not be justified in comparison to the no-action alternative (i.e., it would not provide a positive net benefit). The financial feasibility examined and evaluated the project beneficiaries' ability to repay the Federal Government's investment in the project over a period of time consistent with applicable law. The remaining authorized construction cost ceiling for the San Luis Unit was about $428.7 million, whereas the estimated cost of the selected alternative was $2.69 billion. None of the beneficiaries would be able to repay their share of the project cost if the selected alternative were implemented. The resulting recommendation of the Feasibility Report was to implement the selected alternative, contingent on several required Congressional actions to increase the construction cost ceiling and take associated actions to implement the drainage solution.

wildlife due to selenium in California, Colorado, Nevada, Utah, and Wyoming during Phase 4 of the NIWQP (NIWQP 2009). Planning was coordinated with appropriate Federal, State, and local agencies. The planning process was led by the bureau (Reclamation or Bureau of Indian Affairs) responsible for the irrigation project and governed by a core team made up of representatives from Reclamation or Bureau of Indian Affairs (as appropriate), USGS, and the USFWS. A technical team worked under the direction of the core team to evaluate various technical issues related to the proposed remediation alternatives.

Phase 5 of the NIWQP was implementation of corrective actions identified in Phase 4 (NIWQP 2009). Two projects, or wetland sites within project areas, were in Phase 5 remediation in 2005 (Gunnison/Grand Valley, Colorado; Middle Green River, Utah). The core team for the Gunnison/Grand Valley Project determined that there were over 20 wetland sites that could require remediation. Orchard Mesa Wildlife Area and Colorado River Wildlife Area were remediated by 2005 in the Grand Valley. In the Gunnison area, a lateral piping project in cooperation with the Uncompahgre Valley Water Users Association was initiated and was still in process as of 2005, when further funding for NIWQP ended. The Gunnison/Grand Valley Core Team was also working in cooperation with the Gunnison Basin and Grand Valley Selenium Task Forces.

In 2005, remediation activities in the Middle Green River Project in Utah included extending drains away from Stewart Lake Waterfowl Management Area to the Green River and construction of the seepage collection system in Stewart Lake (NIWQP 2009). Another important source of selenium to Stewart Lake (in addition to irrigation drainage) was seepage from a sewage-lagoon system constructed for the city of Vernal, Utah upgradient from the lake (USGS 2003). This system of lagoons did not directly discharge treated wastewater to Ashley Creek (a tributary of Stewart Lake), but it was constructed on Mancos Shale and seepage from the lagoons mobilized selenium from the fractured shale and discharged it to Ashley Creek in seeps and shallow groundwater. Because of the contamination resulting from lagoon seepage, a new wastewater treatment facility was constructed for Vernal, and the sewage lagoons were decommissioned. The facility was operational in April 2001, and releases of selenium to Ashley Creek declined.

Conservation Gains

Estimates of the numbers of aquatic birds that died in other areas are not available, but I estimated that more than 1,000 birds died due to selenium toxicity at Kesterson Reservoir in 1983–1985 (Ohlendorf 1986a; U.S. General Accounting Office [USGAO] 1987). Discovery of the effects of selenium on birds (and fish) at Kesterson Reservoir, and subsequently at other sites in the western USA where seleniferous agricultural waters were entering wetland habitats, created an awareness of the needs for resource protection (Engberg et al. 1998). Based on experience from the SJVDP and NIWQP, it became evident that the success of selenium management plans depends on partnership between federal, state, and local entities and on strong commitments by all participants to the management efforts.

It is not possible to quantify the area of aquatic or wetland habitats that directly benefitted from the awareness that selenium is a chemical of concern in agricultural drainage in certain parts of the western USA, but the total is almost certainly in the hundreds of thousands of acres. Water management has changed in the agricultural landscape, and there is much more regulatory control to minimize the exposure of aquatic birds to harmful levels of selenium, as noted in the previous section.

Realization that there was a problem with the intended disposal of agricultural drainwater to the San Francisco Bay Estuary derailed the plans for that discharge, and construction of the Drain was never completed. Thus, Kesterson Reservoir was not expanded to the anticipated 5,900 acres, other regulating reservoirs for the temporary storage of San Joaquin Valley drainage along the route of the Drain were not built, and discharge to the Estuary occurs only through regulated discharges via the San Joaquin River (the Grassland Bypass Project mentioned above). Concurrent with our studies at Kesterson, we also found elevated selenium concentrations in San Francisco Bay (Ohlendorf et al. 1986b, White et al. 1987), so increased discharges to the Estuary would have exacerbated the problem (Presser and Luoma 2006).

The results of our studies at Kesterson, in combination with other emerging knowledge about occurrence of selenium in the environment, stimulated research in a number of other exposure scenarios. Field and laboratory selenium research as related to aquatic habitats (especially concerning fish and birds) expanded rapidly in the 1980s and 1990s. As a result, selenium was found to be an issue of concern with regard to mining of coal, phosphate, and metals; disposal of fly ash from coal-fired power plants; and refining of petroleum (see Frankenberger and Engberg 1998; Chapman et al. 2010). Effects of selenium on fish have been documented in many of the same areas where selenium is of concern for birds.

Human Element

My personal experience in studying the effects of selenium on birds at Kesterson Reservoir was a mixture of excitement in making a significant scientific discovery, distress in seeing the devastating effects selenium was causing in bird embryos and hatchlings, a flurry of activities caused by media coverage of the study, and an introduction to the politics of water management in the West. I had anticipated there would be effects on egg hatchability and embryo development, but actually seeing the effects – especially the severely deformed embryos and hatchlings – was disturbing. In most settings, I was able to describe my findings in a matter-of-fact scientific manner, but there were occasions when my self-control broke down, and it was difficult to discuss the results of the study. In those cases, I had to slow down, take a deep breath, and regain my composure before continuing.

I had not anticipated the level of media coverage that would focus on the study as soon as the preliminary results were made public. By nature, I was inclined to focus on quietly and competently doing my work and to publish my findings in the normal course of completing the research and publication process. However, that was not to be. There were frequent requests for interviews from the news media (especially newspapers and television, but also on National Public Radio), and they borrowed the close-up photos of some of the deformed embryos to use in running their stories. Regulatory and resource agencies also borrowed the photos, and they have been widely circulated and printed in various publications. My choice in having those photos taken by a professional photographer was a wise decision, as they provided clear documentation of my observations.

In addition to the local and regional media, the story of Kesterson Reservoir was covered (usually accompanied by some of the photos of deformed embryos) by many conservation organizations in their publications (far too numerous to list) as well as by National Geographic magazine (Madson 1984), a public television documentary (KQED; "Down the Drain" October 10, 1984), 60 minutes (March 10, 1985), a news team from British Broadcasting Corporation (BBC Natural History Unit; Fall 1986, "Vanishing Earth"), plus an investigation by the USGAO (1987), and congressional hearings on the subject.

Initially, my research findings were viewed as a milestone in the Environmental Contaminants Research Program of the Patuxent Wildlife Research Center and the USFWS, and in the mid-1980s I received two USFWS Special Achievement Awards for outstanding performance as a research scientist. This study occurred during a period when there was lowered concern and interest about the effects of environmental contaminants on wildlife than had existed during the hearings about the banning of DDT in the early 1970s and other highly visible events, so my findings spurred interest and a number of other studies related to the effects of selenium in birds (described above).

By the late 1980s, however, I was receiving much less support from my research headquarters, and the "inconvenient" findings of my research, as well as the extensive requests for early use of the research results in making management decisions or for public disclosure, were wearing thin on research administrators' patience. I was advised that it was unlikely I would receive a promotion anytime soon that would put me back to my previous grade (which I had given up to move to California and become a full-time researcher). During one visit by an administrator I was told that I was "more trouble than I was worth." Some have said that comment was intended as a joke; it did not seem that way to me, and by 1990 I was able to find a very interesting, challenging job as an environmental consultant, where I have been happy to be making a difference while continuing my interests in the ecotoxicology of selenium.

Harris (1991) describes some of the background and fallout of the studies at Kesterson Reservoir and other parts of the San Joaquin Valley for those of us who were involved in studying the effects of selenium. I was not the only one who had a career-altering experience in the process. As Felix Smith said as we were examining the severely deformed coot chick that was hatching on June 7, 1983, (described above), "I don't think our lives will ever be the same" and he has right. In the end, it seems that short-term political expediency will prevail over scientific evidence when there is a conflict, especially when the research findings have large implications for interests other than the environment.

In general, our discoveries of the effects of selenium on birds at Kesterson Reservoir have made a significant contribution to environmental stewardship, though there was a personal cost to several of us who were doing the research or disseminating information about the work. The awareness of selenium's potentially toxic effects have expanded to other sources, such as mining (primarily coal, metals, and phosphate), power generation (mainly associated with fly ash), and petroleum extraction or refining. Aquatic birds exposed in environments affected by agricultural irrigation or industrial activities continue to be at some risk, but many advances in selenium evaluation and management have been made over the past 25 years to reduce the levels of exposure and effects. Nevertheless, the problems remain daunting, because of the complex biogeochemistry of selenium in the environment.

In reflecting on my experiences and my career path that changed as a result of the research that I undertook at Kesterson Reservoir in 1983, a couple of logical questions might be "Is there something I would have done differently?", and "Are there critical actions (perhaps toward a particular person with a political agenda, speaking to the press, etc.) that came back to bite me in the end?"

My overall sense is that there is not much I would have done differently, though I realize there is a possibility that might be interpreted as denial of responsibility on my part for what unfolded. Throughout, I did my best to focus on the science and understanding of selenium cycling and behavior in the environment and how selenium affects animals that are exposed to it. By maintaining objectivity and not taking a political or partisan position on what should be done, I believe I was able to focus on discovery of new information that could be applied or used by others toward making management decisions and that the research findings have been broadly accepted as being well documented. When possible, I provided interpretation of the data, because selenium biogeochemistry and ecotoxicology are complex and not easily understood. I do not profess to know everything on those subjects, but I tried to stick to the facts, as I understood them. I believe that is essential if the research findings are to be accepted as being valid and unbiased.

Because of the urgency of having usable information available for making decisions as related to the management of agricultural drainage water in the San Joaquin Valley, there was a lot of pressure for us to get our samples analyzed quickly and to provide interpretation when we were reasonably confident of what the data meant. This sometimes resulted in my speaking to the press or other news media directly or in releasing photos for their use, but I followed protocols in getting clearance for doing so. When I received invitations for presentations at meetings or symposia I routed them up through my administrative line within the USFWS. Unfortunately, some of those requests apparently were viewed by some individuals in the agency as promoting an activist agenda and were denied, which sometimes resulted in extensive back-and-forth exchanges about why presentations were being denied. In the end, that came back to bite me, and it seemed that it would have been safer to focus on issues of lower public interest. Nevertheless, since leaving the USFWS 20 years ago I have continued to focus on selenium-related issues for study, understanding, and management, and have enjoyed the opportunities to make a difference toward our stewardship of the Earth that we are borrowing from future generations.

Summary

Selenium was identified in the 1930s as the cause of embryo mortality and severe embryo deformities when chickens were fed grains grown on seleniferous soils in South Dakota. There had been no documented occurrences of such effects in wild birds before 1983, when we studied the effects of agricultural irrigation drainage water contaminants on birds feeding and nesting at Kesterson Reservoir, which was located within the Kesterson NWR and was used for disposal of subsurface saline drainage waters from agricultural fields. The Reservoir was intended to provide beneficial habitat for wildlife, particularly waterfowl and other aquatic birds.

During 1983–1985, we studied the exposure and effects of contaminants on several species of birds that were common at one or both of our study sites (Kesterson Reservoir and a nearby reference site). Analyses of food-chain biota (plants, aquatic invertebrates, and fish) and bird tissues or eggs showed that selenium was the only chemical found at concentrations high enough to cause the adverse effects on bird health or reproduction we observed.

Results of the studies at Kesterson Reservoir stimulated interest and concern about the effects of selenium in agricultural drainage in other portions of the San Joaquin Valley of California and the western USA where similar scenarios might exist. Those studies showed that selenium-related problems with agricultural drainage were widespread and locally significant. In addition, selenium was found to be an issue of concern with regard to mining of coal, phosphate, and metals; disposal of fly ash from coal-fired power plants; and refining of petroleum. Effects of selenium on fish have been documented in many of the same areas where selenium is of concern for birds.

Management problems of selenium in agricultural drainwater are difficult to solve or mitigate, despite intensive efforts to do so. There is no single action that can be implemented to manage seleniferous drainage; instead, a combination of actions (such as source control, drainage reuse, evaporation systems [designed to minimize bird use], land retirement, controlled discharges, etc.) must be used.

Discovery of the effects of selenium on birds (and fish) at Kesterson Reservoir, and subsequently at other sites in the western USA where seleniferous agricultural waters were entering wetland habitats, created an awareness of the needs for resource protection. Based on experience from the SJVDP and NIWQP, it became evident that the success of selenium management plans depends on partnership between federal, state, and local entities and on strong commitments by all participants to the management efforts. There has been success in some areas, but management of agricultural drainage is still a significant problem in many others.

Our discoveries of the effects of selenium on birds at Kesterson Reservoir also have made a significant contribution to environmental stewardship though the awareness of selenium's potentially toxic effects through industrial activities such as mining (primarily coal, metals, and phosphate), power generation (mainly associated with fly ash), and petroleum extraction or refining. Aquatic birds exposed in environments affected by agricultural irrigation or industrial activities continue to be at some risk, but many advances in selenium evaluation and management have been made over the past 25 years to reduce the levels of exposure and effects. Nevertheless, the problems remain daunting, because of the complex biogeochemistry of selenium in the environment.

Acknowledgments While conducting the initial studies at Kesterson (1983–1985), I was employed by the US Fish and Wildlife Service, which provided funding for the research. Many other USFWS biologists participated in the field studies, conducted the related experimental studies, or provided useful insight concerning interpretation of research findings. For such contributions, I especially acknowledge the help of T.W. Aldrich, D.R. Clark, Jr., G.H. Heinz, D.J. Hoffman, R.L. Hothem, M.K. Saiki, J.P. Skorupa, F.E. Smith, and G.R. Zahm. Biological monitoring at Kesterson Reservoir has been conducted by CH2M HILL (primarily by G.M. Santolo) for Reclamation since 1987. Scientists from other agencies (such as I. Barnes and T.S. Presser from USGS) also provided very helpful contributions to the studies of selenium at Kesterson Reservoir.

References

Albers PH, Green DE, Sanderson CJ (1996) Diagnostic criteria for selenium toxicosis in aquatic birds: dietary exposure, tissue concentrations, and macroscopic effects. J Wildl Dis 32:468–485

Blum D (1983) Mineral is linked to bird deformities. The Fresno Bee, September 21, pp. A1, A14

Byron ER, Ohlendorf HM, Santolo GM, Benson SM, Zawislanski PT, Tokunaga TK, Delamore M (2003) Ecological risk assessment example: waterfowl and shorebirds feeding in ephemeral pools at Kesterson Reservoir, California. In: Hoffman DJ, Rattner BA, Burton GA Jr, Cairns J Jr (eds) Handbook of ecotoxicology, 2nd edn. Lewis, Boca Raton, Florida, pp 985–1014

Central Valley Regional Water Quality Control Board (CVRWQCB) (2001) Total maximum daily load for selenium in the lower San Joaquin River. Sacramento, California. Staff Report. August

CH2M HILL and Lawrence Berkeley National Laboratory (CH2M HILL and LBNL) (2000) Ecological risk assessment for Kesterson Reservoir, Prepared for U.S. Bureau of Reclamation Mid-Pacific Region, Sacramento, CA. December

Chapman PM, Adams WJ, Brooks ML, Delos CG, Luoma SN, Maher WA, Ohlendorf HM, Presser TS, Shaw DP (eds) (2010) Ecological assessment of selenium in the aquatic environment. SETAC, Pensacola, Florida

Davis DE, Hanson CH, Hansen RB (2008) Constructed wetland habitat for American avocet and black-necked stilt foraging and nesting. J Wildl Manage 72:143–151

Deason JP (1986) U.S. Department of the Interior investigations of irrigation-induced contamination problems. In: Summers JB, Anderson SS (eds) Toxic substances in agricultural water supply and drainage – defining the problem – proceedings of regional meeting of the U.S. Committee on Irrigation and Drainage, . September 1986, Boulder, Colorado: U.S Government Printing Office, Washington, DC, pp 201–210

Engberg RA, Westcot DW, Delamore M, Holz DD (1998) Federal and state perspectives on regulation and remediation of irrigation-induced selenium problems. In: Frankenberger WT Jr, Engberg RA (eds) Environmental chemistry of selenium. Marcel Dekker, New York, pp 1–25

Entrix (2008) Grassland Bypass Project, 2010–2019 Environmental Impact Statement and Environmental Impact Report. Draft dated December 2008. Prepared for U.S. Bureau of Reclamation and San Luis & Delta-Mendota Water Authority by Entrix, Concord, California

Evaporation Ponds Technical Committee (EPTC) (1999) Evaporation ponds final report. The San Joaquin Valley Drainage Implementation Program and The University of California Salinity/ Drainage Program. February
Franke KW, Tully WC (1935) A new toxicant occurring naturally in certain samples of plant food-stuffs. V: Low hatchability due to deformities in chicks. Poultry Sci 14:273–279
Frankenberger WT Jr, Engberg RA (eds) (1998) Environmental chemistry of selenium. Marcel Dekker, New York
Garone P (1999) The tragedy at Kesterson Reservoir: a case study in environmental history and a lesson in ecological complexity. Environs 22:107–144
Gilmer DS, Miller MR, Bauer RD, LeDonne JR (1982) California's Central Valley wintering waterfowl: concerns and challenges. Trans N Am Wildl Nat Resour Conf 47:441–452
Gordus AG (1999) Selenium concentrations in eggs of American avocets and black-necked stilts at an evaporation basin and freshwater wetland in California. J Wildl Manage 63:497–501
Hamilton SJ, Buhl KJ, Faerber NL, Wiedmeyer RH, Bullard FA (1990) Toxicity of organic selenium in the diet to Chinook salmon. Environ Toxicol Chem 9:347–358
Hanson BR (1982) A master plan for drainage in the San Joaquin Valley. Calif Agric 36(5&6):9–11
Harris T (1991) Death in the marsh. Island, Washington DC
Heinz GH (1993a) Re-exposure of mallards to selenium after chronic exposure. Environ Toxicol Chem 12:1691–1694
Heinz GH (1993b) Selenium accumulation and loss in mallard eggs. Environ Toxicol Chem 12:775–778
Heinz GH (1996) Selenium in birds. In: Beyer WN, Heinz GH, Redmon-Norwood AW (eds) Environmental contaminants in wildlife: interpreting tissue concentrations. Lewis, Boca Raton, Florida, pp 447–458
Heinz GH, Fitzgerald MA (1993a) Overwinter survival of mallards fed selenium. Arch Environ Contam Toxicol 25:90–94
Heinz GH, Fitzgerald MA (1993b) Reproduction of mallards following overwinter exposure to selenium. Environ Pollut 81:117–122
Heinz GH, Hoffman DJ (1998) Methylmercury chloride and selenomethionine interactions on health and reproduction in mallards. Environ Toxicol Chem 17:139–145
Heinz GH, Hoffman DJ, Krynitsky AJ, Weller DMG (1987) Reproduction in mallards fed selenium. Environ Toxicol Chem 6:423–433
Heinz GH, Hoffman DJ, Gold LG (1988) Toxicity of organic and inorganic selenium to mallard ducklings. Arch Environ Contam Toxicol 17:561–568
Heinz GH, Hoffman DJ, Gold LG (1989) Impaired reproduction of mallards fed an organic form of selenium. J Wildl Manage 53:418–428
Heinz GH, Pendleton GW, Krynitsky AJ, Gold LG (1990) Selenium accumulation and elimination in mallards. Arch Environ Contam Toxicol 19:374–379
Hoffman DJ, Heinz GH (1988) Embryotoxic and teratogenic effects of selenium in the diet of mallards. J Toxicol Environ Health 24:477–490
Hoffman DJ, Heinz GH (1998) Effects of mercury and selenium on glutathione metabolism and oxidative stress in mallard ducks. Environ Toxicol Chem 17:161–166
Hoffman DJ, Ohlendorf HM, Aldrich TW (1988) Selenium teratogenesis in natural populations of aquatic birds in central California. Arch Environ Contam Toxicol 17:519–525
Hoffman DJ, Heinz GH, Krynitsky AJ (1989) Hepatic glutathione metabolism and lipid peroxidation in response to excess dietary selenomethionine and selenite in mallard ducklings. J Toxicol Environ Health 27:263–271
Hoffman DJ, Heinz GH, LeCaptain LJ, Bunck CM, Green DE (1991) Subchronic hepatotoxicity of selenomethionine ingestion in mallard ducks. J Toxicol Environ Health 32:449–464
Hothem RL, Ohlendorf HM (1989) Contaminants in foods of aquatic birds at Kesterson Reservoir, California, 1985. Arch Environ Contam Toxicol 18:773–786
Hothem RL, Welsh D (1994) Contaminants in eggs of aquatic birds from the Grasslands of Central California. Arch Environ Contam Toxicol 27:180–185
Interagency Drainage Program (IDP) (1979) Agricultural drainage and salt management in the San Joaquin Valley. Final Report Including Recommended Plan and First-Stage Environmental

Impact Report. San Joaquin Valley Interagency Drainage Program: U.S. Bureau of Reclamation, California Department of Water Resources, and California State Water Resources Control Board, p 189 + appendices
Jones and Stokes (1977) An evaluation of the feasibility of utilizing agricultural tile drainage water for marsh management in the San Joaquin Valley, California. Prepared for U.S. Fish and Wildlife Service and U.S. Bureau of Reclamation by Jones and Stokes Associates, Inc. Sacramento, California, p 172
Madson J (1984) A lot of trouble and a few triumphs for North American waterfowl. National Geographic November issue, pp 562–599
Moxon AL, Olson OE (1974) Selenium in agriculture. In: Zingaro RA, Cooper WC (eds) Selenium. Van Nostrand Reinhold, New York, pp 675–707
National Academy of Sciences (NAS) (1976) Medical and biologic effects of environmental pollutants: Selenium. Division of Medical Sciences, National Research Council, NAS, Washington, DC, pp 203
National Irrigation Water Quality Program (NIWQP) (2009) National Irrigation Program Projects. http://www.usbr.gov/niwqp/niwqpprojects/index.html. Accessed 19 Sep 2009
O'Toole D, Raisbeck MF (1997) Experimentally induced selenosis of adult mallard ducks: clinical signs, lesions, and toxicology. Vet Pathol 34:330–340
O'Toole D, Raisbeck MF (1998) Magic numbers, elusive lesions: comparative pathology and toxicology of selenosis in waterfowl and mammalian species. In: Frankenberger WT Jr, Engberg RA (eds) Environmental chemistry of selenium. Marcel Dekker, New York, pp 355–395
Ohlendorf HM (1983) Environmental contaminants in aquatic birds and their food chains in marsh habitats created by using San Luis Drain water. Study Plan 906.01.03. Patuxent Wildlife Research Center, U.S. Fish and Wildlife Service, Laurel, Maryland. March 15
Ohlendorf HM (1986) Letter to Mr. Ken Anderson, U.S. General Accounting Office dated May 16
Ohlendorf HM (1989) Bioaccumulation and effects of selenium in wildlife. In: Jacobs LW (ed) Selenium in agriculture and the environment. Soil Science Society of America, Madison, Wisconsin. Special Publication No. 23:133–177
Ohlendorf HM (1996) Selenium. In: Fairbrother A, Hoff GL, Locke LN (eds) Noninfectious diseases of wildlife, 2nd edn. Iowa State University Press, Ames, Iowa, pp 128–140
Ohlendorf HM (2002) The birds of Kesterson Reservoir: a historical perspective. Aquatic Toxicol 57:1–10
Ohlendorf HM (2003) Ecotoxicology of selenium. In: Hoffman DJ, Rattner BA, Burton GA Jr, Cairns JC Jr (eds) Handbook of ecotoxicology, 2nd edn. Lewis, Boca Raton, Florida, pp 465–500
Ohlendorf HM, Heinz GH (2011) Selenium in birds. In: Beyer WN, Meador JP (eds) Environmental contaminants in biota: interpreting tissue concentrations (Final Draft Chapter dated May 1, 2009; update for final draft of this chapter), 2nd edn. Taylor and Francis, Boca Raton, Florida
Ohlendorf HM, Hothem RL (1995) Agricultural drainwater effects on wildlife in central California. In: Hoffman DJ, Rattner BA, Burton GA Jr, Cairns J Jr (eds) Handbook of ecotoxicology. Lewis, Boca Raton, Florida, pp 577–595
Ohlendorf HM, Santolo GM (1994) Kesterson reservoir – past present, and future: an ecological risk assessment. In: Frankenberger WT Jr, Benson S (eds) Selenium in the environment. Marcel Dekker, New York, pp 69–117
Ohlendorf HM, Skorupa JP (1989) Selenium in relation to wildlife and agricultural drainage water. In: Carapella SC Jr (ed) Proceedings, Fourth International Symposium on Uses of Selenium and Tellurium, 8–10 May 1989, Banff, Alberta. Selenium-Tellurium Development Association, Darien, Connecticut, pp 314–338
Ohlendorf HM, Hoffman DJ, Saiki MK, Aldrich TW (1986a) Embryonic mortality and abnormalities of aquatic birds: apparent impacts of selenium from irrigation drainwater. Sci Total Environ 52:49–63
Ohlendorf HM, Lowe RW, Kelly PR, Harvey TE (1986b) Selenium and heavy metals in San Francisco Bay diving ducks. J Wildl Manage 50:64–71
Ohlendorf HM, Hothem RL, Aldrich TW, Krynitsky AJ (1987) Selenium contamination of the Grasslands, a major California waterfowl area. Sci Total Environ 66:169–183

Ohlendorf HM, Kilness AW, Simmons JL, Stroud RK, Hoffman DJ, Moore JF (1988) Selenium toxicosis in wild aquatic birds. J Toxicol Environ Health 24:67–92
Ohlendorf HM, Hothem RL, Welsh D (1989) Nest success, cause-specific nest failure, and hatchability of aquatic birds at selenium-contaminated Kesterson Reservoir and a reference site. Condor 91:787–796
Ohlendorf HM, Hothem RL, Bunck CM, Marois KC (1990) Bioaccumulation of selenium in birds at Kesterson Reservoir, California. Arch Environ Contam Toxicol 19:495–507
Ohlendorf HM, Skorupa JP, Saiki MK, Barnum DA (1993) Food-chain transfer of trace elements to wildlife. In: Allen RG, Neale CMU (eds) Management of irrigation and drainage systems: integrated perspectives. American Society of Civil Engineers, New York, pp 596–603
Ort JF, Latshaw JD (1977) The effect of intermediate levels of selenium in the diets of SCWL laying hens. Poultry Sci 56:1744
Ort JF, Latshaw JD (1978) The toxic level of sodium selenite in the diet of laying chickens. J Nutr 108:1114–1120
Paveglio FL, Bunck CM, Heinz GH (1992) Selenium and boron in aquatic birds from Central California. J Wildl Manage 56:31–42
Paveglio FL, Kilbride KM, Bunck CM (1997) Selenium in aquatic birds from Central California. J Wildl Manage 61:832–839
Presser TS, Luoma SN (2006) Forecasting selenium discharges to the San Francisco Bay-Delta Estuary: ecological effects of a proposed San Luis Drain Extension. Professional Paper 1646. U.S. Department of the Interior, U.S. Geological Survey. Reston, Virginia
Presser TS, Ohlendorf HM (1987) Biogeochemical cycling of selenium in the San Joaquin Valley, California, USA. Environ Manage 11:805–821
Presser TS, Schwarzbach SE (2008) Technical analysis of In-Valley Drainage Management Strategies for the Western San Joaquin Valley, California. Open-File Report 2008–1210. U.S. Department of the Interior, U.S. Geological Survey. Reston, Virginia
Saiki MK (1986) A field example of selenium contamination in an aquatic food chain. In: Proceedings Symposium on Selenium in the Environment. Publication CATI/860201, pp. 67–76. California Agricultural Technology Institute, California State University, Fresno
Saiki MK, Lowe TP (1987) Selenium in aquatic organisms from subsurface agricultural drainage water, San Joaquin Valley, California. Arch Environ Contam Toxicol 16:657–670
Saiki MK, Ogle RS (1995) Evidence of impaired reproduction by western mosquitofish inhabiting seleniferous agricultural drainwater. Trans Am Fish Soc 124:578–587
Saiki MK, Schmitt CJ (1985) Population biology of bluegills, *Lepomis macrochirus*, in lotic habitats on the irrigated San Joaquin Valley floor. Calif Fish Game 71:225–244
San Joaquin Valley Drainage Program (SJVDP) (1990) A management plan for agricultural subsurface drainage and related problems on the Westside San Joaquin Valley. Final Report of the San Joaquin Valley Drainage Program. U.S. Department of the Interior and California Resources Agency, Sacramento, California. September
Santolo GM, Yamamoto JT, Pisenti JM, Wilson BW (1999) Selenium accumulation and effects on reproduction in captive American kestrels fed selenomethionine. J Wildl Manage 63:502–511
Schuler CA, Anthony RG, Ohlendorf HM (1990) Selenium in wetlands and waterfowl foods at Kesterson Reservoir, California, 1984. Arch Environ Contam Toxicol 19:845–853
Seiler RL, Skorupa JP, Peltz LA (1999) Areas susceptible to irrigation-induced selenium contamination of water and biota in the Western United States. U.S. Geological Survey Circular 1180, Denver, Colorado
Skorupa JP, Ohlendorf HM (1991) Contaminants in drainage water and avian risk thresholds. In: Dinar A, Zilberman D (eds) The economics and management of water and drainage in agriculture. Kluwer, Boston, pp 345–368
Skorupa JP, Schwarzbach SE (1995a) Compensation habitat protocol, USFWS, for drainwater evaporation basins. Sacramento Fish and Wildlife Office, USFWS, Sacramento, California. January
Skorupa, JP, Schwarzbach SE (1995b) Alternative habitat protocol, USFWS, for drainwater evaporation basins. Sacramento Fish and Wildlife Office, USFWS, Sacramento, California. March

Smith FE (2007) Statement to the house natural resources subcommittee on water and power, hearing on extinction is not sustainable water policy: the Bay-Delta Crisis and Implications for California Water Management, Vallejo, California. July 2

Smith GJ, Heinz GH, Hoffman DJ, Spann JW, Krynitsky AJ (1988) Reproduction in black-crowned night-herons fed selenium. Lake Reservoir Manage 4:175–180

Stanley TR Jr, Spann JW, Smith GJ, Rosscoe R (1994) Main and interactive effects of arsenic and selenium on mallard reproduction and duckling growth and survival. Arch Environ Contam Toxicol 26:444–451

Stanley TR Jr, Smith GJ, Hoffman DJ, Heinz GH, Rosscoe R (1996) Effects of boron and selenium on mallard reproduction and duckling growth and survival. Environ Toxicol Chem 15:1124–1132

Suter GW II (1993) Ecological risk assessment. Lewis, Boca Raton, Florida

U.S. Bureau of Reclamation (USBR) (1986) Final Environmental Impact Statement, Kesterson Reservoir. USBR, Mid-Pacific Region, in cooperation with USFWS and U.S. Army Corps of Engineers

U.S. Bureau of Reclamation (USBR) (2006) San Luis Drainage Feature Re-evaluation Final Environmental Impact Statement, Mid-Pacific Region, Sacramento, California. May

U.S. Bureau of Reclamation (USBR) (2007) San Luis Drainage Feature Re-evaluation Record of Decision, Mid-Pacific Region, Sacramento, California. March

U.S. Bureau of Reclamation (USBR) (2008) San Luis Drainage Feature Re-evaluation Feasibility Report, Mid-Pacific Region, Sacramento, California. March

U.S. Department of the Interior (USDI) (1989) Submission to California State Water Resources Control Board in Response to Order No. WQ-88-7, Effectiveness of Filling Ephemeral Pools at Kesterson Reservoir, Kesterson Program Upland Habitat Assessment, and Kesterson Reservoir Final Cleanup Plan

U.S. Department of the Interior (USDI) (1998) Guidelines for Interpretation of the Biological Effects of Selected Constituents in Biota, Water, and Sediment. National Irrigation Water Quality Program Information Report No. 3. USDI, Denver, CO. November

U.S. Fish and Wildlife Service (USFWS) (1982) Regional resource plan, Region 1. USFWS, Portland, Oregon

U.S. General Accounting Office (USGAO) (1987) National refuge contamination is difficult to confirm and clean up. Report to the Chairman, Subcommittee on Oversight and Investigations, Committee on Energy and Commerce, House of Representatives. GAO/RCED-87-128. USGAO, Gaithersburg, Maryland

U.S. Geological Survey (USGS) (2003) Selenium contamination and remediation at Stewart Lake Waterfowl Management Area and Ashley Creek, Middle Green River Basin, Utah. USGS Fact Sheet 031–03 (Authored by Rowland RC, Stephens DW, Waddell B, Naftz DL; http://pubs.usgs.gov/fs/fs-031-03/)

White JR, Hofman PS, Hammond D, Baumgartner S (1987) Selenium verification study. A Report to the California State Water Resources Control Board from the California Department of Fish and Game. Stockton and Rancho Cordova, California. May

Wiemeyer SJ, Hoffman DJ (1996) Reproduction in eastern screech-owls fed selenium. J Wildl Manage 60:332–341

Williams ML, Hothem RL, Ohlendorf HM (1989) Recruitment failure in American avocets and black-necked stilts nesting at Kesterson Reservoir, California, 1984–1985. Condor 91:797–802

Yamamoto JT, Santolo GM (2000) Body condition effects in American kestrels fed selenomethionine. J Wildl Dis 36:646–652

Yamamoto JT, Santolo GM, Wilson BW (1998) Selenium accumulation in captive American kestrels (*Falco sparverius*) fed selenomethionine and naturally incorporated selenium. Environ Toxicol Chem 17:2494–2497

Chapter 12
Eliminating Lead from Recreational Shooting and Angling: Relating Wildlife Science to Environmental Policy and Regulation in North America

A.M. Scheuhammer and Vernon G. Thomas

Abstract The manufacture of projectiles for ammunition used in hunting and target shooting, and for terminal tackle (sinkers and jigs) used in recreational angling, comprises a significant continuing commercial use of lead, and a major source of lead deposition into the environment. Thousands of tons of metallic lead are deposited into the North American environment annually from hunting, target-shooting, and recreational angling activities. Numerous symposia and conferences have been held, and hundreds of research papers have been published, addressing lead exposure and toxicosis in wildlife from ingestion of spent lead ammunition and fishing sinkers, but the transition (regulatory or otherwise) to nontoxic substitutes has been slow, impeded in large part by the resistance of hunters, anglers, and their representative organizations to adopt nontoxic products, rather than an inability of the ammunition and tackle industries to manufacture and distribute such products. Here, we present a historical analysis of the interactions between environmental science and regulatory policy development with respect to the use of lead in recreational shooting and angling in North America.

Introduction

Lead is a toxic heavy metal with no known nutritional function in animals. Abundant scientific evidence of lead's detrimental effects on the developing nervous system and cognitive function in children has led to a consideration that there may be no safe level of human lead exposure. The manufacture of projectiles for ammunition used in hunting and target shooting, and for terminal tackle (sinkers and jigs) used

A.M. Scheuhammer (✉)
Environment Canada, National Wildlife Research Centre, Carleton University, Ottawa, ON, K1A 0H3, Canada
e-mail: Tony.Scheuhammer@ec.gc.ca

J.E. Elliott et al. (eds.), *Wildlife Ecotoxicology: Forensic Approaches*, Emerging Topics in Ecotoxicology 3, DOI 10.1007/978-0-387-89432-4_12,

in recreational angling, comprise a significant continuing commercial use of lead, and a major source of lead deposition into the environment. Thousands of tons of metallic lead are deposited into the North American environment annually from hunting, target-shooting, and recreational angling activities. Numerous symposia and conferences have been held, and hundreds of research papers have been published, addressing lead exposure and toxicosis in wildlife from ingestion of spent lead ammunition and fishing sinkers, but the transition (regulatory or otherwise) to nontoxic substitutes has been slow, impeded in large part by the resistance of hunters, anglers, and their representative organizations to adopt non-toxic products, rather than an inability of the ammunition and tackle industries to manufacture and distribute such products. Here, we present a historical analysis of the interactions between environmental science and regulatory policy development with respect to the use of lead in recreational shooting and angling in North America.

Although many of the once-common uses of lead in North America (for example, in water pipes, paints, glazes, solder, and gasoline) have been gradually phased out or dramatically reduced, due mainly to concerns for lead exposure and toxicity in humans, the use of lead in recreational shooting and angling continues (albeit with some restrictions). The manufacture of projectiles for ammunition used in hunting and target shooting, and for terminal tackle (sinkers and jigs) used in recreational angling, comprise a significant continuing commercial use of lead, and a major source of lead deposition into the environment. Thousands of tons of metallic lead are deposited into the North American environment annually from hunting, target-shooting, and recreational angling activities.

Ecosystem health and wildlife management issues surrounding the use of lead ammunition, especially for game bird hunting, have been studied and debated for over a century. Numerous symposia and conferences have been held, and hundreds of research papers have been published, addressing lead exposure and toxicosis in wildlife from ingestion of spent lead ammunition and fishing sinkers, but the transition (regulatory or otherwise) to nontoxic substitutes has been very slow, impeded in large part by the resistance of hunters, anglers, and their representative organizations to adopt nontoxic products, rather than an inability of the ammunition and tackle industries to manufacture and distribute such products. More recently, questions regarding human exposure to lead from consumption of the flesh of hunter-killed game animals have entered the discussion. In May of 2008, we participated in a landmark international conference in Boise, Idaho, where the risks to both wildlife and humans from lead ammunition (gunshot and bullets) were presented and discussed. In the same year, The Wildlife Society, and The American Fisheries Society, published a collaborative Technical Review of the impacts of lead ammunition and fishing sinkers on fish and wildlife (Rattner et al. 2008). Shortly thereafter (July 2009), The Wildlife Society released its Final Position Statement on risks to wildlife arising from use of lead ammunition and fishing tackle, advocating the replacement of lead with nontoxic products (http://joomla.wildlife.org/documents/positionstatements/Lead_final_2009.pdf). Also in 2009, the International Council for Game and Wildlife Conservation (CIC) passed a resolution calling for an expert assessment of the consequences of the use of lead ammunition, and the need to adopt nontoxic substitutes (http://www.cic-wildlife.org/uploads/media/Recommendation_lead_shot_2009_EN_01.pdf). Accordingly, a panel comprising some of the world's foremost scientific experts on the lead ammunition issue, along with representatives of hunting associations, held a workshop in Aarhus, Denmark, November 2009. A Workshop Resolution resulting from this meeting and presented to the CIC indicated that "risks to wildlife, humans and the environment require urgent adoption of the use of nontoxic ammunition"; that "voluntary or partial restrictions on the use of lead ammunition have been largely ineffective"; and recommended "a phase-in of nontoxic ammunition for all hunting and shooting as soon as practicable" (Kanstrup 2010; www.cic-wildlife.org/uploads/media/CIC_Sustainable_Hunting_Ammunition_Workshop_Report_low_res.pdf).

The scientific understanding of this issue is now very extensive, and additional research extends our knowledge only in relatively minor ways, such as providing more information about additional species or places at risk. The use of nontoxic shot by waterfowl hunters in the US, Canada, and some European countries for well over a decade has shown that the sport of hunting is not significantly diminished by the use of lead-free ammunition. The same conclusion can be applied to the use of nontoxic sinkers and jigs in sport angling. Although yet more scientific information may be useful, a greater need is the political will to extend existing regulatory requirements for nontoxic ammunition and tackle to all categories of hunting, shooting, and angling. Here we present a brief history of the scientific, policy, and regulatory responses to these issues, focusing primarily on the US and Canada. Based on an extensive body of scientific research, we discuss options for future policy and regulatory development.

Toxicology of Lead

The main target organs and physiological processes impacted by high lead exposure are: the brain and peripheral nervous system (neurotoxicity); the kidneys (nephropathy); and the red blood cells (hematotoxicity). A rather typical constellation of biochemical and physiological effects is often associated with elevated lead exposure, including inhibition of δ-aminolevulinic acid dehydratase (ALAD) in blood, anemia, low hematocrit, renal impairment, intranuclear lead inclusion bodies in kidney cells, cognitive impairment, weakness, partial paralysis of limb muscles, emaciation, and, if exposure is sufficiently high, death. Ingested lead has also been implicated as a likely carcinogen in animals (Johnson 1998; Silbergeld 2003). The ability of lead (Pb^{++}) to substitute for calcium (Ca^{++}), whether in bone or other cellular sites, and to interfere with calcium-based biochemical processes is well established (Simons 1993; Bressler et al. 1999). Animals consuming low calcium diets absorb relatively more lead, and are more likely to suffer from lead toxicity, than animals consuming high calcium diets (Six and Goyer 1970; Carlson and Nielson 1985; Scheuhammer 1996). Lead poisoning has been documented in individuals of virtually all vertebrate taxa (Pokras and Kneeland 2009). The signs of lead exposure and mechanisms of toxicity are similar for different vertebrate species, including humans, other mammals, and birds (Sanderson and Bellrose 1986; Goyer 1993; Needleman 2004), though some evidence suggests that birds may be generally less sensitive to lead exposure than mammals (Lumeij 1985). A major source of toxic lead exposure in wild birds is the ingestion of metallic lead from spent ammunition (lead gunshot and bullet fragments); and, under certain conditions for some species, ingestion of lost fishing sinkers and jigs.

Lead in Wetland Hunting: Historical Perspectives

Lead shot poisoning of waterfowl was first described over a century ago by Grinnell (1894), who examined ducks, geese, and swans that were found dead or moribund in heavily hunted locations in Texas. These birds had lead shot in their gizzards – a clear indication of shot ingestion – and exhibited pathological changes consistent with lead poisoning. Grinnell stated that "*From these examinations I conclude that the birds dissected died from chronic lead poisoning, the cause of which was sufficiently obvious*" (Grinnell 1901). Although the cause of poisoning may have been obvious, and additional reports of the phenomenon soon followed (e.g. – Bowles 1908; McAtee 1908; Wetmore 1919), the question of what to do about the problem, if anything, has been debated ever since. Although initially considered a rather minor conservation issue, Shillinger and Cottam (1937) concluded that the problem of lead shot poisoning of waterfowl was much more serious and widespread than generally thought; and Quortrup and Shillinger (1941) reported that lead poisoning was a major cause of waterfowl mortality in the USA. However, no viable alternatives to lead shot ammunition were available and from a waterfowl management perspective, no significant actions were undertaken until after the publication of the landmark study by Bellrose (1959), who estimated that 2–3% of the North American waterfowl population, or up to four million individuals, died annually of lead shot poisoning. Although Bellrose (1959) did not advocate "*drastic regulations*" such as the regulatory prohibition of lead shot for waterfowl hunting, his study, more than any previous research, ultimately provided the scientific basis for substantive policy and regulatory development in North America, especially after concerns about declining waterfowl population numbers increased during the 1960s. Although the conclusions reached by Bellrose (1959) appear initially to have been accepted without much debate, his work was strongly challenged following a USFWS (1974) proposal to restrict the use of lead shot for hunting waterfowl in some locations. A number of arguments were presented in an attempt to forestall regulations prohibiting the use of lead shot. The most common of these were (and to some extent, still are): (1) Bellrose's estimates of waterfowl mortality from lead shot poisoning were erroneously high; (2) nontoxic alternatives to lead shot (primarily steel shot) are ballistically inferior to lead and their widespread use will result in increased losses of waterfowl through crippling, negating any small benefit that might be derived from a decrease in lead poisoning; (3) steel shot damages gun barrels; (4) nontoxic ammunition is more expensive than lead; (5) other less onerous means (e.g. – provision of high-calcium grit to mitigate lead poisoning, and active "hazing" to frighten birds away from problem areas) should be employed to solve the lead poisoning problem, without resorting to prohibitive regulations. These objections have, over time, been shown to be largely spurious (although they still receive varying levels of support from opponents of nontoxic ammunition regulations). Later estimates of waterfowl mortality from lead shot poisoning were generally in accord with those of Bellrose (1959) – for example, Clemens et al. (1975), and Sanderson and Bellrose (1986). Coating lead shot with other metals or nonmetallic materials such as plastics failed to reduce the toxicity of the shot when ingested by birds (USFWS 1986). Although some early steel shot ammunition was problematic, both from a ballistic and a barrel damage point of view, these problems were addressed and solved by the ammunition industry (Brister 1992; Colburn 1992). Increasing the size of steel pellets compensated for steel's lower density; and increasing the propellant charge and using a magnum primer ensured that the retained velocity of steel pellets was comparable to that of lead pellets two sizes smaller. Steel shot cartridges were loaded with a greater volume of pellets to compensate for their larger size, and a rigid plastic wad was added to prevent steel pellets from contacting and possibly scoring gun barrels upon discharge. Major arms and ammunition manufacturers confirmed that modern steel shot loads caused no significant reduction in the life of most US full-choke shotguns (USFWS 1986). In none of several lawsuits brought against the US federal government by various opponents of nontoxic shot regulations (mainly hunters and their sponsoring organizations) have courts accepted arguments that steel shot is ballistically inferior to lead, cripples excessively, or damages firearms (Feierabend 1985). Although steel shot cartridges can be slightly more expensive than lead cartridges, the cost increase is minor and certainly not prohibitive for hunters as only a small proportion of a hunter's budget is typically spent on ammunition. Scheuhammer and Norris (1995) estimated that the average waterfowl hunter's overall annual hunting expenditure would increase by only about 2% after a switch from lead to steel ammunition. In addition, over the past 20 years, an increasing number of other nontoxic shot products have been developed that have densities and ballistic properties more similar to those of lead, including various combinations of tungsten, bismuth, tin, and iron. These newer products have given waterfowl hunters an increasingly wide choice of nontoxic shot ammunition. Because lead shot ingestion and poisoning in waterfowl were found to occur over wide geographic areas, labour-intensive habitat manipulation – such as hazing, provision of high-calcium grit, or plowing to bury lead pellets – is impractical as a general control option.

In the US, nontoxic shot zones were first established in the 1970s, after the U.S. Fish and Wildlife Service (USFWS) drafted an environmental impact statement (USFWS 1976) that considered a gradual transition from the use of lead shot to the use of steel shot in areas deemed by the USFWS to have a significant lead shot poisoning problem. A number of state governments, however, were opposed to nontoxic shot zones, and hearings on the issue resulted in the Stevens Amendment, which gave state regulatory authorities the option to not approve the implementation of nontoxic shot zones within their jurisdictions (Anderson 1992; Weyhrauch 1986). Because states did not have to consent to proposed nontoxic shot regulations, the final decision to establish nontoxic shot zones lay with the states and not with the federal government. The Stevens Amendment significantly slowed the transition towards nontoxic shot usage in the USA; but another factor emerged that would ultimately tip the balance in the other direction, towards an even broader prohibition on the use of lead shot than was initially envisioned. This was the issue of so-called "secondary poisoning" of bald eagles (*Haliaeetus leucocephalus*), then a protected species in the USA under the Endangered Species Act. Research showed that predatory raptors – and in particular, bald eagles – were becoming lead poisoned through ingestion of lead shot embedded in the carcasses of waterfowl killed or wounded with lead shot (Pattee and Hennes 1983; USFWS 1986). In 1984, the National Wildlife Federation (NWF) petitioned the USFWS to take immediate action to protect bald eagles from lead shot poisoning. In response, the USFWS developed criteria for nontoxic shot zoning based on protection of eagles as well as protection of waterfowl; however, more than half of the states scheduled for zoning in 1985 under the new criteria for protection of bald eagles did not comply with the proposed regulations, citing the Stevens Amendment. The NWF then sued the federal government, charging that by allowing the continued use of lead shot, the government was in violation of the Endangered Species Act, and the Migratory Bird Treaty Act. During the trial, the government proposed to phase out the use of lead shot for all waterfowl hunting within 5 years, and the case was dismissed. A further court ruling in 1987 determined that the Stevens Amendment did not repeal the authority of the federal government under the Migratory Bird Treaty Act, the Endangered Species Act, or the Bald and Golden Eagle Protection Act (Anderson 1992). By 1991, a general prohibition on the use of lead shot for waterfowl hunting was in effect in the USA (notwithstanding that some hunted migratory species, such as snipe and cranes, were exempted). Thus, almost 100 years had passed between the first North American scientific investigations of lead shot poisoning in waterfowl (Grinnell 1894) and the banning of lead shot for waterfowl hunting throughout the USA. Additional discussion of the transition from lead shot to steel shot for waterfowl hunting in the USA may be found in Friend et al. (2009).

In Canada, the management of the lead shot issue first began to receive attention by the Canadian Wildlife Service (CWS) in the late 1960s and early 1970s, paralleling discussions in the USA regarding the creation of the first nontoxic shot zones for waterfowl hunting. During this time, the CWS in collaboration with the National Research Council conducted studies on the production and ballistic properties of several alternative (non-lead) types of gunshot. A few studies on shot ingestion rates in waterfowl were also undertaken in some locations in Canada during this period. However, a pervasive belief among Canadian waterfowl managers was that, because waterfowl did not generally overwinter in Canada, lead shot ingestion could not be a significant problem in Canadian wetlands; and that a switch to steel shot, the only alternative to lead then available, might cripple far more waterfowl than would be saved by prohibiting the use of lead. Ultimately, Canada took no regulatory action against the use of lead for hunting until the USFWS announced a nation-wide prohibition on the use of lead shot for waterfowl hunting, to come into effect in the USA in 1991. In response, the CWS decided that it should either consider harmonizing with the US regulatory approach, or present persuasive evidence to show that lead shot ingestion was not a significant problem for waterfowl in Canada. There was still a general opinion among waterfowl managers (and certainly among major Canadian hunting associations) that lead shot poisoning in Canadian wetlands was probably a minor issue not requiring regulatory action. However, new research led by the CWS [collated in Kennedy and Nadeau (1993)] indicated that shot ingestion rates in a number of heavily hunted locations across Canada were, in general, comparable to ingestion rates reported for the US. Based on these studies, the CWS judged that there was sufficient evidence of shot ingestion in Canadian wetlands to develop a set of criteria for assessing whether local lead exposure in waterfowl was great enough to require nontoxic shot regulations. Thus as the USFWS had done earlier, the CWS established criteria for the regulatory creation of nontoxic shot zones (Wendt and Kennedy 1992). The CWS criteria were accepted in 1990 by federal and provincial wildlife Ministers as an interim policy for managing the problems associated with the use of lead shot for waterfowl hunting, and in 1991 and 1992 the first nontoxic shot zones were established in British Columbia, Manitoba, and Ontario. However, a subsequent study to determine the pattern of elevated bone-lead concentrations in hatch-year ducks across Canada during the hunting season reported an unexpectedly broad geographic association between elevated bone-lead and waterfowl hunting (Scheuhammer and Dickson 1996). In addition, lead shot poisoning of bald eagles was documented in Canada (Elliott et al. 1992). Scheuhammer and Norris (1995) reviewed the use and environmental impacts of lead shotshell ammunition in Canada, pointing out that because the US had banned lead shot nationally for waterfowl hunting, Canada was now responsible for an increasingly large proportion of the North American lead shot poisoning problem, as well as being a major source of migrating waterfowl carrying embedded lead shot that could potentially pose a hazard of lead poisoning for eagles and other predators and scavengers, both in Canada and USA. Scheuhammer and Norris (1995) also pointed out that neither the federal nor the provincial/territorial governments of Canada had the resources to adequately assess all areas for which nontoxic shot zoning might be appropriate, nor did they have the capacity to effectively enforce bans on the use of lead shot in increasingly numerous nontoxic shot zones. The difficulties associated with enforcing nontoxic shot regulations using a zoning approach, among other considerations, led some provinces (such as British Columbia) to move towards more comprehensive restrictions on the use of lead shot for waterfowl hunting within their jurisdictions. At the same time, the ammunition industry began to produce a wider variety of nontoxic shot products, some of which had ballistic properties close to those of lead, and this practice has continued. Thus, scientific, enforcement, and commercial aspects of the lead shot issue combined to facilitate an evolution towards increasingly broad restrictions on the use of lead shot for migratory bird hunting. Beginning in 1997, Canada prohibited the use of lead shot for the purpose of hunting migratory game birds within 200 m of any watercourse [exempting upland migratory species – American woodcock (*Scolopax minor*), mourning doves (*Zenaida macroura*), and band tailed pigeons (*Columba fasciata*)], and in 1999 this regulation was expanded to include dry land as well as wetland areas (although upland migratory species were, and still are, exempt).

After comprehensive prohibitions against the use of lead shot for most migratory bird hunting were established in USA and Canada, lead shot ingestion rates and tissue-lead concentrations in waterfowl declined substantially in both countries (Anderson et al. 2000; Stevenson et al. 2005). However, lead shot ingestion and poisoning continues to occur in some water bird species, such as swans in southern British Columbia and northern Washington (Degernes et al. 2006; Wilson et al. 2004), in areas where the availability of "old" lead shot may still be high, and/or deposition of lead from other activities such as target shooting and non-migratory game bird hunting, may be significant. Thus, there are aspects of the lead shot issue that continue to require further scientific and regulatory consideration.

Lead in Upland Hunting: Historical Perspectives

As with waterfowl, lead shot ingestion in upland game birds has been documented for over 100 years (Calvert 1876). The annual hunter harvest of upland game birds and small mammals in North America is at least as great as, and probably significantly greater than, that of waterfowl (Scheuhammer and Norris 1995; USFWS 1986). However there has been comparatively less research on lead shot ingestion rates and poisoning in upland birds. Similarly, the development of policy and regulation to protect upland birds from lead shot poisoning has not kept pace with actions to protect waterfowl and their predators. In Canada, migratory upland species such as American woodcock were explicitly exempted from the national regulation requiring nontoxic shot for migratory game bird hunting, based on a lack of information on lead shot ingestion or elevated lead accumulation in these species, and the notion that these species were in general not preyed upon by eagles. However, subsequent research indicated that woodcock in eastern Canada had a high incidence of elevated lead exposure, although the main sources of exposure were uncertain (Scheuhammer et al. 1999). In a follow-up study, it was decided to use stable lead isotope analysis to attempt to differentiate among different possible sources of lead in woodcock. Because the relative abundance of the different stable isotopes of lead (Pb^{204}, Pb^{206}, Pb^{207}, and Pb^{208}) change with different lead ore sources, lead isotope ratios can be used to help distinguish among different potential sources of environmental lead (Komárek et al. 2008). Using this forensic technique, Scheuhammer et al. (2003a) were able to eliminate some potential sources of lead exposure in woodcock, such as lead from Canadian gasoline combustion and lead from Precambrian mining sources; however, stable isotope signatures in woodcock bones with high lead exposure were consistent with lead shot ingestion. Thus, there is now evidence to include upland migratory game birds such as woodcock under nontoxic shot regulations. Other studies have also made use of stable lead isotope ratios to assess the importance of different environmental sources of lead with respect to lead accumulation in wild birds (e.g. – Scheuhammer and Templeton 1998; Pain et al. 2007; Thomas et al. 2009).

In a recent review, Pain et al. (2009) reported that over 60 terrestrial bird species, including Threatened and Near-Threatened species, are documented to ingest lead shot, and are thus at risk for lead poisoning. Kramer and Redig (1997) reported continued lead poisoning mortality of bald eagles well after national nontoxic shot regulations had been established, and suggested that eagles were continuing to be poisoned by ingesting lead bullet fragments while feeding on gut piles from hunter-killed deer. Similarly, on the western Canadian prairies, golden eagles (*Aquila chrysaetos*) feeding exclusively on terrestrial prey die of lead poisoning as frequently as bald eagles that typically feed on more aquatic prey (Wayland and Bollinger 1999); and the shooting of small nuisance mammals, such as gophers, with lead bullets may represent an important source of lead for upland raptors and scavengers (Fisher et al. 2006; Knopper et al. 2006). Lead poisoning from ingestion of lead bullet fragments embedded in carcasses of hunter-killed animals is a major cause of mortality for the California condor (*Gymnogyps californianus*) and has been an important factor limiting the successful reintroduction of this endangered species (Cade 2007; Church et al. 2006). Kendall et al. (1996) concluded that concern for lead shot toxicity in upland game birds and raptors preying upon them was warranted, and that the issue merited continued scrutiny to protect upland game birds and raptors. Replacing lead shot and high velocity lead bullets used in upland hunting with nontoxic alternatives would eliminate the only significant source of high lead exposure and poisoning for large avian predators and scavengers.

Lead in Recreational Angling: Historical Perspectives

As reviewed in Scheuhammer et al. (2003b) for Canada, and USEPA (1994) for USA, millions of North Americans participate in recreational angling each year, and contribute to environmental lead deposition through the (largely unintentional) loss of lead fishing sinkers and jigs. Wildlife, primarily common loons (*Gavia immer*) and other water birds, ingest small fishing sinkers and jigs during feeding, mistaking these items for food or grit stones, or consuming lost bait fish with the line and weight still attached (Fig. 12.1). Unlike lead shot ingestion in wild birds, the ingestion and toxicity of lead fishing weights is a phenomenon that has only relatively recently been reported and investigated. Perhaps the earliest North American report is that of Locke and Young (1973) who described sinker ingestion in a whistling (tundra) swan (*Olor columbianus*). Subsequent North American reports have focused mainly on sinker ingestion and mortality in common loons (e.g. – Locke et al. 1982; Pokras and Chafel 1992; Daoust et al. 1998; Twiss and Thomas 1998; Stone and Okoniewski 2001; Sidor et al. 2003); however, individuals of numerous wildlife species have been documented to ingest sinkers or jigs, including various waterfowl species, bald eagles, pelicans, cormorants, and turtles (Perry 1994; Scheuhammer et al. 2003b). In Britain, mortality from lead sinker ingestion in swans has been the focus of concern and research (e.g. – Birkhead 1982; O'Halloran et al. 1988).

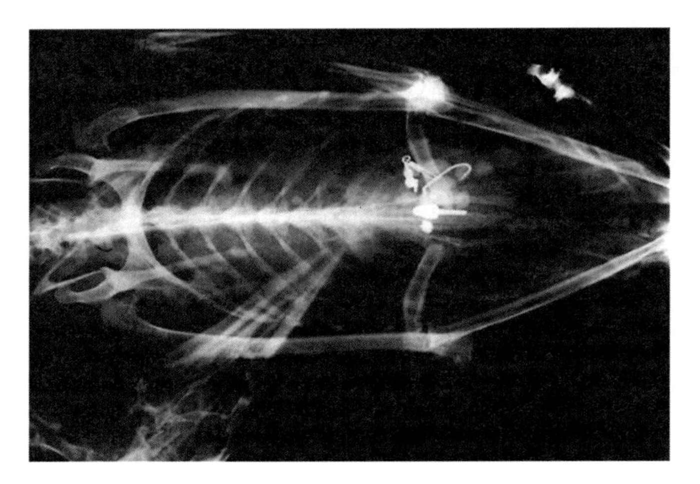

Fig. 12.1 X-ray of dead lead-poisoned common loon showing a hook, swivel, and sinker, as well as several pebbles, in the gizzard (courtesy Avian Care and Research Foundation, Verona, Ontario, Canada)

Lead sinkers and jigs that weigh <50 g and are <2 cm in any dimension are generally the size found to be ingested by wildlife. Ingestion of a single small (<50 g) lead sinker or lead-headed jig is sufficient to expose virtually any water bird to a lethal dose of lead. Lead sinker and jig ingestion is the only significant source of elevated lead exposure and lead toxicity for common loons, and the single most frequently reported cause of death for adult common loons in eastern Canada and USA during the breeding season in habitats where recreational angling is popular, often exceeding deaths associated with other causes of mortality (Scheuhammer et al. 2003b; Sidor et al. 2003). Numerous functional nontoxic materials for producing fishing sinkers and jigs are available including tin, steel, bismuth, tungsten, rubber, ceramics, and clays. Tin, steel, and bismuth sinkers and bismuth jigs are the most common commercially available alternatives. These alternative products tend to be more expensive than their lead counterparts, but not prohibitively so (Scheuhammer et al. 2003b; USEPA 1994).

Science-based regulatory actions have been taken by some nations to restrict lead sinkers and jigs. For example, in 1987, in response to widespread lead poisoning of swans, Britain banned the use of lead fishing sinkers weighing <1 ounce (<28.4 g). USA has banned the use of lead sinkers and jigs in some National Wildlife Refuges, and in Yellowstone National Park. Several individual states have established regulations prohibiting the use and/or sale of lead sinkers (Rattner et al. 2008). However, a USEPA proposal (USEPA 1994) to nationally ban lead sinkers and jigs failed. In 1997, Environment Canada, and Parks Canada, prohibited the possession of lead fishing sinkers or lead jigs weighing less than 50 g in National Wildlife Areas and National Parks under the *Canada Wildlife Act* and the *National Parks Act*, respectively. Environment Canada is currently considering broader controls on the manufacture, import, and/or sale of small lead sinkers and jigs; but to date, as in USA, broad national restrictions on lead sinkers and jigs have yet to be established.

Lead in Recreational Shooting and Angling: Policy and Regulatory Considerations

It is paradoxical that, at the start of the twenty-first century, lead products continue to be used in hunting, target shooting, and angling in light of what is known about the toxicological consequences of lead ingestion in animals. This represents a widespread failure of the policy process to protect wildlife from the recreational use of lead products. Although some nations (Denmark and The Netherlands) have implemented strong legal measures to ban lead shot in all hunting, and other nations have imposed partial bans on lead shot and/or sinkers (e.g. UK, USA and Canada), most nations have not begun to seriously address the issue of lead exposure and toxicosis from these products in a substantive and consistent manner (Mateo 2009; Thomas and Guitart 2010).

The policy process in most countries requires that restrictions on the use of lead ammunition and sinkers be met with evidence of widespread exposure and mortality in wildlife from lead ingestion; that affordable and functional nontoxic substitutes are commercially available; and that their use is enforceable, significantly reducing the risk of lead toxicosis, and does not impact negatively on sport hunting and angling activities. At present, all of these conditions can be met, and none is in serious dispute (Thomas 2009a, b; Thomas and Guitart 2010). Thus the failure of the policy process to establish broader restrictions on lead ammunition and terminal tackle lies elsewhere. As Friend et al. (2009) have indicated, societal/political issues are as important as biological/scientific information in determining policy.

Traditionally, there has been a pronounced separation of human health issues from those of wildlife health in the regulation of lead products; and even within the wildlife community, consideration of the impacts of lost lead sinkers have largely been separated from those of spent shot and rifle bullets (Goddard et al. 2008; Rattner et al. 2008). Thus, while a coherent toxicological syndrome from ingested lead exists (Thomas and Guitart 2003), different jurisdictions have used different criteria for lead exposure and toxicosis to support (or reject) regulations on the use of certain lead products. The regulation of lead in the human environment has occurred in a large number of countries, and especially emphasizes the need to protect children from exposure to environmental lead (Weidenhamer 2009; Braun et al. 2006; Dietrich et al. 2001; Lanphear et al. 2000). Nationally and internationally enforceable standards have been developed to define maximum tolerable levels of lead in drinking water, food, paints, solders, gasoline, glass and glazes, and many other sources of potential exposure in the human environment. These measures are based on the fact that even low-level chronic lead exposure may be attended by serious physiological, developmental, and behavioral impairment that should be prevented by society (CDC 2005). Thus regulations in the human health arena are often precautionary in nature, and are not applied only to geographical areas where the risk to humans from lead exposure is demonstrably highest. Furthermore, as empirical evidence has accumulated on the negative impacts of ingested lead on human health, agencies have lowered the acceptable levels of lead that may exist in the human environment, and the absence of a no-effect exposure threshold for lead is now a serious consideration

(CDC 2005). This relatively advanced state of regulation of lead in the human environment [notwithstanding the concerns of Lanphear (1998)] reflects the preeminent position that human health occupies relative to that of wild animal species.

Thomas and Guitart (2010) have shown that the use of different criteria for lead exposure can lead governments to different conclusions about the impacts of ingested lead shot, sinkers and bullet projectiles on wildlife, as illustrated by very different regulations in USA and Canada versus the European Union; and very different policies among different European nations (Mateo 2009). Similarly, within North America, national regulations on lead product use have arisen from the application of somewhat different criteria (Thomas 2003; Thomas and Guitart 2010). Here, we examine some examples in the context of Canadian and US policy and law.

Phasing Out Lead Ammunition: Science, Policy and Regulation

One of the clearest examples of reliable scientific information driving policy and regulatory change relates to the passage of the Ridley-Tree Condor Preservation Act by the State of California in 2007. A large body of evidence [see papers in Watson et al. (2009)] showed that lead exposure and consequent mortality in California condors (*G. californianus*) was due to the direct ingestion of lead ammunition fragments embedded in the discarded remains of shot game. Green et al. (2009) suggested that lead exposure in this endangered species, if not reduced drastically, would produce further rapid decline of the species. Accordingly, hunters in California are now required to use only nontoxic shot and bullets when hunting in the range of the California condor. This range also extends into Arizona (and Utah), but Arizona has adopted a different policy. Rather than legislate restrictions on the use of lead ammunition, Arizona's policy is to offer hunters free nontoxic bullets and shotgun slugs. The intent is for hunters satisfied with the performance of the nontoxic substitute, and aware of the rationale behind its use, to continue to use lead-free products voluntarily, rather than under the threat of law (Sieg et al. 2009).

Friend et al. (2009) reviewed how extensive scientific, legislative, and socioeconomic approaches interacted to achieve, eventually, a national reduction in lead exposure in waterfowl and their raptorial predators through the regulatory prohibition of lead shot for waterfowl hunting. A large body of evidence revealed that bald eagles contracted lead poisoning from ingesting lead pellets from dead or wounded waterfowl (USFWS 1986). Although this species was protected directly in the US, under four separate federal acts (Anderson 1992), the USFWS chose to use regulations under the Migratory Bird Treaty Act pertaining to the "take" (i.e. hunting) of waterfowl to require the use of nontoxic shot. While this would not prevent bald eagles from ingesting pellets, such (steel) pellets would no longer cause lead poisoning. The objective of the US law was to protect eagles from ingesting lead shot, in addition to protecting waterfowl. Friend et al. (2009) remarked that it was the need to protect eagles in USA that provided the greatest impetus to invoke nontoxic shot use nation-wide in 1991 for waterfowl hunting. Bald eagles are also native to

Canada, and were known to be afflicted by lead shot poisoning (Elliott et al. 1992). However, a parallel legal approach to that of USA was not possible in Canada, mainly because bald eagles were not protected under any dedicated federal legislation (a national Species at Risk Act was not in place in 1991). A further several years would elapse before Canada adopted nontoxic shot for waterfowl hunting nationwide. Initially, passage of the national regulation was to be implemented in 1995, but provincial objections (mainly from Alberta and Saskatchewan) delayed its full implementation for a further 4 years. These provinces contended, unsuccessfully, that they be exempt from the federal regulation because their demonstrated prevalence of lead exposure in waterfowl was low compared to other provinces. Despite the fact that USA and Canada are both Parties to the 1918 Migratory Bird Treaty, and have, for the most part, harmonized legislation under the Treaty, different approaches were required and used to ultimately arrive at an equivalent management position in which nontoxic shot composition and use is mandated for waterfowl hunting, and is similarly regulated in both countries. Now Canadian law complements the 1991 US initiative, rather than thwarting it (Thomas and Guitart 2010), and the risk from lead shot ingestion has been effectively reduced throughout the continental range of most migratory waterfowl species.

Lead exposure and toxicosis is not confined to waterfowl and their predators: it is clearly manifested in many species of upland game birds and their predators as well (Kendall et al. 1996; Pain et al. 2009). It is interesting to examine the continental situation further in the case of North American migratory game bird species that are not included in nontoxic shot regulations in either USA or Canada. In USA, the use of lead shot for hunting "waterfowl and coots" is prohibited, but other migratory wetland birds such as snipe and cranes are not included in the regulation; nor are upland migratory species such as mourning doves. In Canada the regulation is somewhat broader, applying to all migratory game bird species, thus including, for example, snipe and cranes, but specifically exempting three upland species (American woodcock, mourning dove, and band-tailed pigeon). The initial intent of the Government of Canada was to include all migratory game birds under the national nontoxic shot hunting regulations. However, Canadian hunter groups (especially woodcock hunters) objected that the scientific evidence of shot ingestion in upland game birds in Canada was not strong enough to warrant their inclusion under nontoxic shot regulations. Upland migratory species are still not afforded protection from lead shot ingestion in USA or Canada. However, research on lead in woodcock (Scheuhammer et al. 1999, 2003a; Strom et al. 2005) may now support regulations requiring that hunters of this species use nontoxic ammunition.

The hunting of mourning doves in USA is a major recreational event and is still conducted mainly with lead shot in most localities. This species (together with woodcock and snipe) was excluded from the USFWS 1991 listing of species to be hunted with nontoxic shot because the scientific evidence did not implicate these species in the lead-poisoning deaths of bald eagles. However, there is now considerable scientific evidence that widespread lead exposure from shot ingestion is occurring in mourning doves in USA (Schultz et al. 2002, 2006a, b, 2007), and that the extension of federal nontoxic shot regulations to this migratory game bird is warranted (Schultz et al. 2009). It is still uncertain whether the strength of the scientific arguments favoring nontoxic shot adoption to enhance dove conservation will prevail over the objections of hunters.

The accepted inefficiency of recreational hunting may have contributed to the resistance against lead shot restrictions. Average waterfowl crippling (unrecovered harvest) rates are at least 18% of the number of ducks and geese brought to bag in both Canada and USA (Norton and Thomas 1994; USFWS 1986). Considering that the total North American waterfowl harvest is approximately 20 million birds annually, an estimated four million individuals are probably lost to crippling every year. Lead poisoning losses were estimated to be about 1.6–3.8% of the fall population, or about two million individuals annually, prior to the 1991 ban on lead shot use (Feierabend 1983; Bellrose 1959). Insofar as crippling loss has traditionally been accepted in both Canada and USA, and no regulations have been implemented to reduce it, opposition to government proposals to regulate a smaller absolute loss of birds due to lead poisoning can be understood. An additional obstacle to prohibiting the use of lead shot for waterfowl hunting was that it was feared that nontoxic (steel) shot use would cause an increase in wounding loss, exceeding the number of birds saved from lead poisoning by banning lead (e.g. – Wendt and Kennedy 1992). Although this concern was shown over time to be unjustified (Morehouse 1992), the same argument continues to be used by opponents of nontoxic shot regulations for other species, such as mourning doves (Schultz et al. 2006a, 2007). Additionally, some US wildlife professionals have contended that further bans on lead shot use might cause more hunters to give up their sport, resulting in a reduction in the hunting license fees that have traditionally supported much federal and state wildlife management activity (Miller et al. 2009). These views serve to inhibit a science-based conservation policy process. However, The Wildlife Society has issued a policy statement endorsing the use of nontoxic shot (and sinkers) in recreational sports (The Wildlife Society 2009), based on clear evidence that lead is a toxicant to both wildlife and humans. This policy statement by the body representing wildlife professionals across many countries adds legitimacy to initiatives to replace lead ammunition with nontoxic alternatives for use in recreational activities.

While the USFWS has undertaken no nation-wide federal initiatives to require nontoxic ammunition for hunting game other than waterfowl, individual states have implemented varying levels of additional control on the use of lead shot for hunting, whether for state-managed species or migratory species (NSAC 2006; Thomas 2009b). Individual states have heeded the available scientific literature on lead exposure and acted in a precautionary manner, and as of 2009, 26 states have additional regulations restricting the use of lead shot for hunting either migratory or non-migratory birds (Rattner et al. 2008). Despite these progressive actions, opposition to change can still be influential, as in the case of Minnesota, which was prepared to adopt nontoxic shot for hunting all game bird species within the state, but opposition from pro-lead factions defeated the proposal (Thomas 2009b). Nonetheless, some momentum for further change appears to be coming from the state level, rather than from the federal level in USA. By contrast, no initiatives to require the use of lead-free ammunition have been advanced by any Canadian province. However, the efficiency of regulation, as opposed to voluntary compliance, is apparent. Eighteen years have passed since regulations requiring use of nontoxic shot for waterfowl hunting came into effect in USA. The hunting community has had considerable time to accept the need for a transition to nontoxic shot, and to more broadly adopt nontoxic products of its own accord. That this has not happened in any comprehensive manner attests to the importance of entrenched negative opinion within a significant faction of the hunting community about the existing science on lead exposure, the need for change, and the personal implications of such a transition. Such negative opinions continue to have an inhibitory effect on progressive policy development to phase out the use of lead in hunting.

A critical aspect of the policy and regulatory process is to show that transitions to the use of nontoxic products do result in a reduction of lead exposure and a lowering of the mortality rate attributed to lead ingestion (Thomas and Guitart 2010). Evidence of reduced lead exposure in waterfowl following nontoxic shot regulations has been published, but has not been widely publicized and promoted by wildlife agencies. Anderson et al. (2000) reported that, following the 1991 national ban on the use of lead shot for waterfowl hunting, the ingestion of two or more lead pellets by ducks declined by 72%; that an estimated 64% of mallard ducks (*Anas platyrhynchos*) that had ingested four or more pellets were saved from death; and that an estimated 1.4 million ducks across USA were saved from lead poisoning each year. Samuel and Bowers (2000) reported a 44% reduction in the prevalence of high blood-lead levels in black ducks (*Anas rubripes*) following the 1991 US ban on lead shot use. Stevenson et al. (2005) demonstrated that bone-lead levels of black ducks, mallards ducks, and ring-necked ducks (*Aythya collaris*) in Canada declined by at least 50% following introduction of Canadian nontoxic shot regulations. Collectively, these scientific studies reveal a rapid and substantial reduction in lead exposure in waterfowl in the wake of nontoxic shot regulations, and attest to the effectiveness of a regulatory approach to replacing lead ammunition with nontoxic substitutes. These studies could further be used to promote policies for extending nontoxic shot and bullet use to other hunting practices, using a precautionary approach.

Phasing Out Lead Fishing Tackle: Science, Policy and Regulation

The negative impacts of ingested lead fishing weights on the health of water birds has been well established in published research, whether in Canada, USA, or the UK. (Pokras and Chafel 1992; Perry 1994; Twiss and Thomas 1998; Sidor et al. 2003; Scheuhammer et al. 2003b), and the case has been made for the use of nontoxic sinkers and jigs for recreational angling. The extent of the regulatory response to replace lead sinkers with nontoxic alternatives in North America has been less than that to replace lead shot for hunting; however, the study of lead sinker mortality in water birds has a much shorter history than that of lead shot toxicity. The brief history of policy in this area contains examples of marked failures to regulate the use of lead sinkers, but also examples of successful regulatory actions. Parks Canada introduced regulations under the National Parks Act prohibiting the possession of lead sinkers and jigs weighing <50 g in all national parks, and Environment Canada applied the same regulations to all National Wildlife Areas under the Canada Wildlife Act (Scheuhammer 2009). A major basis for these decisions was that the use of small lead sinkers and jigs by anglers, and the risk for lead exposure and toxicity in wildlife when these items were lost or discarded in the environment, was not consistent with ecological integrity, a concept enshrined in the National Parks Act. An important aspect of this initiative is that the regulations were not based on a series of comprehensive scientific studies documenting the extent of lead exposure and risk to wildlife species from lost fishing tackle in each individual National Park or Wildlife Area. The Canadian federal government was thus willing to use a precautionary regulatory approach to prohibit lead sinkers and jigs on lands under its direct control. While this federal regulatory precedent is important, only a small percentage (~1%) of the total Canadian recreational angling effort occurs in National Parks or National Wildlife Areas (Scheuhammer et al. 2003b). Conversely, no Canadian province or territory has yet to launch policy, let alone regulations, to require lead-free fishing tackle in provincially regulated waters. A 2005 initiative to extend nontoxic sinker requirements to all parts of Canada using the regulatory authority of the Canadian Environmental Protection Act was vigorously opposed by some angling organizations and much of the Canadian tackle industry, and was rejected by the parliament of the day (Thomas 2009b), despite having the support of the federal parliamentary advisory committee for the environment (Caccia 1995).

In the USA, the USEPA (1994) attempted to regulate sinkers nationally, but halted this initiative in the mid 1990s (Thomas 2003). However, federal regulations now prohibit use of lead tackle in seven National Wildlife Refuges (Rattner et al. 2008). Five US states (New York, Massachusetts, Vermont, New Hampshire, and Maine) have also introduced nontoxic sinker regulations (Pokras and Kneeland 2009), albeit with conflicting provisions for sale and use (Thomas 2003). The growing role of individual states in advancing the policy process in their jurisdictions is apparent. These states have acted to protect wildlife from lead exposure from sinkers in the same way that some other US states have acted on lead shot use for upland game hunting. Hopefully, the future will see the development of increasingly harmonized policy approaches to restricting both lead shot and lead sinkers throughout North America.

Policy and Regulation: The Path Forward

The US national agencies representing fishing (American Fisheries Society) and wildlife management (The Wildlife Society), while aware of the scientific basis of concern and the need to comment on the issue of regulating the use of lead shot and sinker products, have presented only a variety of possible policy options written in the conditional voice (Goddard et al. 2008; Rattner et al. 2008). This may reflect their dual responsibilities both to conservation of fish and wildlife, and to the human communities that use these resources for recreation.

Although primary lead shot exposure in waterfowl has diminished substantially following broad restrictions on the use of lead shot for waterfowl hunting, many thousands of tons of "old" lead shot still lies accessible in some wetlands (especially hard-bottomed water bodies) and in uplands where hunting has occurred for many years, where it may remain available for ingestion by birds far into the future. Primary lead exposure in wildlife will not disappear quickly in such locations, but will nonetheless diminish over time following prohibitions on lead shot use (Anderson et al. 2000). By contrast, secondary lead exposure and mortality in raptors should diminish very quickly following implementation of nontoxic ammunition regulations in both wetland and upland habitats (assuming a high compliance by hunters). The elimination of secondary lead poisoning constitutes, in itself, a strong policy rationale for extending nontoxic regulations to all types of hunting. Similarly, a requirement for nontoxic sinkers and jigs would result in a rapid lowering of lead poisoning mortality of common loons and other piscivorous species, especially in those situations where live-bait fishing is practiced.

The sport hunting and angling communities are heavily regulated regarding seasons, species, and daily limits, yet are still under-regulated with respect to the use of lead products compared to other segments of society that use lead and release it to the environment (Thomas and Guitart 2010). Lead-using industries in the developed world have had to reduce environmental lead emissions, comply with regulations, internalize the costs of emissions reductions, install new equipment to achieve compliance, and still remain profitable. Society largely expects these diverse industries to adhere to the Polluter Pays and the Precautionary Principles. Hunters and anglers have so far resisted a similar degree of regulation of lead product use within their recreational sports, and governments have not exerted sufficient non-regulatory pressure for these constituencies to voluntarily adopt nontoxic products in a comprehensive manner. That an annual discharge of unrecoverable lead shot and bullets far exceeding 70,000 tonnes for North America should continue is inconsistent with pollution criteria applied to industry, and is also inconsistent with the conservation of high-quality wildlife habitat. The USEPA (2001) estimated that outdoor target shooting alone results in the deposition of 72,600 tonnes of lead annually. Several thousand additional tonnes are contributed by hunting, and recreational angling. Industries outside the shooting and angling fields have had to adapt to producing products containing increasingly lower levels of lead as lead-free nontoxic substitutes became available, such as in paints, glass, solders, glazes, water pipes, and gasoline. Yet, the hunting and angling communities have been largely allowed to continue along their traditional pollution trajectories despite scientific considerations that should require transitions to the use of approved lead-free products.

The European Union, since 2007, has required mandatory registration, evaluation, and authorization of chemicals (REACH) manufactured or imported in quantities exceeding 1 tonne per year (CEC 2006). Lead shot and sinkers are subject to this regulation and are currently under evaluation. Should the European Union rule

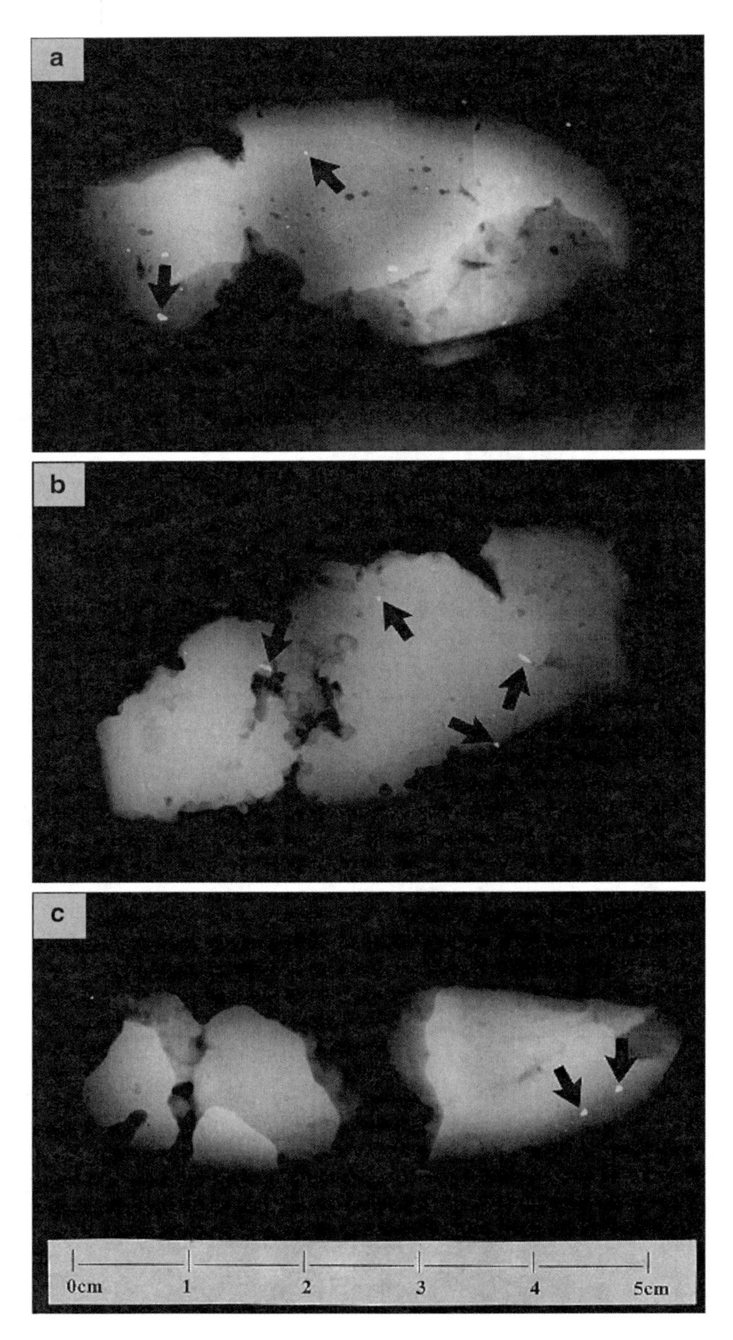

Fig. 12.2 X-ray of breast muscle samples from various game birds collected using lead shot ammunition, showing numerous small fragments of metallic lead (*arrows*) embedded in the tissue (from Scheuhammer et al 1998)

against the use of these products in sporting ammunition and fishing tackle, a strong signal and legal precedent will be sent to North American governments, and especially to those ammunition manufacturers that export to, or import from, Europe.

Humans, especially aboriginal people and other subsistence hunters who consume a high proportion of game animals in their diets, are now the latest group to show marked exposure to lead from ingestion of shot and bullet fragments in hunter-killed game (Johansen et al. 2004, 2006; Mateo et al. 2007; Tsuji et al. 2008; Kosnett 2009). These human exposure examples provide yet another policy rationale for establishing broader regulations requiring the use of nontoxic shot and bullets. A distinction between lead deposition and toxicity in the human environment versus that of wildlife is largely illusory, and in many cases policy makers have sufficient scientific information and precedents to develop regulations that apply equally to wild species and humans. In both North America and Europe, regulations concerning the presence of lead in the human environment explicitly address exposure and are largely precautionary in approach, whereas in the case of wildlife, regulations have been introduced mainly to mitigate large-scale mortality from documented high-exposure scenarios. A shift in the focus of government policy towards a precautionary approach is recommended for all uses of lead shot, bullets, and small sinkers and jigs.

The angling and hunting communities have thus far largely viewed any regulation of lead ammunition and tackle to be an unwarranted government imposition. However, governments should consider that removing lead from the environment is an investment in the health of wildlife species and the quality of their habitats, and that adopting nontoxic ammunition and sinkers is an effective conservation tool that complements the creation and rehabilitation of wildlife habitats (Anderson et al. 2000; Thomas 2009b). Meyer et al. (2008) advocated preventing exposure as the best approach to reducing lead exposure in humans. These authors recommended further that identification of sources, and elimination or control of sources, should be the two primary prevention strategies. In the case of elevated lead exposure in wildlife, the identity of the important lead sources (i.e. spent shot, bullets, and sinkers/jigs) is well-established. The existence of approved, nontoxic lead substitutes enables the elimination of this source of exposure. Regulation at a national level also assures the ammunition and sinker industries of an economically viable market for nontoxic products. A single policy and regulatory vehicle, such as the Canada Environmental Protection Act for Canada, or its equivalent in other nations, could be considered for regulating all categories of lead ammunition and sinker manufacture, sale, and/or use. This would have the greatest regulatory efficiency of any policy proposal, and would better align the policy approaches for both human health and wildlife health with respect to major outstanding issues of lead exposure.

References

Anderson WL (1992) Legislation and lawsuits in the United States and their effects on nontoxic shot regulations. In: Pain DJ (ed) Lead poisoning in waterfowl. IWRB Spec Publ 16. Slimbridge, UK, pp 56–60

Anderson WL, Havera SP, Zercher BW (2000) Ingestion of lead and non-toxic shotgun pellets by ducks in the Mississippi flyway. J Wildl Manage 64:848–857

Bellrose FC (1959) Lead poisoning as a mortality factor in waterfowl populations. Illinois Nat Hist Surv Bull 27:235–288

Birkhead ME (1982) Causes of mortality in the mute swan (*Cygnus olor*) on the River Thames. J Zool (London) 199:59–73

Bowles JM (1908) Lead poisoning in ducks. Auk 25:312–313

Braun JM, Kahn RS, Froelich T et al (2006) Exposures to environmental toxicants and attention deficit hyperactivity disorder in US children. Environ Health Perspect 114:1904–1909

Bressler J, Kim K, Chakraborti T, Goldstein G (1999) Molecular mechanisms of lead neurotoxicity. Neurochem Res 24:595–600

Brister B (1992) Steel shot: ballistics and gunbarrel effects. In: Pain DJ (ed) Lead poisoning in waterfowl. IWRB Spec Publ 16. Slimbridge, UK, pp 26–28

Caccia CL (1995) It's about our health: towards pollution prevention. Report of The House of Commons Standing Committee on Environment and Sustainable Development. Publ Works Gov Serv, Ottawa, Ontario

Cade TJ (2007) Exposure of California condors to lead from spent ammunition. J Wildl Manage 71:2125–2133

Calvert JH (1876) Pheasants poisoned by swallowing shots. The Field 47:189

Carlson BL, Nielson SW (1985) Influence of dietary calcium on lead poisoning in mallard ducks (*Anas platyrhynchos*). Am J Vet Res 46:276–282

CDC. Centers for Disease Control and Prevention (2005) Preventing lead poisoning in young children. CDC, Atlanta, Georgia

CEC. Commission of the European Communities (2006) Regulation (EC) No 1907/2006 of the European Parliament and of the Council. Official J Eur Union L396:1–849

Church ME, Gwiazda R, Risebrough RW et al (2006) Ammunition is the principal source of lead accumulated by California condors re-introduced to the wild. Environ Sci Technol 40:6143–50

Clemens ET, Krook L, Aronson AL, Stevens CE (1975) Pathogenesis of lead shot poisoning in the mallard duck. Cornell Vet 65:248–285

Colburn C (1992) Lead poisoning in waterfowl: the Winchester perspective. In: Pain DJ (ed) Lead poisoning in waterfowl. IWRB Spec Publ 16. Slimbridge, UK, pp 46–50

Daoust P-Y, Conboy G, McBurney S, Burgess N (1998) Interactive mortality factors in Common Loons from Maritime Canada. J Wildl Dis 34:524–531

Degernes L, Heilman S, Trogdon M et al (2006) Epidemiological investigation of lead poisoning in trumpeter and tundra swans in Washington state, USA, 2000–2002. J Wildl Dis 42:345–358

Dietrich KN, Ris MD, Succop PA et al (2001) Early exposure to lead and juvenile delinquency. Neurotoxicol Teratol 23:511–518

Elliott JE, Langelier KM, Scheuhammer AM et al (1992) Incidence of lead poisoning in bald eagles and lead shot in waterfowl gizzards from British Columbia, 1988–91. Can Wildl Serv Prog Note 220, Ottawa, Canada, pp 7

Feierabend JS (1983) Steel shot and lead poisoning in waterfowl. An annotated bibliography of research 1976–1983. National Wildlife Federation Science and Technical Series No. 8, Washington, DC

Feierabend JS (1985) Legal challenges to nontoxic (steel) shot regulations. Southeastern Assoc Fish Wildl Agencies Annu Conf Proc 39:452–458

Fisher IJ, Pain DJ, Thomas VG (2006) A review of lead poisoning from ammunition sources in terrestrial birds. Biol Conserv 131:421–432

Friend M, Franson JC, Anderson WL (2009) Biological and societal dimensions of lead poisoning in birds in the USA. In: Watson RT, Fuller M, Pokras M et al (eds) Ingestion of lead from spent ammunition: implications for wildlife and humans. The Peregrine Fund, Boise, Idaho, pp 34–60

Goddard CI, Leonard NJ, Stang DL et al (2008) Management concerns about known and potential impacts of lead use in shooting and fishing activities. Fisheries 33:228–236

Goyer RA (1993) Lead toxicity: current concerns. Environ Health Perspect 100:177–187

Green RE, Hunt WG, Parish CN et al (2009) Effectiveness of action to reduce exposure of free-ranging California condors in Arizona and Utah to lead from spent ammunition. In: Watson RT, Fuller M, Pokras M et al (eds) Ingestion of lead from spent ammunition: implications for wildlife and humans. The Peregrine Fund, Boise, Idaho, pp 240–253
Grinnell GB (1894) Lead poisoning. Forest and Stream 42:117–118
Grinnell GB (1901) American duck shooting. Field & Stream Publishing, Stackpole Books, Harrisburg, PA
Johansen P, Asmund G, Riget F (2004) High human exposure to lead through consumption of birds hunted with lead shot. Environ Pollut 127:125–129
Johansen P, Pedersen HS, Asmund G et al (2006) Lead shot from hunting as a source of lead in human blood. Environ Pollut 142:93–97
Johnson FM (1998) The genetic effects of environmental lead. Mutat Res 410:123–140
Kanstrup N (2010) 8.2 Workshop Resolution. In: Kanstrup N (ed) Sustainable Hunting Ammunition Workshop Report, CIC Workshop, Aarhus, Denmark, 5–7 November 2009. The International Council for Game and Wildlife Conservation (CIC), p 72
Kendall RJ, Lacher TE Jr, Bunck C et al (1996) An ecological risk assessment of lead shot exposure in non-waterfowl avian species: upland game birds and raptors. Environ Toxicol Chem 15:4–20
Kennedy JA, Nadeau S (1993) Lead shot contamination of waterfowl and their habitats in Canada. Can Wildl Serv Tech Rpt Ser 164, Can Wildl Serv, Ottawa, Ontario
Knopper LD, Mineau P, Scheuhammer AM et al (2006) Carcasses of shot Richardson's ground squirrels may pose lead hazards to scavenging hawks. J Wildl Manage 70:295–299
Komárek M, Ettler V, Chrastný V, Mihaljevič M (2008) Lead isotopes in environmental sciences: a review. Environ Int 34:562–577
Kosnett MJ (2009) Health effects of low dose exposure in adults and children, and preventable risk posed by the consumption of game meat harvested with lead ammunition. In: Watson RT, Fuller M, Pokras M et al (eds) Ingestion of lead from spent ammunition: implications for wildlife and humans. The Peregrine Fund, Boise, Idaho, pp 24–33
Kramer JL, Redig PT (1997) Sixteen years of lead poisoning in eagles, 1980–1995: an epizootiologic view. J Raptor Res 31:327–332
Lanphear BP (1998) The paradox of lead poisoning prevention. Science 281(5383):1617–1618
Lanphear BP, Dietrich K, Auinger P et al (2000) Cognitive deficits associated with blood lead concentrations <10 microg/dl in US children and adolescents. Public Health Rep 115:521–529
Locke LN, Young LT (1973) An unusual case of lead poisoning in a whistling swan. Maryland Birdlife 29:106–107
Locke LN, Kerr SM, Zoromski D (1982) Lead poisoning in common loons (*Gavia immer*). Avian Dis 26:392–396
Lumeij JT (1985) Clinicopathologic aspects of lead poisoning in birds: a review. Vet Q 7:133–138
Mateo R (2009) Lead poisoning in wild birds in Europe and the regulations adopted by different countries. In: Watson RT, Fuller M, Pokras M et al (eds) Ingestion of lead from spent ammunition: implications for wildlife and humans. The Peregrine Fund, Boise, Idaho, pp 71–98
Mateo R, Rodriguez-de-la Cruz M, Vidal D et al (2007) Transfer of lead from shot pellets to game meat during cooking. Sci Total Environ 372:480–485
McAtee WL (1908) Lead poisoning in ducks. Auk 25:472
Meyer PA, Brown MJ, Falk H (2008) Global approach to reducing lead exposure and poisoning. Mutat Res/Revs Mutat Res 659:166–175
Miller DA, Smith MD, Miller JE (2009) Caution before action on lead. Wildl Prof 3(52):66
Morehouse KA (1992) Crippling loss and shot-type: the United States experience. In: Pain DJ (ed) Lead poisoning in waterfowl. IWRB Spec Publ 16. Slimbridge, UK
Needleman H (2004) Lead poisoning. Annu Rev Med 55:209–222
Nontoxic Shot Advisory Committee (NSAC) (2006) Report of the nontoxic shot advisory committee. Fish and Wildlife Division, Minnesota Department Natural Resources, St. Paul, Minnesota
Norton MR, Thomas VG (1994) Economic analyses of crippling losses of North American waterfowl and their policy implications for management. Environ Conserv 21:347–353

O'Halloran J, Myers AA, Duggan PF (1988) Lead poisoning in swans and sources of contamination in Ireland. J Zool (London) 216:211–223

Pain DJ, Carter I, Sainsbury AW et al (2007) Lead contamination and associated disease in captive and reintroduced red kites *Milvus milvus* in England. Sci Total Environ 376:116–127

Pain DJ, Fisher IJ, Thomas VG (2009) A global update of lead poisoning in terrestrial birds from ammunition sources. In: Watson RT, Fuller M, Pokras M et al (eds) Ingestion of lead from spent ammunition: implications for wildlife and humans. The Peregrine Fund, Boise, Idaho, pp 99–118

Pattee OH, Hennes SK (1983) Bald eagles and waterfowl: the lead shot connection. Trans N Am Wildl Nat Res Conf 48:230–237

Perry C (1994) Lead sinker ingestion in avian species. Division of Environmental Contaminants Information Bulletin 94-09-01. US Fish Wildl Serv, Arlington, Virginia, pp 83

Pokras MA, Chafel RM (1992) Lead toxicosis from ingested fishing sinkers in adult Common Loons (*Gavia immer*) in New England. J Zoo Wildl Med 23:92–97

Pokras MA, Kneeland MR (2009) Understanding lead uptake and effects across species lines: a conservation medicine based approach. In: Watson RT, Fuller M, Pokras M et al (eds) Ingestion of lead from spent ammunition: implications for wildlife and humans. The Peregrine, Boise, Idaho, pp 7–22

Quortrup ER, Shillinger JE (1941) 3,000 wild bird autopsies in western lake areas. J Am Vet Med Assoc 99:382–387

Rattner BA, Franson JC, Sheffield SR et al (2008) Sources and implications of lead-based ammunition and fishing tackle to natural resources. Wildlife Society Technical Review, The Wildlife Society, Bethesda, Maryland, pp 62

Samuel MD, Bowers EF (2000) Lead exposure in American black ducks after implementation of non-toxic shot. J Wildl Manage 64:947–953

Sanderson GC, Bellrose FC (1986) A review of the problem of lead poisoning in waterfowl. Illinois Nat Hist Surv. Special Publ. 4, pp 34

Scheuhammer AM (1996) Influence of reduced dietary calcium on the accumulation and effects of lead, cadmium, and aluminum in birds. Environ Pollut 94:337–343

Scheuhammer AM (2009) Historical perspective on the hazards of environmental lead from ammunition and fishing weights in Canada. In: Watson RT, Fuller M, Pokras M et al (eds) Ingestion of lead from spent ammunition: implications for wildlife and humans. The Peregrine Fund, Boise, Idaho, pp 61–67

Scheuhammer AM, Norris SL (1995) A review of the environmental impacts of lead shotshell ammunition and lead fishing weights in Canada. Can Wildl Serv Occas Paper 88, Environment Canada, Ottawa, Canada, p 52

Scheuhammer AM, Dickson K (1996) Patterns of environmental lead exposure in waterfowl in eastern Canada. Ambio 25:14–20

Scheuhammer AM, Templeton DM (1998) Use of stable isotope ratios to distinguish sources of lead exposure in wild birds. Ecotoxicology 7:37–42

Scheuhammer AM, Perrault JA, Routhier E et al (1998) Elevated lead concentrations in edible portions of game birds harvested with lead shot. Environ Pollut 102:251–257

Scheuhammer AM, Rogers CA, Bond D (1999) Elevated lead exposure in American woodcock (*Scolopax minor*) in eastern Canada. Arch Environ Contam Toxicol 36:334–340

Scheuhammer AM, Bond DE, Burgess NM et al (2003a) Lead and stable isotope ratios in soil, earthworms, and bones of American woodcock (*Scolopax minor*) from eastern Canada. Environ Toxicol Chem 22:2585–2591

Scheuhammer AM, Money SL, Kirk DA, Donaldson G (2003b) Lead fishing sinkers and jigs in Canada: review of their use patterns and toxic impacts on wildlife. Can Wildl Serv Occas Paper 108, Environment Canada, Ottawa, Canada, pp 48

Schultz JH, Millspaugh JJ, Washburn BE et al (2002) Spent-shot availability and ingestion on areas managed for Mourning Doves. Wildl Soc Bull 30:112–120

Schultz JH, Padding PI, Millspaugh JJ (2006a) Will Mourning Dove crippling rates increase with nontoxic-shot regulations? Wildl Soc Bull 34:861–865

Schultz JH, Millspaugh JJ, Bermudez AJ et al (2006b) Acute lead toxicosis in Mourning Doves. J Wildl Manage 70:413–421
Schultz JH, Reitz RA, Sheriff SL et al (2007) Attitudes of Missouri small game hunters toward non-toxic shot regulations. J Wildl Manage 71:628–633
Schultz JH, Potts GE, Cornely JE et al (2009) The question of lead: considerations for mourning dove nontoxic shot regulation. Wildl Prof 3(2):46–49
Shillinger JE, Cottam CC (1937) The importance of lead shot in waterfowl. N Am Wildl Conf Trans 2:398–403
Sidor IF, Pokras MA, Major AR et al (2003) Mortality of Common Loons in New England, 1987–2000. J Wildl Dis 39:306–315
Sieg R, Sullivan KA, Parish CN (2009) Voluntary lead reduction efforts within the northern Arizona range of the California condor. In: Watson RT, Fuller M, Pokras M et al (eds) Ingestion of lead from spent ammunition: implications for wildlife and humans. The Peregrine Fund, Boise, Idaho, pp 341–349
Silbergeld EK (2003) Facilitative mechanisms of lead as a carcinogen. Mutat Res 533:121–133
Simons TJ (1993) Lead-calcium interactions in cellular lead toxicity. Neurotoxicology 14:77–85
Six KM, Goyer RA (1970) Experimental enhancement of lead toxicity by low dietary calcium. J Lab Clin Med 76:933–942
Stevenson AL, Scheuhammer AM, Chan HM (2005) Effects of nontoxic shot regulations on lead accumulation in ducks and American woodcock in Canada. Arch Environ Contam Toxicol 48:405–413
Stone WB, Okoniewski JC (2001) Necropsy findings and environmental contaminants in Common Loons from New York. J Wildl Dis 37:178–184
Strom SM, Patnode KA, Langenberg JA et al (2005) Lead contamination in American Woodcock (*Scolopax minor*) from Wisconsin. Arch Environ Contam Toxicol 49:396–402
The Wildlife Society (2009) Final position statement. Lead in ammunition and fishing tackle. The Wildlife Society, Bethesda, Maryland. (http://joomla.wildlife.org/documents/positionstatements/Lead_final_2009.pdf)
Thomas VG (2003) Harmonizing approval of nontoxic shot and sinkers in North America. Wildl Soc Bull 31:292–295
Thomas VG (2009a) Nontoxic shot ammunition: types, availability, and use for upland game hunting. Wildl Prof 3:50–51
Thomas VG (2009b) The policy and legislative dimensions of nontoxic shot and bullet use in North America. In: Watson RT, Fuller M, Pokras M et al (eds) Ingestion of lead from spent ammunition: implications for wildlife and humans. The Peregrine Fund, Boise, Idaho, pp 351–362
Thomas VG, Guitart R (2003) Lead pollution from shooting and angling, and a common regulative approach. Environ Policy Law 33(3/4):143–149
Thomas VG, Guitart R (2010) Limitations of European Union policy and law for regulating use of lead shot and sinkers: comparisons with North American regulation. Environ Policy Gov 20:57–72
Thomas VG, Scheuhammer AM, Bond DE (2009) Bone lead levels and lead isotope ratios in red grouse from Scottish and Yorkshire moors. Sci Total Environ 407:3494–3502
Tsuji JJS, Wainman BC, Martin ID, Sutherland C, Weber JP, Dumas P, Nieboer E (2008) The identification of lead ammunition as a source of lead exposure in First nations: the use of lead isotope ratios. Sci Total Environ 393:291–298
Twiss MP, Thomas VG (1998) Preventing fishing-sinker-induced lead poisoning of common loons through Canadian policy and regulative reform. J Environ Manage 53:49–59
U S Environmental Protection Agency (1994) Lead fishing sinkers; response to citizens' petition and proposed ban; proposed rule. Fed Reg Part III 40(745):11121–11143
U S Environmental Protection Agency (2001) Best management practices for lead at outdoor shooting ranges, EPA-902-B-01-001.USEPA, New York, New York
U S Fish and Wildlife Service (1974) Proposed use of steel shot for hunting waterfowl in the United States. US Dept of the Interior, Draft Environmental Statement, DES 74–76, 79 pp

U S Fish and Wildlife Service (1976) Final environmental impact statement: proposed use of steel shot for hunting waterfowl in the United States. US Dept of the Interior, Washington, DC, pp 276

U S Fish and Wildlife Service (1986) Use of lead shot for hunting migratory birds in the United States: final supplemental environmental impact statement. USFWS, US Dept of the Interior, Arlington, Virginia

Watson RT, Fuller M, Pokras M et al (eds) (2009) Ingestion of lead from spent ammunition: implications for wildlife and humans. The Peregrine Fund, Boise, Idaho

Wayland M, Bollinger R (1999) Lead exposure and poisoning in bald eagles and golden eagles in the Canadian prairie provinces. Environ Pollut 104:341–350

Wendt JS, Kennedy JA (1992) Policy considerations regarding the use of lead pellets for waterfowl hunting in Canada. In: Pain DJ (ed) Lead poisoning in waterfowl. IWRB Spec Publ 16. Slimbridge, UK, pp 61–67

Weidenhamer JD (2009) Lead contamination of inexpensive seasonal and holiday products. Sci Total Environ 407:2447–2450

Wetmore A (1919) Lead poisoning in waterfowl. US Dept Agric Bull 793:12

Weyhrauch BB (1986) Waterfowl and lead shot. Environ Law 16:883–934

Wilson LK, Davison M, Kraege D (2004) Lead poisoning of trumpeter and tundra swans by ingestion of lead shot in Whatcom County, Washington, USA, and Sumas Prairie, British Columbia, Canada. Bull Trump Swan Soc 32:11–13

Chapter 13
Feminized Fish, Environmental Estrogens, and Wastewater Effluents in English Rivers

Charles R. Tyler and Amy L. Filby

Our research journey into endocrine disruption has been a fascinating one. Never did we think that the production of a yolk protein in males would signify the occurrence of such a major phenomenon in wild fish populations – the widespread feminisation of males. Our findings, together with others in the field, have fundamentally changed the way we think about the toxicity and potential health effects of chemicals and this without doubt has been for the good of our environment. We now better appreciate that a chemical does not have to kill an animal to be harmful, but can do so through more subtle effects within the body, altering our hormone signalling systems.

Abstract Our story on the feminization of wild roach (*Rutilus rutilus*) populations in English Rivers is a product of many years of research in collaboration with various colleagues. Over the past 20 years, we have shown that male roach are being feminized as a consequence of exposure to effluents from wastewater treatment works (WwTWs). Here we provide a history of wastewater treatment in England to illustrate how the treatment processes in operation today have been developed and to show why many English Rivers still suffer problems associated with WwTW effluent discharges. We recount the history of finding feminized roach and how, through combined field and laboratory studies, we have identified which chemicals in WwTW effluents cause the intersex condition (principally environmental estrogens) and that feminized males have a reduced breeding capability. We also discuss our work investigating how the feminized effects in male roach might affect the genetics of subsequent populations. The weight of evidence showing adverse effects of estrogenic endocrine disrupting chemicals (EDCs) in WwTW effluents on fish has been a major influence on the development of the so-called national *Endocrine Disrupter Demonstration Programme* (*EDDP*) – an initiative set up by the UK

C.R. Tyler (✉)
School of Biosciences, Hatherly Laboratories, University of Exeter,
Prince of Wales Road, Exeter, Devon, EX4 4PS, UK
e-mail: c.r.tyler@exeter.ac.uk

J.E. Elliott et al. (eds.), *Wildlife Ecotoxicology: Forensic Approaches*,
Emerging Topics in Ecotoxicology 3, DOI 10.1007/978-0-387-89432-4_13,

water companies and the Environment Agency of England and Wales. The objectives of this program have guided our recent research to investigate the most cost-effective methods for EDC removal from effluents. Our journey over the past two decades has been both exciting and challenging but, most importantly, it has helped raise awareness on the need for a better understanding of EDCs and their potential environmental and human health impacts. Finally, in the wider context, we share some of the lessons we have learned from our research on the feminization of roach in English rivers that are relevant to the future protection of aquatic wildlife from discharged chemicals.

The History of Wastewater Treatment in England

Until the late 1800s, sewage was disposed into cesspits and via open above-ground ditches and channels directly into rivers. Then in 1865, following numerous cholera epidemics and the advent of "The Great Stink" in 1858 in London, which occurred as a consequence of pollution of the River Thames, the first of England's sewerage systems was established. In this system, untreated sewage was carried via fully enclosed pipes to large reservoirs and discharged into the tidal reaches of the Thames [reviewed in detail in Lemmoin-Cannon (1912), Humphreys (1930), and Stanbridge (1976)]. Around this time, legislation was also introduced to protect watercourses and the fish they contained from the impacts of raw sewage (The Public Health Act 1875 and Rivers Pollution Prevention Act 1876). In 1884, the Royal Commission designated that solids should be removed from sewage prior to its discharge in the River Thames, and this resulted in England's first WwTWs, constructed in London at the Beckton outfall (1887–1889) and later at the Crossness outfall (1888–1891). In these first WwTWs, sewage was clarified by the addition of chemicals (lime and iron) and the resulting sludge deposited offshore via sludge vessels. This approach was subsequently used elsewhere in England and the liquid effluent portion of the remaining sewage was allowed to pass into the river. In some cases in England, raw sewage was applied to agricultural land, an approach that had the added benefit of improving crop yields. This practice was continued until the 1980s, when it was abandoned due to a combination of reduced land availability, water logging and clogging problems, and insufficient hygiene standards. Today, WwTWs are the only form of sewage treatment in England, most of which discharge into rivers (Cooper 2001). The original WwTW established at Beckton is still in operation (managed by Thames Water) and is the largest in Europe, treating daily sewage from 3.4 million Londoners.

Chemical precipitation techniques, although effective at removing suspended solids, left around one third of the total pollutant load in the effluent, and produced a large quantity of sludge for disposal. Bacterial or biological treatments (which removed dissolved pollutants as well as suspended matter) were the subject of extensive research between 1897 and 1900 while early biological filters were in use at English WwTWs from 1893. In 1916, the use of chemicals (lime and iron) was suspended, partly due to supply shortages of these chemicals during the war and their associated costs, but also because of the advent of the activated sludge process, whereby biological treatment was accelerated via artificial aeration. In 1920, after pilot plants at two WwTWs in Manchester, the first English city (Sheffield) fully implemented the activated sludge process, this also occurred at London's Crossness outfall in 1921. The widespread implementation of activated sludge treatment in England then occurred in earnest in the mid 1900s following the Second World War, and the activated sludge process became the most commonly adopted treatment process for WwTWs, which holds true today (Allen and Prakasam 1983). From the mid 1900s, there was a gradual development of biological filtering processes in England, but these processes do not differ significantly today.

Legislations from the 1970s, such as the Urban Waste Water Treatment Directive, aimed at raising water quality standards and improving environmental protection (Cooper 2001), have most recently driven investments in nutrient removal via processes including denitrification and phosphate removal, and disinfection. As a result, tertiary treatments, including microfiltration by lagoons, sand filters and reed beds, or disinfection by either ultraviolet (UV) light treatment, ozonation or chlorination, have been implemented at some English WwTWs. These have not, however, been adopted widely due to their associated costs and concerns regarding the formation of toxic by-products. Disposal of sewage sludge at sea was banned in Europe in the 1990s and most treated sludge disposal in England is now via application to agricultural land, although incineration and drying/pelletization are becoming increasingly popular (Cooper 2001). A trend over the last 30 years has also been for wastewater treatment to be organized relative to the river basin rather than via municipal councils, and this has improved environmental protection of these areas.

An important consideration in the history of the development of WwTWs that has had a significant influence on the pollution of English rivers today is that the "state-of-the-art" for sewage treatment has advanced more quickly than for sewage conveyance. The combined sewer systems (CSSs) developed in London, which became a model for combined sewers in New York and more widely, were designed with little provision for dealing with periods of heavy rainfall and with underestimations of population growth rates. While CSSs are no longer used in new developments, these ailing sewerage systems developed at the turn of the century are still in use in many older cities and raw sewage at these sites is frequently discharged into English rivers via emergency overflow points. There are approximately 26,000 such events per year in England and Wales (Arthur et al. 2008).

The History of Feminized Responses in Fish in English Rivers and Their Association with WwTW Effluent Discharges

Disruption of sexual development of fish in English Rivers was discovered when wild roach (*Rutilus rutilus*) living in a wastewater treatment works (WwTW) effluent settlement lagoon in the River Thames were reported by local anglers to be intersex – that is, their gonads contained both male (testicular) and female (ovarian) tissue (Sweeting 1981). Although some fish species undergo a sex change during normal development, the roach is a single-sex species that develops as either a male or a female, and remains that sex throughout its life (Jafri and Ensor 1979; Schultz 1996; Paull et al. 2008). The finding of a few intersex roach was, thus, deemed unusual. No further studies followed up on this chance observation of intersex in wild roach, but it prompted a laboratory study (Ensor et al., unpublished) in rats to investigate whether WwTW effluent had endocrine altering properties. Rats were fed concentrates of WwTW effluent and alterations to their reproductive endocrinology were shown, but no studies were undertaken to identify the nature of the contaminants causing these effects.

An independent series of studies conducted by the Centre for Environment, Fisheries and Aquaculture Sciences (CEFAS), and a team of researchers at Brunel University in London where CRT was employed at that time, subsequently established that effluents from WwTWs throughout England were estrogenic to fish. We determined this based on findings that these effluents induced the production of vitellogenin (VTG, a precursor of yolk normally produced in females in response to estrogen stimulation) in male rainbow trout (*Oncorhynchus mykiss*) and carp (*Cyprinus carpio*) caged in discharges at 15 study sites (Purdom et al. 1994). In these caged fish studies, we found up to a one million-fold increase in the amount of circulating VTG in males, which represented more than half of the circulating blood protein, following only a 3-week exposure. This phenomenon of "estrogenic effluents" was subsequently reported in mainland Europe and worldwide. Estrogenic effluents inducing VTG synthesis in fish via controlled exposures and/or in wild native populations have now been reported in Germany (Hecker et al. 2002), Sweden (Larsson et al. 1999), Denmark (Bjerregaard et al. 2006), Portugal (Diniz et al. 2005), Switzerland (Vermeirssen et al. 2005), The Netherlands (Vethaak et al. 2005), the United States (Folmar et al. 1996), Japan (Higashitani et al. 2003), China (Ma et al. 2005), South Africa (Barnhoorn et al. 2004), and Australia (Game et al. 2006).

Some of our (and others) exposure work with fish, including rainbow trout, carp and roach, has demonstrated that WwTW effluent discharges vary widely in their estrogenic potency (Harries et al. 1997; Rodgers-Gray et al. 2000, 2001; Sheahan et al. 2002a; Liney et al. 2005; Tyler et al. 2005). A wide range of factors have been shown to affect the estrogenic potency of WwTW effluent and they include differences in the estrogenic load in the influents received by the works, the nature of the treatment processes in operation within the works, and season and rainfall/influent/effluent dilution rates.

Estrogen loading to WwTWs can vary due to differences in the population sizes/structures served by each WwTW. Most of the steroid estrogens in influents are derived from the resident human population in the catchment area, and the amount derived from women can vary depending on the number of menstruating, pregnant, and menopausal women [e.g. pregnant women produce 120 times more 17β-estradiol (E_2; approx 393 μg/day) compared with non-pregnant women (Johnson and Williams 2004)]. The numbers of women using (and therefore excreting) pharmaceutical estrogens derived from contraceptives or hormone replacement therapy (HRT), will also vary between WwTW catchment areas. Depending on the location of a WwTW, the proportion of the influent derived from industrial sources may also vary, and this can also affect the amount of specific estrogenic chemicals contained within the effluent. As an example, the presence of (estrogenic) alkylphenols is more associated with industrially derived influents (Sheahan et al. 2002a).

Considering the effect of the level and type of treatment employed by a WwTW on the concentrations of estrogens within effluents (Kirk et al. 2002; Johnson et al. 2007; reviewed in Koh et al. 2008), a study of 17 European WwTWs found the highest levels of estrogen in effluents where the WwTWs were using primary treatment (simply a filtration and settlement process) only. In that study, estrogen levels in effluents from WwTWs employing activated sludge treatment typically were at least a factor of 6 lower (Johnson et al. 2005). In general, the use of secondary treatment at WwTWs, and in particular, aerated activated sludge, has been shown to reduce considerably the estrogenic activity of effluents, presumably due to the biological degradation that occurs under these conditions (Kirk et al. 2002; Johnson et al. 2005). Within WwTWs, longer solid and hydraulic retention times have also been associated with increased estrogen removal, since these represent the time allowed for chemical breakdown/sorption to occur (Holbrook et al. 2002; Kirk et al. 2002; Johnson et al. 2005).

Considering season as a factor determining the estrogenic potency of WwTW effluents (Harries et al. 1999; Labadie and Budzinski 2005; Fernandez et al. 2008), temperature has been shown to affect the activity of microbes responsible for chemical breakdown (Fernandez et al. 2008) and even determine the types of microbial fauna present (e.g. Molina-Muñoz et al. 2009; Song et al. 2009). Seasonal, as well as geographic variations in rainfall can also significantly affect estrogen concentrations in effluents, due to dilution effects (Ternes 1998). For example, in one of our studies with our Brunel colleagues, we found that the estrogenic activities were appreciably lower in effluents collected from five English WwTWs in August, after high levels of rainfall, compared with equivalent samples collected from the same WwTW in April and May (Kirk et al. 2002).

Importantly, in some English locations studied, estrogenic activity has been shown to persist in the receiving rivers for many kilometers downstream of the point source of WwTW discharges and with considerable dilution of the effluent (Harries et al. 1995, 1997; Rodgers-Gray et al. 2000, 2001; Williams et al. 2003; Liney et al. 2005), highlighting the potential for river catchment-wide effects. This persistence in the river is likely related to the speed of passage of effluent through the river, rather than necessarily a low rate for biodegradation. As an example, for the River Thames in London, which has a 221-km length to its tidal limit, river residence times have been estimated to range between 7 and 20 days depending on the season, and predicted to increase to up to 34 days with climate change by 2080 (Johnson et al. 2009).

Controlled exposures to WwTW effluents have shown that effluent concentrations as low as 10% can induce a vitellogenic response in juvenile roach and that longer exposure periods reduce the threshold for feminizing effects (Rodgers-Gray et al. 2000; Lange et al. 2009). This is highly relevant for wild roach populations that spend much, even all, of their lives in rivers where the flow is comprised of at least 10% of treated WwTW effluent (Fig. 13.1). Indeed, in some English rivers, during the summer months and periods of low water flow, half of the flow of the river can be made up of WwTW effluent and, in the most extreme cases, the complete flow of a river can be comprised of treated WwTW effluent (Jobling et al. 1998; Fig. 13.1).

It is worth emphasizing that most studies assessing the potential impacts of WwTW effluents on roach and other fish have employed relatively short-term (typically 2–3 week) exposures and this may not accurately reflect the severity of effect that WwTW effluents impose on wild fish living in English rivers. Examples of associations between a reduced threshold for effects on reproduction and/or development in fish and longevity of exposure to environmental estrogens are peppered throughout this chapter. In many countries, dilution rates of effluents in rivers are generally higher

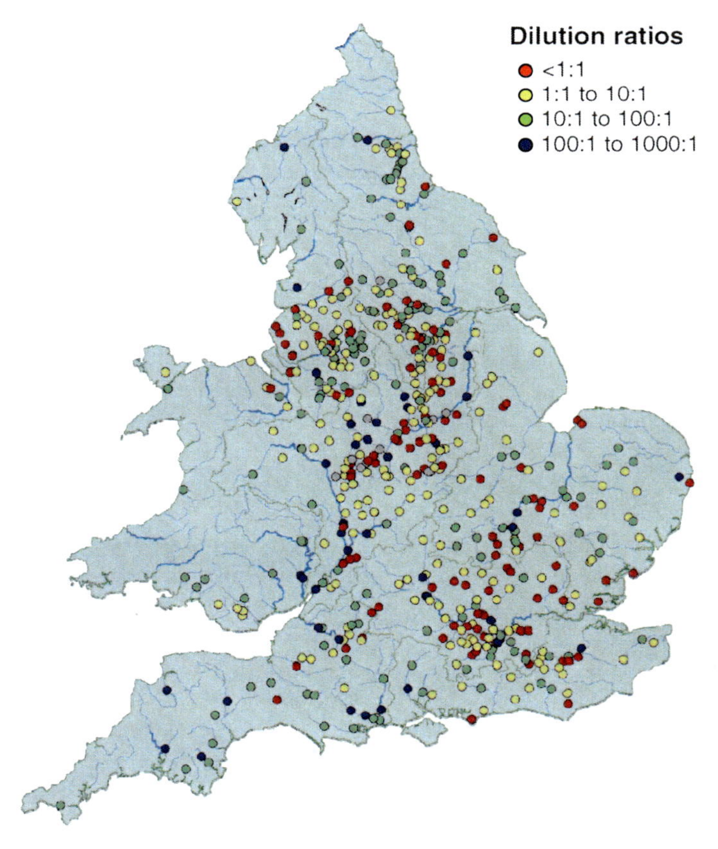

Fig. 13.1 Dilution ratios of wastewater treatment works effluents discharged from major wastewater treatment works (WwTWs) in rivers in England and Wales

than in the England, and this perhaps explains why reported levels of VTG induction in wild carp and bream, members of the same family of fish as the roach, in mainland Europe, and in largemouth bass (*Micropterus salmoides*) in the US, are generally lower compared with in wild roach in English Rivers.

Why the Roach Was Chosen as Our Sentinel Species

For our work on the effects of WwTWs on sexual development in fish in English rivers, we adopted the roach as our sentinel species. There are a number of reasons for this choice including the fact that intersex in fish in English rivers was first identified in this species, thus showing it has a susceptibility for gender disruption. The roach is also one of the most common fish species in lowland rivers in England, sometimes making up to half of the fish biomass, and this has enabled us to undertake the widespread destructive sampling needed in our nationwide surveys to examine for effects of WwTW effluent discharges on gonad development. The roach is also widely distributed across Europe potentially allowing for comparative studies across wide geographical regions. The roach is found in a wide range of habitats (even including brackish waters), and is tolerant of poor water quality, so is, therefore, amenable for WwTW effluent exposures in which oxygen levels are typically low and sediment loads high. For laboratory work, the roach is easily cultured, facilitating controlled chemical exposure studies. Importantly, the roach is a member of the carp (Cyprinidae) family, one of the largest and most important group of freshwater fish worldwide, and, like most other cyprinids, it is a group spawner. As such, the principles derived from work on roach are likely to be widely applicable to other group-spawning species. Finally, the roach is a relatively long-lived species, maturing after 2 or 3 years in the wild and living for up to 10 years, making it highly suited for assessing impacts of long-term exposures to chemicals and their mixtures on sexual development and breeding; in fish, germ cells (which are the most susceptible stages for disruption in sexual programming) undergo proliferation every breeding season.

The Extent of Sexual Disruption in Roach in English Rivers

The extent of sexual disruption in roach in English rivers has now been thoroughly investigated through two major national surveys of wild populations. These were conducted in collaborative efforts between the Environment Agency of England and Wales, Brunel University and ourselves at the University of Exeter, (Jobling et al. 1998, 2006; Nolan et al. 2001). In the first major survey conducted between 1995 and 1996, roach were sampled at locations up- and down-stream of WwTWs on eight rivers, and at five reference sites throughout England and Ireland. Microscopic analysis of the gonads collected revealed that a large proportion (16–100%) of the putative males living immediately downstream of WwTW effluent outfalls were, in fact, intersex (Jobling et al. 1998). Additionally, although intersex was also seen in roach at sites upstream of WwTWs, generally this was observed at lower incidences (11–44%; Jobling et al. 1998) and the severity of the condition was much reduced, compared with their respective downstream sites. It is generally the case that individual rivers in England receive a number of WwTW effluent discharges. In England, there are 862 classified rivers/brooks and a total of 4,422 WwTWs, of which only 271 discharge to sea (Williams et al. 2009). Thus, the observation of some intersex fish at the upstream sites was likely due to the fact that, in most cases the study sites labeled as "upstream" in the first national survey were also downstream from more distant WwTWs. In the second and more extensive survey, conducted 2002–2003, intersex was found in wild roach at 86% of 51 river locations studied receiving WwTW effluent discharges (Jobling et al. 2006). This finding firmly established the widespread nature of sexual disruption in this species in English rivers.

The gonadal phenotypes seen in the wild "feminized male" roach were wide-ranging. In some cases, the gonads affected contained single developing eggs (either oocytes or small nests of oogonia, which are oocytes at a very early stage of development) interspersed throughout an otherwise normal testis. In the most extreme cases, half of the testis was comprised of ovarian tissue (Nolan et al. 2001). In some individuals analyzed, the sperm duct that enables the sperm to be released was absent and replaced by an ovarian cavity, a structural feature associated with an ovary (Jobling et al. 1998; van Aerle et al. 2001; Nolan et al. 2001). Examples of how these gonads appeared visually in histologically prepared sections under a microscope are shown in Fig. 13.2. An analysis of the extensive dataset on intersex in wild populations of roach has shown that both the incidence of intersex in the population and the degree of intersex within those individuals were highly positively correlated with age (Jobling et al. 2006), with oocytes only appearing commonly in the testis when the roach reached 2+ years old. This evidence further supports a growing body of data indicating that longevity of exposure to WwTW effluents can affect the degree of the feminizing effect.

A question that arose for us from the extensive survey work on roach in English rivers was whether similar responses occur in other fish species. We thus conducted a study on the gudgeon (*Gobio gobio*; van Aerle et al. 2001), an allied member of the carp family, but one that lives on the river bed rather than in mid-water like the roach, and we showed a similar effect, albeit at a lower level compared with in the roach. Similarly, altered sexual development has also been reported in mainland Europe in other fish species from the carp family, including the bream (*Abramis brama*; River Elbe, Germany, Hecker et al. 2002), common carp (*C. carpio*; Anoia River, Spain, Solé et al. 2002, 2003) and barbel (*Barbus barbus*; River Po, Italy, Viganò et al. 2006). Not all fish appear to be equally affected by effluents, however, and contrasting with the findings for roach and gudgeon in English rivers, a study we conducted with our Brunel colleagues on the perch (*Perca fluviatilis*) and pike (*Esox lucius*), top predatory fish, found no clear evidence of sexual disruption despite the fact that they were living in the same rivers (Vine et al. 2004). We found

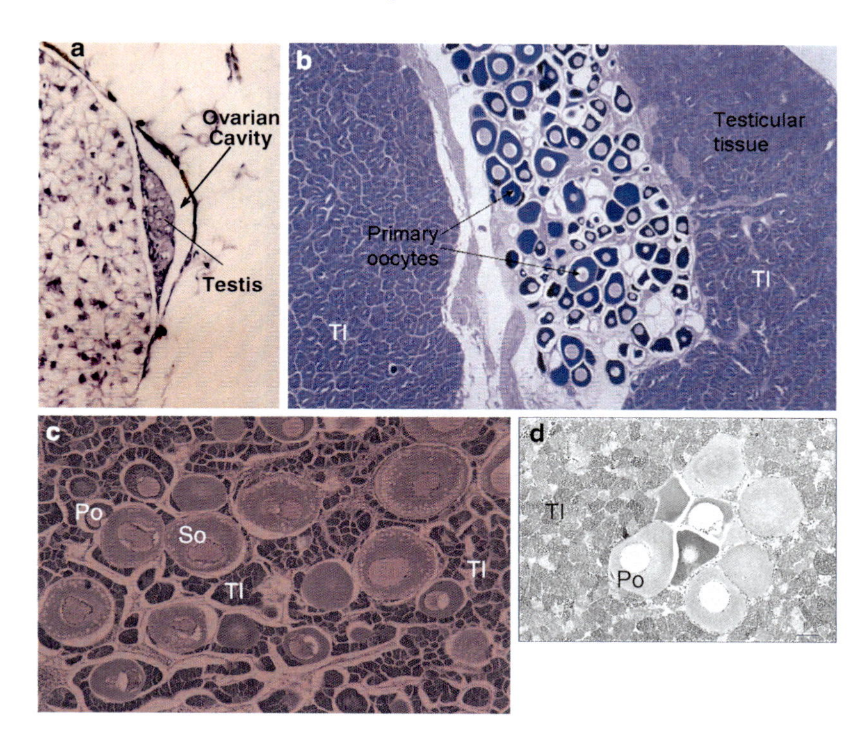

Fig. 13.2 Histological sections showing the range of gonadal abnormalities seen in wild "feminized male" roach. (**a**) Presence of a female-like ovarian cavity in an otherwise normal testis, (**b**) severely intersex gonad showing a testis containing a large number of primary oocytes at a single focus, (**c**) severely intersex gonad with large numbers of both primary and secondary oocytes dispersed throughout the testis, and (**d**) mildly intersex gonad with a small number of primary oocytes found at focal points throughout the testis (the more common condition of roach living in English rivers). *Tl* testis lobule, *Po* primary oocyte, *So* secondary oocyte. The scale bar represents 100 μm

this surprising as, given their position in the food chain and the fact that they readily predate on roach, we envisaged that these species would also have been affected. We thus speculate that some species have a lower susceptibility and/or lower sensitivity to the chemical disruption of sex. Interestingly, in some of our unpublished studies, we found pike to be less responsive to estrogen treatment (as determined by VTG induction) compared with almost all other fish species we have studied.

The prevalence of sexual/endocrine disruption in marine fish in England is beyond the scope of this chapter. However, elevated levels of plasma VTG, the presence of intersex gonads, and/or altered development of secondary sex characters (SSCs; driven by sex hormones) have also been reported in fish living in estuaries. Feminized responses in both the European flounder (*Platichthys flesus*) and goby (*Pomatoschistus* spp.) have been associated with exposure to WwTW effluent discharges (Allen et al. 1999; Kirby et al. 2003, 2004; Kleinkauf et al. 2004).

Whodunit? The Identification of the Causative Chemicals

In some of the original work on wild intersex roach, we established a positive correlation between the proportion of intersex individuals in the populations studied and the concentration of the effluent at the different sampling sites (Jobling et al. 1998). A further study on wild roach by the Brunel University team then provided some evidence for a link between high levels of effluent discharges into a river and an abnormally high proportion of juveniles with a feminized reproductive duct (Beresford et al. 2004). We, however, through a series of controlled exposures to WwTW effluents have now provided definitive evidence that effluents from WwTWs induce sexual disruption in roach, showing they induce all of the feminine characters seen in wild roach, including the presence of an ovarian cavity and oocytes in the testis of males (Rodgers-Gray et al. 2000, 2001; Liney et al. 2005, 2006; Tyler et al. 2005; Gibson et al. 2005; Lange et al. 2011). Very recently in a controlled exposure of roach to full-strength WwTW effluent for a period of 3 years, we have shown that complete gonadal feminization occurs in males, resulting in a functional ovary with no obvious signs of testis tissue (Lange et al. 2011). This finding again highlights that long-term exposures to WwTW effluents can result in more severe, and different, phenotypic outcome (with a more likely functional consequence) compared with shorter-term exposures. Although roach populations move within a catchment as they are able to, and determined in part by river barriers, they tend to reside within a specific location and thus those close to WwTW outfalls will likely have a lifetime of exposure. No studies have yet been forthcoming to assess whether full sex reversal is occurring in wild roach populations, but we cannot yet get a definitive answer to this question as there is no sex probe available for this species.

In theory, intersex wild roach could arise as a consequence of either the feminization of males or the androgenization of females, since exposure of fish to sex hormones (either estrogens or androgens), or other chemicals that interfere with the balance of natural sex hormones, can cause altered sexual development and even complete sex reversal in either direction (see review in Baroiller and Guiguen 2001). The available evidence, however, strongly supports the hypothesis that intersex roach in English rivers are derived from the feminization of males. This evidence includes: (1) we have found that the number of roach with normal testes in the wild populations studied is inversely proportional to the number of intersex roach (Jobling et al. 1998, 2006); (2) WwTW effluent discharges into English rivers are predominantly estrogenic (Purdom et al. 1994; Harries et al. 1997, 1999; Rodgers-Gray et al. 2000, 2001; Jobling et al. 2003) and/or anti-androgenic (which would further enhance any feminization of males; Johnson et al. 2004; Jobling et al. 2009), but rarely androgenic; (3) wild male and intersex roach contain the estrogen-inducible protein VTG in their blood plasma (Jobling et al. 1998, 2002a), and (4) our studies on wild intersex roach have shown that they generally have plasma levels of 11-ketotestosterone (11-KT), the main male sex hormone in fish, and E_2 more similar to those in normal males than in normal females (Jobling et al. 1998). As detailed above, however, as yet for roach there are no genetic sex probes available to provide definitive proof that intersex fish are derived from feminized males.

Based on the evidence that the intersex condition arises from the feminization of males, investigative work undertaken in various laboratories has sought to identify the causative chemicals contained in WwTW effluents with a focus on estrogens. Identifying some of the estrogenic chemicals in WwTW effluents has involved fractionating the effluent and systematically testing the fractions for estrogen activity using a recombinant yeast reporter assay containing the human estrogen receptor [the yeast estrogen screen (YES)]. The specific estrogenic compounds have then subsequently been identified using gas chromatography–mass spectrometry (GC–MS)/liquid chromatography–mass spectrometry (LC–MS) (Desbrow et al. 1998; Rodgers-Gray et al. 2000, 2001; Gibson et al. 2005). A wide range of chemicals with estrogenic activity are now known to be discharged in effluents from WwTWs, albeit with widely varying potencies. However, these fractionation studies have shown that most of the estrogenic activity (up to 80%) is contributed by the natural steroidal estrogens E_2 and estrone (E_1), together with the synthetic estrogen 17α-ethinylestradiol (EE_2; a component of the contraceptive pill) (Desbrow et al. 1998; Rodgers-Gray et al. 2000, 2001). Working with Dr Liz Hill at the University of Sussex, we recently identified horse (equine) estrogens used in HRT in WwTW effluent discharges (Gibson et al. 2005; Tyler et al. 2009). Other estrogenic chemicals identified include alkylphenolic chemicals (APs) derived from the breakdown of industrial surfactants (for example, used in cleaning agents and paints; Waldock et al. 1997). These APs have been shown to be especially prevalent in the effluents from WwTWs receiving significant inputs from the wool scouring and textile industries (Sheahan et al. 2002a, b). The estrogenic potency of APs is, however, significantly lower than for steroidal estrogens. For example, in the YES, nonylphenol was 5,000-fold less potent than E_2 (Gaido et al. 1997), and it showed an approximately 800- to 1,300-fold lower potency than E_2 at inducing vitellogenesis in rainbow trout (Thorpe et al. 2001). A summary of the concentration ranges at which these different classes of environmental estrogens have been observed in English final WwTW effluents is shown in Table 13.1.

Studies adopting a similar fractionation process with fish bile, which focused on chemicals that physically enter and bioconcentrate in exposed fish rather than simply

Table 13.1 Concentration ranges of environmental estrogens measured in English final WwTW effluents

Chemical	Concentration range
Steroidal estrogens	
Estrone (E_1)	0.25–169.0 ng/L
17β-estradiol (E_2)	0.25–20.0 ng/L
17α-ethinylestradiol (EE_2)	0.1–1.5 ng/L
Equine estrogens	0.07–1.32 ng/L
Alkylphenols	
Nonylphenol and ethoxylates	0.62–363 μg/L
Octylphenol	0.03–0.41 μg/L

Data obtained from: Liney et al. (2005, 2006), Rodgers-Gray et al. (2001), Sheahan et al. (2002a, b), Thorpe et al. (2006, 2008, 2009), and Tyler et al. (2009)

measuring those present in the effluent itself, have further supported the hypothesis that steroidal estrogens, and to a lesser extent APs, are the major environmental estrogens accumulating in fish exposed to WwTW effluents (Gibson et al. 2005; Tyler et al. 2009). Indeed, steroidal estrogens and APs can bioconcentrate up to 40,000-fold in fish (Larsson et al. 1999; Gibson et al. 2005), serving to enhance their (potential) feminizing effects.

Laboratory exposures of various fish species to steroidal estrogens and APs found in WwTW effluents, and found to bioconcentrate in exposed fish, have been shown to induce VTG synthesis, gonad duct disruption, and the development of oocytes in the testis (e.g. Blackburn and Waldock 1995; Tyler and Routledge 1998; Metcalfe et al. 2001; Yokota et al. 2001; van Aerle et al. 2002; Hill and Janz 2003), strongly supporting their involvement in the feminization of wild roach. For the most part, however, the concentrations that induced oocytes in the testis have been generally higher than that found in effluents and almost always higher than found in receiving rivers. Nevertheless, recently, long-term exposure of roach to EE_2 at 4 ng/L, a concentration found in some of the more heavily contaminated effluent discharges, was shown to induce complete sex reversal resulting in an all-female population (Lange et al. 2008, 2009). Adding further to the hypothesis that steroidal estrogens play a major role in causing intersex in wild roach, a study on wild roach at 45 sites on 39 English rivers found that both the incidence and severity of intersex were significantly correlated with the modeled concentrations of E_1, E_2, and EE_2 (determined via combining measured steroid concentrations in effluents, with dilution rates and river flow data; Jobling et al. 2006). For some study locations, however, compelling evidence has been presented that APs have been significant contributors to the estrogenic effect of those WwTW effluent discharges. The WwTW discharges in question, discharging into the Rivers Aire and Calder in Northern England, received considerable inputs of nonylphenol and other APs from local wool/textile industries. When a voluntary withdrawal of these APs was adopted, this was accompanied by a much reduced vitellogenic response in caged fish placed in those rivers containing discharged effluents (Sheahan et al. 2002b).

The range of environmental estrogens now identified in English WwTW effluents that induce responses in fish also includes plasticizers, such as phthalates, bisphenols, and various pesticides and herbicides (reviewed in Turner and Sharpe 1997; Sonnenschein and Soto 1998; Oehlmann et al. 2009). Individually, these chemicals, unlike steroidal estrogens, probably do not play a significant role in the disruption of sex in wild roach, given their relatively lower estrogenic potency. However, as part of a mixture they may contribute to the overall effects seen. Indeed, in vivo studies in the roach and other fish species have shown that estrogenic chemicals in combination with each other have additive feminizing effects (Thorpe et al. 2001, 2003, 2006; Brian et al. 2005).

Adding further to the likelihood that chemical causation of sexual disruption in the roach is a mixtures effect, many English WwTW effluents have been reported to be anti-androgenic, as measured in the anti-YAS screen (the YAS is similar to the YES, but has the human androgen receptor incorporated into the yeast system, rather that the estrogen receptor), as well as estrogenic (Johnson et al. 2004). Anti-androgens also induce feminizing effects in males in both fish and higher vertebrates, albeit through different mechanisms than for estrogens (e.g. Filby et al. 2007; reviewed in Kelce and Wilson 1997). Recently, a strong statistical association was shown between the widespread feminization of wild roach and the presence of anti-androgenic activity (Jobling et al. 2009). A wide range of chemicals that are discharged into the environment are now known to have anti-androgenic activity, including some phthalates (e.g. dibutyl phthalate), plasticizers (e.g. bisphenol A; Sohoni and Sumpter 1998), polyaromatic hydrocarbons (PAHs; e.g. benzo[a]anthracene and fluoranthene), APs (e.g. nonylphenol), pesticides (e.g. the organophosphate insecticide fenitrothion and the herbicide linuron), and pharmaceuticals (e.g. flutamide and cyproterone acetate) (Lee et al. 2003; Orton et al. 2009; Weiss et al. 2009) though they are generally only weakly active. We are now working with Dr. Liz Hill (University of Sussex) to try to identify anti-androgens that enter and accumulate in effluent-exposed fish, adopting a similar targeted fractionation process to that used to identify the estrogenic contaminants in fish bile, but employing the YAS to identify the anti-androgenic fractions separated out by high performance liquid chromatography (HPLC) and then GC–MS/LC–MS to identify the chemicals. To date, the anti-androgenic chemicals identified accumulating in bile in WwTW effluent-exposed fish include di(chloromethyl)anthracene and dichlorophene (Hill et al. 2010). Neither of these compounds identified are, however, especially potent as anti-androgens (only 3–6 times more potent than the standard anti-androgen, flutamide, as assessed in the human androgen receptor yeast reporter assay; Hill et al. 2010). The likelihood is, therefore, that fish are exposed to many more environmental anti-androgens and/or that there are some compounds that are more potent as anti-androgens yet to be isolated.

Biological Significance of Feminized Responses in Roach

The ecological importance of these feminized responses in roach in English rivers will depend on whether the reproductive competence of these roach populations is compromised and, if so, how this affects the size, structure, and genetic integrity of the population. Considering first the implications of exposure to environmental estrogens for the health of individual fish, commonly measured VTG induction in wild roach does not necessarily imply an adverse effect on their reproductive health since it can be cleared from the blood following removal of the estrogenic stimulus (Craft et al. 2004). Equally, however, wild roach may live in estrogenic effluent plumes throughout their lives and may be stimulated to produce VTG continuously. High level induction of VTG has been shown to have wider health effects in fish, inducing kidney damage and/or failure (e.g. Herman and Kincaid 1988; Zaroogian et al. 2001), but these levels of VTG induction have not been reported in wild fish in the ambient environment, away from discharge hotspots. Inappropriate VTG induction at lower levels in fish during their early life stages, when energy budgets are critically balanced, could also impact on their survivorship in the wild, but there is no direct evidence for this. Laboratory-based exposures of various fish species have found positive correlations between the induction of VTG and reduced reproductive output, but only at relatively high VTG levels (above 1 mg VTG/mL; Thorpe et al. 2007, 2009). This association may be a statistical one only, rather than an enhanced VTG level having a direct effect on reproduction.

The ability of intersex roach to produce gametes is variable and dependent on the degree of disruption in the reproductive ducts and/or altered germ cell development. In some cases, we have found wild roach that cannot release any gametes at all due to the presence of severely disrupted gonadal ducts (Jobling et al. 2002a, b). In the majority of intersex roach found, male gametes are produced, released and are viable, but for moderately to severely affected fish they are of poorer quality than those from normal males obtained from environments that do not receive WwTW effluent. We established this through in vitro fertilizations of gametes stripped from wild roach (Jobling et al. 2002b) and in fertilization and hatchability studies showed that intersex roach are compromised in their reproductive capacity and produce less offspring than roach from uncontaminated sites under laboratory conditions (Jobling et al. 2002b). In that study, there was an inverse correlation between reproductive performance and the severity of gonadal intersex (Jobling et al. 2002b). More recently, in a study funded by the UK Department for Environment Food and Rural Affairs (DEFRA), as part of a catchment-wide study on the potential impacts on endocrine disrupting chemicals (EDCs) in fish and other aquatic wildlife, we, together with our Brunel colleagues, have made assessments on the ability of intersex roach to breed in laboratory-maintained colonies (thus under conditions that better mimic those that occur in the wild). Using DNA microsatellites to determine parentage, the intersex condition was shown to compromise the individual "reproductive fitness" of more severely affected males. Those "males" with only a few oocytes in their testes were equally as effective as control males in siring offspring (Harris et al. 2010).

Modeling for predicting population-level impacts of intersex in roach requires comprehensive information on their normal population dynamics [ideally over extensive periods of time as they are a reasonably long lived (10+ years) species], and knowledge of their breeding biology. Some of these data, but certainly not all, are now available and recently an impressive attempt has been made to model the impact of the intersex seen in roach on wild populations in English rivers (An et al. 2009). In that work, various parameters were taken into account including the incidence and severity of intersex within the population and used to estimate the reduction in fertilization rate. The outcome was that intersex, when at the levels and severities reported for English roach, would have little or no impact on the viability of wild populations. The authors, however, emphasized that the model was based on an assumption that the roach populations were not subjected to any other additional major environmental stressors and this is very likely not the case. Furthermore, modeling approaches for informing on population sustainability generally work on abundance only (which, in turn, is largely derived as a function of egg/embryo production and early life survivorship) and this should not be the only consideration.

For example, possible effects of chemicals on genetic diversity and selection may have key implications in the long-term maintenance of a viable and healthy population (Bickham et al. 2000).

Other shortfalls in the roach population modeling work include that it is not the intersex condition alone that is induced in roach populations in effluent-contaminated rivers that could compromise their reproductive capability. Male roach living in effluent-contaminated rivers in England also have altered sex steroid hormone profiles, causing altered spawning times and reduced sperm production (Jobling et al. 2002a, b). These behavioural and physiological changes are also likely to impact on the reproductive capability and success of those individuals. The timing of gamete synthesis and release in the appropriate season is essential for reproductive success in annually breeding species. In the recent study where complete gonadal sex reversal was induced in male roach exposed to full-strength WwTW effluent, delayed development of the resulting ovary occurred compared with ovaries in normal females (Lange et al. 2011). This, in turn, would suggest that any potentially viable eggs in sex-reversed males would be produced at an inappropriate time to contribute to the fish population; roach breed once annually in English Rivers between late April and early May. Little study has been given to the effects of effluent exposure in female fish. However, female roach living in rivers heavily contaminated with treated effluent (Jobling et al. 2002b), or exposed for 3.5 years to a WwTW effluent at full-strength (Lange et al. 2011), have been shown to have a higher incidence of ovarian atresia, potentially reducing the number of eggs produced.

Various laboratory studies on a range of fish species continue to show that exposure to environmental estrogens, anti-androgens, and estrogenic effluents can alter normal breeding behavior (e.g. Saaristo et al. 2009; Salierno and Kane 2009; Sebire et al. 2009). We have shown in zebrafish colonies that altered behaviors as a consequence of estrogenic EDC exposure can impact on the reproductive success of individuals (Nash et al. 2004; Coe et al. 2008, 2009). In some of these studies, using DNA microsatellites to track parentage of the offspring produced, we have shown that exposure to an environmental estrogen (EE_2) disrupts the normal pattern of paternity, reducing the reproductive success of the most dominant male and in turn affecting the genetic outcome in subsequent populations (Coe et al. 2008, 2009). Data generated from studies in the zebrafish (Nash et al. 2004) have further implied that reproductively compromised males can also impact negatively on the ability of other healthy males to breed successfully by interfering with normal female interactions.

The Endocrine Disruption Demonstration Programme: A Step Forward

It has been a long journey from those initial studies identifying sexual disruption in fish in English rivers to the generation of a sufficient weight of evidence for causation by EDCs contained in WwTW effluents. On this journey, the Environment Agency has been engaged and supportive throughout. It has been more difficult to convince the water industry of the problem and to persuade them to take up the challenge of EDC removal from WwTW effluents. The science now generated on the feminization of wild fish in English rivers, and even in the absence of proven population level effects, has however lead to the development of sufficient concern at governmental level to establish a so-called *Endocrine Disruption Demonstration Programme* (*EDDP*). This is a programme run by the Environment Agency of England and Wales in collaboration with the UK water industry (Gross-Sorokin et al. 2006; Huo and Hickey 2007), with advisory input from academics, including ourselves. The EDDP, running from 2005 to 2010, at an estimated cost of £40M, has been designed to improve understanding of how current WwTW processes affect the concentrations of EDCs (focused on steroidal estrogens and APs) and via full-scale demonstration trials to identify the most promising advanced treatment technologies for EDC removal. A key aim of the EDDP is to deliver a basis for future decisions on investment costs of improved treatment both in terms of operational and capital expenditure to inform on options for UK investment planning decisions and the forthcoming Water Framework Directive of Measures.

In order to determine a strategy to improve WwTWs in the UK, the EDDP set out a plan involving a large scale pilot study. First they needed to select which WwTWs to study. A risk analysis was conducted on 450 WwTWs (with population equivalents greater than 10,000 people) and their receiving waters based on their predicted estrogen discharges. Predicted impacts in the receiving rivers were supported by our assessments with our Brunel colleagues on sexual disruption in roach at 51 river sites (see Jobling et al. 2006). In the final selection process, 16 WwTWs were chosen for the EDDP covering the range of estrogen discharges (sites were ranked as at low, medium or high risk, representing steroid estrogen equivalents of <1 ng/L, >1 to <10 ng/L and >10 ng/L, respectively), including representation of the different operational treatment types and also considering the geographical location of the WwTWs to ensure the active engagement of all ten water companies in England and Wales. The different treatment processes at these works included primary treatment, secondary treatments [biological filters, and nitrifying and non-nitrifying activated sludge plants (ASPs)], and tertiary treatments [sand filters, lagoons, nitrifying biological aerated flooded filter (BAFF), and UV disinfection].

At all of the selected WwTWs, analytical assessments were made on estrogen removal, for steroid estrogens, nonylphenol and octylphenol and their polyethoxylates, and total estrogenic activity (measured using the YES) to assess the efficacy of the different treatment processes. At two of the selected sites, granular activated carbon (GAC) technology, and at one site ozone (O_3) and chlorine dioxide (ClO_2), were installed to test the efficiency of these technologies at removing steroid estrogens. The potential of these technologies for the removal of EDCs has been highlighted in a number of recent studies: GAC (Zhang and Zhou 2008; Ifelebuegu et al. 2006); ozonation (Huber et al. 2003; Ternes et al. 2003; Deborde et al. 2005; Westerhoff et al. 2005; Kim et al. 2008); and chlorination (Westerhoff et al. 2005; Schiliro et al. 2009) (see also reviews in Larsen et al. 2004; Koh et al. 2008, Chang et al. 2009). Various sampling regimes were undertaken at the different works to measure the estrogenicity and they included bi-weekly samples over 1 year, intensive 7 day samplings (up to 42 events), and hourly samplings over 24 h.

Evaluations of the efficacy of the different conventional treatment processes currently in use at the WwTWs are ongoing as the full data sets are not yet available for a complete assessment (see Butwell et al. 2009). Nevertheless, we can summarise some of the key initial findings. Considering first the APs, although they were found to vary widely in the incoming influents, both APs and their polyethoxylates (APEOs) were substantially removed by all secondary treatment processes. In the resulting effluents, at some sites APs/APEOs were lower, but many were at the analytical detection limits (1–5 μg/L). Based on these AP/APEO levels in the discharging effluents and their known biological potencies, they would not be expected to induce significant feminized responses in wild fish (Kang et al. 2003; Yoon et al. 2008; Jürgens et al. 2009). The only exception perhaps is for very prolonged exposures to nonylphenol, which have been shown to be biologically effective at inducing feminized responses in fish in the low μg/L concentration range (e.g. Ackermann et al. 2002). Due consideration, however, needs to be given to the total summed content of the APs/APEOs in the effluent discharges, as they are known to be additive in their biological effects (Thorpe et al. 2001; Brian et al. 2005).

Overall reported ranges for steroid estrogen levels in discharged effluents assessed via analytical chemistry were 36.2–59.8, 14.3–17.5, and 0.75–1.15 ng/L for E_1, E_2, and EE_2, respectively, similar to those reported previously in the literature (e.g. Ternes et al. 1999). These data reflect "free steroids" (biologically active) and do not account for conjugated steroids that can potentially be de-conjugated to render them biologically active (e.g. D'Ascenzo et al. 2003).

The removal rates of the steroid estrogens and total estrogenic activity observed for the different treatments (compared with levels in the incoming influents to these treatments) are shown in Table 13.2. The data supported previous work demonstrating that primary settlement was relatively ineffective at removing steroid estrogens, with the exception of E_2 where a consistent (but small) reduction was observed at all three sites assessed (Table 13.2). In terms of secondary treatment, it was evident from the data that the nitrifying ASP treatment resulted in a significantly greater removal of the steroid estrogens when compared to the other processes. On average, the biological filter performance exceeded that of the non-nitrifying ASP sites, but considerable variation was observed for this treatment type and some of the biological filter sites were less effective than the non-nitrifying ASP sites (Table 13.2). Tertiary treatment, applied at four of the biological filtration sites, showed variable results, with estrogen concentrations appearing to increase as a result of the settlement lagoon process (although not significantly so), and was unchanged by UV treatment (Table 13.2). Sand filtration resulted in significant reductions in estrogen levels, however, and BAFF (one site only) resulted in a substantial removal of E_1, E_2, and total estrogenic activity although it was less effective at removing EE_2 (Table 13.2).

At the two study sites in the EDDP that are employing the advanced treatment technologies, in addition to the analytical chemistry undertaken, biological effects data have been gathered using two test exposure scenarios with fathead minnow (*Pimephales promelas*) as a model species (Thorpe et al. 2008, 2009). We have undertaken the exposures and biological effect measures for one of these two EDDP study sites [see Filby et al. (2010) for full details]. In one of these test scenarios,

Table 13.2 Effects of different WwTW treatments on the removal of steroid estrogens from influents at English WwTWs

WwTW treatments	Number of sites	E_1 (%)	E_2 (%)	EE_2 (%)	Total estrogenic activity (%)
Primary	3	4.7 (–2.7 to 16.1)	22.3 (9.4–41.8)	16.0 (0–28.6)	–0.475 (–26.5 to 21.2)
Secondary					
Nitrifying ASP	4	93.7 (89.0–98.2)	97.6 (95.5–99.4)	53.9 (20.0–84.6)	94.9 (91.7–99.0)
Biological filter	10	40.2 (–40.5 to 88.1)	75.5 (34.4–93.6)	–24.1 (–233 to 56.4)	48.8 (–53.4 to 90.7)
Non-nitrifying ASP	2	17.9 (13.8 to 21.9)	62.5 (59.9–65.1)	–10.8 (–13.2 to 28.3)	18.5 (15.5–21.5)
Tertiary					
Sand filter	3	77.6 (73.9–79.7)	78.3 (73.9–81.0)	18.3 (10.0–25.0)	75.5 (67.5–77.0)
Lagoon	1	–11.7	–16.0	0.0	–19.0
BAFF	1	94.8	90.2	38.5	88.4
UV disinfection	1	7.7	0.0	25.0	–13.8

Data were collected as part of the Endocrine Disruption Demonstration Programme (EDDP) and are shown as the mean (and range) percentage removal for the steroid estrogens, estrone (E_1), 17β-estradiol (E_2), and 17α-ethinylestradiol (EE_2), and total estrogenic activity [E_2 equivalent; as determined by the yeast estrogen screen (YES)]. For full details, see Butwell et al. (2009)

maturing male fathead minnows were exposed to the different effluents for 21 days and effects quantified on VTG induction (as a biomarker of estrogen exposure), and on the development of secondary sex characters. SSCs (which include nuptial tubercles and a dorsal fatpad in male fathead minnows; see Fig. 13.3) are androgen-mediated in their development (Smith 1978). However, following exposure to EDCs, their development can be suppressed (by estrogens/anti-androgens) or enhanced (by androgens) (e.g. Pawlowski et al. 2004). SSCs are therefore a useful endpoint for screening for chemicals that act both via the androgen and/or estrogen mediated pathways. Importantly, SSCs in fish are believed to function in spawning, maintenance of territories and, in fathead minnows, also in parental care of the developing embryos. Furthermore, there are data indicating that chemically induced effects on SSCs in fish may have consequences for individual reproductive success (Salierno and Kane 2009). Sexually mature male roach develop well-defined nuptial tubercles, similarly to those in male fathead minnows. In the second test scenario adopted to test the biological effects of advanced treatments on fish (here only for GAC treatment), we paired mature fathead minnows and made assessments on their breeding output, as well as effects on VTG induction and SSCs in males.

For the study site for which we conducted the biological effect studies with fathead minnows, the standard final effluent without any advanced treatments contained E_1, E_2 and EE_2 at mean concentrations spanning 4.5–6.7, 0.1–0.7 and 0.6–0.8 ng/L, respectively. The GAC treatment of this effluent reduced its total estrogenicity (defined as E_2 equivalent, as assessed by the in vitro YES assay) by 70–86%, and concentrations of E_1 by 80–98%, E_2 by 79–100% and EE_2 by 53–73%. Similarly, O_3 treatment of the standard effluent reduced its total estrogenicity by 88% and concentrations of E_1 by 74%, E_2 by 77%, and EE_2 by 64%. ClO_2 reduced the total estrogenicity and EE_2 concentrations to below the limit of detection of the

Fig. 13.3 Photographs of adult male and female fathead minnow (*Pimephales promelas*) showing the secondary sex characters (SSCs) present in the male: the nuptial tubercles and fatpad. The fatpad is separated into two regions: the head pad and the dorsal pad

assays and concentrations of E_1 and E_2 by 96 and 90%, respectively. Induction of VTG that occurred in male fish exposed to the standard effluent was removed by the GAC and ClO_2 treatments. In the O_3 treatment too, there was a reduction in the vitellogenic response compared with the standard effluent. In this treatment however, there was still a statistically significant induction of VTG compared with the dilution water controls. Development of the SSCs in males was affected by exposure to the standard effluent and some (but not all) of these effects were remediated by the advanced treatment processes. Data on the effects of effluent exposure on egg production in the pair-breeding fathead minnows showed that the standard effluent inhibited reproduction at the two highest test concentrations: there were 46 and 34% reductions in mean cumulative egg production over the 21-day exposure period compared to the 21-day pre-exposure period for breeding pairs exposed to 50 and 100% effluent concentrations, respectively (Fig. 13.4). Interestingly, exposure to the same full-strength effluent after GAC treatment did not remediate for the effluent's effects on reproduction and there was similarly a 40% decrease in egg production in these fish (Fig. 13.4). This last finding, together with the fact that a parallel exposure to EE_2 (at 10 ng/L) similarly had no effect on reproductive output in the pair-breeding fathead minnows, strongly suggests that the poor egg production resulting from

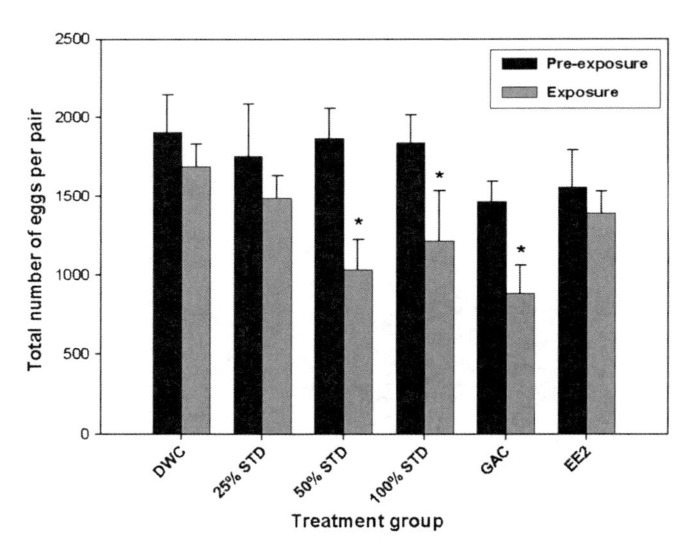

Fig. 13.4 The effect of exposure to treated WwTW effluents on egg production by pair-breeding fathead minnows. The data are shown as the total number of eggs (as assessed daily) produced per pair of fish over a 21-day pre-exposure period (during which time the fish were held in clean water) compared to a 21-day exposure period. Fish were exposed to a dilution water control (DWC), graded concentrations (25–100%) of standard-treated effluent (STD), full-strength STD effluent treated with granular activated carbon (GAC), or an estrogen reference (10 ng/L 17α-EE_2) ($n = 7$–8 pairs per treatment group). Significant differences in the number of eggs produced by the fish between the pre-exposure period and the exposure period for each treatment group are denoted with an asterisk ($P < 0.05$; determined using the Kolmogorov–Smirnov (KS) test)

exposure to standard effluent was not related to the estrogenic content of the effluent and/or that the GAC treatment process changed the physicochemical characteristics of the water making them less optimal for fish reproduction.

The full outcome and subsequent recommendations from the EDDP are eagerly awaited, but initial findings have shown that various advanced treatment processes, including GAC, ClO_2 and O_3, are highly effective at removing steroid estrogens and their associated estrogenic responses in exposed fish. Numerous factors need to be carefully considered to establish whether the widespread implementation of these technologies into WwTWs would be justified. Of these factors, the EDDP needs to weigh the relatively high cost of some of these technologies, their environmental impacts (increased energy consumption and CO_2 emissions), the dependency of their success on various operating parameters, concerns about their effectiveness at removing other EDCs or pharmaceuticals or inefficiency in the presence of certain compounds, and/or concerns about the formation, fate and toxicity of chemical byproducts (e.g. disinfection by-products) (Kosaka et al. 2000; Snyder et al. 2003; Westerhoff et al. 2005; Jones et al. 2007; Koh et al. 2008). Of particular importance is to understand how these advanced treatment processes might affect the wider health of fish, including reproduction, rather than simply their estrogenic activity per se.

Lessons Learned and Our Concluding Thoughts

We now understand many aspects of the feminization process in roach in English rivers, including some of the causative chemicals. However, even after 25 years of study we are not able to say with any degree of certainty that there are substantive changes in the abundance or composition of wild populations due to EDC exposure. Recent modeling work, that considered fish abundance only, has concluded that intersex in roach, when considered alone and in the absence of any other environmental stressors, is unlikely to have significant impacts on populations. The reality, however, is that roach (and other fish) populations in English rivers have (and continue to be) subjected to a wide range of environmental stressors, ranging from habitat alterations to climate change. Furthermore, exposure to estrogenic WwTW effluents can induce other effects on the reproductive physiology of wild fish, including alterations in their endocrinology and timing of sexual development, and also effects on gamete quality, including increased atresia (egg resorption) in females, none of which have yet been given due consideration. It is also the case that genetics can have a major bearing on population sustainability. Some EDCs are now known to be able to affect behavior and breeding dynamics, impacting on the parentage and fitness of subsequent populations in laboratory-maintained model fish species. It is not known whether this also applies to wild fish populations in effluent-contaminated rivers. Again, these factors need to be considered in future models that attempt to predict chemical impacts on population viability.

Our research into the feminization of roach in English rivers, as part of a wider body of literature on endocrine disruption in wildlife populations, has heightened awareness on (and the importance of understanding) the more subtle effects of environmental pollutants discharged into the environment. Our work on estrogens also illustrates that long-term exposures to relatively low levels of pollutants can result in substantially different phenotypic effects compared with short-term exposure effects, and with potentially highly significant functional outcomes. The rich vein of data developed over the past 20 years on the feminization of roach in English rivers provides an opportunity to address some of the most pertinent questions in endocrine disruption, namely whether roach can (indeed, are) adapting to the environmental estrogens present in effluent-contaminated rivers and, if so, how their endocrine physiology is changed to accommodate this adaptation, and whether estrogen exposure is impacting on the sexual selection process. These are extremely difficult questions to address but of considerable importance in developing our understanding of the long-term impacts of exposure to EDCs (and other pollutants) in wild populations.

Wildlife are (in our option) not only important to protect in their own right, but provide us with sentinels that reflect environmental health more widely. This, of course, has a bearing on our own health too. Much of the vertebrate endocrine system is highly conserved and, in many (most) cases, it is highly likely that a chemical that is hormonally active in fish will also be active in humans. The issue is not perhaps so much as to whether established EDCs can have adverse effects in humans, as has been clearly established for studies in various forms of vertebrate wildlife spanning fish, amphibians, reptiles, birds, and mammals [for review see: Goodhead and Tyler (2008)], but rather whether exposure scenarios in humans are sufficient to do so, although there is an increasing body of evidence to suggest that this may be the case for some EDCs (e.g. Louis et al. 2008). We would also emphasize, however, that our fresh- and marine-water environments act as sinks for a very wide range of chemicals and many EDCs are most concentrated in these environmental compartments. Thus, exposure of aquatic organisms, like fish, will generally be significantly higher than for many other animals, including humans.

The EDDP, arising as a consequence of the feminization of roach, has brought together academics, government agencies, and the water companies in England and Wales with a focus on combating the issue of estrogenic discharges into rivers. Through this partnership, significant resources and expertise have been harnessed to great effect to progress our understanding on the efficacies of the various available treatment technologies employed in WwTWs for EDC removal and to consider (and trial) other more advanced treatment technologies. The scale of this work would simply not have been possible without direct industry collaboration. The full outcome of the EDDP has yet to be realized, however, the initial findings that some of the advanced technologies are effective in estrogen removal are encouraging. We would hope that the partnership this work has created between the water industry, government bodies and academics in the UK has paved the way for the development of improved practices in wastewater treatment, and an enhanced exchange of relevant knowledge and information between the different parties to help reduce harmful estrogen discharges and improve the quality of English rivers for fish and other wildlife.

Acknowledgments Much of the research on roach reported in this review was funded on grants from the UK Research Councils (NERC and EPSRC), the Environment Agency of England and Wales, and the Department for Environment, Food and Rural Affairs (DEFRA). The authors would like to thank research colleagues, post-docs and Ph.D. students who have contributed to the work reviewed in this chapter. Special thanks to Alan Henshaw and his team at the Environment Agency's Fish Production unit for their invaluable help and advice over the years on roach breeding and husbandry.

References

Ackermann GE, Schwaiger J, Negele RD, Fent K (2002) Effects of long-term nonylphenol exposure on gonadal development and biomarkers of estrogenicity in juvenile rainbow trout *Oncorhynchus mykiss*. Aquat Toxicol 60:203–221

Allen JE, Prakasam TBS (1983) Reflections on seven decades of activated sludge history. J Water Pollut Control Fed 55:436–443

Allen Y, Matthiessen P, Scott AP, Haworth S, Feist S, Thain JE (1999) The extent of oestrogenic contamination in the UK estuarine and marine environments – further surveys of flounder. Sci Total Environ 233:5–20

An W, Hu J, Giesy JP, Yang M (2009) Extinction risk of exploited wild roach (*Rutilus rutilus*) populations due to chemical feminization. Environ Sci Technol 43(20):7895–7901

Arthur S, Crow H, Pedezert L (2008) Understanding blockage formation in combined sewer networks. Proc ICE Water Manage 161:215–221

Barnhoorn IE, Bornman MS, Pieterse GM, van Vuren JH (2004) Histological evidence of intersex in feral sharptooth catfish (*Clarias gariepinus*) from an estrogen-polluted water source in Gauteng, South Africa. Environ Toxicol 19:603–608

Baroiller JF, Guiguen Y (2001) Endocrine and environmental aspects of sex differentiation in gonochoristic fish. EXS 91:177–201

Beresford N, Jobling S, Williams R, Sumpter JP (2004) Endocrine disruption in juvenile roach from English rivers: a preliminary study. J Fish Biol 64:580–586

Bickham JW, Sandhu S, Hebert PD, Chikhi L, Athwal R (2000) Effects of chemical contaminants on genetic diversity in natural populations: implications for biomonitoring and ecotoxicology. Mutat Res 463:33–51

Bjerregaard LB, Korsgaard B, Bjerregaard P (2006) Intersex in wild roach (*Rutilus rutilus*) from Danish sewage effluent-receiving streams. Ecotoxicol Environ Saf 64:321–328

Blackburn MA, Waldock MJ (1995) Concentrations of alkylphenols in rivers and estuaries in England and Wales. Water Res 29:1623–1629

Brian JV, Harris CA, Scholze M, Backhaus T, Booy P, Lamoree M, Pojana G, Jonkers N, Runnalls T, Bonfà A, Marcomini A, Sumpter JP (2005) Accurate prediction of the response of freshwater fish to a mixture of estrogenic chemicals. Environ Health Perspect 113:721–728

Butwell T, Gardner M, Johnson I, Leverett D, Rockett L (2009) Removal of endocrine disrupting chemicals: the National Demonstration Programme. The Chartered Institution of Water and Environmental Management (CIWEM) Annual Conference, London, 29–30 April 2009. Presentation. Available at http://www.ciwem.org/events/annual_conference/2009_outputs.asp

Chang HS, Choo KH, Lee B, Choi SJ (2009) The methods of identification, analysis, and removal of endocrine disrupting compounds (EDCs) in water. J Hazard Mater 172(1):1–12

Coe TS, Hamilton PB, Hodgson D, Paull GC, Stevens JR, Sumner K, Tyler CR (2008) An environmental estrogen alters reproductive hierarchies, disrupting sexual selection in group-spawning fish. Environ Sci Technol 42:5020–5025

Coe TS, Hamilton PM, Hodgson D, Paull GC, Tyler CR (2009) Parentage outcomes in response to estrogen exposure are modified by social grouping in zebrafish. Environ Sci Technol 43:8400–8405

Cooper PF (2001) Historical aspects of wastewater treatment. In: Lens P, Zeeman G, Lettinga G (eds) Decentralised sanitation and reuse: concepts, systems and implementation. IWA Publishing, London, pp 11–36

Craft JA, Brown M, Dempsey K, Francey J, Kirby MF, Scott AP, Katsiadaki I, Robinson CD, Davies IM, Bradac P, Moffat CF (2004) Kinetics of vitellogenin protein and mRNA induction and depuration in fish following laboratory and environmental exposure to oestrogens. Mar Environ Res 58:419–423

D'Ascenzo G, Di Corcia A, Gentili A, Mancini R, Mastropasqua R, Nazzari M, Samperi R (2003) Fate of natural estrogen conjugates in municipal sewage transport and treatment facilities. Sci Total Environ 302:199–209

Deborde M, Rabouan S, Duguet JP, Legube B (2005) Kinetics of aqueous ozone-induced oxidation of some endocrine disruptors. Environ Sci Technol 39:6086–6092

Desbrow C, Routledge EJ, Brighty GC, Sumpter JP, Waldock M (1998) Identification of estrogenic chemicals in STW effluent. 1. Chemical fractionation and in vitro biological screening. Environ Sci Technol 32:1549–1558

Diniz MS, Peres I, Magalhaes-Antoine I, Falla J, Pihan JC (2005) Estrogenic effects in crucian carp (*Carassius carassius*) exposed to treated sewage effluent. Ecotoxicol Environ Saf 62:427–435

Fernandez MP, Buchanan ID, Ikonomou MG (2008) Seasonal variability of the reduction in estrogenic activity at a municipal WWTP. Water Res 42:3075–3081

Filby AL, Thorpe KL, Maack G, Tyler CR (2007) Gene expression profiles revealing the mechanisms of anti-androgen- and estrogen-induced feminization in fish. Aquat Toxicol 81:219–231

Filby AL, Shears JA, Drage BE, Churchley JH, Tyler CR (2010) Effects of advanced treatments of wastewater effluents on estrogenic and reproductive health impacts in fish. Environ Sci Technol 44(11):4348–4354

Folmar LC, Denslow ND, Rao V, Chow M, Crain DA, Enblom J, Marcino J, Guillette LJ Jr (1996) Vitellogenin induction and reduced serum testosterone concentrations in feral male carp (*Cyprinus carpio*) captured near a major metropolitan sewage treatment plant. Environ Health Perspect 104:1096–1101

Gaido KW, Leonard LS, Lovell S, Gould JC, Barbai D, Portier CJ, McDonnell DP (1997) Evaluation of chemicals with endocrine modulating activity in a yeast-based steroid hormone receptor gene transcription assay. Toxicol Appl Pharmacol 143:205–212

Game C, Gagnon MM, Webb D, Lim R (2006) Endocrine disruption in male mosquitofish (*Gambusia holbrooki*) inhabiting wetlands in Western Australia. Ecotoxicology 15:665–672

Gibson R, Smith MD, Spary C, Tyler CR, Hill EM (2005) Mixtures of oestrogenic contaminants in bile in fish exposed to wastewater treatment works effluents. Environ Sci Technol 39:2461–2471

Goodhead RM, Tyler CR (2008) Endocrine disrupting chemicals and their environmental impacts. In: Walker C (ed) Organic pollutants. CRC Press, London, pp 265–292

Gross-Sorokin MY, Roast SD, Brighty GC (2006) Assessment of feminization of male fish in English rivers by the Environment Agency of England and Wales. Environ Health Perspect 114(Suppl 1):147–151

Harries JE, Jobling S, Matthiesen P, Sheahan DA, Sumpter JP (1995) Effects of trace organics on fish – Phase 2. Report to the Department of the Environment, Foundation for Water Research, Marlow, UK

Harries JE, Sheahan DA, Jobling S, Matthiessen P, Neall P, Sumpter JP, Tylor T, Zaman N (1997) Estrogenic activity in five United Kingdom rivers detected by measurement of vitellogenesis in caged male trout. Environ Toxicol Chem 16:534–542

Harries JE, Janbakhsh A, Jobling S, Matthiessen P, Sumpter JP, Tyler CR (1999) Estrogenic potency of effluent from two sewage treatment works in the United Kingdom. Environ Toxicol Chem 18:932–937

Harris CA, Hamilton PB, Jobling S, Vinciotti V, Tyler CR, Sumpter JP (2010) Breeding capabilities of fish with pollution-induced intersex. In preparation

Harris CA, Hamilton PB, Runnalls TJ, Vinciotti V, Henshaw A, Hodgson DJ, Coe TS, Jobling S, Tyler CR, Sumpter JP (2010b) The consequences of feminisation in breeding groups of wild fish. Environ Health Perspect 119:306–311
Hecker M, Tyler CR, Maddix S, Karbe L (2002) Plasma biomarkers in fish provide evidence for endocrine modulation in the Elbe River, Germany. Environ Sci Technol 36:2311–2321
Herman RL, Kincaid HL (1988) Pathological effects of orally-administered estradiol to rainbow-trout. Aquaculture 72:165–172
Higashitani T, Tamamoto H, Takahashi A, Tanaka H (2003) Study of estrogenic effects on carp (*Cyprinus carpio*) exposed to sewage treatment plant effluents. Water Sci Technol 47:93–100
Hill J, Janz DM (2003) Developmental estrogenic exposure in zebrafish (*Danio rerio*): I. Effects on sex ratio and breeding success. Aquat Toxicol 63:417–429
Hill EM, Evans KL, Horwood J, Rostkowski P, Gibson R, Shears JA, Tyler CR (2010) Bioconcentration and some initial identifications of (anti)androgenic compounds in fish exposed to wastewater treatment works effluents. Environ Sci Technol 44:1137–1143
Holbrook RD, Novak JT, Grizzard TJ, Love NG (2002) Estrogen receptor agonist fate during wastewater and biosolids treatment processes: a mass balance analysis. Environ Sci Technol 36:4533–4539
Huber MM, Canonica S, Park GY, Von Gunten U (2003) Oxidation of pharmaceuticals during ozonation and advanced oxidation processes. Environ Sci Technol 37:1016–1024
Humphreys GW (1930) Main drainage of London. London County Council, London, p 54
Huo CX, Hickey P (2007) EDC demonstration programme in the UK – Anglian Water's approach. Environ Technol 28:731–741
Ifelebuegu AO, Lester JN, Churchley J, Cartmell E (2006) Removal of an endocrine disrupting chemical (17a-ethinyloestradiol) from wastewater effluent by activated carbon adsorption: effects of activated carbon type and competitive adsorption. Environ Technol 27:343–1349
Jafri SIH, Ensor DM (1979) Occurrence of an intersex condition in the roach *Rutilus rutilus* (L.). J Fish Biol 4:547–549
Jobling S, Nolan M, Tyler CR, Brighty G, Sumpter JP (1998) Widespread sexual disruption in wild fish. Environ Sci Technol 32:2498–2506
Jobling S, Beresford N, Nolan M, Rodgers-Gray T, Brighty GC, Sumpter JP, Tyler CR (2002a) Altered sexual maturation and gamete production in wild roach (*Rutilus rutilus*) living in rivers that receive treated sewage effluents. Biol Reprod 66:272–281
Jobling S, Coey S, Whitmore JG, Kime DE, Van Look KJ, McAllister BG, Beresford N, Henshaw AC, Brighty G, Tyler CR, Sumpter JP (2002b) Wild intersex roach (*Rutilus rutilus*) have reduced fertility. Biol Reprod 67:515–524
Jobling S, Casey D, Rodgers-Gray T, Oehlmann J, Pawlowski S, Baunbeck T, Turner AP, Tyler CR (2003) Comparative responses of molluscs and fish to environmental oestrogens and an oestrogenic effluent. Aquat Toxicol 65:205–220
Jobling S, Williams R, Johnson A, Taylor A, Gross-Sorokin M, Nolan M, Tyler CR, van Aerle R, Santos EM, Brighty G (2006) Predicted exposures to steroid oestrogens in UK Rivers correlate with widespread sexual disruption in wild roach populations. Environ Health Perspect 114:32–39
Jobling S, Burn RW, Thorpe K, Williams R, Tyler CR (2009) Statistical modeling suggests that antiandrogens in effluents from wastewater treatment works contribute to widespread sexual disruption in fish living in English rivers. Environ Health Perspect 117:797–802
Johnson AC, Williams RJ (2004) A model to estimate influent and effluent concentrations of estradiol, estrone, and ethinylestradiol at sewage treatment works. Environ Sci Technol 38:3649–3658
Johnson I, Hetheridge M, Tyler CR (2004) Assessment of the (anti-)oestrogenic and (anti-)androgenic activity of sewage treatment works effluents. R&D Technical Report, Environment Agency
Johnson AC, Aerni HR, Gerritsen A, Gibert M, Giger W, Hylland K, Jürgens M, Nakari T, Pickering A, Suter MJ, Svenson A, Wettstein FE (2005) Comparing steroid estrogen, and nonylphenol content across a range of European sewage plants with different treatment and management practices. Water Res 39:47–58

Johnson AC, Williams RJ, Simpson P, Kanda R (2007) What difference might sewage treatment performance make to endocrine disruption in rivers? Environ Pollut 147:194–202

Johnson AC, Acreman MC, Dunbar MJ, Feist SW, Giacomello AM, Gozlan RE, Hinsley SA, Ibbotson AT, Jarvie HP, Jones JI, Longshaw M, Maberly SC, Marsh TJ, Neal C, Newman JR, Nunn MA, Pickup RW, Reynard NS, Sullivan CA, Sumpter JP, Williams RJ (2009) The British river of the future: how climate change and human activity might affect two contrasting river ecosystems in England. Sci Total Environ 407:4787–4798

Jones OAH, Green P, Voulvoulis N, Lester JN (2007) Questioning the excessive use of advanced treatment to remove organic micropollutants from wastewater. Environ Sci Technol 41:5085–5089

Jürgens MD, Johnson AC, Pottinger TG, Sumpter JP (2009) Do suspended sediments modulate the effects of octylphenol on rainbow trout? Water Res 43:1381–1391

Kang IJ, Yokota H, Oshima Y, Tsuruda Y, Hano T, Maeda M, Imada N, Tadokoro H, Honjo T (2003) Effects of 4-nonylphenol on reproduction of Japanese medaka, *Oryzias latipes*. Environ Toxicol Chem 22:2438–2445

Kelce WR, Wilson EM (1997) Environmental antiandrogens: developmental effects, molecular mechanisms, and clinical implications. J Mol Med 75:198–207

Kirby MF, Bignell J, Brown E, Craft JA, Davies I, Dyer RA, Feist SW, Jones G, Matthiessen P, Megginson C, Robertson FE, Robinson C (2003) The presence of morphologically intermediate papilla syndrome in United Kingdom populations of sand goby (*Pomatoschistus* spp): endocrine disruption? Environ Toxicol Chem 22:239–251

Kirby MF, Allen YT, Dyer RA, Feist SW, Katsiadaki I, Matthiessen P, Scott AP, Smith A, Stentiford GD, Thain JE, Thomas KV, Tolhurst L, Waldock MJ (2004) Surveys of plasma vitellogenin and intersex in male flounder (*Platichthys flesus*) as measures of endocrine disruption by estrogenic contamination in United Kingdom estuaries: temporal trends, 1996 to 2001. Environ Toxicol Chem 23:748–758

Kim SE, Park NS, Yamada H, Tsuno H (2008) Modeling of decomposition characteristics of estrogenic chemicals during ozonation. Environ Technol 29:287–296

Kirk LA, Tyler CR, Lye CM, Sumpter JP (2002) Changes in estrogenic and androgenic activities at different stages of treatment in wastewater treatment works. Environ Toxicol Chem 21:972–979

Kleinkauf A, Scott AP, Stewart C, Simpson MG, Leah RT (2004) Abnormally elevated VTG concentrations in flounder (*Platichthys flesus*) from the Mersey estuary (UK) – a continuing problem. Ecotoxicol Environ Saf 58:356–364

Koh YKK, Chiu TY, Boobis A, Cartmell E, Scrimshaw MD, Lester JN (2008) Treatment and removal strategies for estrogens from wastewater. Environ Technol 29:245–267

Kosaka K, Yamada H, Matsui S, Shishida K (2000) The effects of the co-existing compounds on the decomposition of micropollutants using the ozone/hydrogen peroxide process. Water Sci Technol 42:353–361

Labadie P, Budzinski H (2005) Determination of steroidal hormone profiles along the Jalle d'Eysines River (near Bordeaux, France). Environ Sci Technol 39:5113–5120

Lange A, Katsu Y, Ichikawa R, Paull GC, Chidgey LL, Coe TS, Iguchi T, Tyler CR (2008) Altered sexual development in roach (*Rutilus rutilus*) exposed to environmental concentrations of 17α-ethinylestradiol and associated expression dynamics of aromatases and estrogen receptors. Toxicol Sci 106:113–123

Lange A, Paull GC, Coe TS, Katsu Y, Urushitani H, Iguchi T, Tyler CR (2009) Sexual reprogramming and estrogenic sensitization in wild fish exposed to ethinylestradiol. Environ Sci Technol 43:1219–1225

Lange A, Paull GC, Hamilton PB, Tyler CR (2010) Long-term effluent exposure impacts on sexual development and breeding capabilities in roach (*Rutilus rutilus*). In preparation

Lange A, Paull GC, Hamilton PB, Iguchi T, Tyler CR (2011) Implications of persistent exposure to treated wastewater effluent for breeding in wild roach (*Rutilus rutilus*) populations. Environ Sci Technol 45(4):1673–1679

Larsen TA, Lienert J, Joss A, Siegrist H (2004) How to avoid pharmaceuticals in the aquatic environment. J Biotechnol 113:295–304

Larsson DGJ, Adolfsson-Erici M, Parkkonen J, Pettersson M, Berg AH, Olsson PE, Forlin L (1999) Ethinyloestradiol – an undesired fish contraceptive? Aquat Toxicol 45:91–97
Lee HJ, Chattopadhyay S, Gong EY, Ahn RS, Lee K (2003) Antiandrogenic effects of bisphenol A and nonylphenol on the function of androgen receptor. Toxicol Sci 75:40–46
Lemmoin-Cannon H (1912) Sewage disposal in the United Kingdom. St. Bride's Press, London
Liney KE, Jobling S, Shears J, Simpson P, Tyler CR (2005) Assessing the sensitivity of different life stages for sexual disruption in roach (*Rutilus rutilus*) exposed to effluents from wastewater treatment works. Environ Health Perspect 113:1299–1307
Liney KE, Hagger JA, Tyler CR, Depledge MH, Galloway TS, Jobling S (2006) Health effects in fish after long-term exposure to effluents from wastewater treatment works. Environ Health Perspect 114:81–89
Louis GMB, Gray LE Jr, Marcus M, Ojeda SR, Pescovitz OH, Witchel SF, Sippell W, Abbott DH, Soto A, Tyl RW, Bourguignon JP, Skakkebaek NE, Swan SH, Golub MS, Wabitsch M, Toppari J, Euling SY (2008) Environmental factors and puberty timing: expert panel research needs. Pediatrics 121:S192–S207
Ma TW, Wan XQ, Huang QH, Wang ZJ, Liu JK (2005) Biomarker responses and reproductive toxicity of the effluent from a Chinese large sewage treatment plant in Japanese medaka (*Oryzias latipes*). Chemosphere 59:281–288
Metcalfe CD, Metcalfe TL, Kiparissis Y, Koenig BG, Khan C, Hughes RJ, Croley TR, March RE, Potter T (2001) Estrogenic potency of chemicals detected in sewage treatment plant effluents as determined by in vivo assays with Japanese medaka (*Oryzias latipes*). Environ Toxicol Chem 20:297–308
Molina-Muñoz M, Poyatos JM, Sánchez-Peinado M, Hontoria E, González-López J, Rodelas B (2009) Microbial community structure and dynamics in a pilot-scale submerged membrane bioreactor aerobically treating domestic wastewater under real operation conditions. Sci Total Environ 407:3994–4003
Nash JP, Kime DE, van derven LTM, Wester PW, Brion F, Maack G, Stahlschmidt-Allner P, Tyler CR (2004) Long-term exposure to environmental concentrations of the pharmaceutical ethinyloestradiol causes reproductive failure in fish. Environ Health Perspect 112:1725–1733
Nolan M, Jobling S, Brighty G, Sumpter JP, Tyler CR (2001) A histological description of intersexuality in the roach. J Fish Biol 58:160–176
Oehlmann J, Schulte-Oehlmann U, Kloas W, Jagnytsch O, Lutz I, Kusk KO, Wollenberger L, Santos EM, Paull GC, Van Look KJ, Tyler CR (2009) A critical analysis of the biological impacts of plasticizers on wildlife. Philos Trans R Soc Lond B Biol Sci 364:2047–2062
Orton F, Lutz I, Kloas W, Routledge EJ (2009) Endocrine disrupting effects of herbicides and pentachlorophenol: in vitro and in vivo evidence. Environ Sci Technol 43:2144–2150
Paull GC, Lange A, Henshaw AC, Tyler CT (2008) Ontogeny of sexual development in the roach (*Rutilus rutilus*) and its interrelationships with growth and age. J Morphol 269:884–895
Pawlowski S, van Aerle R, Tyler CR, Braunbeck T (2004) Effects of 17a-ethinylestradiol in a fathead minnow (*Pimephales promelas*) gonadal recrudescence assay. Ecotoxicol Environ Saf 57:330–345
Purdom CE, Hardiman PA, Bye VJ, Eno NC, Tyler CR, Sumpter JP (1994) Estrogenic effects of effluents from sewage treatment works. Chem Ecol 8:275–285
Rodgers-Gray TP, Jobling S, Morris S, Kelly C, Kirby S, Janbakhsh A, Harries JE, Waldock MJ, Sumpter JP, Tyler CR (2000) Long-term temporal changes in the estrogenic composition of treated sewage effluent and its biological effects on fish. Environ Sci Technol 34:1521–1528
Rodgers-Gray TP, Jobling S, Kelly C, Morris S, Brighty G, Waldock MJ, Sumpter JP, Tyler CR (2001) Exposure of juvenile roach (*Rutilus rutilus*) to treated sewage effluent induces dose-dependent and persistent disruption in gonadal duct development. Environ Sci Technol 35:462–470
Saaristo M, Craft JA, Lehtonen KK, Lindström K (2009) Sand goby (*Pomatoschistus minutus*) males exposed to an endocrine disrupting chemical fail in nest and mate competition. Horm Behav 56:315–321
Salierno JD, Kane AS (2009) 17a-ethinylestradiol alters reproductive behaviors, circulating hormones, and sexual morphology in male fathead minnows (*Pimephales promelas*). Environ Toxicol Chem 28:953–961

Schiliro T, Pignata C, Rovere R, Fea E, Gilli G (2009) The endocrine disrupting activity of surface waters and of wastewater treatment plant effluents in relation to chlorination. Chemosphere 75:335–340

Schultz H (1996) Drastic decline of the proportion of males in the roach (*Rutilus rutilus* L.) population of Bautzen reservoir (Saxony, Germany): result of direct and indirect effects of biomanipulation. Limnologica 26:153–164

Sebire M, Scott AP, Tyler CR, Cresswell J, Hodgson DJ, Morris S, Sanders MB, Stebbing PD, Katsiadaki I (2009) The organophosphorous pesticide, fenitrothion, acts as an anti-androgen and alters reproductive behavior of the male three-spined stickleback, *Gasterosteus aculeatus*. Ecotoxicology 18:122–133

Sheahan DA, Brighty GC, Daniel M, Kirby SJ, Hurst MR, Kennedy J, Morris S, Routledge EJ, Sumpter JP, Waldock MJ (2002a) Estrogenic activity measured in a sewage treatment works treating industrial inputs containing high concentrations of alkylphenolic compounds – a case study. Environ Toxicol Chem 21:507–514

Sheahan DA, Brighty GC, Daniel M, Jobling S, Harries JE, Hurst MR, Kennedy J, Kirby SJ, Morris S, Routledge EJ, Sumpter JP, Waldock MJ (2002b) Reduction in the estrogenic activity of a treated sewage effluent discharge to an English river as a result of a decrease in the concentration of industrially derived surfactants. Environ Toxicol Chem 21:515–519

Smith RJF (1978) Seasonal changes in the histology of the gonads and dorsal skin of the fathead minnow *Pimephales promelas*. Can J Zool 56:2103–2109

Snyder SA, Westerhoff P, Yoon Y, Sedlak DL (2003) Pharmaceuticals, personal care products, and endocrine disruptors in water: implications for the water industry. Environ Eng Sci 20:449–469

Sohoni P, Sumpter JP (1998) Several environmental oestrogens are also anti-androgens. J Endocrinol 158:327–339

Solé M, Barceló D, Porte C (2002) Seasonal variation of plasmatic and hepatic vitellogenin and EROD activity in carp, *Cyprinus carpio*, in relation to sewage treatment plants. Aquat Toxicol 60:233–248

Solé M, Raldua D, Piferrer F, Barceló D, Porte C (2003) Feminization of wild carp, *Cyprinus carpio*, in a polluted environment: plasma steroid hormones, gonadal morphology and xenobiotic metabolizing system. Comp Biochem Physiol C Toxicol Pharmacol 136:145–156

Song Z, Ren N, Zhang K, Tong L (2009) Influence of temperature on the characteristics of aerobic granulation in sequencing batch airlift reactors. J Environ Sci 21:273–278

Sonnenschein C, Soto AM (1998) An updated review of environmental estrogen and androgen mimics and antagonists. J Steroid Biochem Mol Biol 65:143–150

Stanbridge HH (1976) History of sewage treatment in Britain. Institute of Water Pollution Control, Maidstone

Sweeting RA (1981) Hermaphrodite roach in the river Lee. Thames Water, Lea Division

Ternes TA (1998) Occurrence of drugs in German sewage treatment plants and rivers. Water Res 32:3245–3260

Ternes TA, Stumpf M, Mueller J, Haberer K, Wilken RD, Servos M (1999) Behavior and occurrence of estrogens in municipal sewage treatment plants-I. Investigations in Germany, Canada and Brazil. Sci Total Environ 225:81–90

Ternes TA, Stuber J, Herrmann N, McDowell D, Ried A, Kampmann M, Teiser B (2003) Ozonation: a tool for removal of pharmaceuticals, contrast media and musk fragrances from wastewater? Water Res 37:1976–1982

Thorpe KL, Hutchinson TH, Hetheridge MJ, Scholze M, Sumpter JP, Tyler CR (2001) Assessing the biological potency of binary mixtures of environmental estrogens using vitellogenin induction in juvenile rainbow trout (*Oncorhynchus mykiss*). Environ Sci Technol 35:2476–2481

Thorpe KL, Cummings RI, Hutchinson TH, Scholze M, Brighty G, Sumpter JP, Tyler CR (2003) Relative potencies and combination effects of steroidal estrogens in fish. Environ Sci Technol 37:1142–1149

Thorpe KL, Gross-Sorokin M, Johnson I, Brighty G, Tyler CR (2006) An assessment of the model of concentration addition for predicting the estrogenic activity of chemical mixtures in wastewater treatment works effluent. Environ Health Perspect 114:90–97

Thorpe K, Benstead R, Hutchinson TH, Tyler CR (2007) Associations between altered vitellogenin concentrations and adverse health effects in fathead minnow (*Pimephales promelas*). Aquat Toxicol 85:176–183
Thorpe KL, Benstead R, Eccles P, Maack G, Williams T, Tyler CR (2008) A practicable laboratory flow-through exposure system for assessing the health effects of effluents in fish. Aquat Toxicol 88:164–172
Thorpe KL, Maack G, Benstead R, Tyler CR (2009) Estrogenic wastewater treatment works effluents reduce egg production in fish. Environ Sci Technol 43:2976–2982
Turner KJ, Sharpe RM (1997) Environmental oestrogens – present understanding. Rev Reprod 2:69–73
Tyler CR, Routledge E (1998) Oestrogenic effects in roach in English rivers with evidence for causation. Pure Appl Chem 70:1795–1804
Tyler CR, Spary C, Gibson R, Shears J, Santos E, Sumpter JP, Hill EM (2005) Accounting for differences in the vitellogenic responses of rainbow trout (*Oncorhynchus mykiss*) and roach (*Rutilus rutilus*: Cyprinidae) exposed to oestrogenic effluents from wastewater treatment works. Environ Sci Technol 39:2599–2607
Tyler CR, Filby AL, Bickley LK, Cumming RI, Gibson R, Labadie P, Katsu Y, Liney KE, Shears JA, Silva-Castro V, Urushitani H, Lange A, Winter MJ, Iguchi T, Hill EM (2009) Environmental health impacts of equine estrogens derived from hormone replacement therapy. Environ Sci Technol 43:3897–3904
van Aerle R, Nolan TM, Jobling S, Christiansen LB, Sumpter JP, Tyler CR (2001) Sexual disruption in a second species of wild cyprinid fish (the gudgeon, *Gobio gobio*) in United Kingdom freshwaters. Environ Toxicol Chem 20:2841–2847
van Aerle R, Pounds N, Hutchinson TH, Maddix S, Tyler CR (2002) Window of sensitivity for the estrogenic effects of ethinylestradiol in early life-stages of fathead minnow, *Pimephales promelas*. Ecotoxicology 11:423–434
Vermeirssen EL, Burki R, Joris C, Peter A, Segner H, Suter MJ, Burkhardt-Holm P (2005) Characterization of the estrogenicity of Swiss midland rivers using a recombinant yeast bioassay and plasma vitellogenin concentrations in feral male brown trout. Environ Toxicol Chem 24:2226–2233
Vethaak AD, Lahr J, Schrap SM, Belfroid AC, Rijs GB, Gerritsen A, de Boer J, Bulder AS, Grinwis GC, Kuiper RV, Legler J, Murk TA, Peijnenburg W, Verhaar HJ, de Voogt P (2005) An integrated assessment of estrogenic contamination and biological effects in the aquatic environment of The Netherlands. Chemosphere 59:511–524
Viganò L, Mandich A, Benfenati E, Bertolotti R, Bottero S, Porazzi E, Agradi E (2006) Investigating the estrogenic risk along the river Po and its intermediate section. Arch Environ Contam Toxicol 51:641–651
Vine E, Shears JA, van Aerle R, Tyler CR, Sumpter JP (2004) Endocrine (sexual) disruption is not a prominent feature in pike (*Esox lucius*), a top predator, living in English waters. Environ Toxicol Chem 24:1436–1443
Waldock M, Sheahan D, Routledge EJ, Sumpter JP, Brighty G, Kennedy J (1997) Endocrine disruption in fish: a case study in identification of the problem, control measures and recovery in the UK. Proceedings of the SETAC Conference, San Francisco, USA
Weiss JM, Hamers T, Thomas KV, van der Linden S, Leonards PE, Lamoree MH (2009) Masking effect of anti-androgens on androgenic activity in European river sediment unveiled by effect-directed analysis. Anal Bioanal Chem 394:1385–1397
Westerhoff P, Yoon Y, Snyder S, Wert E (2005) Fate of endocrine-disruptor, pharmaceutical, and personal care product chemicals during simulated drinking water treatment processes. Environ Sci Technol 39:6649–6663
Williams RJ, Johnson AC, Smith JJL, Kanda R (2003) Steroid estrogens profiles along river stretches arising from sewage treatment works discharges. Environ Sci Technol 37:1744–1750
Williams RJ, Keller VDJ, Johnson AC, Young AR, Holmes MGR, Wells C, Gross-Sorokin M, Benstead R (2009) A National risk assessment for intersex in fish arising from steroid estrogens. Environ Toxicol Chem 28:220–230

Yokota H, Seki M, Maeda M, Oshima Y, Tadokoro H, Honjo T, Kobayashi K (2001) Life-cycle toxicity of 4-nonylphenol to medaka (*Oryzias latipes*). Environ Toxicol Chem 20:2552–2560

Yoon SH, Itoh Y, Kaneko G, Nakaniwa M, Ohta M, Watabe S (2008) Molecular characterization of Japanese sillago vitellogenin and changes in its expression levels on exposure to 17β-estradiol and 4-tert-octylphenol. Mar Biotechnology 10:19–30

Zaroogian G, Gardner G, Horowitz DB, Gutjahr-Gobell R, Haebler R, Mills L (2001) Effect of 17β-estradiol, o, p′-DDT, octylphenol and p, p′-DDE on gonadal development and liver and kidney pathology in juvenile male summer flounder (*Paralichthys dentatus*). Aquat Toxicol 54:101–112

Zhang Y, Zhou JL (2008) Occurrence and removal of endocrine disrupting chemicals in wastewater. Chemosphere 73:848–853

Chapter 14
South Asian Vultures in Crisis: Environmental Contamination with a Pharmaceutical

J. Lindsay Oaks and Richard T. Watson

My late night arrival in November of 2000 to Pakistan was filled with much anticipation and extensive planning for locating vultures, and collecting those that were dead or dying to try and discover what was killing them. My suitcase was filled with medical and sampling supplies, and my luggage included a liquid nitrogen container – all of which held great interest for and led to several hours of questions by customs officials. In the end we were all cleared, after which I could then focus on my expectations of traveling the Pakistani countryside searching for wild vultures with my colleagues Munir Virani and Pat Benson. But my first vulture sighting occurred much closer than expected. The next morning, we simply took a taxi to the historic Jinnah's Garden (formerly known as Lawrence Gardens) in downtown Lahore where there were vultures nesting in the trees within the garden, completely unperturbed by the people below. That was the moment when I had my first inkling of how closely intertwined the lives of humans and vultures had become in Southern Asia.

Abstract In the late 1990s an unprecedented decline in the population of two of the world's most abundant raptors, the Oriental White-backed vulture (*Gyps bengalensis*) and the Long-billed vulture (*Gyps indicus*), was noticed in India. By the early 2000s, similar catastrophic declines followed in neighboring Pakistan. Ecological and forensic studies ultimately found that a non-steroidal anti-inflammatory pharmaceutical, diclofenac, was responsible. Diclofenac, long used in human medicine, had found its way into the veterinary market as a safe, inexpensive, and very popular drug for livestock in Southern Asia. Unfortunately, diclofenac residues caused kidney failure in *Gyps* vultures that fed on treated carcasses. And the loss of breeding adult vultures had a profound impact on the population, leading to declines on the

R.T. Watson (✉)
The Peregrine Fund, 5668 West Flying Hawk Lane, Boise, ID, 83709, USA
e-mail: rwatson@peregrinefund.org

J.E. Elliott et al. (eds.), *Wildlife Ecotoxicology: Forensic Approaches*,
Emerging Topics in Ecotoxicology 3, DOI 10.1007/978-0-387-89432-4_14,

order of 30% per year. In 2004, a series of meetings were held with government officials to inform them of this discovery. The extensive lobbying efforts that followed successfully led in 2006 to a ban on the manufacture of veterinary diclofenac in India, Pakistan, and Nepal. Sadly, in 2010, diclofenac still appears to be readily available and widely used in veterinary medicine, leaving the fate of wild *Gyps* vultures in doubt.

History and Background of the Decline

While the charismatic Peregrine Falcon was the "canary in the coal-miner's cage" of the 1950s and 1960s warning of the insidious effects of environmental contamination by persistent organochlorine pesticides, the new millennium dawned with signs of environmental disaster from a "canary" of a very different sort. Misunderstood, often ignored, sometimes reviled, but ecologically important vultures were declining in number at a catastrophic rate in South Asia. Beginning in the mid-1990s, there were increasing newspaper reports and anecdotal accounts of vultures disappearing from Northern India. The first scientific data were collected in 1997, when biologists from the Bombay Natural History Society (BNHS) documented the decline in Keoladeo National Park in the western Indian state of Rajasthan (Prakash 1999). In the park, the breeding population of Oriental White-backed vultures (*Gyps bengalensis*) had gone from about 250 breeding pairs in the mid-1980s to none in 1999. These alarming data from Keoladeo, fueled by continued anecdotal reports of disappearing vultures from other parts of India, led to an expanded survey of vulture populations in Northern, Western, and Eastern India. Road-transect surveys were conducted in 2000 by the BNHS with support from The Royal Society for the Protection of Birds (RSPB), and when compared to similar surveys from 1991 to 1993, found a staggering decline of at least 96% for White-backed vultures and 92% for the Long-billed vultures (*Gyps indicus*) and Slender-billed vultures (*Gyps tenuirostris*)[1] (Prakash et al. 2003). Ongoing surveys in 2002, 2003, and 2007 further documented the decline on an annual basis, with an average annual mortality between 2002 and 2007 of 44% for White-backed vultures and 16% for Long-billed vultures (Prakash et al. 2007). In September of 2000, the first official meeting to address this problem was organized in New Delhi by the BNHS, and was attended by government and non-governmental representatives from India, as well as biologists from other countries. From that meeting came the first plan of action to identify and address the cause of the vulture decline in India, a joint effort initially between the BNHS, the state government of Haryana, the RSPB, the Zoological Society of London, and the National Birds of Prey Trust (Pain et al. 2008).

[1] Initial surveys did not distinguish between the Long-billed and Slender-billed vultures, as these were not officially separated into different species until 2001.

Also attending the initial meeting in New Delhi were representatives of The Peregrine Fund (TPF), a non-governmental birds of prey conservation organization based in the USA. Upon their return to the USA, an internal meeting was quickly held to review the information and to decide if TPF was in a position to assist in the investigation. The outcome of this meeting at TPF was to establish a complementary study of the vulture decline in neighboring Pakistan. The reasons that TPF chose to work in Pakistan included for one that a large investigational effort was already underway in India by well-established organizations, and felt that an overlapping effort there would be counterproductive. TPF also had contacts with the Ornithological Society of Pakistan (OSP), and while OSP ornithologists had indicated that there did appear to be many dying vultures in Pakistan, there were still nesting colonies of vultures with large numbers of birds. That was possibly very important, as it was a potential opportunity to study the decline as it began to affect the population – as opposed to India where the decline appeared to have been well underway and much of the population already gone. TPF also hypothesized that vultures, and thus the vulture decline, were unlikely to recognize human-defined international boundaries. One of the TPF/OSP study sites, at the Changa Manga forest plantation south of Lahore, was deliberately selected for its proximity to the Indian border. And finally, critically importantly from a logistical standpoint, Pakistan was willing to allow export of diagnostic samples if in-country testing was not available. In contrast, India had extremely strict regulations controlling the export of any viable genetic or biological materials that could potentially delay testing.[2]

Both TPF and the RSPB also established contacts with Bird Conservation Nepal, where ornithologists had reported a situation similar to Pakistan with the White-backed vultures in the Terai lowland areas of Nepal bordering India. However, due to the relative inaccessibility of that area and the difficult terrain, along with ongoing political instability, intensive diagnostic work was not carried out there. Nepal's political neutrality, however, would later play a major role in the response phase of this story.

At that stage, the majority of investigators considered the most logical explanation for the decline, and the predominant working hypothesis, to be either a new infectious disease or a known infectious disease that had gained access to the vulture populations. Although no one knew what that disease might be, its effects were clearly devastating. In less than ten years, what were three of the most abundant birds of prey in the world suddenly were all listed as critically endangered by the IUCN (Birdlife International 2001). The loss was estimated to be on the order of tens of millions of vultures. Moreover, the Eurasian Griffon vulture (*G. fulvus*) had a contiguous population from south Asia to the African savannahs. If a contagious

---

[2]India is a party to the Convention on Biological Diversity (1992), which recognizes the sovereign rights of states to use their own biological resources, and provides the basis for these regulations. To protect these resources, India has enacted an umbrella legislation called the Biological Diversity Act of 2002 which is administered by the National Biodiversity Authority (http://www.nbaindia.org). All export of genetic material requires permits issued by the National Biodiversity Authority.

disease were involved, it raised the specter of the disaster spreading throughout Asia, into Europe, and ultimately into Africa. At the beginning of the twenty first century, the events in India had the full attention of ornithologists throughout the world.

The Forensic Investigation in Pakistan

The diagnostic investigation began in November, 2000 in Pakistan. The objectives were to identify active breeding colonies of White-backed vultures in the Punjab province of Pakistan,[3] begin ecological studies of the vultures in Pakistan with a special focus on documenting a population decline and abnormal mortality, and to collect samples from dead vultures for diagnostic testing. Those studies were a joint effort between TPF, the OSP, and the Bahauddin Zakariya University (BZU) in Multan, Pakistan. Critically important to the success of the effort was the availability of manpower, which in addition to staff from the TPF and the OSP, was supplied by five MSc students[4] from the BZU Institute of Pure and Applied Biology.

Ecological Studies

We felt very strongly from the beginning that tackling this problem would be done most effectively with a multidisciplinary approach, and that understanding the ecology of vultures in Pakistan would be essential to understanding both normal and abnormal mortality. Initial surveys soon identified three large and active breeding colonies in the Pakistani Punjab, which were selected for ecological and diagnostic studies (Gilbert et al. 2006). Those included Dholewala in the Layyah and Muzaffargarh districts, Toawala in the Muzaffargarh, Multan and Khanewal districts, and the Changa Manga forest plantation in the Kasur district (Fig. 14.1). The Dholewala and Toawala study sites were both linear plantations of sheesham (*Dalbergia sisoo*) and acacia (*Acacia* spp.) trees, lining earthen flood control barriers and canal banks, respectively. The Dholewala breeding colony was approximately 24 km in length, of which 5.2 km was studied intensively. The Toawala breeding colony was approximately 19 km in length, of which 6.4 km were studied intensively. The Changa Manga forest site was also a large breeding site, but nonlinear

[3] Due to the lack of suitable nesting cliffs, Long-billed vultures were not present, or expected to be present, in this part of Pakistan.

[4] The students were Shakeel Ahmed, Muhammad Jamshed Chaudhry, Muhammad Arshad, Shahid Mahmood and Ahmad Ali. Their MSc supervisor was Dr. Aleem Ahmed Khan. TPF staff that also assisted with the field work was Dr. Pat Benson, a volunteer biologist from South Africa, and Mr. Muhammad Asim (Mr. Asim, also known as "Awesome Asim", was selected for the 2005 Disney Conservation Hero Award for Asia for extraordinary efforts).

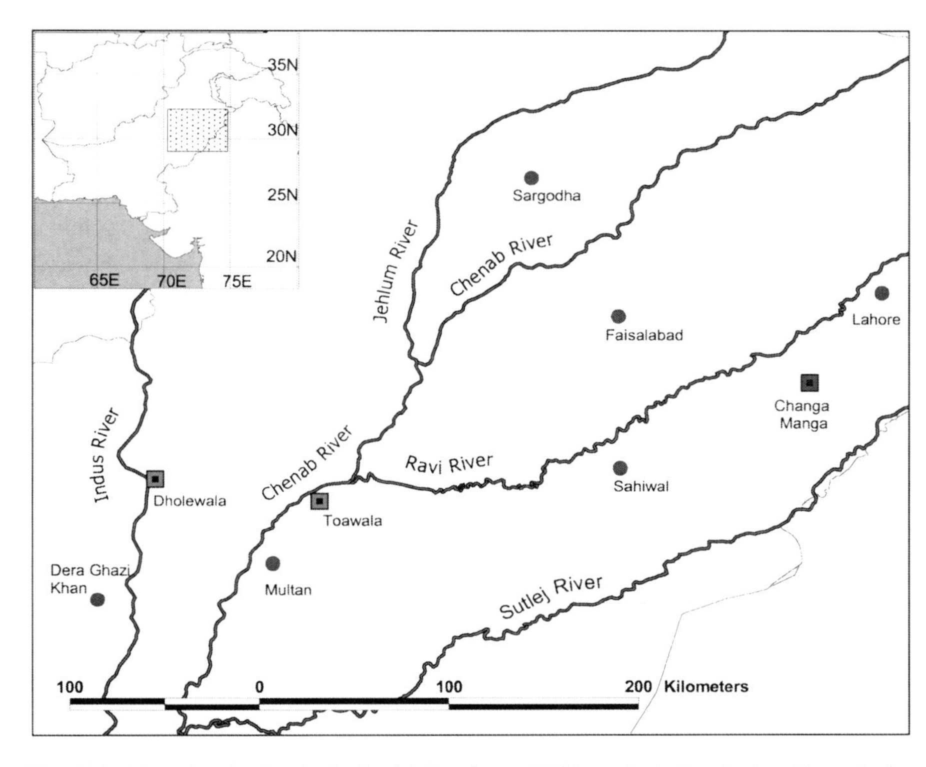

Fig. 14.1 Map of study sites in the Punjab Province of Pakistan, including the breeding colonies at Dholewala, Toawala, and Changa Manga. Inset shows location of main figure. Reprinted with permission from Cambridge University Press, Gilbert et al. (2006). Rapid population declines and mortality clusters in three Oriental white-backed vulture *Gyps bengalensis* colonies due to diclofenac poisoning. Oryx 40:388–399

in configuration, and the surveyed area was based on the highest nesting density identified at the start of the study. The Changa Manga site was also selected for its proximity to the border with India, since it was anticipated that if the disease were to spread from India it would most likely be detected first nearest the border. The typical breeding season for the White-backed vultures began with courtship and breeding as early as August, with peak egg laying occurring in November, followed by hatching of a single egg in about January, and fledging from about March to May. The ecological studies were started in December of 2000 and continued until the end of the 2003 breeding season for Dholewala and Changa Manga, and the 2004 breeding season for Toawala.

The ecological studies were aimed at determining the size of the breeding population at the start of the study, breeding success, movement during the breeding season, and vulture mortality parameters. It was hoped that conducting the study at multiple breeding colonies spread over a large area and time-span of years would provide a good representation of what was happening with the overall vulture population in Pakistan. It was immediately obvious at the start of the study that there were large numbers of vultures at all of the sites. There were approximately 760, 420 and 445 active nests at the Changa Manga, Dholewala and Toawala sites, respectively (Gilbert et al. 2006). When the non-breeding adults, subadults, and immature birds were included, the total vulture population at each of these sites was well over 1,000 individuals. It is difficult to describe the fantastic spectacle that occurred daily at mid-morning. As the air began to warm and thermals formed, a major segment of the vulture population began to soar upward in great kettles to disperse in the search for food. It is also especially sad that this may have been the last significant congregation of these magnificent birds in Southern Asia. The timing for the work in Pakistan was indeed fortuitous, as our studies began early in the decline, or at least at a time when a decline could be characterized. And the decline, as in India, was spectacular and to our knowledge unprecedented. Within three years, there was not a single active nest in either the Dholewala or Changa Manga sites, and the number of active nests at Toawala had declined by 54% (Gilbert et al. 2006). Counts of total numbers of vultures mirrored the decline in active nests (Fig. 14.2). During the 2007–2008 breeding season, the colony at Toawala was also completely extirpated when the last two breeding pairs disappeared. The situation was the same at other non-study sites around the Punjab. For all practical purposes, White-backed vultures were extinct in Pakistan (Johnson et al. 2008).

From those ecological studies we learned a vitally important element, which was that the primary driving force for the decline was mortality of the breeding adults and concomitant nesting failures (Gilbert et al. 2006). That also helped to establish a critical parameter for a case definition that could guide diagnostic sampling. Although that seems readily apparent in retrospect, at the time identifying who was dying in association with the decline was problematic and at times even misleading. For example, when asking the local citizens[5] if they had noticed abnormal numbers of dead vultures, invariably the answer was "yes". The locals also noted that most of the mortality occurred during the hot months, suggesting that the mortality was seasonal. When pressed as to why it may have begun recently, local people invariably and emphatically attributed the problem to increasingly hot weather, somehow associated with testing of nuclear weapons by India and Pakistan in 1998! As it turns out, there really were many more dead vultures in the summer, but more careful analysis revealed that the observed mortality was primarily associated with the fledgling vultures – mortality that is considered normal, and not relevant to the decline. Other important findings from the ecological studies were that, despite the assertions (sometimes very loudly) from people peripheral to the investigation, there was no

[5]The Oriental White-backed vultures in the Pakistani Punjab live in intimate association with the local farmers, often nesting in trees next to houses. One time we had been invited for tea with some local villagers, and in the course of our discussion had branches literally falling on us from nest-building efforts going on overhead! Vultures even nested in major cities, as evidenced by the presence of active nests in Lawrence Gardens within the city of Lahore.

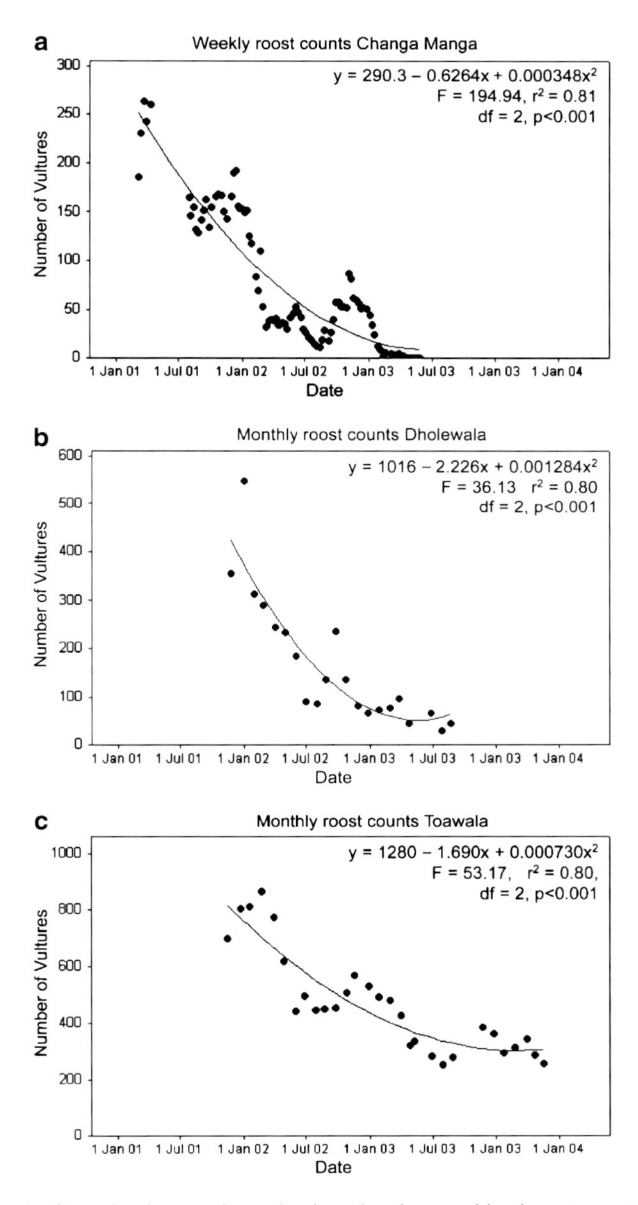

Fig. 14.2 Graphs from the three main study sites showing combined roost counts of adult and subadult Oriental White-backed vultures made weekly at Changa Manga (**a**), and monthly at Dholewala (**b**) and Toawala (**c**) colonies. Lines of best fit are given as quadratic equations with regression statistics. For the purposes of the quadratic equations, dates were expressed numerically with 1 January 2001 equating to day 1. Reprinted with permission from Cambridge University Press, Gilbert et al. (2006). Rapid population declines and mortality clusters in three Oriental white-backed vulture *Gyps bengalensis* colonies due to diclofenac poisoning. Oryx 40:388–399

solid evidence that the actual decline was related to reproductive problems, persecution,[6] loss of food or loss of habitat (Pain et al. 2008).

Other findings from the ecological studies would also prove to be very important. For example, we learned that domestic livestock was virtually the sole food source for vultures in Pakistan (Oaks et al. 2004b; Watson et al. 2004). At the same time, data from satellite telemetry showed that, not unexpectedly, the vultures ranged very widely to locate livestock food. They could cover up to 225 km/day, which over several months constituted a minimum convex polygon area of up to 68,930 km^2 (Gilbert et al. 2007a). From early on it was also evident that mortality measured in the field was clustered temporally and spatially, which suggested a highly lethal point source most compatible with a toxin (Gilbert et al. 2006). Those findings became very important in identifying the cause of the decline, as well as in attempts at mitigation.

Diagnostic Studies

Another key objective from the outset was to collect good quality samples from representative dead vultures to aid in diagnosing specific cause(s) of death. Determining what was "representative", through a case definition, was extremely important. Mortality within a large vulture population would presumably occur from a number of causes, some natural, some anthropogenic, and many would be "normal" and not have serious detrimental effects on the stability of the population. Thus, an appropriate case definition was essential to differentiate those cases that died of the decline-related disease from those cases that died from other more incidental, or normal causes. In addition, typical causes for a large scale mortality event would include infectious or toxic diseases, for which proving causality would require a proper set of control cases. In the very beginning, before there were enough data to be able to firmly define a case, our plan was to simply collect as many dead vultures as possible, including those in all age groups and where the cause of death was likely incidental (e.g. a traumatic injury such as hit by a car). Later, after a good case definition was established, we could then assign the cases we had collected to their proper groups.

Another important objective was to perform comprehensive post mortem examinations and testing. Each dead bird, whether it was likely to have died of the disease responsible for the decline (a "case") or from some other problem (a "control") would have all relevant testing needed to establish the cause of death. Each case would begin with a complete histopathologic examination to identify the types of lesions, if any, that were present. The microscopic lesions in tissues would provide important

[6]There was evidence for limited persecution by the military and civil aviation authorities in India, who were killing vultures in the vicinities of airports to decrease the incidence of bird-strikes on aircraft. However, the scope of this type of persecution was very limited, and unlikely to have any impact on the overall vulture populations.

clues as to the type(s) of disease processes in the dead bird, and would be important for guiding the types of, and in many cases for interpreting the results of, ancillary testing such as microbiology and toxicology. That kind of guidance would help prioritize additional testing, which could be time consuming and expensive. It would make better use of limited numbers of tissue samples and restricted financial budgets. Meaningful, comprehensive testing would also require a multidisciplinary approach, with input from general, wildlife, and avian (including poultry) pathologists, microbiologists, and toxicologists, and ultimately from any other relevant specialty that was determined to be appropriate.

Implicit in a comprehensive testing strategy was collection of complete sets of tissues for examination and analysis. Our plan was to include a sample of major organ systems fixed in formalin for histopathology, and replicate samples of frozen tissues from major organ systems for any additional testing deemed necessary, such as toxicology or microbiology. To ensure that each post mortem was complete, and that historical data and sample collection from each case remained consistent, a form was created which requested all the relevant data and provided a check off box for samples required. Also implicit in our strategy was collection of samples from freshly dead vultures so that the tissue samples were optimal for diagnostic testing. Proper storage and handling also were very important to maintain tissue quality for diagnostic work. The highest priority was a place to freeze and store samples, which was accommodated by the laboratory at BZU.

The other major decision was how and where to have the tissues tested. It soon became clear that the appropriate facilities and expertise for comprehensive testing were either not present, or more typically, not available to us in a time frame appropriate for the urgency of the problem. So the decision was made to export all of the samples to Washington State University from which the various samples would be distributed to the relevant laboratories and experts, and where the data would be collated and integrated. The various Pakistani government agencies involved in this venture recognized local constraints with testing, and demonstrating a refreshing level of institutional common sense, were extremely helpful in efficiently providing the necessary permissions and permits. The final major hurdle was to identify a means of shipping the tissues from Pakistan to the USA. Any mode of shipment would take several days, at best, to get from door-to-door, undermining the ability to keep the samples cool or frozen with ice or dry-ice in coolers, respectively. So we purchased several "dry shippers" – liquid nitrogen containers specifically designed for air transport, which would store samples frozen for some weeks, and which ought to have solved all transportation problems. But more perplexing problems were yet to be encountered: we discovered to our dismay that the same airlines that would carry the shippers *to* Pakistan would not carry them *out* of Pakistan! Liquid nitrogen, as a hazardous material, can only be carried on commercial aircraft after appropriate verification by hazardous material inspectors that the material is appropriately contained. While the airlines had such inspectors at major airports in North America and Europe, they did not have any such inspectors in Pakistan. It was thus a long arduous process to arrange a system whereby private inspectors could examine the shipments to the satisfaction of the airlines. We finally made our first shipment of samples at the

Fig. 14.3 Dead Oriental White-backed vulture on the bank of a canal, Punjab Province, Pakistan, in 2001. Photo credit: The Peregrine Fund

end of the first breeding season in 2001. Then, after the terrorist events of September 11, 2001, our carrier, along with many other airlines, suspended their service to Pakistan. Thus the process had to be repeated all over again.

On our initial visits to the nesting colonies, it was clear, based on the number of carcasses and body parts strewn around the area, that there were large numbers vultures dying (Fig. 14.3). Those birds had been dead for days to months. Our first assumption was that in the face of such a massive die-off, there would be no shortage of freshly dead vultures for necropsy examination and testing. This was soon proven to be wrong, and our first visit was met with enormous frustration! In spite of the many dead birds, finding an intact carcass that was fresh enough to collect was something of a challenge. We were clearly not the only scavengers in the Punjab; we had considerable competition from feral dogs, jackals, mongoose, various corvids, and even humans. In other cases, apparently freshly-dead birds were caught in the branches high in the treetops making them inaccessible.[7] After 3 weeks of searching, only two possible cases and two likely controls (traumatic injuries) were collected. And none of those cases was less than several days old. But one of these cases, found at the Changa Manga plantation, gave us our first important clue as to the cause of the decline. That bird had obvious visceral gout, which in birds, is the manifestation of renal failure. Birds metabolize dietary nitrogen (from protein in their food) in the form of uric acid, rather than as urea, as do mammals. Uric acid is

[7]We later circumvented this problem by enlisting the aid of a "professional" climber whose usual job was to collect honey.

Fig. 14.4 Photograph showing visceral gout as a white, pasty membrane covering the surface of the liver (*arrows*). The dark areas just past the arrows show normal liver where the membrane was peeled away. Photo credit: The Peregrine Fund

then excreted by the kidneys to give the white pasty parts of the typical avian dropping and which is the avian form of urine. So when there is renal failure, uric acid in the blood is no longer filtered out adequately by the kidneys, and the blood levels build up until the uric acid actually precipitates as solid white crystals. That leads to the gross appearance of visceral gout, which is a white pasty deposit that covers the surface of the internal organs including the liver, heart, kidneys, and intestinal tract – and is immediately evident upon opening the carcass (Fig. 14.4).

It was soon obvious that conducting post mortems and collecting samples was going to be a long-term effort. TPF recruited a veterinarian, Dr. Martin Gilbert, who along with Dr. Munir Virani, a TPF biologist based in Kenya, were to be present in Pakistan throughout much of the vulture breeding season, and coordinate the post mortems, sample collection, and ecological studies. Those studies were done in collaboration with the Ornithological Society of Pakistan (OSP) and the MSc students from BZU. With that large investment in manpower, and daily monitoring of transects, the picture of mortality and a case definition began to emerge. In retrospect we can now recognize that the collection of good quality samples was only possible with a systematic search for cases and controls, which clearly merited the very large investment of researcher-hours in the field (Fig. 14.5).

Fig. 14.5 Peregrine Fund veterinarian Dr. Martin Gilbert (*left*) and BZU students examine a dead vulture in the field. Photo credit: The Peregrine Fund

Initial Diagnostic Testing

The obvious nature of visceral gout was fortuitous. Gout is very apparent on visual inspection, even in birds that have been dead for some time and otherwise decomposed beyond being useful for many types of tests. By the end of the 2002 breeding season, there was a very strong correlation emerging between adult mortality and the presence of visceral gout. Gross post mortem examinations done on 259 adult and subadult White-backed vultures revealed that 219 (85%) of those had visceral gout. Visceral gout was also present in 17 (39%) of 44 immature and juvenile vultures examined (Oaks et al. 2004b). In the younger birds we could partly attribute the lower rate of gout to a higher incidence of other causes of mortality commonly associated with immaturity, including normal fledgling mortality and traumatic injuries. Thus, it became increasingly apparent that visceral gout was likely the disease responsible for the population crash, and renal failure associated with visceral gout became the case definition. Yet, while it was clear that the actual cause of death was renal failure, we were very aware that many different disease factors can cause the underlying renal pathology. Therefore, identifying the specific cause of renal disease became the focus of the diagnostic investigation.

The two most common or likely causes of renal pathology in wild birds would be toxicity or infectious disease. Other common causes, frequently cited in the limited literature, include high protein diets and dehydration. However, that literature generally refers to domestic fowl, and is likely of limited relevance to free living vultures. For example, for an obligate carnivore it does not seem likely that their diet could be too high in protein. If dehydration were the cause, this was potentially problematic from a diagnostic perspective, since dehydration can be a nonspecific consequence of many diseases. Fortunately, while dehydration is commonly cited in text books as a possible cause of avian visceral gout, a closer examination of the primary literature reveals a lack of any direct evidence for this (Meteyer et al. 2005). Thus, it was most likely that we were searching for a specific disease process.

This is where the approach of letting histopathology guide the testing proved very valuable. Histopathology was performed on 28 vultures with visceral gout (cases) and 14 cases without gout (controls). In all the cases, and with complete consensus from all the examining pathologists,[8] the lesion was severe, acute renal tubular necrosis and resultant uric acid crystal deposition in many tissues (Meteyer et al. 2005; Oaks et al. 2004b) (Fig. 14.6). There was no evidence that the lesions were chronic, which was consistent with the finding that the vultures that died of gout were in good body condition. Also, and very importantly, there were no significant inflammatory cellular infiltrates, which would be expected with an infectious disease. Thus, the lesions were most compatible with an acute toxic insult to the kidneys, and focused the investigation on toxicological causes (Oaks et al. 2004b).

We started by testing tissue samples for known toxic causes of avian renal disease, starting with the heavy metals cadmium, lead, and mercury. Tissue concentrations for those metals fell within normal limits, with the exception of a single bird without gout which had toxic levels of lead – which was not surprising given the potential for a scavenger of this type to acquire fragments of lead projectiles or other sources of lead (Church et al. 2006). We also tested tissues for abnormal levels of arsenic, copper, iron, manganese, molybdenum and zinc, all of which were unremarkable. We did extensive testing for possible poisoning by organophosphate, carbamate, and organochlorine pesticides. Although those chemicals are not commonly associated with renal failure, they certainly can cause acute mortality, and organophosphates were widely used in the agricultural habitats used by vultures. While it appeared that accidental and intentional organophosphate toxicity was fairly common in humans and livestock in the area, we only found a single case in one of the non-renal failure vultures. We did find traces of DDE, lindane, and dieldrin in some birds, but those

[8]The primary pathologists for this study were Drs. Bruce Rideout (from the San Diego Zoo) and Carol Meteyer (from the National Wildlife Health Center in Madison, WI) who specialize in wildlife, and Dr. HL Shivaprasad (from the University of California) who specializes in poultry and exotic birds.

Fig. 14.6 Histopathology section of kidney from an affected vulture showing necrotic epithelial cells lining urate-filled tubule (*arrows*). The tubule is filled with urate crystals (*star*)

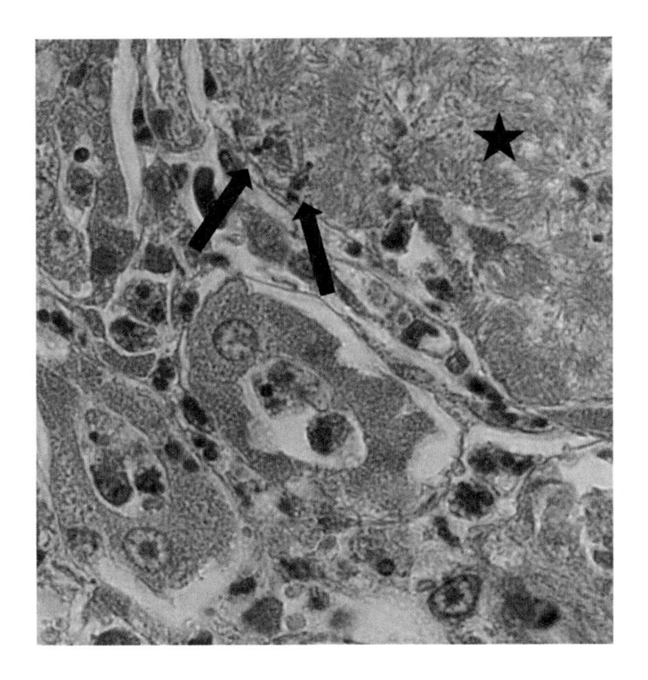

were all at levels below those associated with acute toxicity, and were present at similar levels in both gout and control cases. An avicide (3-chloro-p-toluidine) that works by causing renal failure was also considered, but we could find no evidence that chemical was used in Pakistan. Moreover, it has no other known uses, and has little potential for causing secondary poisoning in raptors.

Although the histopathology results did not support the diagnosis of an infectious cause, in the interest of being thorough and with the possibility of a novel and/or unusual infectious disease being present, we performed extensive microbiology studies, including bacteriology and virus isolation. However, no pathogens that correlated with gout were detected. Polymerase chain reaction (PCR) studies were performed for avian influenza and infectious bronchitis viruses, agents recognized as renal pathogens in poultry, but those tests were all negative.

In the course of this research, a novel mycoplasma bacterium, now named *Mycoplasma vulturii*, was isolated and characterized (Oaks et al. 2004a). However, while that new mycoplasma was quite interesting to microbiologists, its prevalence in vultures was about 30% in both cases of renal failure and nonrenal failure. Thus, this agent was not associated with gout or the decline. The identification of that organism was very instructive and a good reminder about the distinction between 'new' and 'biologically important'. Whenever a species suddenly becomes the focus of intensive investigation, new things are inevitably going to be discovered. But care needs to be taken to ensure that those new insights are relevant to the question at hand. Another example of an interesting finding that proved not directly relevant to identifying the cause of the vulture decline was the role of "head-drooping". Early on it was noted that in areas where vultures were disappearing and/or ill, vultures were hanging their heads. This behavior had not been noted previously in Asian *Gyps* vultures, and was proposed to be an indicator of sick birds and affected populations. However, upon further investigation, it was discovered that anecdotally this behavior had been described and appeared to be quite common in African *Gyps* vultures (where there was no evidence of the problem), and that it also correlated with ambient environmental temperatures and was most likely a thermoregulatory behavior (Gilbert et al. 2007b). As such, although it is likely that sick birds would head-droop as they weaken, this behavior by itself was not a specific indicator of birds affected with gout. Interestingly, we also observed an association of head-drooping posture by birds on the ground with proximity to human observers, in which the sick-looking behavior may normally serve as a deterrent to closer investigation (Watson et al. 2008) (Fig. 14.7).

Discovery of Diclofenac Toxicity

In mid-2002, with extensive and very expensive testing completed but no diagnosis, and the most likely and usual suspects effectively ruled out, our frustration levels began to grow quite rapidly. The level of frustration, and alarm, grew even more when a group of ten captive vultures in our care died unexpectedly and suddenly with visceral gout, most within a week. Obviously, we needed to expand our thinking and develop new hypotheses as to the possible cause. And once again, the coordinated diagnostic and ecological studies proved extremely useful. At that point, although we had no idea what it might be, we were quite certain the cause was a toxic agent of some kind. But in wild birds, a toxin is most likely something that is ingested and thus coming through the food supply. In Pakistan, our ecological studies indicated that the vultures fed almost exclusively on domestic livestock, as large wild ungulates had long been extirpated from this region of Asia. Therefore, we realized that we should be looking at the chemicals that would either go on, or in, the livestock – and to look at this we designed a survey to query the owners about what types of chemicals they might be using. This survey focused on what veterinary care might be provided, as well as other supplements and agricultural chemicals being used. We assumed that most of the small shareholder farmers, the predominant type of livestock ownership in that part of Pakistan, would themselves administer most, if any, veterinary treatments. Like many assumptions, that one turned out to be false, and the reply to the question about veterinary/health care products was essentially, "I don't know. Ask my veterinarian!" So we revised the survey to focus on veterinarians and the ubiquitous veterinary retail stores in the area.

In late 2002, we conducted a survey of 74 veterinarians and veterinary drug retailers in the region, and compiled a list of 34 commonly used drugs. Those identified drugs were then screened for candidates that could be responsible for kidney disease in vultures. We applied the following criteria: known to be nephrotoxic in other birds or mammals, absorbed orally (since the presumed route of exposure was ingestion), and to be compatible with the recent and widespread decline of the vultures the drug should also be commonly used and new to the market. Remarkably enough, one drug and only one drug, diclofenac, met all those criteria. Diclofenac is a nonsteroidal anti-inflammatory drug (NSAID) that has been widely used for decades in human medicine, primarily to treat rheumatoid and other types of arthritis. However, in the early 1990s in India, and in about 1998 in Pakistan, diclofenac entered the veterinary

Fig. 14.7 From top to bottom: (**a**) A wing-tagged Oriental White-backed vulture is observed standing on a trail with wings held out in a drying posture after bathing in a nearby pool. (**b**) The vulture stands alert as a man and donkeys approach.

Fig. 14.7 (continued) From top to bottom: (**c**) Rather than taking flight, the wet vulture adopts a head-drooping posture as the man and donkeys investigate. (**d**) In this and other situations of human proximity, vultures in Pakistan adopted a head-drooping posture to deceive the human and deter closer investigation (Watson et al. 2008). The man and donkeys continued, leaving the vulture unmolested to resume drying. Photo credit: The Peregrine Fund

market (Pain et al. 2008). And thanks to its low cost, and apparent safety and efficacy in livestock, it was a very successful veterinary product. But NSAIDs as a group are well known for potential renal toxicity, and other related anti-inflammatory drugs such as indomethacin and flunixin had been shown to cause renal failure and visceral gout in other bird species (Paul-Murphy and Ludders 2001).

Based on that new information, kidney samples from 23 vultures with renal failure and 13 control cases without renal failure (causes of death shown to be from trauma, lead poisoning, organophosphate poisoning, and intestinal foreign bodies) were tested by high performance liquid chromatography and mass spectroscopy[9] for residues of diclofenac. Amazingly, every single one of the renal failure cases had tissue residues of diclofenac, while none of the nonrenal failure control cases were positive (Oaks et al. 2004b). Those results obviously implicated diclofenac as the cause of renal disease in vultures. We also tested tissues from the captive vultures that had died of gout. All were positive for diclofenac, as was the buffalo meat which had been obtained from a local butcher and was being fed to the vultures when they began dying.

We also felt it was crucial to experimentally verify the toxicity of diclofenac to white backed vultures, so we conducted a dosing trial. Veterinary diclofenac was administered orally to two non-releasable juvenile vultures at 2.5 mg/kg, the standard veterinary dose recommended for mammals. Two other vultures were given a lower dose of 0.25 mg/kg. Within 58 h of dosing, both of the high dose birds and one of the low dose vultures died of visceral gout (Oaks et al. 2004b). On postmortem, the three dead vultures had kidney lesions identical to those found in the field cases. Subsequent experiments also showed that visceral gout was reproducible in vultures fed meat from buffalo treated with labeled doses of veterinary diclofenac. Those controlled dosing experiments clearly demonstrated the toxicity of diclofenac to be dose-dependent (Oaks et al. 2004b), with a median lethal dose of about 0.098 mg kg^{-1} (Swan et al. 2006a). At that point we were elated; after 3 years of grueling field work and painstaking diagnostic analysis, we had discovered that a pharmaceutical drug, recently introduced for use in domestic livestock, was responsible for the visceral gout syndrome seen in the Pakistani vultures.

Diclofenac as the Cause of the Vulture Decline

The analytical and experimental work with diclofenac was being completed in mid-May 2003, just as many of the researchers working on the vulture problem in India and Pakistan were gathering for the VI World Conference on Birds of Prey and Owls in Budapest, Hungary. That meeting provided us an opportunity to present our findings almost immediately after their discovery. Our data were met with considerable

[9]Numerous people assisted with the toxicological studies. The diclofenac analysis was performed by Todd Taruscio (from the University of Idaho). Key assistance with assays and interpretation of pesticide and heavy metal analyses included Drs. Patricia Talcott (University of Idaho) and Val Beasley (University of Illinois).

excitement at the conference; many discussions soon followed concerning the important questions that still needed resolving, and the best ways forward. One of the most direct and important questions was to determine the extent that diclofenac was used in India[10] and if vulture deaths in India also could be tied to diclofenac. Regarding use of diclofenac in India, a casual search of the internet and visits to veterinarians and veterinary pharmacies showed the drug was indeed quite commonly used. India alone had about 50 companies that manufactured veterinary diclofenac (Pain et al. 2008). Researchers from the BNHS and RSPB then documented the presence of visceral gout in India and Nepal in approximately 72% of the dead vultures examined, and were able to show that 100% of the gout cases tested had detectable residues of diclofenac, while none of the non-gout cases were positive (Schultz et al. 2004). Cases of diclofenac poisoning were recorded from the far Eastern parts of India, some 2,000 km from Pakistan. Over a broad spatial scale, diclofenac poisoning was, therefore, occurring wherever vulture declines had been documented. Sadly, it also showed that diclofenac poisoning was occurring over most of the current known range of these birds, which raised the specter that this problem was causing extinction of three species of vultures at a staggering rate.

We also questioned how diclofenac could have such a devastating effect on the entire population of these vultures. While the type of poison might be new and even unique, certainly poisoning of large raptors and other wildlife by ingesting a tainted carcass is hardly novel. Apart from malicious poisonings of scavengers and predators through carcasses laced with poison, most chemicals that have led to population declines are those that have unintentionally accumulated in the environment and ultimately caused toxicity in animals at all trophic levels. In contrast, diclofenac, like the other nonsteroidal anti-inflammatory drugs (NSAIDs), does not persist in a treated animal. Its half-life is approximately 12.2 h and tissue residues become undetectable after about three days in cattle and two days in goats (Taggart et al. 2006). Thus, a toxic exposure was a random event. Intuitively it would seem that such a massive population effect would require some unreasonable proportion of the livestock population to perish very soon after treatment. Part of the solution to that paradox was supplied by our learning of the exquisite sensitivity of vultures to diclofenac. The toxicity of NSAIDs to the mammalian species for which they were developed generally occurs at abusive, higher doses, while the median lethal dose for *Gyps* vultures was approximately 10% of the usual therapeutic dose for mammals.

The question of scale was effectively put to rest by some elegant ecological modeling. Rhys Green and colleagues (2004) used a simulation model based on the demographics and feeding behavior of *Gyps* vultures to show that an incidence rate as low as one contaminated carcass in a total of 130–760 carcasses (Fig. 14.8) was sufficient to drive the observed vulture population decline of about 30% per year. Key features of vulture behavior that would predispose them to such profound effects

[10] It was surprising to us the number of people peripheral to the vulture work, many of which were totally uninformed – and in some cases who had never been to India! – who were outright dismissive of diclofenac saying a human drug was simply not used in veterinary medicine, and/or that there was no significant level of veterinary care in a "poor" country like India.

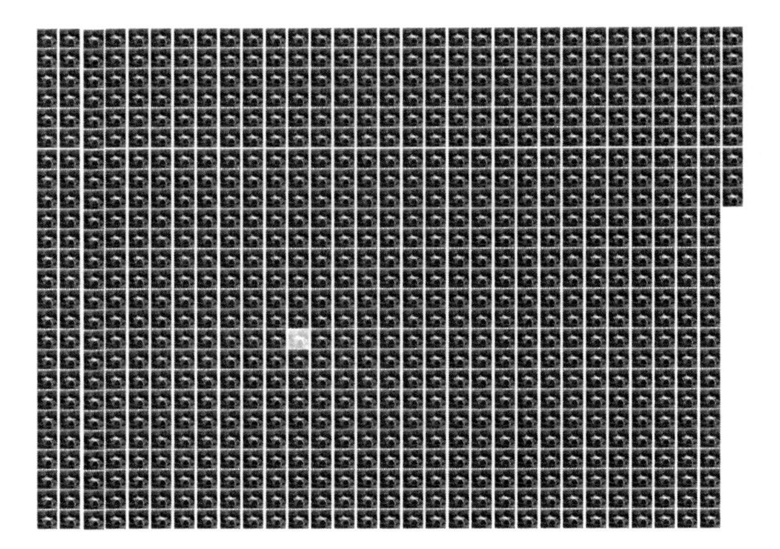

Fig. 14.8 An array of 759 photos of buffalo skulls, with one single different one to illustrate the very low level of frequency (1 in 759) of diclofenac contaminated carcasses required to drive rapid (as high as 30% per annum) population declines in vulture populations

would include the fact that, in addition to slow reproductive rates, they congregate and feed communally at carcasses, and that they can cover vast areas in their search for food. Thus, a single toxic carcass could affect multiple vultures at multiple breeding colonies. Moreover, those modeling studies showed that 71–100% of the excess mortality required for the observed population decline could be attributed to diclofenac. Later studies of the temporal and spatial patterns of mortality in Pakistan reached similar conclusions (Gilbert et al. 2006). There, regression analysis of observed annual mortality and annual breeding population declines supported the conclusion that diclofenac was the primary cause of the decline, and further that a ratio as low as 1–3% contaminated carcasses could cause the decline. Subsequent analyses of liver samples from over 1,800 animals at 67 sites within 12 states of India documented an overall contamination rate of about 10% (range by state from 0% to 22.3%) (Taggart et al. 2007). Collectively, those studies showed that a surprisingly rare event, in the form of a toxic carcass, had the potential to cause rapid and catastrophic effects at the population level. Stated another way, it was quite plausible for these vultures to have been victims of what was essentially a mass poisoning event with a transient environmental contaminant. In this case, the paradigm of an ecological disaster being related to environmental persistence and bioaccumulation of a toxin was simply overcome by the scope and scale of use of a commercial, inexpensive, and therefore popular pharmaceutical. More recently, similar effects have been observed on local populations of Bald eagles (*Haliaeetus leucocephalus*) exposed to non-persistent oganophosphate and carbamate insecticides through feeding on poisoned ducks (Elliott et al. 2008). It should now be recognized that populations of efficient scavengers where relatively large numbers of individuals will congregate at a single toxic food source are highly vulnerable.

The Regulatory and Policy Response

By the end of 2003, the scientists involved with the vulture problem were convinced beyond a reasonable doubt that the cause of the decline was diclofenac. If the decline was going to be stopped, and the populations of at least three vulture species were to have any chance at recovery, diclofenac needed to be removed from the vultures food supply. We felt that this discovery alone should have been sufficient to warrant an immediate ban on the veterinary use of this drug. That was especially so, considering NSAIDs do not 'cure' anything (as would, for example, an antimicrobial drug that directly eliminates a bacterial infection), but are simply part of supportive or ancillary care mainly to relieve pain or discomfort in animals, and because there were alternative NSAIDs potentially available (although at this stage it was unknown if other NSAIDs would be equally toxic to vultures). Loss of vultures could potentially also do significant damage to public health, economy and culture (Markandya et al. 2008), further warranting immediate restrictions on the use of the drug. The task was then to get this evidence into the hands of people who could make a difference.

Informing Policy Makers

TPF and Bird Conservation Nepal convened the Kathmandu Diclofenac Summit in February 2004, a meeting in politically neutral Nepal (Pakistan and India being openly at war over Kashmir at the time, which made it almost impossible for any government officials to travel to the other country). That meeting of authorities and interested non-governmental organizations (NGOs) from India, Nepal, and Pakistan was to present the scientific evidence in one place and time to policy makers. The key researchers from across the subcontinent who had been involved in the study of the vulture decline all participated, providing a strong and unified message. The ecological and diagnostic studies were presented first. That was followed by the ecological modeling work in anticipation of arguments that there was insufficient diclofenac use to cause the observed rate of decline. The meeting concluded with expert accounts on how similar problems with raptors had been resolved elsewhere, and finally with proposed solutions to the problem. Those solutions included firstly, an immediate ban on the veterinary use of diclofenac, and secondly, the collection of a representative sample of healthy vultures from each of the affected species for safe-keeping. The healthy birds could also be used for captive breeding and/or release in the future to restore the seriously depleted populations. The scientific session was followed by a workshop to develop a response plan, aiming for some level of agreement if not coordination across the three countries involved. The workshop also provided an opportunity to gauge whether our message was understood and if it was likely to be acted upon. We gained a majority consensus, and we were encouraged by the feedback during the workshop, and the declaration of intent to act by the majority of agencies represented.

A second, similar meeting was held in India a few days later, the International South Asia Recovery Plan Workshop, at which a vulture recovery plan for India was developed with key state agencies involved. Again, the recommendations from that meeting were to ban the use of veterinary diclofenac in all the range states, and to establish captive populations of the three affected species of *Gyps* vultures (Pain et al. 2008).

The Regulatory Response: Legislation to Ban Diclofenac

In principle, the most direct approach to solving the problem was a ban on the use of veterinary diclofenac. More specifically, since the manufacture and sale of veterinary drugs required a license, there was a proposal to revoke those licenses in the vulture range states. Such a process would certainly require an enormous amount of advocacy, since until that time no pharmaceutical had ever been banned for environmental reasons, and considerable opposition from the pharmaceutical industry was expected. In November of 2004, a formal program was initiated by the BNHS and RSPB, whose goal was to advocate a ban on the veterinary use of diclofenac, and to educate and inform the relevant government agencies about ongoing research (Pain et al. 2008). The program was extremely effective and ultimately successful, with a series of meetings and recommendations from the Indian National Board for Wildlife and Indian Ministry of Environment and Forests, which ultimately led to a directive by the Drug Controller General of India in May of 2006 to withdraw manufacturing licenses for veterinary diclofenac (Pain et al. 2008). In August of 2006, there was a similar ban in Nepal. Meanwhile, a single legislator and a few passionate individuals carried the cause forward in Pakistan, resulting in a ban there later in 2006.

Officially banning the veterinary use of diclofenac occurred at a pace that seemed agonizingly slow, given that continued use of the drug was causing the vultures to disappear at an extremely rapid rate. In retrospect, however, the ban was a remarkable achievement, and a tremendous feat of persuasion by many individuals and organizations in each country. It was accomplished within 3 years of the actual discovery of diclofenac toxicity, and two years from the time when the case for diclofenac was fully developed and presented to policy makers. In contrast, it took more than a decade to bring about even limited restrictions on use of DDT.

A significant factor in that regulatory success was the unprecedented joint effort by the BNHS and RSPB, with collaborators in South Africa, which resulted in the identification of a safe alternative to diclofenac. In this way, the environmentalists were not just part of the problem (at least from the veterinary and pharmaceutical industry perspective), but also part of the solution. It was entirely unknown at the time whether *Gyps* vultures were uniquely sensitive to diclofenac and/or other NSAIDs, or whether diclofenac was uniquely toxic to birds in general. However, the range of side effects among different species to NSAIDs was well recognized, suggesting that alternatives could be found which would be safe for vultures.

Alternatively, if *Gyps* were highly sensitive to all NSAIDs, that information also would have important implications for their conservation. The data was initially collected from surveys of avian and zoo veterinarians regarding their clinical experiences with NSAID use in raptors (Cuthbert et al. 2006). While the results of the survey did not address diclofenac, as it was not being used in avian veterinary medicine, the results did suggest that meloxicam was a safe alternate. In addition to documenting the successful use of meloxicam in some 700 individual birds belonging to 54 different species, it had also been used safely in 39 individual *Gyps* vultures belonging to six species. In contrast, other NSAIDs, including flunixin,[11] carprofen, ibuprofen, and phenylbutazone have all been associated with visceral gout.

The next step was to experimentally demonstrate the safety of meloxicam in *Gyps* vultures under controlled conditions. That was accomplished in a series of phased studies performed by an international consortium involving groups from Africa (Pretoria University, DeWildt Cheetah and Wildlife Trust, and Rare and Endangered Species Trust), India (BNHS, Indian Veterinary Research Institute) and the UK (RSPB, Cambridge University, and Aberdeen University). In preliminary studies, done in part to avoid experiments with the now highly endangered Asian *Gyps* species, diclofenac toxicity studies were performed in African White-backed vultures (*Gyps africanus*) and Eurasian griffons (*Gyps fulvus*) – in whom it was found that diclofenac was as toxic as it was in their Asian cousins (Swan et al. 2006a). Thus those birds made suitable non-endangered surrogates for potentially dangerous studies into the safety of meloxicam to vultures. Fortunately, these studies found that meloxicam, in very high doses, as well as doses that exceeded the calculated maximum likely exposure from feeding on a treated animal, was safe in 40 African White-backed vultures (Swan et al. 2006b). Follow-up studies demonstrated that meloxicam was also safe at doses well in excess of calculated maximum exposure levels in a total of 41 Oriental White-backed vultures and Long-billed vultures (Swan et al. 2006b, Swarup et al. 2007).

Unfortunately, meloxicam does not appear to have gained widespread acceptance or use in Indian veterinary medicine due to higher cost and perceived lack of efficacy in comparison to diclofenac. Nonetheless, it is another NSAID that is already licensed and available for use by Indian veterinarians. In a carcass survey conducted in 2006, meloxicam was found to be, after diclofenac, the second most commonly used NSAID in livestock (Taggart et al. 2009). Ketoprofen was another candidate veterinary NSAID that might potentially replace diclofenac, based on apparent lack of toxicity in the clinical survey (Cuthbert et al. 2006). However, it has more recently been shown that ketoprofen, although less toxic than diclofenac, can cause renal failure in African White-backed vultures and Cape vultures (*G. coprotheres*) at doses found in carcasses (Naidoo et al. 2010).

Much more alarming than poor acceptance of meloxicam, however, is that as of early 2010 it appears, based on numerous anecdotal reports, that diclofenac is still

[11] Flunixin has long been recognized in other birds as significantly toxic (Klein et al. 1994).

routinely used in Indian livestock. One loophole was that the original ban only specified that it was illegal to manufacture diclofenac, not sell it. The latter problem was addressed in a 2008 letter from the Drugs Controller General of India to the Manufacturer's Associations (Taggart et al. 2009). Other likely, and unresolved, reasons for the continued illicit use of diclofenac include lack of effective enforcement, lack of incentives either to veterinarians or industry to switch to meloxicam, diversion of human diclofenac products to the veterinary market, illegal manufacture and/or importation, and lack of awareness among the veterinary community (Pain et al. 2008; Taggart et al. 2009). Interestingly, and somewhat surprisingly, given the success of diclofenac sales, the pharmaceutical industry put up little resistance to the ban on diclofenac. The reasons for that are unclear, but probably include the comprehensive nature and high quality of the science exposing its toxicity which was published in a timely manner in peer-reviewed literature. The evidence was overwhelming and gave opponents little basis for disputing the connection between diclofenac and the vulture decline. Another possible factor was the lack of direct attacks on or assigning blame to the pharmaceutical industry. As of the mid-2000s, that was probably appropriate as the industry would have had no reason to suspect their product would affect vultures in this way. Even in countries that require environmental safety testing of pharmaceuticals, current assessment standards would not have detected or predicted the effect of diclofenac on vulture populations. Of course, with current knowledge, any individuals or companies that continue to sell diclofenac will be judged much more harshly, and one hopes, prosecuted within the civil and criminal legal systems.

Additional Conservation Strategies: Captive Breeding Populations

In addition to the ban on diclofenac, the other primary conservation strategy was the establishment of captive holding and breeding programs for the three most severely affected vulture species. That move was imperative given the rate of decline, and the realistic time frame for removing diclofenac from the environment.

In Pakistan, TPF pursued a plan to hold Oriental White-backed vultures in safekeeping at an existing, empty facility in United Arab Emirates until the Pakistan authorities and local NGOs could build suitable facilities and secure long-term funding. However, the plan broke down due to politics, self-interest, and government bureaucracy. In the end, only 11 White-backed vultures have been put into safe-keeping – an inconsequential fraction of the minimum 75 individuals recommended. Sadly, because almost no vultures are left, the window of opportunity to collect these birds in Pakistan has passed.

In India, the BNHS has developed a vulture conservation breeding program recognized by the World Association of Zoos and Aquariums (WAZA), and with support from state Forestry Departments (http://www.vulturedeclines.org/). The nidus for this program was an existing facility in Haryana, which was originally designed as a care center to provide medical treatment and provide diagnostic services for sick vultures.

After the discovery of the diclofenac-driven population crash of vultures, it was clear that breeding facilities rather than care centers were needed, so this facility was quickly converted and expanded to accommodate a breeding population. Similar holding and breeding facilities were subsequently built in West Bengal and Assam. Collectively, as of 2008, those facilities had 83 White-backed, 71 Long-billed and 28 Slender-billed vultures, with the eventual goal of 25 pairs of each species at each of the three facilities (Pain et al. 2008). The first White-backed vulture was produced in 2008, and the first Slender-billed vultures were produced in 2009 (http://www.vulturedeclines.org/).

Nepal has also recently acquired permission and resources to construct a breeding facility in Chitwan National Park, and in 2008 obtained 14 White-backed vulture chicks, and 30 more in 2009, with the eventual goal of maintaining 25 pairs each of White-backed and Slender-billed vultures (http://www.vulturedeclines.org/).

Additional Conservation Strategies: Food Provisioning

Another strategy that has been discussed is the provision of safe, diclofenac-free food at sites known as "vulture restaurants". Vulture restaurants have shown some promise where food supplies were limiting factors in *Gyps* vulture populations, such as for Eurasian griffons in France and Cape vultures in South Africa. TPF tested the effects on foraging behavior of one vulture feeding station in Pakistan where diclofenac-free food was provided ad-libitum (Gilbert et al. 2007a) (Fig. 14.9). The restaurant did reduce the vulture foraging range and potential rates of exposure, but only during the breeding season. Between breeding seasons, vultures dispersed widely and over large areas, rarely fed at the restaurant, and were still subject to increasing rates of diclofenac poisoning. That restaurant ultimately failed to prevent the extirpation of White-backed vultures in the area, and was closed at the end of the 2005–2006 breeding season for lack of customers.

Conservation Gains and Lessons Learned

The discovery that environmental contamination with a pharmaceutical compound was the cause of the vulture declines, and the relatively rapid regulatory response, were significant and positive accomplishments. Those successes also illustrate the power of cooperative and multidisciplinary problem solving, as well as the ability of solid science to facilitate policy change. Although perhaps unique to this situation, the ability to help provide solutions – in this case finding alternative drugs that were safe – was also extremely helpful in bringing about policy change.

In theory, the science and regulatory response should result in a very significant conservation gain for south Asian *Gyps* vultures.

Fig. 14.9 Vultures gather at a 'vulture restaurant' while a Peregrine Fund assistant prepares diclofenac-free food in an attempt to reduce vulture poisoning. Photo credit: The Peregrine Fund

Sadly, almost 4 years after veterinary diclofenac has been banned, this drug continues to be used widely and illegally. Because modeling studies suggest that even limited use of the drug can have a significant impact on vulture populations, mere reductions in diclofenac use will not prevent ongoing declines. So this story is very far from over – and the outlook for vulture survival looks grim. Measures to more effectively implement the ban on veterinary diclofenac are desperately needed. This should include the traditional approaches of increased enforcement and harsher penalties for offenders. In addition, more creative and positive strategies should be explored and/or enhanced. These include incentives to use safe alternatives and awareness campaigns targeted at veterinarians who should be given personal responsibility for the drugs carried by the veterinary offices in their districts, since many of those local veterinary offices are run by people with no veterinary or science education. Public education campaigns should be used to improve the general image of vultures among the public (a hard sell, but one which has been done successfully in South Africa), and vulture safe zones (http://www.vulturedeclines.org/). The BNHS, RSPB, and Bird Conservation Nepal have started significant work on these initiatives.

The stakes are high. The failure to effectively control carcass contamination by diclofenac, will likely lead to extinction of these magnificent birds which, through their scavenger role, have controlled the spread of infectious disease for millennia, as well as provided other important ecological services. The breeding programs, although they have had some initial success, are not yet proven. These programs are also still well short of producing the number of functional breeding pairs necessary for a successful breeding and restoration program. Meanwhile, the opportunity to obtain more vultures from the wild is dwindling rapidly. Large or non-threatened populations of any of the three affected *Gyps* species do not exist outside of Pakistan, India, and Nepal. Smaller populations have been found in Cambodia, where diclofenac is not used; but those birds have their own environmental threats including lack of food, habitat loss, persecution, and other types of poisoning (http://www.wcs.org/saving-wildlife/birds/white-rumped-vulture.aspx).

The vulture decline is to some extent vindication to scientists who have expressed concern about environmental contamination by pharmaceuticals and other personal care products. Vast quantities of these compounds and their metabolites, which are by their very design biologically active, find their way into the environment via treated and untreated sewage (Daughton and Ternes 1999). Although concern has been voiced for some time about the potential effects of these compounds on the ecosystem, poisoning of the vultures was the first time that a pharmaceutical has been associated with massive and widespread ecological damage. In particular, Daughton and Ternes postulated that ecological damage did not require chemical persistence if it was continually being introduced (Daughton and Ternes 1999). And this was exactly what was happening with diclofenac and the vultures.

What other pharmaceutical-related environmental disasters await us? It is impossible to say, but it bears remembering that the exquisite sensitivity of *Gyps* vultures to residual levels of diclofenac in carcasses was never anticipated. Most NSAIDs are toxic to *Gyps* vultures at therapeutic or higher doses. In fact, although recent research has identified mechanisms of diclofenac-associated renal damage, the basis for the particular sensitivity of *Gyps* vultures is still unknown (Meteyer et al. 2005; Naidoo and Swan 2009). The effect cannot be generalized to all birds, as New World Turkey vultures (which are in a different family, the Cathartidae) are not affected by 100 times the median lethal dose of diclofenac for *Gyps* vultures (Rattner et al 2008). It also bears noting that the vulture decline was not at all subtle: it involved a single drug and was relatively easily detected once the source of the die-off was known. And at least in principle, the problem was detected early enough for action to be taken. In contrast, the effects on human and/or ecosystem health by chronic exposure to low levels of bioactive compounds, or mixtures of compounds, may not be so easily detected or repaired.

References

Birdlife International (2001) Threatened birds of Asia: The BirdLife International red data book. BirdLife International, Cambridge

Church ME, Gwiazda R, Risebrough RW et al (2006) Ammunition is the principal source of lead accumulated by California condors re-introduced to the wild. Environ Sci Technol 40:6143–6150

Cuthbert R, Parry-Jones J, Green RE et al (2006) NSAIDs and scavenging birds: potential impacts beyond Asia's critically endangered vultures. Biol Lett 3:90–93

Daughton CG, Ternes TA (1999) Pharmaceuticals and personal care products in the environment: agents of subtle change? Environ Health Perspect 107(Suppl 6):907–938

Elliott JE, Birmingham A, Wilson LK et al (2008) Fonofos poisons raptors and waterfowl several months after labeled application. Environ Toxicol Chem 27:452–460

Gilbert M, Watson RT, Virani MZ et al (2006) Rapid population declines and mortality clusters in three oriental white-backed vulture *Gyps bengalensis* colonies due to diclofenac poisoning. Oryx 40:388–399

Gilbert M, Watson RT, Ahmed S et al (2007a) Vulture restaurants and their role in reducing diclofenac exposure in Asian vultures. Bird Conserv Int 17:63–77

Gilbert M, Watson RT, Virani MZ et al (2007b) Neck-drooping posture in oriental white-backed vultures (*Gyps bengalensis*): an unsuccessful predictor of mortality and its probable role in thermoregulation. J Raptor Res 41:35–40

Green RE, Newton I, Schultz S et al (2004) Diclofenac poisoning as a cause of vulture population declines across the Indian subcontinent. J Appl Ecol 41:793–800

Johnson JA, Gilbert M, Virani MZ et al (2008) Temporal genetic analysis of the critically endangered oriental white-backed vulture in Pakistan. Biol Conserv 141:2403–2409

Klein PN, Charmatz K, Langenberg J (1994) The effect of flunixin meglumine (Banamine®) on the renal function in northern bobwhite quail (*Colinus virginianus*): an avian model. Proc Am Assoc Zoo Vet 1994:128–131

Markandya A, Taylor T, Longo A et al (2008) Counting the cost of vulture decline – an appraisal of the human health and other benefits of vultures in India. Ecol Econ 67:194–204

Meteyer CU, Rideout BA, Gilbert M et al (2005) Pathology and pathophysiology of diclofenac poisoning in free-living and experimentally exposed oriental white-backed vultures (*Gyps bengalensis*). J Wildl Dis 41:707–716

Naidoo V, Swan GE (2009) Diclofenac toxicity in Gyps vultures is associated with decreased uric acid excretion and not renal portal vasoconstriction. Comp Biochem Physiol C Toxicol Pharmacol 149:269–74

Naidoo V, Wolter K, Cromarty D et al (2010) Toxicity of non-steroidal anti-inflammatory drugs to Gyps vultures: a new threat from ketoprofen. Biol Lett. doi:10.1098/rsbl.2009.0818

Oaks JL, Donahoe SL, Rurangirwa FR et al (2004a) Identification of a novel mycoplasma species from an oriental white-backed vulture (*Gyps bengalensis*). J Clin Microbiol 42:5909–5912

Oaks JL, Gilbert M, Virani MZ et al (2004b) Diclofenac residues as the cause of vulture population decline in Pakistan. Nature 427:630–633

Pain DJ, Bowden CGR, Cunningham AA et al (2008) The race to prevent the extinction of South Asian vultures. Bird Conserv Int 18:S30–S84

Paul-Murphy J, Ludders JW (2001) Avian analgesia. Vet Clin North Am Exot Anim Pract 4:35–45

Prakash V (1999) Status of vultures in Keoladeo National Park, Bharatpur, Rajasthan, with special reference to population crash in Gyps species. J Bombay Nat Hist Soc 96:365–378

Prakash V, Pain DJ, Cunningham AA et al (2003) Catastrophic collapse of Indian white-backed *Gyps bengalensis* and long-billed *Gyps indicus* vulture populations. Biol Conserv 109:381–390

Prakash V, Green RE, Pain DJ et al (2007) Recent changes in populations of resident Gyps vultures in India. J Bombay Nat Hist Soc 104:129–135

Rattner BA, Whitehead MA, Gasper G et al (2008) Apparent tolerance of turkey vultures (*Cathartes aura*) to the non-steroidal anti-inflammatory drug diclofenac. Environ Toxicol Chem 27:2341–2345

Schultz S, Baral HS, Charman S et al (2004) Diclofenac poisoning is widespread in declining vulture populations across the Indian subcontinent. Proc Roy Soc Lond B (Supplement) 271(Suppl 6):S458–S460

Swan GE, Cuthbert R, Quevedo M et al (2006a) Toxicity of diclofenac to Gyps vultures. Biol Lett 2:279–282

Swan GE, Naidoo V, Cuthbert R et al (2006b) Removing the threat of diclofenac to critically endangered Asian vultures. PLoS Biol 4:e66

Swarup D, Patra RC, Prakash V et al (2007) The safety of meloxicam to critically endangered Gyps vultures and other scavenging birds in India. Anim Conserv 10:192–198

Taggart MA, Cuthbert R, Das D et al (2006) Diclofenac disposition in Indian cow and goat with reference to Gyps vulture population declines. Environ Pollut 147:60–65

Taggart MA, Senacha KR, Green RE et al (2007) Diclofenac residues in carcasses of domestic ungulates available to vultures in India. Environ Int 33:759–765

Taggart MA, Senacha KR, Green RE et al (2009) Analysis of nine NSAIDs in ungulate tissues available to critically endangered vultures in India. Environ Sci Technol 43:4561–4566

Vulture Rescue website. http://www.vulturerescue.org/page17.html. Accessed 1 Mar 2010

Watson RT, Gilbert M, Oaks JL et al (2004) The collapse of vulture populations in South Asia. Biodiversity 5:3–7

Watson RT, Gilbert M, Virani M (2008) Neck-drooping posture of oriental white-backed vultures (*Gyps bengalensis*) in close proximity to human observers. J Raptor Res 42:66–67

Wildlife Conservation Society website. http://www.wcs.org/saving-wildlife/birds/white-rumped-vulture.aspx. Accessed 2 Mar 2010

Chapter 15
Where Science Stops, and Action Starts

Tim Lougheed

Introduction

Economists have traditionally cast wildlife in the role of an "externality", a neutral background that does not formally interact with the day-to-day business of human affairs. Yet by the time Apollo astronauts had returned with pictures of a lonely earth hanging in space, many of us concluded that there are no externalities. We inhabit a natural world that can be all too relevant to our everyday lives, however much we might try to ignore this fact. The species sharing this world can become susceptible to poisons of our making, toxic agents that might carry profound economic significance for us, but fatal consequences for a variety of plants and animals.

Research and development efforts are bringing such agents into a global marketplace at an unprecedented pace. Various jurisdictions have responded by mounting an equally unprecedented galaxy of regulations. Yet the organizations charged with administering those regulations – which can range from elaborate international bodies to the tiniest of town councils – are creatures of history rather than scientific endeavour. They represent political, economic, and social considerations, sometimes butting heads with scientists whose only consideration is the data in front of them.

Such conflicts can result in glaring inconsistencies. A particular chemical can be declared toxic in one place but not another, even though administrators in both places may be looking at exactly the same evidence for their judgement. As a few examples from Europe, the USA, and Canada will illustrate, different attitudes toward risk can cause scientific insight to be implemented in highly distinctive ways.

T. Lougheed (✉)
Science Writer, Ottawa, ON, K1K 2A5, Canada
e-mail: Stormchild@sympatico.ca

J.E. Elliott et al. (eds.), *Wildlife Ecotoxicology: Forensic Approaches*,
Emerging Topics in Ecotoxicology 3, DOI 10.1007/978-0-387-89432-4_15,

Regulatory Roots

Our outlook on environmental regulation has been shaped by the outcry of passionate observers who tend to pepper their ideas with gloom and trepidation. We remain grateful for their efforts, but you would be forgiven for omitting them from your list of imaginary dinner companions. Aldo Leopold, a towering figure in this regard, might have some rollicking adventures to relate at such a meal. But he could be just as likely to lapse into a more sombre tone when talking about preservation of the wild or a novel land ethic. Likewise, Rachel Carson could bring a stirring intellectual energy to the table, but her enthusiasm might well be tempered by dark musings about the fate of baby birds.

Yet even if the views offered by these pioneers occasionally strike us as bleak, they embraced an optimism that is today in short supply. They held not only to a faith in technology, but a belief that the more we would come to understand the natural world, the better we would be able to preserve the most outstanding features of that world and limit the damage we were inflicting upon it.

> By and large, our present problem is one of attitudes and implements. We are remodeling the Alhambra with a steam-shovel, and we are proud of our yardage. We shall hardly relinquish the shovel, which after all has many good points, but we are in need of gentler and more objective criteria for its successful use. (Leopold 1949)

Carson was even blunter. She may have had no difficulty with using innovative chemicals to help feed the hungry, but she resented any lazy, myopic view that went along with that use. As far as she was concerned, we simply know better than to apply our modern tools indiscriminately.

> Much of the necessary knowledge is now available, but we do not use it. We train ecologists in our universities and even employ them in our government agencies, but we seldom take their advice. We allow the chemical death rain to fall as though there were no alternative, whereas in fact there are many, and our ingenuity could soon discover more if given opportunity. (Carson 1962)

Less than a decade later, international industrialists formed the Club of Rome to answer Carson's complaint in the form of the celebrated Limits to Growth report. But even their pessimistic pronouncements were shot through with a beacon of hope.

> Man possesses, for a small moment in his history, the most powerful combination of knowledge, tools, and resources the world has ever known. He has all that is physically necessary to create a totally new form of human society – one that would be built to last for generations. (Meadows et al. 1972)

The rest should have been easy. As our knowledge grew, so too should our ability to govern how that knowledge would be applied. The legacy of thoughtful observers like Carson has been a popular conception that science should logically inform the regulations to make our progress more civilized than ever before.

Science, Regulation and Precaution

Small wonder, then, that many contemporary observers remain frustrated by the ungainly relationship that has grown up between science and regulation. Representatives at the local, regional, national, and even international level can be found wading through dense reports on the latest environmental hazard confronting them. If they are new to the job, those individuals may be especially eager for the grail promised by Carson, some clear scientific indication of what steps they must take to safeguard the well being of particular places and their inhabitants.

Unfortunately, such guidance can be vague at best, and sometimes altogether absent. What qualifies as a problem can vary from one instance to the next. The regulatory agency of one country may dub toxic the same agent that another country's regulator has approved, even as the people framing these respective policies take their cues from the same research findings.

Individuals who have worked in the belly of these administrative beasts will gladly blame such incongruity on the necessary tension that exists between science and politics. Just as scientists might dream of findings that transcend the rough and tumble of party infighting, so too do politicians dream of science that unarguably defends the position they were going to take anyway. Neither side winds up completely satisfied, yet they can serve each other's purposes.

Daniel Sarewitz would like to improve this interaction. As director of the Consortium for Science, Policy, and Outcomes, an independent unit within the Arizona State University, he has looked for ways of helping scientific researchers to better inform public policy for the benefit of society. As a former Congressional Science Fellow and science consultant who worked on Capitol Hill for five years, he knows how elusive this goal can be. Whenever the uncertainty of a scientific conclusion rears its head, the ensuing debate masks what would otherwise have been just a painful political decision to act or not to act on a particular problem.

> The point is not that stripping away the overlay of scientific debate must force politicians to take action. But if they choose not to act they can no longer claim that they are waiting for the results of the next round of research – they must instead explain their allegiance to inaction in terms of their own values and interests, and accountability now lies with them, not with science or scientists. To the extent that our democratic political fora are incapable of enforcing that accountability, the solution must lie in political reform, not more and better scientific information. (Sarewitz 2004).

A Canadian Compromise

"Most people ask the question 'How can a chemical substance be toxic in Canada but not in the United States or vice versa?'" notes George Enei, director general of Environment Canada's Science and Risk Assessment Directorate. "The basic answer is that, while each country is following the same scientific methods and protocols, our respective legislation is different."

Whereas the USA combines various processes of environmental health assessment and management, he explains, Canada has separated them into three steps under the Canadian Environment Protection Act (CEPA). The first is a risk assessment, asking if the agent in question could or does bring harm to human beings or the environment. If so, the second step places that agent on a designated list of toxic substances. At that point, the third stage offers a range of options, starting with rules for how the agent can be employed and ending up at an outright ban on it.

According to Enei: "We conduct a risk assessment that assembles all of the available science to determine if a substance is harmful or has the potential to cause harm to humans or the environment. If the answer is yes, the substance is added to the list of toxic substances. Once the substance is on the list, we have access to a variety of options under CEPA to manage the risks that the assessment has identified. These include pollution prevention plans, regulations, or, in extreme cases, eliminating the substance from being used or brought into Canada."

Originally passed by Parliament in 1988, then updated and expanded in 1999, CEPA set priorities for two federal government departments, Health Canada and Environment Canada. They became responsible for a kind of "cradle to grave" oversight of materials labeled toxic, managing such goods from the initial point of their manufacture or importation right through to their final disposal or destruction.

The scope of CEPA grew considerably when the federal government launched its Chemicals Management Plan in 2006. This initiative targeted the constituents of a huge number of potentially problematic commodities, including cosmetics, pharmaceuticals, and processed foods. Some of these goods became entrenched in the industrial landscape throughout the twentieth century, with little review of their environmental implications. Tucked away within commercial processes that were often secret or proprietary, few people might even be able to pronounce the names of these chemicals, much less cite their impact. Nor, in the absence of any calamity to highlight the link between a chemical and a specific outcome, would there have been much incentive to seek more information.

All that changed with the drafting of the original CEPA legislation, which featured a definitive list of more than 23,000 chemicals that were in use in Canada between 1984 and 1986. By the time the Chemicals Management Plan was introduced, some 4,000 of these agents had been placed into a special category requiring "further attention". By way of getting the ball rolling, the government singled out 200 chemicals as a "challenge to industry", flagging them as the most likely suspects behind environmental or health problems.

The USA and European Union have since embraced their own versions of the Chemicals Management Plan. The US Environmental Protection Agency's (EPA) Chemicals Assessment and Management Program (ChAMP) and the European Commission's Registration, Evaluation, Authorisation and Restriction of Chemical Substances (REACH) system are no less ambitious in their objectives, tackling lists with tens of thousands of agents old and new. The easy regulatory ride that many products enjoyed during the twentieth century is coming to an abrupt end in the twenty first century.

The Precautionary Approach

The creation of lists of chemicals, however, will not ensure identical outcomes in different jurisdictions. Further distinctions revolve around a phrase that typically precedes key decisions: "better safe than sorry". This familiar rhetoric embodies the precautionary principle, a mainstay of regulatory practice and a justification for action even as opponents of that action call for more evidence.

Bandied about by governments as early as the 1930s, the idea began to acquire legal force toward the end of the century. When representatives of countries bordering the North Sea met in 1987 to discuss their mutual responsibility for protecting these waters, they accepted the need to reduce persistent, accumulating toxic emissions at the source.

> This applies especially when there is reason to assume that certain damage or harmful effects on the living resources of the sea are likely to be caused by such substances, even where there is no scientific evidence to prove a causal link between emissions and effects. (Ministerial Declaration 1987)

Similarly, the 1992 United National Conference on Environment and Development in Rio de Janeiro – hailed as the "Earth Summit" – offered this key recommendation: "In order to protect the environment, the precautionary approach shall be widely applied by States according to their capabilities. Where there are threats of serious or irreversible damage, lack of full scientific certainty shall not be used as a reason for postponing cost-effective measures to prevent environmental degradation." (UNEP 1992).

By 2000, members of the European Union were convinced of the need for taking precautionary measures under certain circumstances, even when cause-and-effect could not be confirmed scientifically. The nature of such circumstances was laid out in an elaborate statement imposing the juridical equivalent of "guilty until proven innocent" on questionable chemicals and their manufacturers, squarely placing public protection from potential harm ahead of commercial interests. Soon afterward, Canada published "A Framework for the Application of Precaution in Science-based Decision Making about Risk" (Government of Canada 2003). The document offers ten guiding principles, generally assigning the greatest weight to whatever scientific information can be brought to bear on a precautionary measure. When science comes up short, however, the text defends the legitimacy of "society's chosen level of protection against risk", as already defined in various pieces of legislation or international agreements.

This Framework also distributes the burden of scientific proof somewhat more widely than do European dictates, establishing a decidedly Canadian take on "guilty until presumed innocent".

> Overall, the responsibility for providing the sound scientific basis should rest with the party who is taking an action associated with a risk of serious harm (e.g., the party engaged in marketing a product, employing a process or extracting natural resources). However, when faced with a concrete scenario, there should be an assessment of who would be in the best position to provide the information base. This could depend upon which party holds the responsibility or authority, and could also be informed by such criteria as who has the capacity to produce timely and credible information. (Government of Canada 2003)

A Question of Quality

Recently the approach to precaution has taken quite a different path in the USA, prompted by two sentences added to a congressional spending bill in 2000. This seemingly inadvertent piece of legislation had no official name, but came to be cited as either the Information Quality Act or the Data Quality Act. It directs the U.S. Office of Management and Budget to issue government-wide guidelines that "provide policy and procedural guidance to Federal agencies for ensuring and maximizing the quality, objectivity, utility, and integrity of information (including statistical information) disseminated by Federal agencies." (Treasury and General Government Appropriations Act 2001)

The implications of this directive came to the fore in 2003, just as the EPA was re-evaluating the herbicide atrazine. Researchers were investigating whether this agent was responsible for altering sexual development in frog species (Hayes et al. 2002). However, some critics invoked the Data Quality Act to challenge the experimental methods that were being employed. "Publication of a research article in a peer-reviewed scientific journal does not mean that the research has been accepted as valid by the scientific community and that it should be considered reliable for regulatory purposes," argued members of the Center for Regulatory Effectiveness (CRE) in a letter to the editor of *Environmental Health Perspectives*, a publication of the National Institute of Environmental Health Sciences (Tozzi et al. 2004).

A self-styled regulatory watchdog, the CRE had welcomed the Data Quality Act as an essential brake against what the organization cast as overzealous acceptance of preliminary or incomplete research findings. The organization's posture on atrazine sparked a direct response in the same journal, from representatives of the Natural Resources Defense Council, a nonprofit group dedicated to strengthening the regulation of toxic chemicals.

> On the surface, the CRE's call for validated tests sounds innocuous, even responsible. On closer inspection, the CRE seems to be arguing that a federal agency may not base any regulatory action on scientific research unless it has been performed in accordance with a pre-existing, government-approved test protocol. However, the government lacks standard protocols to assess many health effect end points and many types of studies. For example, there is no accepted government benchmark for data from epidemiologic research, for the use of pharmacokinetic models, or for most molecular methodologies. Accidental poisoning data are likewise useful to a risk assessor looking at a given chemical, but they are obviously not the result of experiments carried out under government-approved test conditions. If accepted, the CRE's arguments could jeopardize the government's ability to consider most published scientific research. (Sass and Devine 2004)

Journalist Chris Mooney added conspiratorial undertones to the rhetoric surrounding this subject. "The atrazine story provides a perfect example of how industry-friendly groups meddle in science and how the Data Quality Act – just one tool among many, but definitely a powerful one – facilitates the process," he argued (Mooney 2005).

In fact, Mooney regards this attack as just the latest volley in an ongoing campaign to wear down the unprecedented regulatory authority that was invested in the EPA with its creation in the 1970s. Since then, he insists, opponents of regulation have become skilled at marshalling expertise and undermining methodology in order to keep the agency's authority in check.

> Now, each regulatory decision seems to descend into a 'science' fight, not because we don't have enough qualified scientists in federal agencies, or because they are doing a poor job, but rather because those seeking to avoid regulation constantly try to raise the burden of proof required for action. (Mooney 2005)

Precaution Aplenty

Elsewhere these fights over proof can follow a significantly different course. For example, European sensitivities toward the limitations of scientific knowledge remain heightened in the wake of the bovine spongiform encephalopathy (BSE) outbreak of the 1990s. As thousands of livestock were diagnosed with this condition, politicians in the UK and elsewhere initially found themselves defending the safety of the meat that had been approved for sale in the market. Their assurances were founded on claims that the best science of the day had shown no implications for human health.

When further research discovered that BSE was caused by a strange class of misfolded proteins called prions, many of these same advocates were forced to recant, offering a stinging public admission that the problem might indeed extend to human beings. The economic and political impact of this revelation continues to reverberate, as can be seen in Europe's highly critical stance toward the use of genetically modified organisms (GMOs) in food and pharmaceutical processing.

That attitude verges on a profound suspicion, and perhaps nowhere was it better showcased than in Switzerland. There, in 1998, voters were presented with a binding referendum, dubbed the "Gene Protection Initiative", aimed at prohibiting transgenic animals, banning field release of transgenic crops, and limiting patents on biotechnology inventions. The effect could well have crippled the local operations of Swiss-based drug giants such as Novartis and Hoffmann-La Roche. Had the vote approved this measure, these scientific activities and the people conducting them would undoubtedly have migrated elsewhere, hollowing out a mainstay of the Swiss economy.

Nor did the defeat of the "Gene Protection Initiative" put the matter to rest. Even before the vote, the country's parliament committed itself to a more strict regulatory framework for any work with GMOs. By 2005, another referendum had successfully installed a 5-year moratorium on the use of GMO products in Swiss agriculture. While there was no attempt to curtail any kind of laboratory work, organizations such as the Swiss Biotech Association and Swiss Trade Association subsequently questioned the extent to which the country might be willing to restrict the freedom of researchers, building a reputation for being unfriendly to science.

Mooney argues that the USA has already acquired just such a reputation, though not because of any affection for the precautionary principle. He harkens back to Carson, bemoaning political interference in matters better left in the hands of scientists.

Most recently, he has weighed in on the fate of a $500 million pool of research funds provided by BP in the wake of the company's 2009 oil spill in the Gulf of Mexico. In principle, the money was to establish a 10-year program to track the ecological consequences of this major event, but institutions and governments in the affected region complained that their bids for a share of this support would lose out to higher profile scientific players in places such as Massachusetts and California.

According to him, such wrangling delayed the data collection in a rapidly changing situation that called for a more timely response, a critique he also applies to the reception of Carson's definitive polemic against the use of DDT. Her assertions were castigated by those who that had the most to lose from her call for a ban on this chemical, although this regulatory milestone was ultimately imposed in 1972, a full decade after the publication of her book.

DDT Revisited

In light of the ever tighter restrictions placed on this pesticide, it is worth wondering if Carson would have compromised with other, non-pecuniary interests that continue to regret this outcome. Some 100 countries have signed the Stockholm Convention on Persistent Organic Pollutants, committing them to eliminate the use of the most threatening of these chemicals, including DDT. Although its days as an open-field pesticide are long over, this agent is permitted for the specific purpose of controlling malaria. The exception reflects DDT's highly effective role in the ongoing struggle with this disease, which continues to present hundreds of millions of cases annually, leading to more than 800,000 deaths. As organizations just try to keep up with that toll, some of their representatives have flatly rejected the possibility of doing so without DDT.

Roger Bate and Richard Tren, who founded the non-profit group Africa Fighting Malaria, have been among the most outspoken of these representatives. According to them, Carson's critique of DDT set the standard for a model of sustainable development that dismisses chemicals or other techniques that could threaten lasting damage to the environment. The precautionary principle (dubbed "PP" by them) makes it possible to act in anticipation of such damage, but not necessarily on the basis of scientific information.

> Fundamentally the way that the PP, as often interpreted, fails to recognise that every activity that man undertakes involves risk of some sort and the only way to reduce risk to zero is to die. It is impossible to prove that any particular technology will not do harm to the environment as it is always possible to overlook a potential harm, even after the most thorough analysis. Putting these objections to the PP aside however, if one were to apply the PP to DDT, the conclusion would unequivocally be in favour of its continued use and promotion. (Tren and Bate 2001)

They cite the preservation of human health and saving of human lives as virtues that trump the environmental shortcomings of DDT. Moreover, even those shortcomings have been minimized through a steady refining of the techniques for spraying the chemical indoors in very limited amounts. Alternatives, such as combinations of drugs and insecticide-laced bed netting, can go further, eliminating this risk of toxic contamination. However, Bate and Tren are eager to point out that this strategy is much more logistically demanding, calling for significant levels of expertise and organizational co-ordination to achieve the same success as the straightforward application of DDT. In many of the poor places besieged by malaria, the harsh reality is that this well ordered approach is all but impossible.

Others agree, including observers who acknowledge DDT's effectiveness as well as how much is still unknown about its effects.

> Few studies of health outcomes have been conducted in populations where indoor residual spraying with DDT is occurring. These populations likely have much higher exposures to DDT and may differ from those previously studied in ways that might affect susceptibility (e.g., genetics, diet, health status, and social class). Research is needed to determine the exposure and health risks associated with DDT used for indoor residual spraying in the relevant communities. (Eskenazi et al. 2009)

For many, the demonstrated benefits of DDT outweigh its demonstrated hazards, even as we continue to examine how significant those hazards might be and what alternatives might exist. For Carson, this would not have sufficed to redeem the chemical, but the data behind this argument might have persuaded her to tolerate it for now. As testament to just how difficult it can be to regulate or ban even the most objectionable of toxins, then, it would be worth inviting her to dinner to hear her answer.

Scotch Guarding Arctic Wildlife

It would be no less interesting to solicit Carson's opinion on the case of a corporate giant that quickly demolished one of its most lucrative products for environmental reasons. The product in question was the internationally distributed fabric protectant Scotchgard®, manufactured by 3M. It occupied a class known as perfluorinated chemicals, defined by chains of carbon atoms bonding to fluorine atoms. Although the company had been turning out mass quantities of such agents for decades, it was not until the 1990s that innovations in mass spectrometry made it possible to analyse these complex fluorine compounds in a straightforward way.

Among those who applied this technology was John Giesy, who eventually took on a Canada Research Chair in the University of Saskatchewan's Department of Veterinary Biomedical Sciences. He spent several years working closely with the EPA, and later with 3M, to examine the global impact of the chemicals that went into Scotchgard®. He vividly recalls the day in 2000 when he was summoned to a formal meeting of the multinational firm's CEO and Board of Directors; what he had to tell them would rattle a market worth billions of dollars.

Giesy had started on what should have been a routine environmental audit using the new spectrographic approach. The findings turned up anomalous fluorine-based compounds in isolated wildlife samples that should have been far removed from any human influence. Expanding the survey to include no fewer than 3,500 samples from terrestrial and marine species around the world, the primary source of these errant compounds was identified as Scotchgard®.

"The technical people literally were stunned," he says. "Six months earlier they didn't think any of their chemicals were a problem, and suddenly I'm there saying 'you've got a problem'."

When Giesy was invited to present his findings to the company's top brass, they were equally stunned. Here were capable investigators telling them that traces of their company's output had wound up in the bloodstream of everything from albatrosses to polar bears, not to mention human beings on every continent.

Any health implications remained unclear, but the weight of the evidence prompted these managers to immediately shut down the manufacture of Scotchgard® and order a re-engineering of its formula to prevent the escape of the suspect compounds. It was an expensive decision, with direct costs and lost revenues that could be measured in the tens of millions of dollars. The economic disruption extended to major enterprises such as Mcdonald's, which was wrapping most of its food in paper coated with Scotchgard®.

Nevertheless, the effects of the move were apparent in short order. Just a few years after Giesy outlined the parameters of the issue (Giesy et al. 2001), other researchers were able to offer specific measurements (Butt et al. 2006). The latter were able to take advantage of a valuable baseline that was established in the 1970s, when Canadian wildlife officials had thoughtfully archived some Arctic seal liver samples. Within 4 years, animals in the same location were beginning to show lower levels of perfluorooctanesulfonyl fluoride, the signature chemical that had started all the fuss.

That news perhaps provided some cold comfort to bean counters at 3M, who might otherwise have good reason to think their firm hard done by. And yet their experience would be the envy of political and corporate leaders around the world. For in this particular instance, decision-makers were confronted by scientific feedback that guided them in clear, unambiguous terms. It should have been an occasion to cheer the likes of Rachel Carson, a painful step taken on the basis of "necessary knowledge".

References

Butt C et al (2006) Rapid response of Arctic ringed seals to changes in perfluoroalkyl production. Environ Sci Technol 41(1):42–49

Butterworth-Heinemann et al (2001) The precautionary principle: a critical appraisal. Cato Institute, Washington

Carson R (1962) Silent spring. Haughton Mifflin, Boston

Commission of the European Communities (2000) Communication from the commission on the precautionary principle. Commission of the European Communities, Brussels

Eskenazi B et al (2009) The pine river statement: human health consequences of DDT Use. Environ Health Perspect 117(9):1359–1367

Giesy J et al (2001) Global distribution of perfluorooctane sulfonate in wildlife. Environ Sci Technol 35(7):1339–1342
Government of Canada (2003) A framework for the application of precaution in science-based decision making about risk
Hayes T et al (2002) Hermaphroditic, demasculinized frogs after exposure to the herbicide atrazine at low 808 ecologically relevant doses. Proc Natl Acad Sci U S A 99:5476–5480
Leopold A (1949) A sand county Almanac. Oxford University Press, New York
Meadows D et al (1972) The limits to growth. The New American Library, New York
Ministerial Declaration of the Second International Conference on the Protection of the North Sea, ILM, 835, 25 November 1987; found on-line at http://www.seas-at-risk.org/1mages/1987%20London%20Declaration.pdf and in Morris, Julian (2000) Rethinking risk and the precautionary principle
Mooney C (2005) The republican war on science. Basic Books, New York
Sarewitz D (2004) How science makes environmental controversies worse. Environmental Sci Policy 7:385–403
Sass J, Devine J (2004) The center for regulatory effectiveness invokes the data quality act to reject published studies on atrazine toxicity. Environ Health Perspect 112(1):A18
Tozzi J et al (2004) Data quality act: response from the center for regulatory effectiveness. Environ Health Perspect 112(1):A18–A19
Treasury and General Government Appropriations Act for Fiscal Year 2001 (Public Law 106–554)
Tren R, Bate R (2001) Malaria and the DDT story. The Institute of Economic Affairs, London
UNEP (1992) Report of the United Nations Conference on Environment and Development, Rio de Janeiro, 3–14 June 1992, A/CONF.151/26 (Vol. I)
vom Saal F et al (2008) Baby's toxic bottle. The Work Group for Safe Markets
Welshons W et al (2006) Large effects from small exposures. III. Endocrine mechanisms mediating effects of bisphenol A at levels of human exposure. Endocrinology 147(Suppl 6):S56–S69

About the Editors

John E. Elliott is a research scientist with Environment Canada at the Pacific Wildlife Research Centre, Delta, British Columbia. Beginning with his BSc research at Carleton University on impacts of urban runoff on aquatic invertebrates to an MSc on toxicity of PCBs at the University of Ottawa and his PhD on the ecotoxicology of bald eagles from the University of British Columbia, he has always been interested in the application of science to understand and devise solutions to environmental problems. He has actively engaged in regulatory proceedings around topics such as lead projectiles, pulp mill pollutants, pesticides and brominated flame retardants. As adjunct professor at the University of British Columbia and Simon Fraser University, he continues to supervise graduate students engaged in field and laboratory ecotoxicological research, particularly with top predators. He has lectured and given courses in wildlife toxicology in North America and Europe. His publication record includes more than 150 peer reviewed papers, book chapters and reports.

J.E. Elliott et al. (eds.), *Wildlife Ecotoxicology: Forensic Approaches*,
Emerging Topics in Ecotoxicology 3, DOI 10.1007/978-0-387-89432-4,

Christine Bishop is a Research Scientist with the Canadian Federal Dept. of the Environment and adjunct professor at Simon Fraser University and at University of British Columbia. Her research focuses on the effects of multiple stressors on amphibian, reptilian and avian populations and the recovery of Species at Risk. Her doctoral studies at McMaster University examined the effects of pesticide use on birds nesting in apple orchards. Her masters degree research at York University was the study of the effects of organochlorine contaminants on common snapping turtles (*Chelydra serpentina serpentina*). She received her Bachelor's degree in Agricultural Science (Honors) at the University of Guelph. She has combined her research interests with conservation projects involving habitat restoration and preservation in Ontario and British Columbia, Canada. She co-founded the Canadian Amphibian and Reptile Conservation Network, has published more than 70 peer-reviewed scientific articles, and has co-edited three previous books.

Christy Morrissey is an Assistant Professor in the Department of Biology and School of Environment and Sustainability at the University of Saskatchewan. Her academic background, research experience and teaching interests are in avian population ecology, wildlife ecotoxicology, freshwater ecology, ecophysiology, and wildlife conservation. Since completion of her PhD from Simon Fraser University in 2003, she has completed 3 prestigious postdoctoral fellowships in Canada and in the U.K. (NSERC, Royal Society and Leverhulme Trust), building on a theme of how ecological processes impact contaminant exposure and effects to individuals, populations and communities. While still early in her career with a publication record of 16 papers, her current and past projects involve examining effects of endocrine disrupting chemicals on avian life cycles, pesticide impacts on forest bird communities and in agricultural habitats, and tracing nutrient and contaminant inputs in freshwater food webs.

Index

T

U

V